Handbook of Hydrology

Handbook of Hydrology

Edited by Allison Sergeant

MURPHY & MOORE
www.murphy-moorepublishing.com

Published by Murphy & Moore Publishing,
1 Rockefeller Plaza,
New York City, NY 10020, USA
www.murphy-moorepublishing.com

Handbook of Hydrology
Edited by Allison Sergeant

International Standard Book Number: 978-1-63987-284-8 (Hardback)

Cataloging-in-Publication Data

Handbook of hydrology / edited by Allison Sergeant.
p. cm.
Includes bibliographical references and index.
ISBN 978-1-63987-284-8
1. Hydrology. 2. Hydrography. 3. Water. 4. Earth sciences. I. Sergeant, Allison.
GB661.2 .H36 2021
551.48--dc23

TABLE OF CONTENTS

PREFACE

This book has been a concerted effort by a group of academicians, researchers and scientists, who have contributed their research works for the realization of the book. This book has materialized in the wake of emerging advancements and innovations in this field. Therefore, the need of the hour was to compile all the required researches and disseminate the knowledge to a broad spectrum of people comprising of students, researchers and specialists of the field.

Hydrology is the field of science concerned with the study of the movement, management, and distribution of Earth's water. Studies on watershed sustainability, water cycle, and water resources as well as drainage-basin and water quality management fall under the umbrella of this field. Hydrology has three major subdivisions - marine hydrology, groundwater hydrology and surface water hydrology. Advances in hydrology serve to further the solution to natural disasters and aid in environmental preservation and water management. This book contains some path-breaking studies in the field of hydrology. It outlines the processes, applications, and recent advances in hydrology in detail. This book will serve as a reference to a broad spectrum of readers.

At the end of the preface, I would like to thank the authors for their brilliant chapters and the publisher for guiding us all-through the making of the book till its final stage. Also, I would like to thank my family for providing the support and encouragement throughout my academic career and research projects.

Editor

1

Electrical resistivity dynamics beneath a fractured sedimentary bedrock riverbed in response to temperature and groundwater–surface water exchange

Colby M. Steelman, Celia S. Kennedy, Donovan C. Capes, and Beth L. Parker

G360 Institute for Groundwater Research, College of Engineering and Physical Sciences, University of Guelph, Guelph, Ontario, Canada

Correspondence to: Colby M. Steelman (csteelma@uoguelph.ca)

Abstract. Bedrock rivers occur where surface water flows along an exposed rock surface. Fractured sedimentary bedrock can exhibit variable groundwater residence times, anisotropic flow paths, and heterogeneity, along with diffusive exchange between fractures and rock matrix. These properties of the rock will affect thermal transients in the riverbed and groundwater–surface water exchange. In this study, surface electrical methods were used as a non-invasive technique to assess the scale and temporal variability of riverbed temperature and groundwater–surface water interaction beneath a sedimentary bedrock riverbed. Conditions were monitored at a semi-daily to semi-weekly interval over a full annual period that included a seasonal freeze–thaw cycle. Surface electromagnetic induction (EMI) and electrical resistivity tomography (ERT) methods captured conditions beneath the riverbed along a pool–riffle sequence of the Eramosa River in Canada. Geophysical datasets were accompanied by continuous measurements of aqueous specific conductance, temperature, and river stage. Time-lapse vertical temperature trolling within a lined borehole adjacent to the river revealed active groundwater flow zones along fracture networks within the upper 10 m of rock. EMI measurements collected during cooler high-flow and warmer low-flow periods identified a spatiotemporal riverbed response that was largely dependent upon riverbed morphology and seasonal groundwater temperature. Time-lapse ERT profiles across the pool and riffle sequence identified seasonal transients within the upper 2 and 3 m of rock, respectively, with spatial variations controlled by riverbed morphology (pool versus riffle) and dominant surficial rock properties (competent versus weathered rock rubble surface). While the pool and riffle both exhibited a dynamic resistivity through seasonal cooling and warming cycles, conditions beneath the pool were more variable, largely due to the formation of river ice during the winter season. We show that surface electrical resistivity methods have the capacity to detect and resolve electrical resistivity transience beneath a fractured bedrock riverbed in response to porewater temperature and specific conductance fluctuations over a complete annual cycle.

1 Introduction

Fractured sedimentary bedrock represents an important source of water for many communities around the world. Although the effective porosity of rock is low relative to unconsolidated sediment, the existence of dense networks of interconnected fractures, dissolution-enhanced conduits, and karst features can result in productive yet heterogeneous and anisotropic flow systems. An exposed bedrock surface may exhibit greater variability in flow and transport properties as it is subjected to weathering and erosional processes, leading to very complicated groundwater recharge and discharge patterns. Fractured sedimentary rock is best conceptualized as a dual porosity system where fractures dominate flow, but remain connected to water stored in the porous matrix through advection and diffusion. Such conceptualizations of fracture flow and transport are routinely applied to groundwater resource (e.g., Novakowski and Lapcevic, 1998; Lemieux et al., 2009; Perrin et al., 2011) and contaminant transport studies (e.g., Zanini et al., 2000; Meyer et al., 2008; McLaren et al., 2012). Recent studies have extended these concepts to fluvial river environments (e.g., Singha et al., 2008; Toran et al., 2013a).

Groundwater–surface water interactions on the reach scale are conceptualized through gaining, losing, and flow-through interactions (Woessner, 2000). On the channel scale, micro-to-macro bedform variations result in variably scaled zones of surface water downwelling (recharge) and groundwater upwelling (discharge) (e.g., Binley et al., 2013; Käser et al., 2013). Groundwater temperature measurements are frequently used to monitor spatiotemporal variations in groundwater–surface water exchange or flux across riverbeds (e.g., Anderson, 2005; Irvine et al., 2016). Yet, very little is known about the existence and nature of hyporheic and groundwater–surface water mixing zones in fractured sedimentary bedrock, largely because these systems are very difficult to instrument using direct methods (e.g., drive point monitoring wells, seepage meters, thermistors), and the scale of the interaction may be very small or heterogeneous relative to unconsolidated sediment.

Hydrologic processes along a fractured bedrock river were explored by Oxtobee and Novakowski (2002), who concluded that groundwater–surface water interaction was restricted by poor vertical connectivity and limited exposure of horizontal bedding plane fractures. A subsequent numerical sensitivity analysis by Oxtobee and Novakowski (2003) confirmed that groundwater–surface water connectivity through discrete fractures would be highly variable in space and time, and would largely depend on fracture size or aperture, river stage, and the distribution of hydraulic head within the flow system. Fan et al. (2007) numerically explored the influence of larger-scale fracture orientations and geometries on the groundwater-flow system near a stream; they concluded that the base flow to a stream would be higher for streams aligned with fracture dip than those aligned with fracture strike. Therefore, groundwater–surface water interaction in a fractured bedrock environment will depend on stream-fracture alignment. Although these previous studies offered valuable insights into the magnitude of groundwater–surface water exchange, they were based on idealized fracture network conceptualizations, and did not consider the role of matrix porosity and potential exchanges between fractures and the porous matrix.

Electrical and electromagnetic methods such as ground-penetrating radar, electromagnetic induction, and electrical resistivity imaging are commonly used to characterize fluvial deposits (e.g., Naegeli et al., 1996; Gourry et al., 2003; Froese et al., 2005; Sambuelli et al., 2007; Rucker et al., 2011; Orlando, 2013; Doro et al., 2013; Crosbie et al., 2014). The capacity of time-lapse electrical resistivity imaging for conceptualization of groundwater transients in sediment is well-documented in the literature (e.g., Nyquist et al., 2008; Miller et al., 2008; Coscia et al., 2011, 2012; Cardenas and Markowski, 2011; Musgrave and Binley, 2011; Dimova et al., 2012; Wallin et al., 2013). Electrical imaging of natural river systems perturbed by solute tracers has resulted in unprecedented visualizations of fluid flow (e.g., Ward et al., 2010a, b, 2012; Doetsch et al., 2012; Toran et al., 2013a, b; Harrington et al., 2014). More recent applications of electrical resistivity in karst undergoing surface water transients have shown how surface geophysics can unravel complex hydrologic processes in sedimentary bedrock environments (e.g., Meyerhoff et al., 2012, 2014; Sirieix et al., 2014), especially when site conditions limit the use of more invasive direct measurement methods.

While a variety of geophysical tools and techniques can measure flow and water chemistry in space and time (Singha et al., 2015), the most appropriate tool and approach will depend on the scale of interest. The vast majority of geophysical work within shallow river environments has utilized discrete temperature monitoring below the riverbed to detect vertical fluxes (e.g., White et al., 1987; Silliman and Booth, 1993; Evans et al., 1995; Alexander and Caissie, 2003; Conant Jr., 2004; Anderson, 2005; Hatch et al., 2006; Keery et al., 2007; Schmidt et al., 2007; Constantz, 2008). Recent advancements in distributed fiber-optic cables have improved spatial and temporal resolution of groundwater–surface water interactions (e.g., Slater et al., 2010; Briggs et al., 2012; Johnson et al., 2012).

Groundwater and surface water interaction can be monitored through changes in thermal gradient or electrolytic concentration (e.g., Norman and Cardenas, 2014), yet the scale and magnitude of these interactions will vary as a function of riverbed architecture and subsurface hydraulic conditions (Crook et al., 2008; Boano et al., 2008; Ward et al., 2012; Tinkler and Wohl, 1998), resulting in spatially dynamic exchange. These processes are further complicated by diel (e.g., Swanson and Cardenas, 2010) and seasonal (e.g., Musgrave and Binley, 2011) temperature fluctuations across a range of spatial scales, local transients such as precipitation events (e.g., Meyerhoff et al., 2012), river stage fluctuations (e.g., Bianchin et al., 2011), and controlled dam releases (e.g., Cardenas and Markowski, 2011). Relative to other non-invasive geophysical methods, electrical resistivity methods are more robust in their ability to provide information about temperature and solute fluctuations beneath actively flowing surface water bodies (e.g., Nyquist et al., 2008; Cardenas and Markowski, 2011; Ward et al., 2012; Meyerhoff et al., 2014) particularly in a time-lapse manner. Unlike conventional hydrogeological methods (e.g., screened or open "coreholes" – continuously cored boreholes), which may bias conduction in the fractures, surface electrical methods are sensitive to the bulk electrical conductivity of the formation, making them more suited for detection of processes between the open fractures or conduits and the porous matrix.

A hypothetical groundwater–surface water mixing zone in a porous fractured rock system (Fig. 1) will exhibit transience in water temperature and specific conductance as a result of changes in groundwater flow (recharge versus discharge), mixing of surface water and groundwater, seasonal

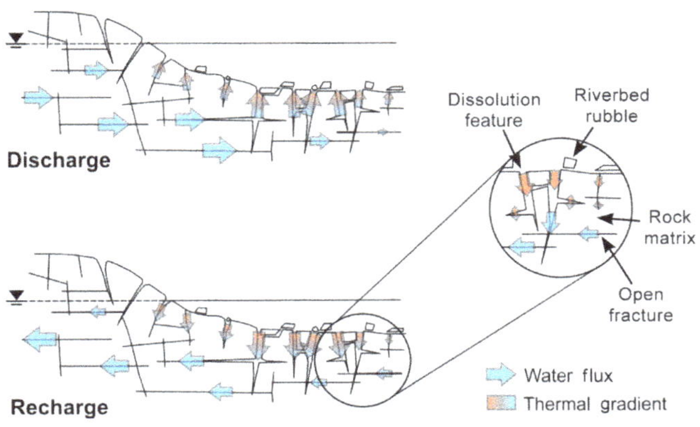

Figure 1. General conceptual model of the groundwater flow system beneath a fractured bedrock river. Groundwater–surface water mixing is controlled by open fractures and dissolution-enhanced features, with secondary exchanges (flux or diffusion) occurring between fractures and rock matrix.

atmospheric temperature fluctuations, and/or biogeochemical changes within the riverbed. Ward et al. (2010a) demonstrated how surface electrical methods sensitive to changes in water conductivity could be used to detect and quantify diffusive mass transport (exchange) between a mobile and immobile storage zone in a shallow riverbed. Our study uses a similar geophysical monitoring approach to assess the magnitude and scale of groundwater–surface water transience beneath a fractured bedrock riverbed based on the detection and characterization of geoelectrical dynamics over a complete annual cycle. Seasonal freezing and thawing of surface water bodies is an important process in mid-latitude climates; winter freeze-up will reduce base-flow contributions, while spring snow melt will result in a sudden and large increase in base flow to a river. Therefore, understanding the transient behaviours of the hydrogeologic system, including water-phase transformations, over a complete annual cycle will be critical to our understanding of the magnitude and spatial extent of groundwater–surface water exchange along fractured bedrock rivers. Given the challenges and costs associated with installing sensors within rock and heterogeneous flow system characteristics, minimally invasive surface and borehole geophysical methods offer an ideal alternative, and possibly a more effective approach for long-term groundwater–surface water monitoring of bedrock environments.

Our study examines the capacity of electrical imaging methods (i.e., electromagnetic induction and electrical resistivity) to monitor geoelectrical transients within a fractured sedimentary bedrock river to better understand groundwater–surface water transience over a complete annual cycle. To achieve this, seasonal variations in electrical resistivity distribution were measured across a 200 m reach of a bedrock river using a ground conductivity meter and time-lapse electrical resistivity measurements along two fixed transects intersecting a pool–riffle sequence. These geophysical surveys were supported by continuous measurements of groundwater and surface water temperature, specific conductance, and river stage. Our study shows that spatiotemporal resistivity dynamics were largely controlled by riverbed morphology in combination with seasonal changes in water temperature, and to a lesser degree electrolytic concentration. The formation of ground frost and basal ice strongly affected the electrical resistivity beneath the riverbed compared to intraseasonal dynamics (spring, summer, and fall). Observed geoelectrical changes beneath the riverbed appear primarily dependent on seasonal temperature trends exhibiting varying zones of influence (vertical and horizontal) across the pool and riffle section of the river. The riverbed was strongly susceptibility to seasonal atmospheric temperature fluctuations, which might have implications for biogeochemical processes or benthic activity.

2 Background

The Eramosa River – a major tributary of the Speed River within the Grand River watershed in Ontario, Canada – resides upon a regional bedrock aquifer of densely fractured dolostone with dissolution-enhanced conduits and karst features (e.g., Kunert et al., 1998; Kunert and Coniglio, 2002; Cole et al., 2009). Although this aquifer represents the sole source of drinking water for the region, the potential effects of increased groundwater pumping on the overlying bedrock river and surrounding ecosystems are not yet understood. This is largely due to a gap in our conceptual understanding of groundwater–surface water interaction in rivers that flow directly along sedimentary bedrock surfaces with exposed fracture networks. The fractured sedimentary bedrock exhibits a complex flow system due to variably connected fracture networks, dissolution-enhanced features, and variable bedrock exposure (Steelman et al., 2015a, b). The bulk of the flow will occur along the fracture networks, with highly variable head distributions; matrix storage could support equally complex biogeochemical processes and thermal dynamics through convective or diffusive exchange with open fractures or dissolution-enhanced features.

A focused geophysical investigation was carried out along a 200 m reach of the Eramosa River (Fig. 2). The study area was positioned at a bend in the river with relatively cleared vegetation along the southern shoreline and adjacent floodplain with exposed rock at the surface. A network of coreholes and streambed piezometers were installed across the site. Locally, the water table elevation corresponds to the surface water or stage elevation, resulting in vadose zone thicknesses from <0.5 to 2.0 m along the shorelines. The temperate southern Ontario climate subjects the river to a wide range of seasonal conditions, including high precipitation periods in spring and fall, hot and dry summers, and variable degrees of ground frost and surface water freeze-up during the winter months (Fig. 3a).

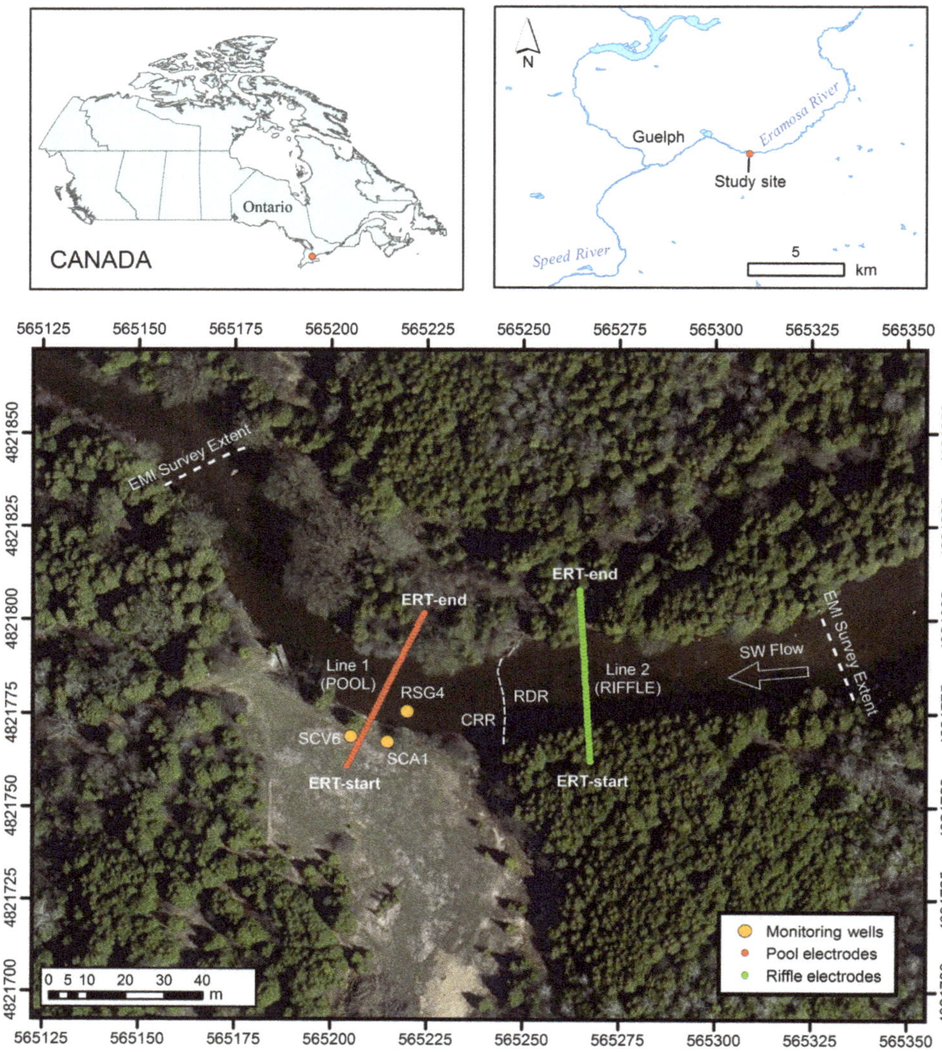

Figure 2. Field site located along the Eramosa River near Guelph, Ontario, Canada. The spatial extent of the electromagnetic induction (EMI) surveys, coreholes and groundwater–surface water monitoring points, and fixed electrical resistivity tomography (ERT) transects are shown relative to the riffle–pool sequence. The riverbed is described as either rubble-dominated riverbed (RDR) or competent rock riverbed (CRR) surface.

Locally, the river incises the Eramosa Formation by 2 to 3 m, exposing abundant vertical and horizontal fractures with little to no alluvial sediment deposited along the riverbed (Fig. 3b). Regionally the Eramosa acts as a discontinuous aquitard unit (Cole et al., 2009); however, core logs collected at the study site show bedding plane and vertical joint set fractures spanning the entire 11 m sequence of Eramosa. This upper formation is underlain by approximately 3 m of the cherty, marble-like Goat Island formation, which exhibits high-angle fractures along cherty nodules near the Eramosa contact. The Goat Island is underlain by more than 15 m of Gasport Formation, which exhibits coral reef mounds of variable morphology. The rock matrix of the Gasport is visually more porous, with well-defined vugs, dissolution-enhanced features, and fewer fractures than the overlying Goat Island and Eramosa. A full description of these bedrock sequences can be found in Brunton (2009).

In this region, the winter season may be accompanied by the formation of ground frost and surface-water freezing. Seasonal freeze-up will consist of an ice crust layer on the surface of the water and the possible formation of basal ice along the riverbed (Stickler and Alfredsen, 2009). The latter phenomenon can occur during extreme atmospheric cooling over turbulent water bodies, resulting in super-cooled water ($< 0\,°C$) that rapidly crystalizes to form frazil (i.e., tiny ice particles with adhesive characteristics); these crystals can flocculate to form slush, which adheres and accumulates on the substrate, forming a basal ice layer.

Figure 3. (a) Images of the river during monitored study period. **(b)** Examples of vertical and horizontal fracturing within pool and rubble covered portions of the riverbed bedrock.

3 Methods

3.1 Electrical properties

Electrical resistivity methods are based upon Ohm's Law ($R = \Delta V / I$). In the case of a homogeneous half-space, the electrical resistance (R) of the subsurface is determined by measuring the potential difference (ΔV) across a pair of "potential" electrodes due to an applied current (I) across a separate pair of "current" electrodes a certain distance away. The measured R (Ω) across a unit volume of the earth can be converted to apparent resistivity (Ωm) using a specific geometric factor that compensates for varying electrode array geometry (Reynolds, 2011). Apparent resistivity measurements are commonly interpreted using tomographic inversion techniques, whereby measured data are reconstructed from forward models of an optimized physical parameter distribution (Snieder and Trampert, 1999; Loke et al., 2013). Although data inversion techniques are standard practice in the interpretation of most geophysical data, the model that best matches the measured data is not necessarily an exact representation of the subsurface. Here, the inversion process ultimately yields a smoothed representation of the true parameter distribution.

The bulk electrical resistivity (i.e., inverse of conductivity) of a porous medium can be calculated through an empirical relationship known as Archie's law (Archie, 1942):

$$\rho_b = \rho_w \phi^{-m} S_w^{-n}, \tag{1}$$

where the resistivity of the bulk formation (ρ_b) is related to the resistivity of the pore water (ρ_w), porosity of the medium (ϕ) raised to the negative power (m) that represents the degree of pore cementation, and fraction of pores containing water (S) raised to the negative power (n) that accounts for the connectedness of the pore water (Glover, 2010). This relationship carries a number of simplifying assumptions: the most significant being that the current flow is entirely electrolytic. While more sophisticated formulations of Archie's Law incorporating interfacial conduction can be found in the literature (e.g., Rhoades et al., 1976; Waxman and Smits, 1968; Glover, 2010), Eq. (1) is considered to be a reasonable approximation for this type of environment. Equation (1) is used in this study to evaluate the impact of observed groundwater and surface water aqueous conductivity variations on the bulk formation resistivity. Here, a value of 1.4 was used for the constant m, which is considered reasonable for fractured dolostone (Doveton, 1986), while S is considered to be equal to unity (i.e., fully saturated). It should be noted that the relative impact of aqueous conductivity changes on the bulk formation resistivity may vary with clay content and pore connectivity due to intrinsic deviations in the m value (Worthington, 1993). Furthermore, orientated fracture networks may result in an anisotropic resistivity response (Steelman et al., 2015b); however, these static properties of the rock will not impact relative changes in resistivity at a fixed location.

The electrolytic (fluid phase) resistivity will depend on the concentration and composition of dissolved ions, and viscosity of the pore water (Knight and Endres, 2005). Increasing ion concentrations and temperature will lead to a reduction in formation resistivity. Empirical evidence has shown that resistivity can decrease anywhere from 1 to 2.5 % per °C (Campbell et al., 1948; Keller, 1989; Brassington, 1998). The effects of temperature on our resistivity signals were determined using Arps' law (Arps, 1953):

$$\rho_{w2} = \frac{\rho_{w1}(T_1 + 21.5)}{(T_2 + 21.5)}, \tag{2}$$

where ρ_w (Ωm) and T (°C) represent the resistivity and temperature of the water at two points. This formulation was developed from a least-squares fit to the conductivity of a NaCl solution ranging from 0 to 156 °C; however, the exact relationship between fluid conductivity and temperature will depend on the composition of the electrolytic solution (Ellis, 1987).

3.2 Bedrock lithology, fractures, and porosity

The geology was characterized through a series of vertical and angled coreholes along the southern shoreline that were advanced into upper Gasport Formation. These drilling activities were part of a broader hydrogeological investigation of groundwater flow and fluxes along the river. A network of riverbed piezometers, bedrock stage gauges, and flux measurement devices were installed between 2013 and 2014 within the pool. Locally, the riverbed morphology can be distinguished in terms of the amount of bedrock rubble or weathered rock fragments covering the exposed rock surface. Figure 2 shows the transition from a rubble-dominated riverbed (RDR) to a more competent rock riverbed (CRR); this boundary roughly corresponds to the riffle–pool transition.

Geophysical measurements were supported by temperature, specific conductance of the fluid, and river stage elevation collected at nearby monitoring points (Fig. 2). The geologic and hydrogeologic data were obtained from a river stage gauge (RSG4), a vertical corehole (SCV6) drilled to a depth of 10.9 m, and an angled corehole (SCA1) drilled to a vertical depth of 31.8 m. The angled corehole plunges at 60° and is orientated at 340°, and therefore, plunges beneath the river with a lateral displacement of 21 m. Coreholes were drilled using a small-diameter portable Hydracore Prospector™ drill with a diamond bit (NQ size: 47.6 mm core and 75.7 mm corehole diameter) and completed with steel casings set into concrete to a depth of 0.6 m below ground surface (b.g.s.). All coreholes were sealed using a flexible impermeable liner filled with river water (FLUTe™ Flexible Liner Underground Technologies, Alcalde, New Mexico, USA; Keller et al., 2014).

The SCA1 rock core was logged for changes in lithology, vugs, and fracture characteristics, intensity and orientation, including bedding plane partings. Rock core subsamples were extracted for laboratory measurements of matrix porosity using the following procedure: sample was oven dried at 40 °C; dimensions and dry mass were recorded; samples were evacuated in a sealed chamber and imbibed with deionized water; the sample chamber was pressurized to 200 to 300 psi for 15 min; and samples were blotted and weighed to obtain saturated mass. Open coreholes were logged using an acoustic (QL40-ABI) and an optical (QL40-OBI) borehole imager (Advanced Logic Technologies, Redange, Luxembourg) to characterize the fracture network.

3.3 Pressure, temperature, specific conductance, and river flow

Temperature, specific conductance, and hydraulic pressure data were recorded using a CTD-Diver™ (Van Essen Instruments, Kitchener, Canada) deployed within RSG4 (surface water) and SCV6 (groundwater) at a depth of 10.5 m b.g.s. The transducer in SCV6 was placed near the bottom of the open corehole prior to being sealed with an impermeable liner, thereby creating a depth-discrete groundwater monitoring point. Surface water data were recorded through the full study period while deeper bedrock conditions were recorded from early September 2014 through late May 2015. All measurements were collected at 15 min intervals.

Vertical temperature profiles were additionally collected along the inclined sealed corehole water column of SCA1 from 4 September 2014 to 22 May 2015 using an RBR*solo*™ temperature logger paired with a RBR*solo*™ pressure logger (RBR Limited, Ottawa, Canada). These data were recorded at 0.5 s intervals while the sensors were manually lowered into the water column using a fiberglass measuring tape at a rate of 0.02 to 0.03 m s^{-1}. Barometric pressure was collected at the site using a Baro-Diver™ (Van Essen Instruments, Kitchener, Canada). These temperature sensors can resolve changes to 0.5×10^{-4} °C with a full-scale response time of 99 % in 2 s.

Rainfall was recorded at the University of Guelph Turfgrass Institute, located 2 km northwest of the site, while snowfall accumulation was obtained from the Region of Waterloo Airport roughly 18 km southwest of the site. Hourly mean river flux was recorded 900 m upstream at the Watson Road gauge operated by the Grand River Conservation Authority. A summary of the weather and river flux data are provided in Fig. 4.

3.4 Riverbed electrical resistivity

3.4.1 Spatial electrical resistivity mapping

Spatial riverbed resistivity distribution was measured using a Geonics EM-31 ground conductivity meter (Geonics, Mississauga, Canada) during a seasonally cool and warm period: mid-summer (low-stage) conditions on 7 July 2014 and early spring (high-stage) conditions on 3 April 2013. Measurements were collected at a rate of 3 readings per second with the device operated in vertical dipole mode held ~ 1 m above the riverbed. The effective sensing depth of this instrument in vertical dipole mode is approximately 6 m (cumulative depth), and it is minimally sensitive to conditions above the ground surface (McNeil, 1980). Data were recorded along roughly parallel lines spaced ~ 1.75 m apart orthogonal to the river orientation, with the coils aligned parallel to surface water flow direction. Water depths over the investigated reach varied from < 0.1 m in the riffle during low-flow to nearly 1 m in the pool during high-flow conditions. Datasets were

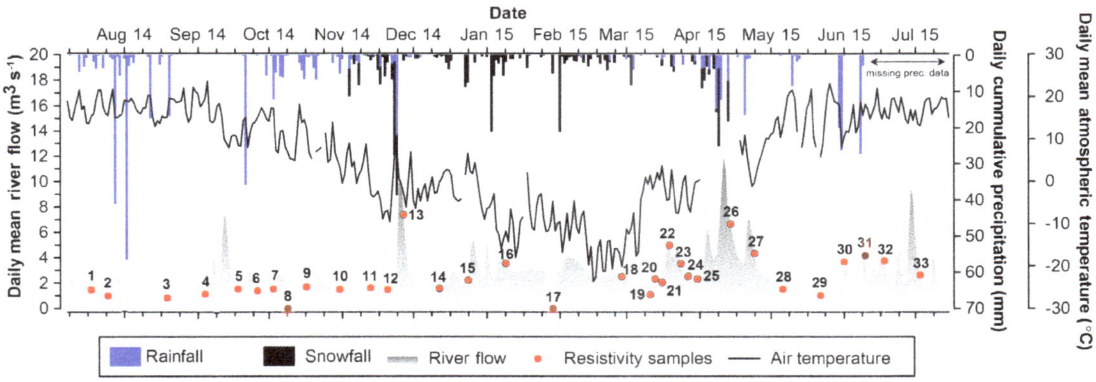

Figure 4. Continuously monitored atmospheric conditions and river flow from Watson Gauge during the study period, with superimposed resistivity geophysical measurement events between 18 July 2014 and 3 July 2015. Note: snowfall is presented as snow water equivalent.

filtered for anomalous outliers prior to minimum curvature gridding.

3.4.2 Time-lapse electrical resistivity imaging

Surface electrical resistivity measurements were collected along two fixed transects orientated orthogonal to the river (Fig. 2), capturing conditions within a pool and riffle sequence (Fig. 3). Line 1 was positioned downstream over a deeper pool section with more substantial bedrock incision into a competent bedrock surface (Fig. 3b.i and b.ii), while line 2 was situated upstream over a shallower riffle section blanked by bedrock rubble fragments with less bedrock incision (Fig. 3b.iii).

For this study, resistivity cables were constructed using a pair of 25 multicore cables (22 gauge strained wire, 600 V rating) wound within a PVC jacket. The PVC jacket was split open every metre to expose and cut out a single wire that was connected to an audio-style banana plug. Spliced sections of outer PVC jacket were resealed using heat shrink tubing and silicon. This process yielded two 24-channel cables, each connected to a single multi-pin connector for direct data logger communication. Electrodes were constructed from half-inch diameter stainless steel rod cut to 15 cm lengths. A hole was drilled on one end of the electrode to receive the banana plug connector. Given the exposed bedrock across the site, a half-inch hole was drilled into the rock at 1 m intervals along the ground surface. In some cases, electrodes were buried beneath a rubble zone of the riverbed, or were pushed into a thin layer of sediment. On the shorelines, electrodes were fully implanted into the rock along with a few teaspoons of bentonite clay to minimize contact resistance. Each monitoring line was instrumented with dedicated electrodes and cables that remained in place for the duration of the study.

Resistivity measurements were recorded using a Syscal Junior Switch 48 (Iris Instruments, Orléans, France) resistivity meter. A Wenner array was selected for its higher S/N ratio. A dipole–dipole array was tested but found to be very susceptible to noise (i.e., bad data points resulting from low potentials); this can be attributed to the high-contact resistances with rock combined with the instrument's more moderate power capability (max 400 V, 1.3 A). Although the Wenner array geometry resulted in a stronger signal (i.e., where potentials are measured across a pair of electrodes located between the current electrodes), it was less sensitive to lateral variations across the riverbed and did not permit the collection of a reciprocal dataset, which could have been used to assess potential measurement errors (e.g., Slater et al., 2000). Surface resistivity data were recorded on a semi-daily to semi-weekly interval from 18 July 2014 to 3 July 2015 covering a complete annual cycle, which included a seasonal freeze–thaw cycle, and numerous wetting–drying events accompanied by large river stage fluctuations. The timing of resistivity measurement events are shown along the river flow data in Fig. 4. Resistivity measurements were generally recorded between 08:00 LT and 13:00 LT.

Measured apparent resistivity data were manually filtered to remove erroneous data points prior to being inverted using RES2DINV v.3.59 (Geotomo Software, Malaysia), which uses the Gauss–Newton least-squares method (Loke and Dahlin, 2002). For this study, a robust inversion scheme was used with a moderate dampening factor applied at the riffle and a slightly higher dampening factor at the pool. These dampening factors were chosen based on the resistivity contrasts observed along the rock surface, along with intermittently noisy measurements at the pool; the width of the model cells were set to half the electrode spacing (i.e., model refinement) to help suppress the effects of large surface resistivity variations on the inversion process. All other parameters within the program were optimized to compensate for high noise and large resistivity contrasts, while achieving the lowest possible model root mean squared (RMS) error (Loke, 2002). Each model was independently inverted with a defined surface water boundary (i.e., stage height above the submerged electrodes) and true aqueous resistivity, both of which were fixed for each model inversion. The same set of initial inversion parameters was applied to all datasets collected along a given line. Model convergence typically occurred within 8 iterations.

Temporal variations in bedrock resistivity were assessed within 5 m wide × 2.5 m deep zones beneath the riverbed based on a resistivity index (RI) calculation. This enabled comparison of datasets with different magnitudes of resistivity variation. The RI was defined as follows:

$$\mathrm{RI}_{i,j} = \frac{\mathrm{MZR}_{i,j} - \mathrm{MAR}}{\mathrm{MAR}}, \tag{3}$$

where $\mathrm{RI}_{i,j}$ = resistivity index for the ith zone on the jth sample date; $\mathrm{MZR}_{i,j}$ = mean zone resistivity for the ith zone on the jth sample date; MAR = mean annual resistivity of the entire profile across the full time series for the pool or riffle.

4 Results

4.1 Bedrock fracture network, temperature, and specific conductance

Formation contacts of the Eramosa–Goat Island and the Goat Island–Gasport formations were identified in the core at depths of 8.6 and 13.0 m b.g.s., respectively (Fig. 5a). Fractures beneath the river were predominantly horizontal to slightly dipping (< 10°), and most abundant in the Eramosa and Goat Island formations. Although vertical and sub-vertical fractures (> 10°) were relatively less abundant than bedding plane fractures, they were more uniformly distributed with depth. These high-angle fractures terminate at the surface as vertical joint sets along two regional orientations: 10 to 20° north-northeast and 280 to 290° west-northwest (Fig. 3b.ii). Matrix porosities from the corehole ranged from 0.5 to 5 % with the lowest porosities observed along the highly weathered riverbed surface and lower portion of the Eramosa Formation. Hydraulic head data collected in the river and at the base of SCV6 (10.5 m b.g.s.) suggest a seasonally sustained downward gradient ranging from 0.21 to 0.26 at the pool; however, this does not necessarily reflect conditions proximal to the streambed.

Vertical temperature profiling within the static water column of the FLUTe™ lined SCA1 corehole from 4 September 2014 to 22 May 2015 captured seasonal fluctuations in ambient groundwater temperature to depths up to 20 m (Fig. 5b), thereby delineating the vertical extent of the heterothermic zone. Temperatures inside the liner ranged from 18 °C in late summer to 5 °C in mid-winter. Although fluctuations were observed along the entire 20 m profile, the bulk of the variations (short- and long-period) were observed in the upper 10 m b.g.s.

Previous studies using ambient temperature profiling in lined coreholes (Pehme et al., 2010, 2014) examined the effects of active groundwater flow around static water columns.

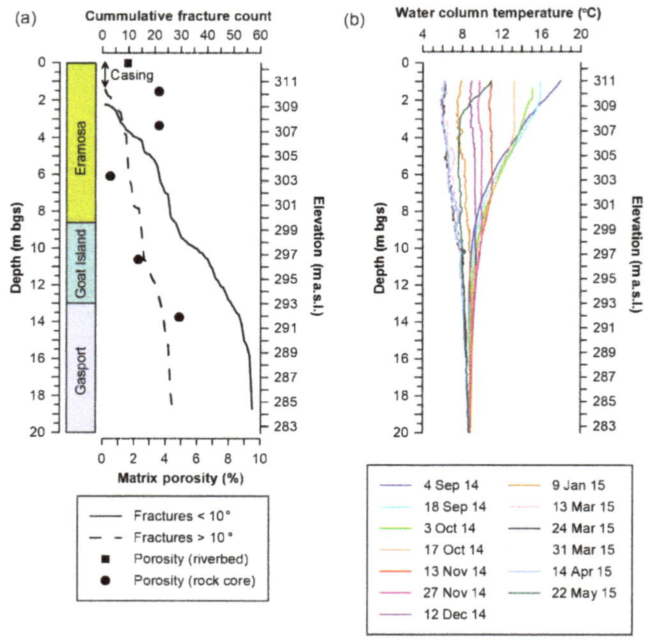

Figure 5. (a) Interpreted rock core from SCA1 (angled corehole plunging at 60° with an azimuth of 340°). Fracture frequency and orientations were obtained using an acoustic televiewer log, while matrix porosity measurements were obtained from subsamples of the continuous core. **(b)** Corehole temperature profiles of the SCA1 Flute™ sealed water column.

Pehme et al. (2010) demonstrated how a lined water-filled corehole in thermal disequilibrium with the surrounding formation would exhibit more short-period temperature perturbations along its vertical profile than an equilibrated water column within zones of active groundwater flow. Here, the onset of winter seasonal conditions (9 January through 31 March 2015) cooled the corehole water column near the ground surface, resulting in density-driven convection within the column and leading to thermal disequilibrium with respect to the surrounding bedrock which then results in subtle temperature perturbations as the water column cooled toward 5 °C. The magnitude and frequency of the perturbations observed in Fig. 5b during these cooler periods correspond to areas of increased fractures (Fig. 5a) supporting active groundwater flow.

Specific conductance and temperature of surface water (RSG4) and groundwater (SCV6) corresponding to geophysical sampling events (Fig. 4) are presented in Fig. 6. These data indicate that surface water specific conductivity varied within a much narrower range than the actual (uncompensated) conductivity, which includes the effects of temperature. While the overall impact of temperature and ionic concentration on the specific conductance of surface water were similar (i.e., equivalently dynamic), variations associated with ionic concentration appear more erratic, exhibiting sharper fluctuations over shorter periods of time. For in-

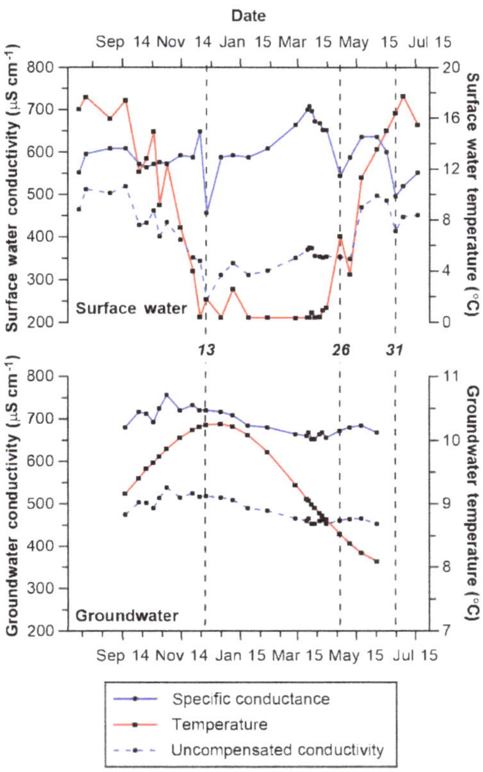

Figure 6. Specific conductance, temperature, and uncompensated aqueous conductivity of surface water at RSG4 and groundwater at the bottom of SCV6 (10.5 m b.g.s.). Uncompensated conductivity represents the actual conductivity of the porewater after re-incorporating the effect of temperature using the sensor's internal temperature–conductivity correction factor.

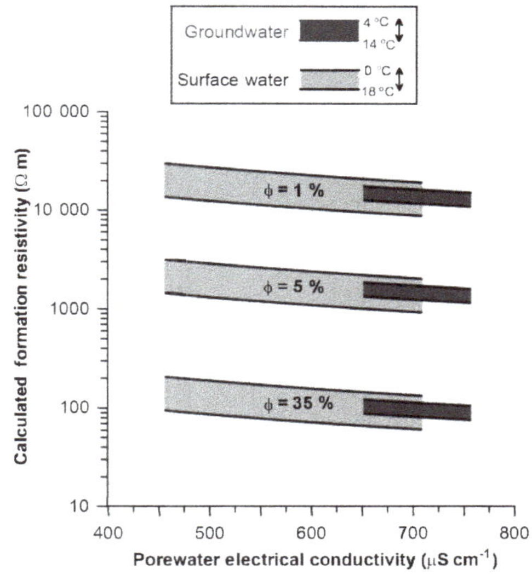

Figure 7. Calculated formation resistivity based on Eq. (1) and measured variations in surface water and groundwater electrical conductivity, including effects of temperature based on Eq. (2). A cementation factor of 1.4 was used to represent the fractured dolostone bedrock. Measured water conductivity and temperature were obtained from CTD-Diver™ sensors deployed in RSG4 (surface water) and SCV6 (groundwater at a depth of 8 m b.g.s.), and the continuous RBR™ temperature profiles shown in Fig. 5b. These data show the potential range in formation resistivity based on the measured range in specific conductance, along with the potential impact of temperature for three different porosity values. Porosities of 1 and 5 % correspond to the range measured in the core, while a value of 35 % would be representative of a rubble zone.

stance, major precipitation events coinciding with measurement events 13, 26, and 31 (refer to Fig. 4) were accompanied by short-period reductions in surface water conductivity and increases in temperature. Seasonal atmospheric temperature trends resulted in more gradual, yet sustained reductions in aqueous conductivity. In comparison, the groundwater specific conductance at 10.5 m b.g.s. was comparatively stable during the study period, exhibiting a moderate temperature-driven decline superimposed by shorter-period fluctuations associated with ion concentration.

Figure 7 shows the potential impact of these observed specific conductivity and temperature variations (based on Figs. 5b and 6) on the bulk formation resistivity using Eqs. (1) and (2) for three representative porosity values. Porosities of 1 and 5 % correspond to the values obtained in the core, while a porosity of 35 % is assumed to represent the maximum porosity of a weathered or broken rubble zone (i.e., unconsolidated porous sediment). These calculations indicate that variations in temperature will likely be the primary driver in formation resistivity dynamics. For instance, water temperature could affect the formation resistivity by as much as 46 % (between isotherms), based on the observed range in groundwater and surface water temperatures. In comparison, measured aqueous conductivity ranges (along an isotherm) for groundwater and surface water would affect the formation resistivity by 18 and 36 %, respectively. These maximum temperature effects represent end-member conditions for a given specific conductance and effective porosity estimate.

4.2 Sub-riverbed electrical resistivity distribution

Two ground conductivity surveys were conducted across the riverbed surface to assess spatial variability in bulk formation resistivity and its relationship to riverbed morphology (e.g., pool versus riffle): resistivity snapshots were collected on 7 July 2014 during low-flow conditions (1.30 $m^3\,s^{-1}$) and on 3 April 2013 during high-flow conditions (6.81 $m^3\,s^{-1}$) (Fig. 8a). The percentage change in resistivity from low- to high-stage periods is shown in Fig. 8b. The daily average river flows for the years 2013 and 2014 were 3.5 and 3.3 $m^3\,s^{-1}$, respectively.

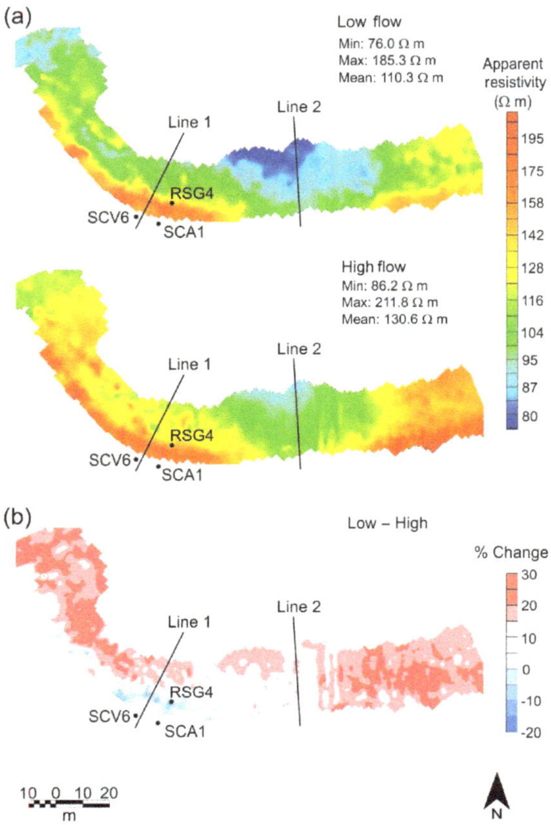

Figure 8. Riverbed resistivity obtained using an EM-31 ground conductivity meter during low-flow (low-stage) conditions on 7 July 2014 and high-flow (high-stage) conditions on 3 April 2013. **(b)** Percentage change in apparent resistivity from low- to high-stage periods.

Two main observations can be made from the changes observed between warmer low-flow and cooler high-flow conditions. First, the southern shoreline exhibited a 10 to 15 % reduction in resistivity along the southern shoreline within the pool and along the cut bank (outer edge of the elbow). These areas were characterized by more competent rock with exposed bedding plane fractures and vertical joints (Fig. 3b.i). Secondly, a broader and more variable increase in resistivity upwards of 20 to 25 % was observed northward into the thalweg, along the slip-off slope (inner edge of the elbow) and the eastern and western portions of the reach; the rock surfaces in these areas were more weathered with large irregular rock fragments and dissolution features, as well as limited exposure of horizontal bedding plane fractures. A lower resistivity zone (blue area) was identified upstream within the northern portion of the riffle section (Fig. 3b.iii). The riffle portion of the river was also accompanied by a break in the high resistivity trend observed along the southern shoreline. A lower average resistivity was observed during warmer low-flow conditions, indicating that a portion of the response may be dependent on formation temperature (i.e., 5 to 20 °C fluctuations). However, the reduction in resistivity along the cut bank during cooler high-stage periods indicates an increase in the specific conductance of the pore fluid, which would be consistent with increased baseflow and groundwater discharge to the river.

4.3 Time-lapse electrical resistivity imaging

4.3.1 Electrical resistivity models

Figure 9 provides a summary of the inverted model results at the pool (Fig. 9a) and riffle (Fig. 9b) sections for the full study period. The mean inverted model resistivity and data range for each sample event is presented along with the number of apparent resistivity data points removed from the dataset prior to inversion, and the root mean squared error of the inverted model. The partially frozen and frozen ground conditions were accompanied by higher signal noise due to a systematic increase in electrode contact resistance and reduction in pore fluid connectivity (liquid water saturation near the surface); here, lower potentials resulted in more frequent failed measurement, which had to be removed from the dataset prior to inversion. Therefore, the noisier datasets collected in these periods were accompanied by higher model RMS errors. A subset of the inverted resistivity models over the annual cycle (i.e., Fig. 9a–h) are shown in Fig. 10. These snapshots capture the spatiotemporal evolution of predominant geoelectrical conditions beneath the riverbed.

Spatial electrical resistivity data were highly variable across the pool (Fig. 10a–h). The highest resistivities were observed along the southern shoreline, which coincided with the presence of competent bedrock (Fig. 3b.i) with bedding plane fractured and vertical joint sets. Similarly resistive conditions extended southward onto the floodplain. Subsurface conditions became less resistive toward the northern shoreline, which coincided with the presence of increased vertical fracturing and dissolution-enhanced features, mechanically broken or weathered bedrock, and a thin layer of organic rich sediment alongside the northern shoreline and floodplain. Initial surveys conducted across the pool on 25 July 2014 identified a relatively low resistivity zone ($< 1000\,\Omega$m) extending 2 m beneath the riverbed that spanned the full width of the river. Measurements on 26 September through 24 December 2014 captured the retraction of this zone toward the northern shore. During this period the resistivity across the full transect increased only slightly. The onset of frozen ground and river conditions on 29 January 2014 resulted in an abrupt shift in the resistivity distribution. A high resistivity zone formed above the water table across the southern floodplain and was accompanied by an increase in resistivity across the full river profile. It is important to note that these frozen periods were accompanied by higher model RMS errors, and thus our interpretation of these data focus on long-period trends. The formation of river ice (visually observed basal and surface ice) may have altered the true geometry of

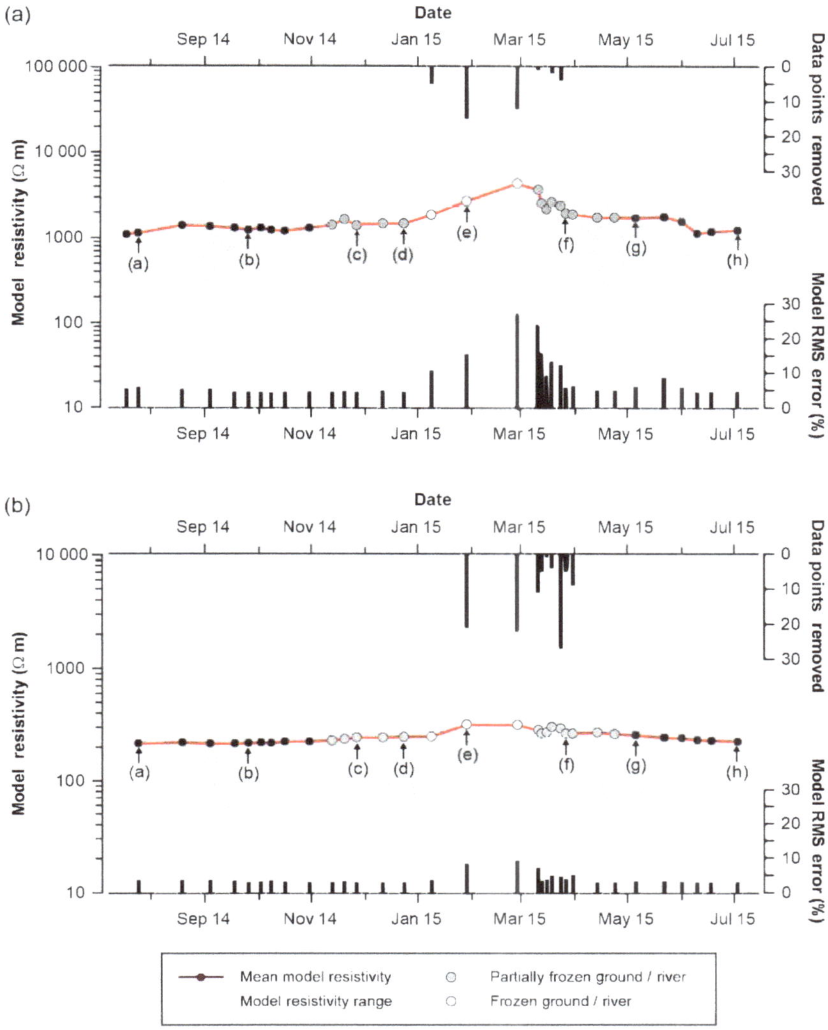

Figure 9. Temporal variations in inverted resistivity models for **(a)** pool and **(b)** riffle. Black dots represent unfrozen conditions and grey dots indicate partially frozen conditions, while white dots indicate completely frozen river conditions. Select resistivity models along the time series are shown in Fig. 10a–h.

the surface water body represented in the model, potentially contributing to the higher RMS errors along with the overall noisier measurements during this time (Fig. 9). The arrival of seasonal thaw conditions on 27 March 2015 was accompanied by reduced resistivities across the river as rock and river ice progressively thawed and was mobilized by spring freshet. Further seasonal warming on 6 May and 3 July 2015 resulted in a systematic decrease in riverbed resistivity from the northern to southern shoreline.

Riverbed resistivity across the riffle portion of the river (Fig. 10a–h) was markedly different with respect to the distribution and magnitude of resistivity fluctuations. The riffle exhibited a zone of comparatively low resistivity ($< 100\ \Omega$m) that extended slightly deeper than that at the pool, to a depth of 3 m. The initial survey on 26 July 2014 identified a zone of very low resistivity that progressively became more resistive over time (26 September through 24 December 2014). Much like the pool, however, this low resistivity zone reverted back toward the northern shoreline. The onset of seasonally frozen river conditions was accompanied by an increase in resistivity across a significant portion of the riverbed. Inverse models during frozen water conditions were again accompanied by higher RMS errors, which we attribute to the formation of river ice. Unlike the pool, which experienced the formation of a substantial zone of ground frost along the southern shore, less ground frost was observed at depth along the riverbanks bounding the riffle. Spring thaw brought reduced resistivities across the riverbed with subtle lateral variations, followed by the beginnings of a less-resistive riverbed zone emanating southward from the northern shoreline. The bedrock resistivity below 3 m depth remained relatively constant throughout the monitoring period.

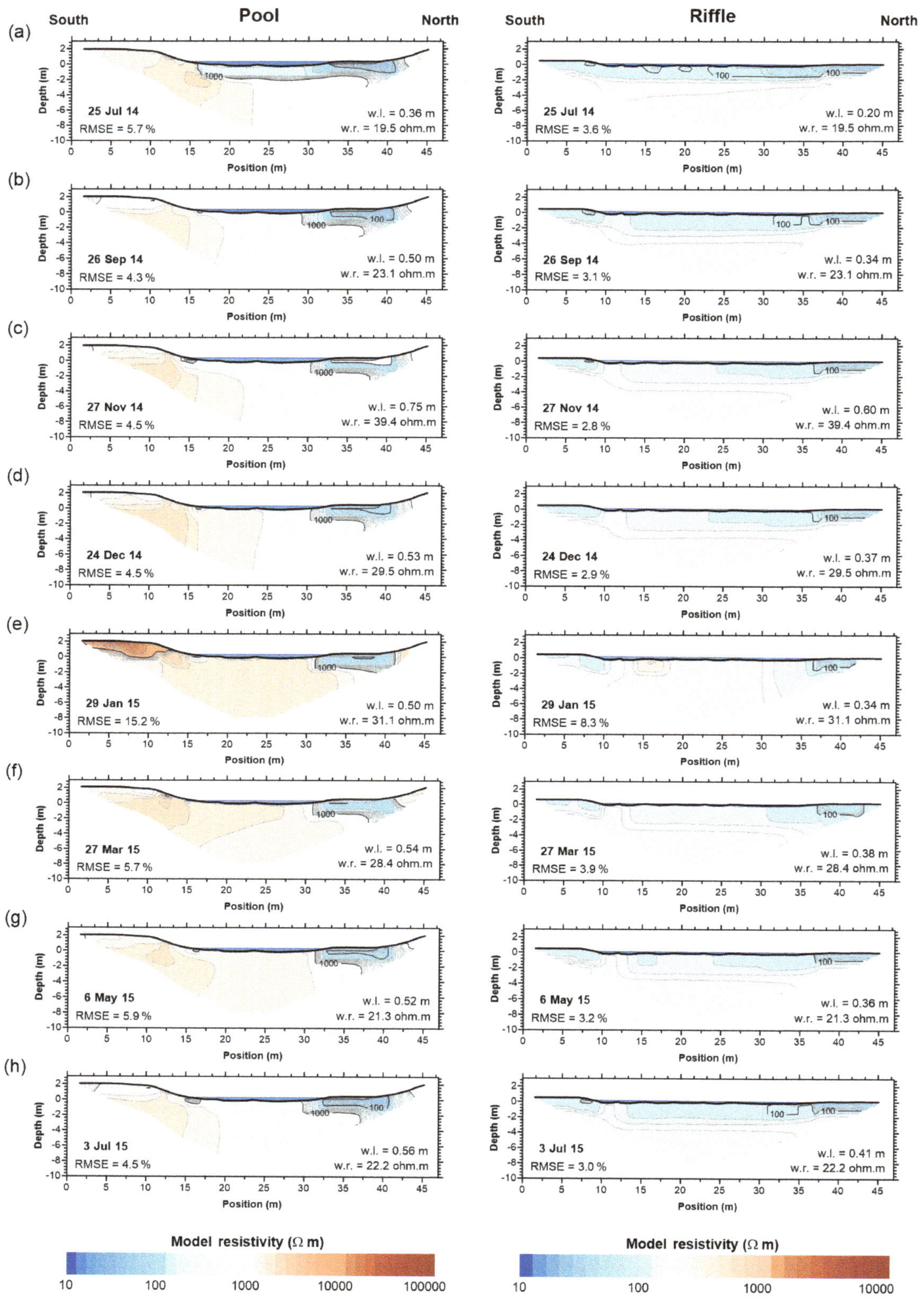

Figure 10. Representative inverse resistivity models across the pool and riffle, orientated from south to north. Datasets **(a–h)** are identified in Fig. 9. River stage (w.l.) and surface water resistivity (w.r.) values were fixed in the inverse model. A marked increase in resistivity was observed beneath the river during colder seasonal conditions (November through March), while lower resistivities were observed during warmer seasonal conditions (July).

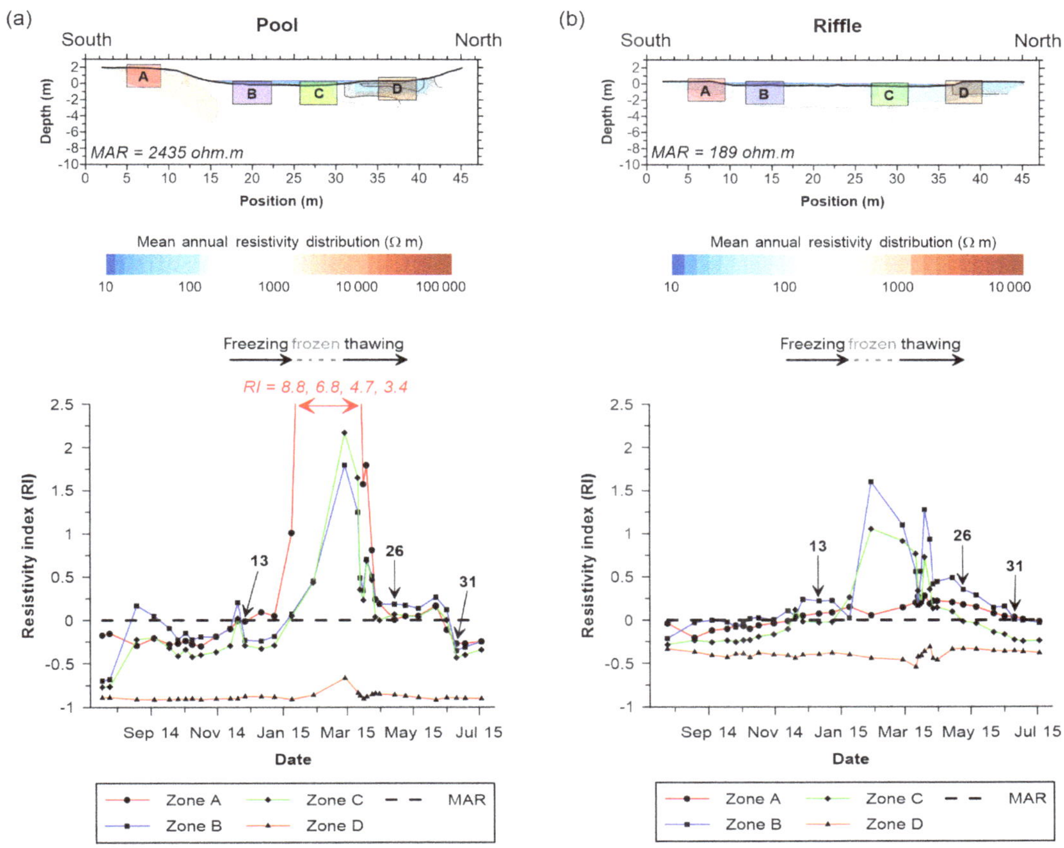

Figure 11. Spatiotemporal fluctuations in resistivity within the focused monitoring zones A, B, C, and D. The resistivity index (RI) was calculated using Eq. (3), using the mean zone resistivity (MZR) for a given measurement date and the mean annual resistivity (MAR) of the whole profile.

4.3.2 Spatiotemporal resistivity trends

A resistivity index was calculated using Eq. (3) to compare spatiotemporal variations in electrical resistivity within predefined zones of the bedrock beneath the river (Fig. 11); zones A and D represent conditions along the south and north riverbank, while zones B and C represent conditions within the southern and northern portions the river. These zones were defined based on their representative areas and the magnitude of the temporal fluctuations observed over the full monitoring period (Fig. 10). A RI of zero indicates a mean zone resistivity (MZR) that is equal to the mean annual resistivity (MAR) of the whole profile. An index of $+1$ indicates a resistivity that is twice the annual mean, while an index of -0.5 indicates a resistivity that is half the annual mean.

The RI time series for the pool (Fig. 11a) and riffle (Fig. 11b) capture the magnitude and frequency of the temporal variability observed within these four zones. Relative to the MAR, the pool exhibited larger and more frequent fluctuations in resistivity compared to the riffle. The southern shoreline (zone A) at the pool was more dynamic than the corresponding zone at the riffle; zone A at the pool encompasses a larger unsaturated zone, which is more likely impacted by changes in temperature and saturation, especially during the freezing and thawing period. The northern shoreline (zone D) at the pool and riffle exhibited lower-than-average resistivities with relatively minimal transience over the study period, with the exception of the mid-to-late-winter freeze-up. Here, a variable layer of sediment and organic matter with higher water content likely moderated freeze–thaw fluctuations relative to sections of exposed rock. Conditions below the riverbed (zones B and C) exhibited both longer-period (seasonal) and shorter-period (intraseasonal) fluctuations. While the relative changes observed at the pool were larger than the riffle, similar seasonal trends were observed at each location. Zones B and C at the pool were mutually consistent, while those at the riffle were less consistent.

Although perturbations were observed in the resistivity beneath the riverbed before and after winter freeze-up (e.g., zones B and C), the responses were dampened relative to the winter period. Events 13, 26, and 31 (Fig. 6), which correspond to periods of increase precipitation, may coincide with observed perturbations in the RI; however, based on these data it is not clear whether the riverbed resistivity and surface water responses are mutually consistent. Here, a limited relationship may suggest that groundwater–surface water interaction does not occur in this section of the river or that groundwater discharge is strong enough to limit potential groundwater–surface water mixing at this particular location. Therefore, these observed geophysical dynamics within the riverbed may be more associated with seasonal temperature transience, with secondary influence due to solute-based fluctuations.

5 Discussion

5.1 Influence of water properties on formation resistivity

Riverbed electrical resistivity mapping during low and high stage periods identified a spatiotemporal response within the upper 6 m of rock, with the southern cut bank exhibiting a reduction in resistivity while the remaining portions of the river increased in resistivity. This spatiotemporal response, together with observed bedrock surface conditions, indicates that riverbed morphology strongly impacts groundwater dynamics below the riverbed.

Long-term resistivity monitoring along the fixed profiles over the pool and riffle portions of the river revealed a transient zone within the upper 2 and 3 m of bedrock, respectively. In particular, the formation of a low resistivity zone (high electrical conductivity) was observed during the warmer summer period that diminished as the environment cooled. While pore water conductivity depends on electrolytic concentration and temperature, their independent variability could not be decoupled across the entire study area given in situ sensor deployment limitations in a bedrock environment. Although this uncertainty in the driving mechanism of observed electrical changes below the riverbed hindered our ability to definitively define the vertical extent of a potential groundwater–surface water mixing zone, our geophysical dataset does suggest that a groundwater–surface water mixing in a bedrock environment may be more limited due to the lower effective porosity of rock and heterogeneous and anisotropic fracture distributions.

Aqueous temperature and specific conductance measurements collected in the river stage gauge (RSG1) and shallow bedrock well (SCV6) provided end-member conditions (Fig. 6). These data were used to assess the influence of aqueous conditions on bulk formation resistivity (Fig. 7). While some degree of overlap was observed between groundwater and surface water properties, they were generally differentiable across the study area. That being said, aqueous temperature fluctuations dominated the bulk electrical response over the full annual cycle. Given the impact of temperature on the bulk formation resistivity, observed bedrock resistivity dynamics are attributed to changes in formation water temperature, with secondary effects caused by changes in electrolytic concentration. These findings are consistent with Musgrave and Binley (2011), who concluded that temperature fluctuations over an annual cycle within a temperate wetland environment, with groundwater electrical conductivities ranging from 400 to 850 $\mu S\,cm^{-1}$, dominated formation resistivity transience. Of course, as our annual temperature range was more extreme than that of Musgrave and Binley, we examined electrical dynamics within a less-porous, more heterogeneous and anisotropic medium and captured a broader range of seasonal conditions including ground frost and riverbed ice formation.

Measurements collected with a shorter time step (diurnal) and shorter electrode spacing may have captured more transient rainfall or snowpack melt episodes focused along discrete fracture pathways, possibly leading to the identification of electrolytic-induced transients beneath the riverbed, which would be more indicative of a groundwater–surface water mixing zone. Based on the short period of intraseasonal fluctuations observed in Fig. 11, and the timing and duration of major precipitation or thawing events (5 to 7 day cycles; Fig. 4), it is reasonable to assume that our geophysical time step (days to weeks) was accompanied by some degree of aliasing. Finally, it is possible that shallower sections of rock within the river exposed to direct sunlight during the day, which can vary depending on cloud cover (daily) and the Sun's position in the sky (seasonally), may have exhibited a wider range, or more transient temperature fluctuations, than those areas beneath or adjacent to a canopy. A closer inspection of the unfrozen temporal response in zone B reveals a wider range in resistivity relative to the more northern zone C. At this latitude in the Northern Hemisphere the southern shore will receive more direct sunlight; therefore, it is possible that the shallow rock on this side of the river experienced greater fluctuations in temperature (both seasonally and diurnally), thereby contributing to the observed geoelectrical dynamics. Although diurnal fluctuations in surface water temperature can be significant ($> 10\,°C$; e.g., Constantz et al., 1994; Constantz, 1998), the relative effects of such transient temperature fluctuations on our geophysical measurements cannot be assessed given our more seasonally scaled measurement sampling interval.

5.2 Formation of ground frost and anchor ice

A dramatic increase in bedrock resistivity was observed with the onset of freezing ground conditions; this can impact a wide range of infrastructure (e.g., dams, hydropower generation), ecologic (e.g., alteration of fish and benthic habitats), and hydraulic functions (e.g., river storage, baseflow) (Beltaos and Burrell, 2015). The formation of a highly resistive zone consistent with a seasonal frost front within the unsaturated portion of the riverbank (Fig. 10e; zone A in Fig. 11a), as well as the accumulation of river ice, resulted in marked changes in resistivity. These winter-season effects are readily evident in Fig. 11 at the pool and riffle. Here, the magnitude of the resistivity increase may reflect a potential reduction in the hydraulic connectivity between surface water and groundwater during the winter months.

Ground frost primarily formed along the riverbank over the southern floodplain at the pool. This topographically higher area was relatively devoid of large shrubs and trees (Fig. 2), and thus likely experienced more severe weather conditions (e.g., higher winds resulted in less snow pack to insulate the ground). These conditions would promote the formation of a thicker frost zone which propagated to the water table (pool in Fig. 10e). The adjacent northern riverbank and those up-gradient at the riffle were topographically lower (i.e., thinner unsaturated zone) and were sheltered by large trees. The formation of a seasonal frost zone along the riverbank may have implications to baseflow dynamics during the winter months and the early-thaw period.

A simple sensitivity analysis of the inversion process using different constraints on surface water geometry and aqueous electrical resistivity indicated that model convergence was highly sensitive to surface water geometry and aqueous resistivity. For instance, setting an aqueous resistivity of one-half the true value led to poor model convergence and unrealistic resistivity values for bedrock. The riffle was relatively less sensitive to surface water properties, likely because of its overall lower river stage compared to the pool and, hence, relatively lower impact of the surface water body on the apparent resistivity measurement. This sensitivity to surface water properties is a consequence of the high electrical conductivity of the surface water relative to high resistivity bedrock.

Anchor ice reduced the electrical connectivity across the riverbed, while the ice crust along the surface of the water altered the effective geometry of the water body, further influencing the inverse solution. While the formation of river ice was accompanied by higher RMS errors at the pool ($> 6\%$) and riffle ($> 4\%$; Fig. 9), direct measurements of river ice thickness and spatial extent could not be collected or explicitly incorporated into the surface water layer geometry during the winter months.

5.3 Implications to the conceptualization of groundwater–surface water exchange in bedrock rivers

The fractured dolostone in this study consists of a visible orthogonal joint network approximately orientated at 10 to 20° north-northwest and 280 to 290° south-northwest, consistent with the regional joint orientations, with frequencies ranging from centimetre to sub-metre scale. Streambed resistivity measurements indicate a seasonally dynamic groundwater zone within the upper 2 to 3 m of riverbed. A less-resistive zone ($< 1000\,\Omega$m) was observed beneath the pool emanating from the northern shoreline during warmer low-flow periods (July and August 2014). This zone retracted in late summer but showed signs of reappearance in early July 2015. A similarly evolving low-resistivity zone ($< 100\,\Omega$m) was observed across the riffle, but was more variable across the river transect. Although dynamic fluctuations in temperature and aqueous conductivity support the potential existence of a groundwater–surface water mixing zone, it is not yet clear how these geoelectrical dynamics were influenced or enhanced by fluid flow in the discrete fractures, exchange between the mobile and immobile pore phase, and seasonal thermal gradients across the riverbed.

Discrete fracture networks and dissolution-enhanced features will result in a more heterogeneous and anisotropic groundwater–surface water mixing zones compared to porous unconsolidated sediment. Swanson and Cardenas (2010) examined the utility of using heat as a tracer of groundwater–surface water exchange across a pool–riffle–pool sequence. Observed thermal patterns and zones of influence (i.e., effective mixing zones) in their study were consistent with conceptual models depicting a pool–riffle–pool sequence. While similar temperature dynamics may be expected across the pool–riffle–pool sequence in a bedrock environment, our coarser temporal sampling interval (days to weeks) combined with our smoothed resistivity models limited our ability to capture subtle diel temperature transience across discrete fractures or flow features. Although the electrical resistivity method was not able to definitively differentiate between groundwater and surface water, our geophysical measurements do provide insight into the magnitude, lateral extent, and spatiotemporal scale of geoelectrical transience, which are largely driven by temperature fluctuations within the upper few metres of rock.

6 Conclusions

Time-lapse resistivity measurements were performed across a 200 m reach of the Eramosa River during low and high-stage periods. Our results showed high spatiotemporal variability within the riverbed that could be attributed to its exposure of the bedding plane and vertical fracture, as well as the morphology or competency of the rock surface. Complementary fracture and temperature profiling within the open and lined borehole revealed abundant active groundwater flow zones spanning the upper 8–10 m of bedrock, with strong intra-seasonal and seasonal temperature variations along horizontal fracture sets. While surface resistivity profiles captured geoelectrical dynamics within the upper few metres of rock, these data represent indirect evidence of riverbed transience resulting from changes in groundwater temperature and specific conductance. Geoelectrical transience was primarily governed by seasonal temperature trends, with secondary effects arising from porewater conductance; however, spatiotemporal variations in temperature and specific conductance could not be decoupled beyond the fixed monitoring points.

Time-lapse electrical resistivity imaging of the pool and riffle portion of the river, sampled on a semi-daily to semi-weekly interval, showed consistently higher resistivity at the pool, with more elevated resistivities along the southern shoreline. Seasonal cooling was accompanied by the formation of a higher-resistivity zone emanating from the southern shore to northern shore in both the pool and riffle. This resistivity trend reversed during the seasonal warming cycle, becoming less resistive toward the southern shoreline as seasonal temperatures increased and river stage decreased. The formation of ground frost and basal ice along the riverbed had a strong impact on the seasonal resistivity profiles during the winter months. Intraseasonal (i.e., spring, summer, fall) geoelectrical changes beneath the riverbed were strongly dependent upon seasonal temperature trends, with the pool and riffle exhibiting variable horizontal and vertical zones of influence. Geoelectrical transience associated with major precipitation events, which were accompanied by short-period perturbations in surface water temperature and specific conductance, had a relatively small impact on sub-riverbed resistivity. This could be explained by a seasonally sustained groundwater discharge zone across this reach of the river, which would have limited or moderated electrical resistivity changes associated with surface water mixing with groundwater.

This study demonstrated that time-lapse resistivity measurements have the capacity to image the magnitude and scale of transience within a bedrock riverbed. Although riverbed resistivity dynamics were largely a function of seasonal atmospheric temperature trends, the ability to map the geometrical extent of such heterothermic zones beneath a river would be relevant to understanding biogeochemical processes, benthic activity, and macro-scale hyporheic zone processes. Imaging the magnitude and scale of transience within the riverbed will be critical to the advancement of our understanding of mechanisms controlling groundwater–surface water exchange within fractured bedrock rivers. Given the challenges and costs associated with installing sensors within rock and effectively sampling a heterogeneous flow system, minimally invasive surface and borehole geophysical methods offer an ideal alternative and possibly a more effective approach for long-term groundwater–surface water monitoring of bedrock environments, by reducing instrumentation costs and impacts to ecosensitive environments.

Competing interests. The authors declare that they have no conflict of interest.

Acknowledgements. This research was made possible through funding by the Natural Sciences and Engineering Research Council of Canada in the form of a Banting Fellowship to Colby M. Steelman, and an Industrial Research Chair (grant no. IRCPJ363783-06) to Beth L. Parker. The authors are very appreciative of the field contributions of staff and students at the University of Guelph and University of Waterloo, particularly field technicians Dan Elliot and Bob Ingleton. This work would not have been possible without land-access agreements with Scouts Canada and the Ontario Ministry of Natural Resources and river-flow data provided by the Grand River Conservation Authority.

Edited by: Mauro Giudici

References

Alexander, M. D. and Caissie, D.: Variability and comparison of hyporheic water temperatures and seepage fluxes in a small Atlantic Salmon stream, Ground Water, 41, 72–82, 2003.

Anderson, M. P.: Heat as a ground water tracer, Ground Water, 43, 951–968, 2005.

Archie, G. E.: The electrical resistivity log as an aid in determining some reservoir characteristics, T. Am. Inst. Mineral. Metall. Petrol. Eng., 146, 54–62, 1942.

Arps, J. J.: The effect of temperature on the density and electrical resistivity of sodium chloride solutions, Trans. A.I.M.E., 164, 54–62, 1953.

Beltaos, S. and Burrell, B. C.: Hydrotechnical advances in Canadian river ice science and engineering during the past 35 years, Can. J. Civil. Eng., 42, 583–591,https://doi.org/10.1139/cjce-2014-0540, 2015.

Bianchin, M. S., Smith, L., and Beckie, R. D.: Defining the hyporheic zone in a large tidally influenced river, J. Hydrol., 406, 16–29, https://doi.org/10.1016/j.jhydrol.2011.05.056, 2011.

Binley, A., Ullah, S. Heathwaite, A. L., Heppell, C., Byrne, P., Lansdown, K., Trimmer, M., and Zhang, H.: Revealing the spatial variability of water fluxes at the groundwater-surface water interface, Water Resour. Res., 49, 3978–3992, https://doi.org/10.1002/wrcr.20214, 2013.

Boano, F., Revelli, R., and Ridolfi L.: Reduction of the hyporheic zone volume due to the stream-aquifer interaction, Geophys. Res. Lett., 35, L09401, https://doi.org/10.1029/2008GL033554, 2008.

Brassington, R.: Field Hydrogeology, John Wiley & Sons, Inc, Chichester, 1998.

Briggs, M. A., Lautz, L. K., McKenzie, J. M., Gordon, R. P., and Hare, D. K.: Using high-resolution distributed temperature sensing to quantify spatial and temporal variability in vertical hyporheic flux, Water Resour. Res., 48, W02527, https://doi.org/10.1029/2011WR011227, 2012.

Brunton, F. R.: Update of revisions to the Early Silurian stratigraphy of the Niagara Escarpment: integration of sequence stratigraphy, sedimentology and hydrogeology to delineate hydrogeologic units, in: Summary of Field Work and Other Activities 2009, Ontario Geologic Survey, Open File Report 6240, 25-1–25-20, 2009.

Campbell, R. B., Bower, C. A., and Richards, L. A.: Change of electrical conductivity with temperature and the relation of osmotic pressure to electrical conductivity and ion concentration for soil extracts, Soil Sci. Soc. Am. Proc., 12, 66–69, 1948.

Cardenas, M. B. and Markowski, M. S.: Geoelectrical imaging of hyporheic exchange and mixing of river water and groundwater in a large regulated river, Environ. Sci. Technol., 45, 1407–1411, https://doi.org/10.1021/es103438a, 2011.

Cole, J., Coniglio, M., and Gautrey, S.: The role of buried bedrock valleys on the development of karstic aquifers in flat-lying carbonate bedrock: insights from Guelph, Ontario, Canada, Hydrogeol. J., 17, 1411–1425, https://doi.org/10.1007/s10040-009-0441-3, 2009.

Conant Jr., B.: Delineating and quantifying ground water discharge zones using streambed temperature, Ground Water, 42, 243–257, 2004.

Constantz, J.: Interaction between stream temperature, streamflow, and groundwater exchanges in alpine streams, Water Resour. Res., 34, 1609–1615, 1998.

Constantz, J.: Heat as a tracer to determine streambed water exchanges, Water Resour. Res., 44, W00D10, https://doi.org/10.1029/2008WR006996, 2008.

Constantz, J., Thomas, C. L., and Zellweger, G.: Influence of diurnal variations in stream temperature on streamflow lows and groundwater recharge, Water Resour. Res., 30, 3253–3264, 1994.

Coscia, I., Greenhalgh, S. A., Linde, N., Doetsch, J., Marescot, L., Günther, T., Vogt, T., and Green, A. G.: 3D crosshole ERT for aquifer characterization and monitoring of infiltrating river water, Geophysics, 76, G49–G59, https://doi.org/10.1190/1.3553003, 2011.

Coscia, I., Linde, N., Greenhalgh, S., Vogt, T., and Green, A.: Estimating traveltimes and groundwater flow pattern using 3D time-lapse crosshole ERT imaging of electrical resistivity fluctuations induced by infiltrating river water, Geophysics, 77, E239–E250, https://doi.org/10.1190/GEO2011-0328.1, 2012.

Crook, N., Binley, A., Knight, R., Robinson, D. A., Zarnetske, J., and Haggerty, R.: Electrical resistivity imaging of the architecture of substream sediments, Water Resour. Res., 44, W00D13, https://doi.org/10.1029/2008WR006968, 2008.

Crosbie, R. S., Taylor, A. R., Davis, A. C., Lamontagne, S., and Munday, T.: Evaluation of infiltration from losing-disconnected rivers using a geophysical characterisation of the riverbed and a simplified infiltration model, J. Hydrol., 508, 102–113, https://doi.org/10.1016/j.jhydrol.2013.07.045, 2014.

Doveton, J. H.: Log analysis of subsurface geology: concepts and computer methods, Wiley-Interscience, New York, 1986.

Dimova, T. N., Swarzenski, P. W., Dulaiova, H., and Glenn, C. R.: Utilizing multichannel electrical resistivity methods to examine the dynamics of the fresh water-seawater interface in two Hawaiian groundwater systems, J. Geophys. Res., 117, C02012, https://doi.org/10.1029/2011JC007509, 2012.

Doetsch, J., Linde, N., Vogt, T., Binley, A., and Green, A. G.: Imaging and quantifying salt-tracer transport in a riparian groundwater system by means of 3D ERT monitoring, Geophysics, 77, B207–B218, https://doi.org/10.1190/GEO2012-0046.1, 2012.

Doro, K. O., Leven, C., and Cirpka, O. A.: Delineating subsurface heterogeneity at a loop of River Steinlach using geophysical and hydrogeological methods, Environ. Earth Sci., 69, 335–348, https://doi.org/10.1007/s12665-013-2316-0, 2013.

Ellis, D. V.: Well logging for earth scientists, Elsevier Science Publishing Company, Inc., New York, 1987.

Evans, E. C., Greenwood, M. T., and Petts, G. E.: Thermal profiles within river beds, Hydrol. Process., 9, 19–25, 1995.

Fan, Y., Toran, L., and Schlische, R. W.: Groundwater flow and groundwater-stream interaction in fractured and dipping sedimentary rocks: insights from numerical models, Water Resour. Res., 43, W01409, https://doi.org/10.1029/2006WR004864, 2007.

Froese, D. G., Smith, D. G., and Clement, D. T.: Characterizing large river history with shallow geophysics: middle Yukon River, Yukon Territory and Alaska, Geomorphology, 67, 391–406, https://doi.org/10.1016/j.geomorph.2004.11.011, 2005.

Glover, P. W. J.: A generalized Archie's law for n phases, Geophysics, 75, E247–E265, https://doi.org/10.1190/1.3509781, 2010.

Gourry, J.-C., Vermeersch, F., Garcin, M., and Giot, D.: Contribution of geophysics to the study of alluvial deposits: a case study in the Val d'Avaray area of the River Loire, France, J. Appl. Geophys., 54, 35–49, https://doi.org/10.1016/j.jappgeo.2003.07.002, 2003.

Harrington, G. A., Gardner, W. P., and Munday, T. J.: Tracking groundwater discharge to a large river using tracers and geophysics, Ground Water, 52, 837–852, https://doi.org/10.1111/gwat.12124, 2014.

Hatch, C. E., Fisher, A. T., Revenaugh, J. S., Constantz, J., and Ruehl, C.: Quantifying surface water-groundwater interactions using time series analysis of streambed thermal records: Method development, Water Resour. Res., 42, W10410, https://doi.org/10.1029/2005WR004787, 2006.

Irvine, D. J., Briggs, M. A., Lautz, L. K., Gordon, R. P., McKenzie, J. M., and Cartwright, I.: Using diurnal temperature signals to infer vertical groundwater-surface water exchange, Ground Water, 55, 10–26, https://doi.org/10.1111/gwat.12459, 2016.

Johnson, T. C., Slater, L. D., Ntarlagiannis, D., Day-Lewis, F. D., and Elwaseif, M.: Monitoring groundwater-surface water interaction using time-series and time-frequency analysis of transient three-dimensional electrical resistivity changes, Water Resour. Res., 48, W07506, https://doi.org/10.1029/2012WR011893, 2012.

Käser, D. H., Binley, A., and Heathwaite, A. L.: On the importance of considering channel microforms in groundwater models of hyporheic exchange, River Res. Appl., 29, 528–535, https://doi.org/10.1002/rra.1618, 2013.

Keery, J., Binley, A., Crook, N., and Smith, J. W. N.: Temoral and spatial variability of groundwater-surface water fluxes: Development and application of an analytical method using temperature time series, J. Hydrol., 336, 1–16, https://doi.org/10.1016/j.jhydrol.2006.12.003, 2007.

Keller, G. V.: Section V: Electrical properties, in: CRC Practical handbook of physical properties of rock and minerals, edited by: Carmichael, R. S., CRC Press, 361–427, 1989.

Keller, C., Cherry, J. A., and Parker, B. L.: New method for continuous transmissivity profiling in fractured rock, Ground Water, 52, 352–367, https://doi.org/10.1111/gwat.12064, 2014.

Knight, R. J. and Endres, A. L.: Chapter 3: An Introduction to Rock Physics Principles for Near-Surface Geophysics, in: Near-Surface Geophysics, Society of Exploration Geophysics, edited by: Butler, D. K., Tulsa, Oklahoma, 31–70, 2005.

Kunert, M. and Coniglio, M.: Origin of vertical shafts in bedrock along the Eramosa River valley near Guelph, southern Ontario, Can. J. Earth Sci., 39, 43–52, https://doi.org/10.1139/E01-053, 2002.

Kunert, M., Coniglio, M., and Jowett, E. C.: Controls and age of cavernous porosity in Middle Silurian dolomite, southern Ontario, Can. J. Earth Sci., 35, 1044–1053, 1998.

Lemieux, J.-M., Kirkwood, D., and Therrien, R.: Fracture network analysis of the St-Eustache quarry, Quebec, Canada, for groundwater resources management, Can. Geotech. J., 46, 828–841, 2009.

Loke, M. H.: Tutorial: 2-D and 3-D electrical imaging surveys, Geotomo Software, Malaysia, 2002.

Loke, M. H. and Dahlin, T.: A comparison of the Gauss-Newton and quasi-Newton methods in resistivity imaging inversion, J. Appl. Geophys., 49, 149–162, 2002.

Loke, M. H., Chambers, J. E., Rucker, D. F., Kuras, O., and Wilkinson, P. B.: Recent developments in the direct-current geoelectrical imaging method, J. Appl. Geophys., 95, 135–156, https://doi.org/10.1016/j.jappgeo.2013.02.017, 2013.

McLaren, R. G., Sudicky, E. A., Park, Y.-J., and Illman, W. A.: Numerical simulation of DNAPL emissions and remediation in a fractured dolomitic aquifer, J. Contam. Hydrol., 136–137, 56–71, https://doi.org/10.1016/j.jconhyd.2012.05.002, 2012.

McNeill, J. D.: Electromagnetic terrain conductivity measurement at low induction numbers, Geonics Ltd. Technical Note, TN-6, Mississauga, Ontario, 1980.

Meyer, J. R., Parker, B. L., and Cherry, J. A.: Detailed hydraulic head profiles as essential data for defining hydrogeologic units in layered fractured sedimentary rock, Environ. Geol., 56, 27–44, https://doi.org/10.1007/s00254-007-1137-4, 2008.

Meyerhoff, S. B., Karaoulis, M., Fiebig, F., Maxwell, R. M., Revil, A., Martin, J. B., and Graham, W. D.: Visualization of conduit-matrix conductivity differences in a karst aquifer using time-lapse electrical resistivity, Geophys. Res. Lett., 39, L24401, https://doi.org/10.1029/2012GL053933, 2012.

Meyerhoff, S. B., Maxwell, R. M., Revil A., Martin, J. B., Karaoulis, M., and Graham, W. D.: Characterization of groundwater and surface water mixing in a semiconfined karst aquifer using time-lapse electrical resistivity tomography, Water Resour. Res., 50, 2566–2585, https://doi.org/10.1002/2013WR013991, 2014.

Miller, C. R., Routh, P. S., Brosten, T. R., and McNamara, J. P.: Application of time-lapse ERT imaging to watershed characterization, Geophysics, 73, G7–G17, https://doi.org/10.1190/1.2907156, 2008.

Musgrave, H. and Binley, A.: Revealing the temporal dynamics of subsurface temperature in a wetland suing time-lapse geophysics, J. Hydrol., 396, 258–266, https://doi.org/10.1016/j.jhydrol.2010.11.008, 2011.

Naegeli, M. W., Huggenberger, P., and Uehlinger, U.: Ground penetrating radar for assessing sediment structures in the hyporheic zone of a prealpine river, J. N. Am. Benthol. Soc., 15, 353–366, 1996.

Norman, F. A. and Cardenas, M. B.: Heat transport in hyporheic zones due to bedforms: An experimental study, Water Resour. Res., 50, 3568–3582, https://doi.org/10.1002/2013WR014673, 2014.

Novakowski, K. S. and Lapcevic, P. A.: Regional hydrogeology of the Silurian and Ordovician sedimentary rock underlying Niagara Falls, Ontario, Canada, J. Hydrol., 104, 211–236, 1998.

Nyquist, J., Freyer, P. A., and Toran, L.: Stream bottom resistivity tomography to map ground water discharge, Ground Water, 46, 561–569, https://doi.org/10.1111/j.1745-6584.2008.00432.x, 2008.

Orlando, L.: Some consideration on electrical resistivity imaging for characterization of waterbed sediments, J. Appl. Geophys., 95, 77–89, https://doi.org/10.1016/j.jappgeo.2013.05.005, 2013.

Oxtobee, J. P. A. and Novakowski, K.: A field investigation of groundwater/surface water interaction in a fractured bedrock environment, J. Hydrol., 269, 169–193, 2002.

Oxtobee, J. P. A. and Novakowski, K. S.: Ground water/surface water interaction in a fractured rock aquifer, Ground Water, 41, 667–681, 2003.

Pehme, P. E., Parker, B. L., Cherry, J. A., and Greenhouse, J. P.: Improved resolution of ambient flow through fractured rock with temperature logs, Ground Water, 48, 191–205, https://doi.org/10.1111/j.1745-6584.2009.00639.x, 2010.

Pehme, P. E., Parker, B. L., Cherry, J. A., and Blohm, D.: Detailed measurement of the magnitude and orientation of thermal gradients in lined coreholecoreholes for characterizing groundwater flow in fractured rock, J. Hydrol., 513, 101–114, https://doi.org/10.1016/j.jhydrol.2014.03.015, 2014.

Perrin, J., Parker, B. L., and Cherry, J. A.: Assessing the flow regime in a contaminated fractured and karstic dolostone aquifer supplying municipal water, J. Hydrol., 400, 396–410, https://doi.org/10.1016/j.jhydrol.2011.01.055, 2011.

Reynolds, J. M.: An Introduction to Applied and Environmental Geophysics, 2nd Edn., John Wiley & Sons, Chichester, 2011.

Rhoades, J. D., Raats, P. A. C., and Prather, R. J.: Effect of liquid-phase electrical conductivity, water content, and surface conductivity on bulk soil electrical conductivity, Soil Sci. Soci. Am. J., 40, 651–655, 1976.

Rucker, D. F., Noonan, G. E., and Greenwood, W. J.: Electrical resistivity in support of geological mapping along the Panama Canal, Eng. Geol., 117, 121–133, 2011.

Sambuelli, L., Leggieri, S., Calzoni, C., and Porporato, C.: Study of riverine deposits using electromagnetic methods at a low induction number, Geophysics, 72, B113–B120, https://doi.org/10.1190/1.2754249, 2007.

Schmidt, C., Conant Jr., B., Bayer-Raich, M., and Schirmer, M.: Evaluation and field-scale application of an analytical method to quantify groundwater discharge using mapped streambed temperatures, J. Hydrol., 347, 292–307, https://doi.org/10.1016/j.jhydrol.2007.08.022, 2007.

Silliman, S. E. and Booth, D. F.: Analysis of time-series measurements of sediment temperature for identification of gaining vs. losing portions of Juday Creek, Indiana, J. Hydrol., 146, 131–148, 1993.

Singha, K., Pidlisecky, A., Day-Lewis, F. D., and Gooseff, M. N. Electrical characterization of non-Fickian transport in groundwater and hyporheic systems, Water Resour. Res., 44, W00D07, https://doi.org/10.1029/2008WR007048, 2008.

Singha, K., Day-Lewis, F. D., Johnson, T., and Slater, L. D.: Advances in interpretation of subsurface processes with time-lapse electrical imaging, Hydrol. Process., 29, 1549–1576, https://doi.org/10.1002/hyp.10280, 2015.

Sirieix, C., Riss, J., Rey, F., Prétou, F., and Lastennet, R.: Electrical resistivity tomography to characterize a karstic Vauclusian spring: Fontaine d'Orbe (Pyrénées, France), Hydrogeol. J., 22, 911–924, https://doi.org/10.1007/s10040-013-1095-8, 2014.

Slater, L., Binley, A. M., Daily, W., and Johnson, R.: Cross-hole electrical imaging of a controlled saline tracer injection, J. Appl. Geophys., 44, 85–102, 2000.

Slater, L. D., Ntarlagiannis D., Day-Lewis, F. D., Mwakanyamale, K., Versteeg, R. J., Ward, A., Strickland, C., Johnson, C. D., and Lane Jr., J. W.: Use of electrical imaging and distributed temperature sensing method to characterize surface water-groundwater exchange regulating uranium transport at the Hanford 300 Area, Washington, Water Resour. Res., 46, W10533, https://doi.org/10.1029/2010WR009110, 2010.

Snieder, R. and Trampert, J.: Inverse problems in geophysics, in: Wavefield Inversion, edited by: Wirgin, A., Springer-Verlag, New York, 119–190, 1999.

Stickler, M. and Alfredsen, K. T.: Anchor ice formation in streams: a field study, Hydrol. Process., 23, 2307–2315, https://doi.org/10.1002/hyp.7349, 2009.

Steelman, C. M., Kennedy, C. S., and Parker, B. L.: Geophysical conceptualization of a fractured sedimentary bedrock riverbed using ground-penetrating radar and induced electrical conductivity, J. Hydrol., 521, 433–446, https://doi.org/10.1016/j.jhydrol.2014.12.001, 2015a.

Steelman, C. M., Parker, B. L., and Kennedy, C. S.: Evaluating local-scale anisotropy and heterogeneity along a fractured sedimentary bedrock river using EM azimuthal resistivity and ground-penetrating radar, J. Appl. Geophys., 116, 156–166, https://doi.org/10.1016/j.jappgeo.2015.03.003, 2015b.

Swanson, T. E. and Cardenas, M. B.: Diel heat transport within the hyporheic zone of a pool-riffle-pool sequence of a losing stream and evaluation of models for fluid flux estimation using heat, Limnol. Oceanogr., 55, 1741–1754, https://doi.org/10.4319/lo.2010.55.4.1741, 2010.

Tinkler, K. J. and Wohl, E. E.: Rivers over rock: fluvial processes in bedrock channels, Geophysical Monograph 17, American Geophysical Union, Washington, D.C., 1998.

Toran, L., Nyquist J. E., Fang, A. C., Ryan, R. J., and Rosenberry, D. O.: Observing lingering hyporheic storage using electrical resistivity: variations around stream restoration structures, Crabby Creek, PA, Hydrol. Process., 27, 1411–1425, https://doi.org/10.1002/hyp.9269, 2013a.

Toran, L., Hughes, B., Nyquist, J., and Ryan, R.: Freeze core sampling to validate time-lapse resistivity monitoring of the hyporheic zone, Ground Water, 51, 635–640, https://doi.org/10.1111/j.1745-6584.2012.01002.x, 2013b.

Wallin, E. L., Johnson, T. C., Greenwood, W. J., and Zachara, J. M.: Imaging high stage river-water intrusion into a contaminated aquifer along a major river corridor using 2-D time-lapse surface electrical resistivity tomography, Water Resour. Res., 49, 1693–1708, https://doi.org/10.1002/wrcr.20119, 2013.

Ward, A. S., Gooseff, M. N., and Singha, K.: Imaging hyporheic zone solute transport using electrical resistivity, Hydrol. Process., 24, 948–953, https://doi.org/10.1002/hyp.7672, 2010a.

Ward, A. S., Gooseff, M. N., and Singha, K.: Characterizing hyporheic transport processes – interpretation of electrical geophysical data in coupled stream-hyporheic zone systems during solute tracer studies, Adv. Water. Resour., 33, 1320–1330, https://doi.org/10.1016/j.advwatres.2010.05.008, 2010b.

Ward, A. S., Fitzgerald, M., Gooseff, M. N., Voltz, T. J., Binley, A. M., and Singha, K.: Hydrologic and geomorphic controls on hyporheic exchange during base flow recession in a headwater mountain stream, Water Resour. Res., 48, W04513, https://doi.org/10.1029/2011WR011461, 2012.

Waxman, M. H. and Smits, L. J. M.: Electrical conductivities in oil-bearing shaly sands, Soc. Petrol. Eng. J., 8, 107–122, 1968.

White, D. S., Elzinga, C. H., and Hendricks, S. P.: Temperature patterns within the hyporheic zone of a northern Michigan river, J. N. Am. Benthol. Soc., 6, 85–91, 1987.

Woessner, W. W.: Stream and fluvial plain ground water interactions: rescaling hydrogeologic thought, Ground Water, 38, 423–429, 2000.

Worthington, P. F.: The uses and abuses of the Archie equations, 1: The formation facture-porosity relationship, J. Appl. Geophys., 30, 215–228, 1993.

Zanini, L., Novakowski, K. S., Lapcevic, P., Bickerton, G. S., Voralek, J., and Talbot, C.: Ground water flow in a fractured carbonate aquifer inferred from combined hydrogeological and geochemical measurements, Ground Water, 38, 350–360, 2000.

2

Analytical flow duration curves for summer streamflow in Switzerland

Ana Clara Santos[1,2], **Maria Manuela Portela**[2], **Andrea Rinaldo**[1,4], **and Bettina Schaefli**[1,3]

[1]School of Architecture, Civil and Environmental Engineering (ENAC),
École Polytechnique Fédérale de Lausanne (EPFL), Lausanne, Switzerland
[2]Instituto Superior Técnico, Technical University of Lisbon, Lisbon, Portugal
[3]Faculty of Geosciences and Environment, University of Lausanne, Lausanne, Switzerland
[4]Dipartimento di Ingegneria Civile Edile e Ambientale, Universitá degli studi di Padova, Padua, Italy

Correspondence: Ana Clara Santos (anaclara.santos@epfl.ch)

Abstract. This paper proposes a systematic assessment of the performance of an analytical modeling framework for streamflow probability distributions for a set of 25 Swiss catchments. These catchments show a wide range of hydroclimatic regimes, including namely snow-influenced streamflows. The model parameters are calculated from a spatially averaged gridded daily precipitation data set and from observed daily discharge time series, both in a forward estimation mode (direct parameter calculation from observed data) and in an inverse estimation mode (maximum likelihood estimation). The performance of the linear and the nonlinear model versions is assessed in terms of reproducing observed flow duration curves and their natural variability. Overall, the nonlinear model version outperforms the linear model for all regimes, but the linear model shows a notable performance increase with catchment elevation. More importantly, the obtained results demonstrate that the analytical model performs well for summer discharge for all analyzed streamflow regimes, ranging from rainfall-driven regimes with summer low flow to snow and glacier regimes with summer high flow. These results suggest that the model's encoding of discharge-generating events based on stochastic soil moisture dynamics is more flexible than previously thought. As shown in this paper, the presence of snowmelt or ice melt is accommodated by a relative increase in the discharge-generating frequency, a key parameter of the model. Explicit quantification of this frequency increase as a function of mean catchment meteorological conditions is left for future research.

1 Introduction

Knowledge of the availability and variability of daily discharges in a given stream section proves useful for many engineering applications (e.g., the design of hydropower plants or water supply systems) and for studies about stream ecology alterations and sediment transport or about water quality and allocation (Basso et al., 2015; Ceola et al., 2010; Searcy, 1959; Vogel and Fennessey, 1995). For many such applications, knowledge of the probability distribution of daily discharges rather than of their exact temporal occurrence is sufficient.

In hydrology, the probability distribution of daily discharges is traditionally not represented as a probability density function (pdf), but in terms of flow duration curves (FDCs) that assign an exceedance probability to each discharge value (Vogel and Fennessey, 1994) and that correspond to the complement of the cumulative distribution function (cdf).

Different methods exist to estimate FDCs (ie. to estimate their shape), the most straightforward method being the assignment of empirical probabilities to observed ranked data (yielding empirical FDCs; Vogel and Fennessey, 1994). FDCs can also be obtained from statistical methods that relate the FDC shape to catchment characteristics (Castellarin et al., 2013).

An important category of FDC models are process-based models that combine climate controls and catchment characteristics to estimate the shape of FDCs. Such models describe the shape of FDCs either based on long-term simulations of the system behavior or based on a direct parameterization of the FDC shape as a function of key hydrological controls. One such model is the model developed by Botter et al. (2007c), who derived an analytical description of streamflow distributions as the result of subsurface flow pulses triggered by stochastic rainfall and censored by the soil moisture dynamics. The resulting streamflow distribution is characterized by only a few parameters: the mean rainfall depth and the frequency of rainfall events that produce discharge, the area of the catchment and the mean residence time of the catchment.

This modeling framework has been applied successfully for a range of case studies in Italy (Botter et al., 2007c, 2009; Ceola et al., 2010; Schaefli et al., 2013), Switzerland (Basso et al., 2015; Doulatyari et al., 2017; Schaefli et al., 2013) and the US (Botter et al., 2007a, 2013; Ceola et al., 2010). Müller et al. (2014) expanded the framework to seasonally dry climates with an application in Nepal. According to Müller and Thompson (2016), the benefits of such a process-based approach, as opposed to purely statistical or empirical methods, can be summarized as follows: (i) it provides an explicit link between the FDC shape, rainfall characteristics and catchment recession characteristics rather than an empirical or statistical link to regional FDC shapes; (ii) the method is applicable to periods characterized by different meteorological conditions, thanks to the explicit treatment of rainfall and evapotranspiration characteristics.

The original model framework was developed for rainfall-driven catchments that show a linear recession behavior. Besides the aforementioned extension to seasonally dry climates, the framework has namely been extended to nonlinear recessions (Botter et al., 2009) and to the description of winter low flow resulting from seasonal snow accumulation (Schaefli et al., 2013).

In the previous applications of the model, the focus was generally on the study of signatures of discharge regimes under different climates and landscape conditions (Botter et al., 2007a, 2013), where the shape of the pdf was more important than the accuracy of the predicted discharge probabilities. Furthermore, all previous applications deliberately excluded all catchments or seasons that were snowmelt affected (Botter et al., 2007a, 2013; Ceola et al., 2010; Doulatyari et al., 2015).

The objective of this research is to assess and compare the performance of the model in its linear and nonlinear forms for summer streamflows for a range of Alpine discharge regimes. The selected set of case studies covers all Swiss catchments that have a natural (unperturbed) discharge regime and long-term discharge monitoring. Compared to existing studies (e.g., Basso et al., 2015; Ceola et al., 2010; Doulatyari et al., 2017), this paper provides a systematic analysis of all model parameters and of their seasonality, and a comprehensive analysis of a wide range of discharge regimes, including namely rainfall-driven and snowfall-influenced regimes. This allows for the first detailed view of the suitability of the modeling framework for Alpine summer discharges (influenced by rain and snow) and an assessment of the model performance as a function of the discharge regime.

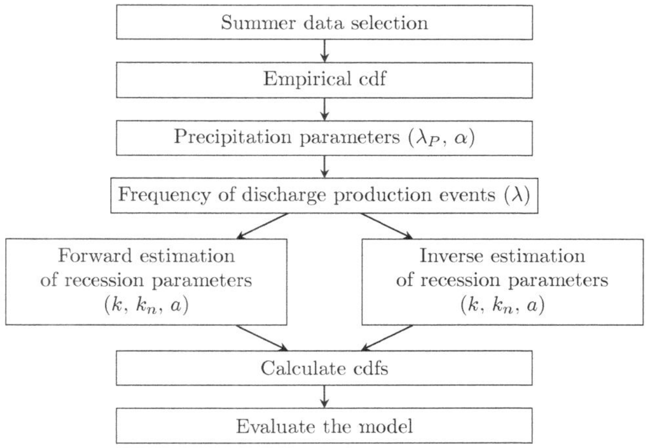

Figure 1. Sketch of the adopted workflow for model parameter estimation and performance assessment.

The paper is organized as follows: Sect. 2 provides a description of the analytical model, together with the methods adopted in this paper to estimate the model parameters and to assess the model performance, followed by a presentation of the Swiss case studies (Sect. 3). The obtained results for the linear and nonlinear model versions (Sect. 4) are discussed in Sect. 5 with a particular focus on the model performance under different hydrological regimes. The conclusions are summarized in Sect. 6.

2 Methods

Here, we first give a short overview of the used analytic modeling framework, followed by the two different methods adopted for parameter estimation and for model performance assessment. All methods are applied only to the summer season (1 June to 31 August; see also Sect. 3). The model evaluation framework adopted here is synthesized in Fig. 1, starting from the empirical cdfs as references for performance evaluation. Next, the precipitation frequency λ_p (Sect. 2.1) is estimated from precipitation and the discharge-producing frequency λ from observed discharge (Eq. 7, Sect. 2.2). The recession parameters are obtained in forward mode (Sect. 2.2) or inverse mode (Sect. 2.3). Based on these parameters, the model cdf is calculated from the linear model (Eq. 4) or from the nonlinear model (Eq. 6). The model performance is evaluated based on two classical performance indicators and by comparison to the natural variability of the observed cdfs (Sect. 2.4).

2.1 Model framework

The analytical modeling framework of Botter et al. (2007c) for probabilistic characterization of rainfall-driven daily discharges is based on a soil moisture model originally proposed by Rodriguez-Iturbe et al. (1999). This point-scale model represents the dynamics of soil moisture as the result of a deterministic, state-dependent loss function, combined with stochastic increments triggered by rainfall events. Rodriguez-Iturbe et al. (1999) showed that the corresponding spatially averaged soil moisture $s(t)$ can be obtained from the water balance equation as follows:

$$\frac{\mathrm{d}s(t)}{\mathrm{d}t} = -\rho[s(t)] + \xi_{\mathrm{t}}, \tag{1}$$

where $-\rho[s(t)]$ is the loss function, due to evapotranspiration, surface runoff and deep percolation, and where ξ_{t} represents the stochastic instantaneous increments due to infiltration from rainfall.

Botter et al. (2007c) proposed describing the dynamics of daily streamflow with a similar stochastic differential equation, supposing that rainfall acts as a stochastic forcing for discharge production and that, at the catchment scale, the water is released following a linear decay:

$$\frac{\mathrm{d}Q(t)}{\mathrm{d}t} = -kQ(t) + \xi_{\mathrm{t}}'', \tag{2}$$

where Q is the daily streamflow, k is the inverse of the time constant associated with the loss function and ξ_{t}'' is the stochastic process associated with discharge-producing precipitation events (i.e., the sequence of events that trigger a flow response in the river).

It is assumed here that discharge Q is the result of a series of rainfall inputs (ξ_{t}'') that deliver enough water to fill the water deficit in the soil, i.e., that deliver enough water to raise the soil moisture level above its retention capacity. The excess water is removed from the soil as subsurface runoff and becomes river discharge. This description of the streamflow response neglects any direct surface flow.

The overall rainfall forcing ξ is modeled as a marked Poisson process with frequency λ_{p} and exponentially distributed rainfall depths with average α (average rainfall on rain days). Nevertheless, not all the rainfall events trigger streamflow responses, and the discharge-producing precipitation events ξ" are also modeled as a marked Poisson process with a decreased frequency $\lambda < \lambda_{\mathrm{p}}$ and the same α. λ is influenced by the soil storage capacity and soil drying time and can be written as (Botter et al., 2007a)

$$\lambda = \eta \frac{\exp(-\gamma)\gamma^{\frac{\lambda_{\mathrm{p}}}{\eta}}}{\Gamma(\lambda_{\mathrm{p}}/\eta, \gamma)}, \tag{3}$$

where $\Gamma(a,b)$ is a lower incomplete Gamma function with parameters a and b, $\eta = E/(nZ_{\mathrm{r}}(s_1 - s_{\mathrm{w}}))$, $\gamma = \gamma_{\mathrm{p}} nZ_{\mathrm{r}}(s_1 - s_{\mathrm{w}})$ and $\gamma_{\mathrm{p}} = 1/\alpha$. E is the maximum evapotranspiration rate and $nZ_{\mathrm{r}}(s_1 - s_{\mathrm{w}})$ synthesizes the soil volume liable to be filled by water before drainage starts; n is the porosity of the soil, Z_{r} is the effective soil depth, s_1 is the retention capacity and s_{w} the permanent wilting point.

As discussed in detail by Botter et al. (2007c), this framework results in the following probability distribution of daily discharges at the catchment scale:

$$p(Q, t \to \infty) = \frac{1}{\Gamma\left(\frac{\lambda}{k}\right)} \frac{1}{Q} \left(\frac{Q}{\alpha k A}\right)^{\frac{\lambda}{k}} \exp\left(-\frac{Q}{\alpha k A}\right), \tag{4}$$

where A is the catchment area. This corresponds to a Gamma distribution with shape parameter λ/k and a scale parameter $\alpha k A$. The corresponding expected mean discharge equals $\overline{Q} = \lambda\alpha$. The model is suitable for steady-state conditions, at the annual or seasonal scale, depending on the temporal variability of the model parameters (Botter et al., 2007a).

Nonlinear storage–discharge relationships at the catchment scale are commonly observed (Botter et al., 2009; Brutsaert and Nieber, 1977; Mutzner et al., 2013). Accordingly, Botter et al. (2009) proposed an extension of the above modeling framework assuming that

$$\frac{\mathrm{d}Q(t)}{\mathrm{d}t} = -k_{\mathrm{n}} Q(t)^a + \xi_{\mathrm{t}}'', \tag{5}$$

where k_{n} and a are the constants of the nonlinear recession. As for the linear model, it is possible to obtain an equation for the pdf of the daily discharges:

$$p(Q, t \to \infty) = C\left\{\frac{1}{Q^a} \exp\left[-\frac{Q^{2-a}}{\alpha k_{\mathrm{n}}(2-a)} + \frac{Q^{1-a}\lambda}{k_{\mathrm{n}}(1-a)}\right]\right\}, \tag{6}$$

where C is a normalizing constant (Botter et al., 2009).

2.2 Parameter estimation 1: forward estimation

We use the term "forward parameter estimation" to emphasize that the parameters are estimated directly from observed data, without calibration. This method is generally used in the context of this modeling framework for the estimation of the parameters related to the stochastic inputs (λ_{p}, α, λ), and this method is always used for these parameters in the present paper. However, the recession parameters (k, k_{n} and a) are either estimated in a forward mode (Botter et al., 2007c, 2009; Ceola et al., 2010; Schaefli et al., 2013) or in an inverse mode (Ceola et al., 2010) (see Sect. 2.3).

The computation of the precipitation parameters first involves the computation of a reference catchment-scale precipitation time series (here obtained from gridded data; see Sect. 3). Then interception losses (I) are subtracted from the observed daily precipitation depths. These losses are in fact evaporated (or sublimated in case of snow) before participating to soil moisture dynamics. Following Rodriguez-Iturbe et al. (1999), previous model applications generally assumed that these losses are accounted for when the frequency of precipitation events is corrected to the frequency of discharge-producing events. In view of understanding how the model

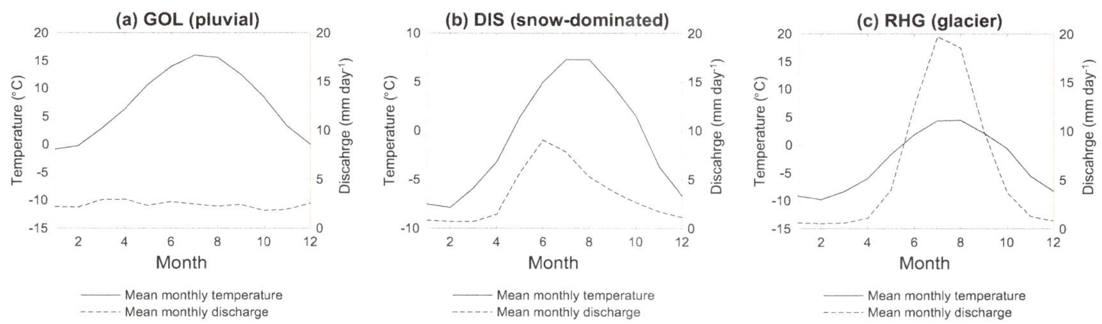

Figure 2. Annual cycle of discharge and air temperature for three selected catchments representing three different hydrologic regimes (pluvial, snow-dominated and glacier). The mean monthly values computed over the entire observation period for each catchment are shown (see Table 1).

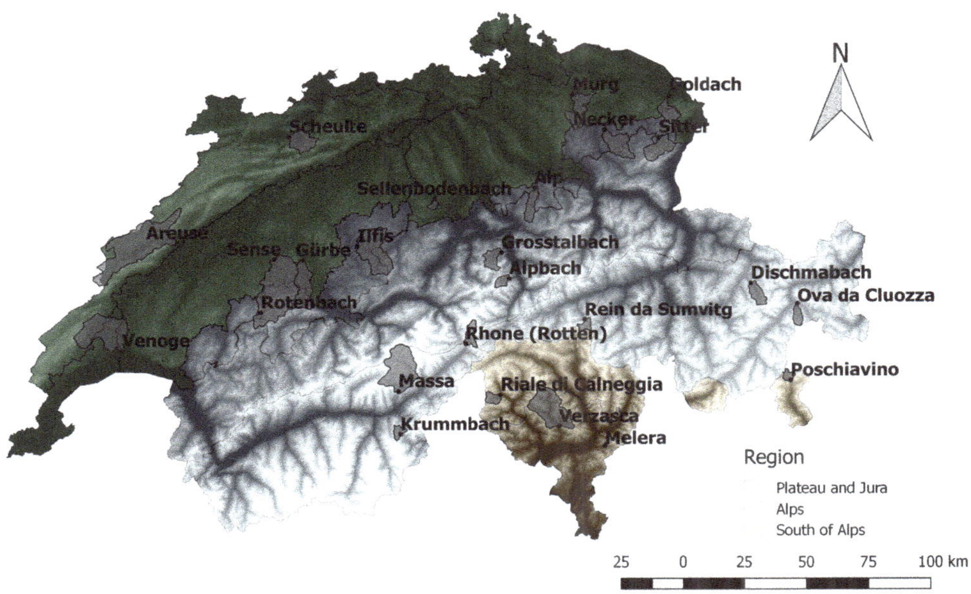

Figure 3. Location of the case study catchments in Switzerland. The six biogeographical regions of Switzerland (Federal Office for the Environment, 2004) are summarized here into three main regions. Data source as follows: digital elevation model, SwissTopo (2005); catchments, Helbling (2016).

parameters vary in space, it was decided here to treat interception losses explicitly with minimal assumptions about this process: different maximum interception depths are attributed to four different land covers: 4 mm for forests, 2 mm for low vegetation, 1 mm for impervious areas and 0 mm for water bodies (Gerrits, 2010). The catchment-scale maximum interception depth is obtained as the land-use-weighted average of these values, but a minimum interception depth of 1 mm is imposed. This catchment-scale interception depth is subtracted from daily precipitation depths, assuming that at a daily time step, all intercepted water re-evaporates during the same time step.

Instead of correcting the frequency of precipitation events λ_p according to Eq. (3), the frequency of discharge-producing events λ is estimated directly from the theoretical relationship between the mean discharge and the precipitation parameters, $\overline{Q} = \lambda\alpha$ (see Eq. 4). Replacing the mean modeled discharge $\overline{Q}$ with the mean observed discharge $\overline{\overline{Q}}$, it follows that

$$\lambda = \frac{\alpha}{\overline{\overline{Q}}}. \tag{7}$$

Estimating λ from the above equation rather than directly from the soil properties as in Eq. (3) has been shown by Ceola et al. (2010) to provide much better results, and this method has been used by the majority of studies since then (e.g., Ceola et al., 2010; Botter et al., 2013; Basso et al., 2015).

The recession parameter for the linear model is calculated directly from observed daily discharge based on a classical

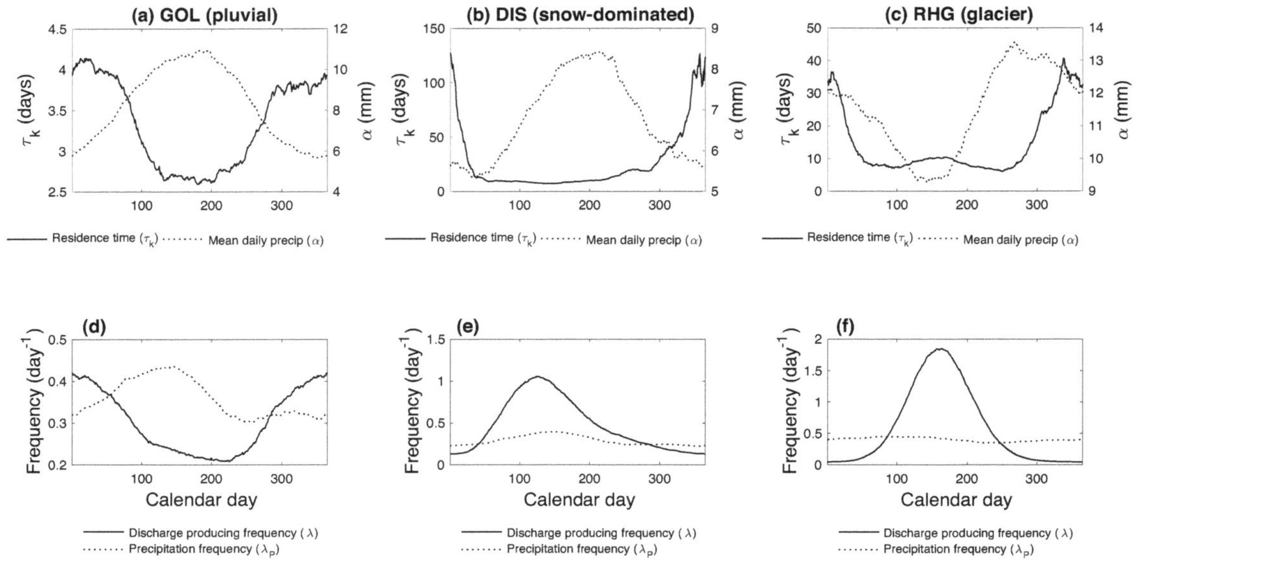

Figure 4. Examples of the temporal variation of the model parameters over the course of a year. The parameters are calculated for 90-day intervals beginning at the calendar day for which the value is plotted; for a given time window, the data points corresponding to this window in all available civil years are pooled together. **(a–c)** Residence time τ_k and mean daily precipitation depth α. **(d–f)** Precipitation frequency λ_p and discharge-producing frequency λ.

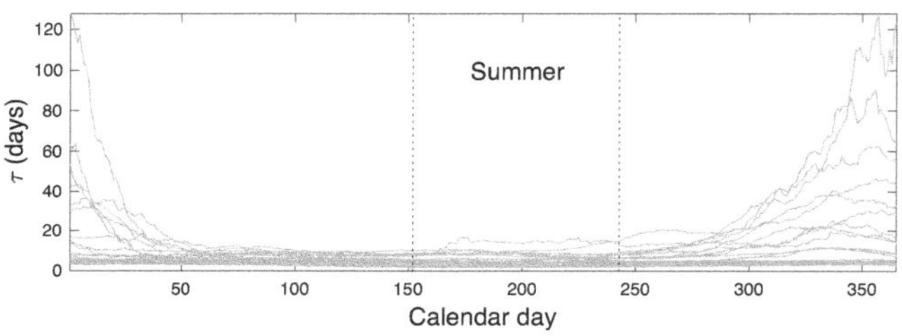

Figure 5. Temporal variation of the residence time ($\tau_k = k^{-1}$) for the 25 catchments. The temporal variation is obtained as in Fig. 4.

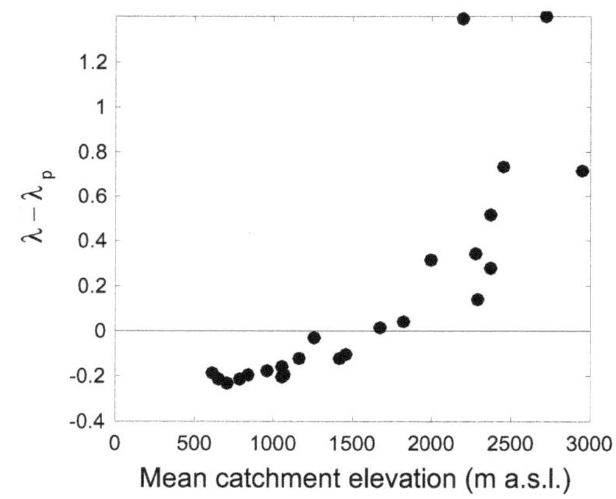

Figure 6. Difference between λ and λ_p as a function of mean catchment elevation.

Brutsaert–Nieber recession analysis (Brutsaert and Nieber, 1977; Biswal and Marani, 2010, 2014; Mutzner et al., 2013), considering, however, only discharges below a certain threshold, fixed to 95 %. The nonlinear recession parameters k_n and a are also obtained from a recession analysis, using the same discharge threshold via linear regression of the logarithm of $(-dQ/dt)$ versus the logarithm of Q, where a is the slope and k_n the intercept.

2.3 Parameter estimation 2: inverse estimation

To objectively compare the potential of the linear and the nonlinear model formulations to capture observed flow-duration curves, the recession parameters for the linear model (k) and for the nonlinear model (k_n, a) are also estimated in a classical inverse estimation mode where the model parameters are obtained by maximizing the likelihood function formulated for the model. For the linear model, the likelihood function is obtained from the model as follows:

$$\mathcal{L}(k|\widetilde{Q},\theta) = \prod_{j=1}^{N} p(\widetilde{Q}_j|k,\theta), \tag{8}$$

where the probability $p(.)$ is obtained from Eq. (4) and $\boldsymbol{\theta} = [\alpha, A, \lambda]$ is the parameter vector containing all parameters that are estimated directly from observed data (i.e., not maximized). For the nonlinear case, the likelihood is obtained analogously by replacing k with k_n and a and using $p(.)$ from Eq. (6).

2.4 Model evaluation criteria

To objectively compare different models, we propose using the Kolmogorov–Smirnov distance between the cdfs corresponding to different models (Ceola et al., 2010; Schaefli et al., 2013), i.e., the maximum difference between the values of the empirical and the modeled cumulative distributions:

$$c^{\mathrm{KS}} = \sup_x |F(\widetilde{Q}) - F(Q)|, \tag{9}$$

where $F(\widetilde{Q})$ corresponds to the empirical cumulative distribution of the discharges and $F(Q)$ to the modeled cumulative distribution of the discharges. A good model should have a low c^{KS} value.

This comparison of the cdfs overcomes an important limitation inherent in the comparison of analytic pdfs and empirical pdfs. In fact, the choice of the number of classes for the calculation of the empirical pdf from observed data (via a so-called frequency polygon; Naghettini, 2016) can change the shape of the empirical pdf. The problem does not arise for cdfs given their cumulative nature.

Since the nonlinear model formulation has an additional parameter, the linear and the nonlinear models are also compared based on the Akaike information criterion (Burnham and Anderson, 2004; Laio et al., 2009; Ceola et al., 2010):

$$c^{\mathrm{AIC}} = 2n - \ln(\hat{\mathcal{L}}), \tag{10}$$

where n is the number of parameters of the model and $\ln(\hat{\mathcal{L}})$ is the logarithm of the maximum likelihood function obtained by maximizing Eq. (8). As for c^{KS}, a good model should have a low c^{AIC} value.

Based on the above criterion, we measure the relative performance increase from the linear to the nonlinear model as follows:

$$r^{\mathrm{AIC}} = -\left(\frac{c_{\mathrm{n}}^{\mathrm{AIC}} - c_{\mathrm{l}}^{\mathrm{AIC}}}{c_{\mathrm{l}}^{\mathrm{AIC}}}\right), \tag{11}$$

where ${c_{\mathrm{n}}}^{\mathrm{AIC}}$ is the Akaike criterion for the nonlinear model and $c_{\mathrm{l}}^{\mathrm{AIC}}$ for the linear model. Taking the opposite of the relative difference between the Akaike criteria ensures that a higher r^{AIC} value indicates a stronger performance increase (recall that the Akaike criterion is to be minimized).

In addition, to assess the performance difference between different models, the obtained models are compared to the natural variability of the observed discharge cdfs. Therefore, an empirical long-term cdf is constructed, obtained by ranking the observed data in ascending order and dividing the rank numbers by the total sample size. Furthermore, to assess the natural yearly variability, individual cdfs are constructed for each summer season of each civil year (Vogel and Fennessey, 1994). From this collection of annual cdfs, envelopes are based on the maximum and minimum values of discharge for each probability class of the annual cdfs. A reliable model should yield a cdf contained between these curves and should be as close as possible to the long-term cdf.

3 Case studies

In this paper, we analyze 25 Swiss catchments with areas ranging from 1.05 to 377 km^2 and with mean elevations ranging from 615 m a.s.l. (meters above sea level) to 2945 m a.s.l. (Table 1, Fig. 3). These catchments correspond to all streamflow gauging stations run by the Swiss Federal Office for the Environment (FOEN, 2017) and that have unperturbed discharges (i.e., minimal anthropogenic influence).

The average precipitation at the country scale is around 1300 mm yr^{-1} (Blanc and Schädler, 2013). The complex topography leads to a high diversity of hydrologic regimes (Weingartner and Aschwanden, 1992), which can be grouped into (i) pluvial or rainfall-driven regimes, (ii) snow-dominated regimes and (iii) glacier regimes. Pluvial regimes are rainfall-dominated with sporadic snowfall events during winter; these regimes occur on the Swiss Plateau and in the Jura region (Fig. 3). Snow-dominated regimes result from seasonal snow cover, roughly at elevations above 900 m a.s.l. In these catchments, solid precipitation accumulates over several weeks up to several months during the cold season (winter) and is released entirely in the following spring and early summer. Glacier regimes result from perennial snow and ice accumulation at elevations roughly beyond 3000 m a.s.l. Most snow-dominated and glacier regimes are located in the Alps region (Fig. 3), few of them are located south of the Alps, which overall has a warmer climate and presents higher precipitation than the other two regions.

Most Swiss discharge regimes show a strong seasonality (Weingartner and Aschwanden, 1992), illustrated in Fig. 2 for typical examples of the three regime main types; air temperature is shown here as a proxy for snow and evapotranspiration processes. The pluvial Goldach River (GOL) shows the typical summer low flow resulting from evapotranspiration; the Dischmabach shows a snow regime with high summer flows resulting from the release of snowmelt stored in the subsurface during the main snowmelt period (spring) and from residual snowmelt during summer. The Rhône River (RHG) with its 50 % glacier cover shows a glacier regime, with significant ice melt during summer, and with monthly discharge peaking for the same month as air temperature (July).

It is noteworthy that surface runoff processes definitely can play a certain role in extreme events in all regions of Switzerland (Bernet et al., 2017), but all hydrologic regimes are dominated by subsurface runoff processes, a precondition for the application of the modeling framework developed by Botter et al. (2007c).

Besides observed daily discharge, the model requires catchment-scale daily precipitation as input. Most of the previous applications of the models used precipitation from one

Table 1. Characteristics of Swiss case study catchments as given in the FOEN database, including the FOEN identification code (ID), the catchment name, the Swiss coordinates of the gauging station, the drainage area, the mean elevation of the catchment and the gauging station elevation, the percentage of glacier-cover of the catchment, the mean annual precipitation, the mean annual temperature and the period of data acquisition. The 16 regime classes of Weingartner and Aschwanden (1992) are regrouped here into three classes (details are available in the Supplement).

	ID	Name	Coordinates (CH1903)	Area (km^2)	Mean elevation (m a.s.l.)	Station elevation (m a.s.l.)	Glaciation (%)	P (mm)	$\overline{T}$ (°C)	Data period	16 regime classes	3 regime classes
1	2430	Rein da Sumvitg – Sumvitg, Encardens	718810/167690	21.8	2450	1490	6.7	1707	−1.19	15 Sep 1977 to 31 Dec 2014	b-glacio nival	glacier
2	2327	Dischmabach – Davos, Kriegsmatte	786220/183370	43.3	2372	1668	2.1	1021	−0.62	24 Jul 1961 to 31 Dec 2014	b-glacio nival	glacier
3	2308	Goldach – Goldach, Bleiche	753190/261590	49.8	833	399	0	1446	7.39	1 Jan 1974 to 31 Dec 2014	pluvial supérieur	pluvial
4	2374	Necker – Mogelsberg, Aachsäge	727110/247290	88.2	959	606	0	1777	6.47	1 Jan 1972 to 31 Dec 2014	nivo-pluvial préalpin	snow-dominated
5	2112	Sitter – Appenzell	749040/244220	74.2	1252	769	0.08	1904	5.10	1 Jan 1961 to 31 Dec 2014	nival de transition	snow-dominated
6	2126	Murg – Wängi	714105/261720	78.9	650	466	0	1357	7.90	1 Jan 1961 to 31 Dec 2014	pluvial inférieur	pluvial
7	2610	Scheulte – Vicques	599485/244150	72.8	785	463	0	1325	7.27	1 Jan 1992 to 31 Dec 2014	nivo-pluvial jurassien	snow-dominated
8	2159	Gürbe – Belp, Mülimatt	604810/192680	117	837	522	0	1295	7.21	1 Jan 1961 to 31 Dec 2014	pluvial supérieur	pluvial
9	2251	Rotenbach – Plaffeien, Schwyberg	587980/170590	1.65	1454	1275	0	1910	5.81	1 Sep 1961 to 31 Dec 2014	nivo-pluvial préalpin	snow-dominated
10	2179	Sense – Thörishaus, Sensematt	593350/193020	352	1068	553	0	1479	6.29	1 Jan 1961 to 31 Dec 2014	nivo-pluvial préalpin	snow-dominated
11	2480	Areuse – Boudry	554350/199940	377	1060	444	0	1531	5.41	1 Jan 1961 to 31 Dec 2014	pluvial jurassien	pluvial
12	2603	Ilfis – Langnau	627320/198600	188	1051	685	0	1719	6.22	1 Apr 1989 to 31 Dec 2014	nivo-pluvial préalpin	snow-dominated
13	2608	Sellenbodenbach – Neuenkirch	658530/218290	10.5	615	515	0	1230	8.72	12 Sep 1980 to 31 Dec 2014	pluvial inférieur	pluvial
14	2299	Alpbach – Erstfeld, Bodenberg	688560/185120	20.6	2200	1022	27.7	1645	0.68	1 Jan 1961 to 31 Dec 2014	b-glaciaire	glacier
15	2276	Grosstalbach – Isenthal	685500/196050	43.9	1820	767	9.3	1801	2.22	1 Jan 1961 to 31 Dec 2014	nival alpin	snow-dominated
16	2609	Alp – Einsiedeln	698640/223020	46.4	1155	840	0	2005	5.43	27 Feb 1991 to 31 Dec 2014	nivo-pluvial préalpin	snow-dominated
17	2268	Rhone – Gletsch	670810/157200	38.9	2719	1761	52.2	2066	−2.98	1 Jan 1961 to 31 Dec 2014	a-glaciaire	glacier
18	2161	Massa – Blatten bei Naters	643700/137290	195	2945	1446	65.9	2423	−3.18	1 Jan 1961 to 31 Dec 2014	a-glaciaire	glacier
19	2432	Venoge – Ecublens, Les Bois	532040/154160	231	700	383	0	1181	9.29	1 Jan 1979 to 31 Dec 2014	pluvial jurassien	pluvial
20	2206	Melera – Melera (Valle Morobbia)	726988/114670	1.05	1419	944	0	1716	4.74	1 Jan 2005 to 31 Dec 2014	nivo-pluvial méridional	snow-dominated
21	2605	Verzasca – Lavertezzo, Campiòi	708420/122920	186	1672	490	0	2051	4.37	1 Sep 1989 to 31 Dec 2014	nivo-pluvial méridional	snow-dominated
22	2356	Riale di Calneggia – Cavergno, Pontit	684970/135960	24	1996	890	0	1918	2.54	1 Jan 1967 to 31 Dec 2014	nival méridional	snow-dominated
23	2244	Krummbach – Klusmatten	644500/119420	19.8	2276	1795	3	1475	1.92	1 Jan 1995 to 31 Dec 2014	nival méridional	snow-dominated
24	2366	Poschiavino – La Rösa	802120/142010	14.1	2283	1860	0.35	1512	0.02	1 Jan 1970 to 31 Dec 2014	nival méridional	snow-dominated
25	2319	Ova da Cluozza – Zernez	804930/174830	26.9	2368	1509	2.2	963	−1.36	24 Jul 1961 to 31 Dec 2014	nivo glaciaire	snow-dominated

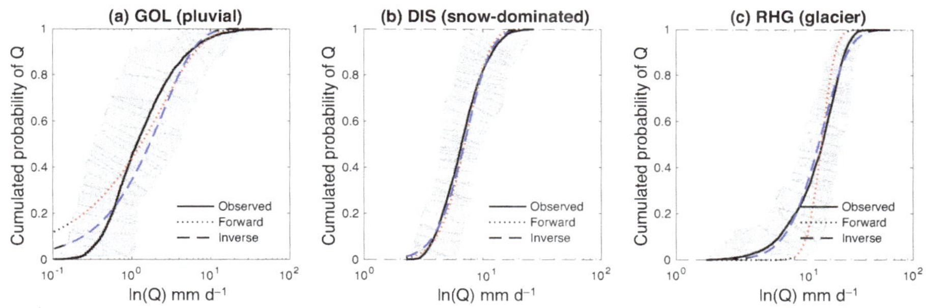

Figure 7. Modeled cdfs with forward and inverse parameter estimation for the three selected catchments. The shaded area is located between the cdf envelopes and represents the natural variability of the daily discharges.

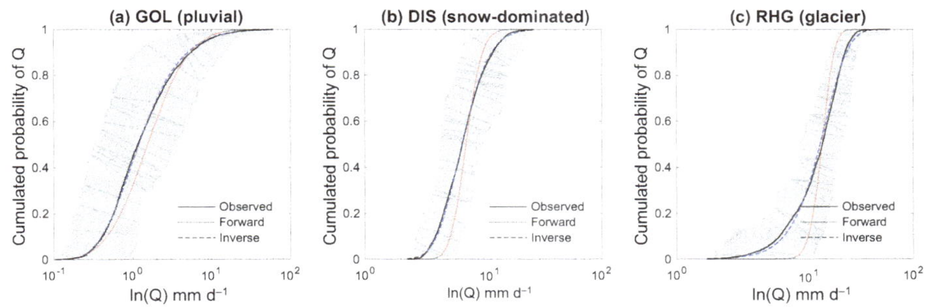

Figure 8. As Fig. 7 but for the nonlinear model.

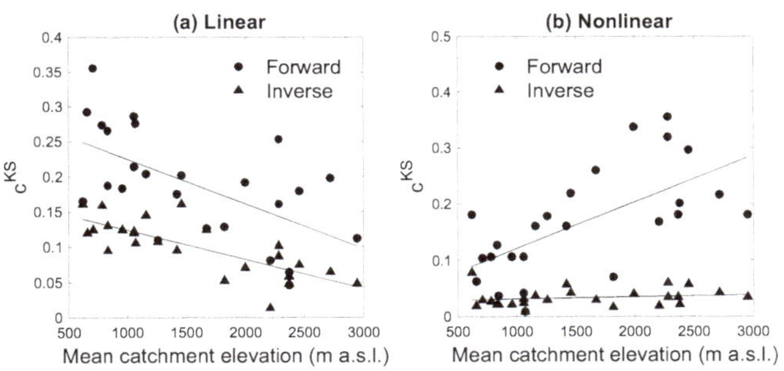

Figure 9. Performance of the linear model and nonlinear model as a function of mean catchment elevation. The performance measure shown, c^{KS}, is zero for a perfect model.

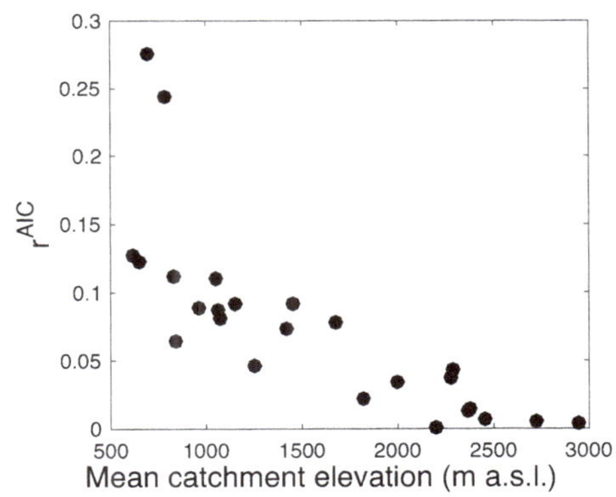

Figure 10. Relative increase in the performance of the nonlinear model with respect to the linear model (as measured by r^{AIC}) as a function of mean catchment elevation. All model parameters are estimated in inverse mode.

or several meteorological stations as input (Botter et al., 2007c, a, 2013; Ceola et al., 2010; Basso et al., 2015; Schaefli et al., 2013), which is potentially limiting for the model performance since good area-averaged input estimates are critical. Recent progress in spaceborne precipitation observation, and in particular the Global Precipitation Measurement (GPM) mission, potentially offers an interesting new data source for area-averaged precipitation estimates, even in such complex terrain as the Swiss Alps (Gabella et al., 2017), with the drawback of covering only short historical periods. Here, we use the relatively new spatial precipitation data set of MeteoSwiss with a nominal resolution of 2.2 km and an effective resolution between 15 and 20 km and extending back to 1961 (MeteoSwiss, 2011a). This data set can be assumed to give relatively good estimates of area-averaged precipitation (Paschalis et al., 2014; Addor and Fischer, 2015), even in mountainous areas where there are only few meteorological stations.

Corresponding catchment-scale average precipitation time series per case study catchment are obtained by averaging the daily precipitation time series of all pixels contained in the catchment (a list of pixels per catchment is included in the Supplement). In addition, we also used the corresponding gridded temperature data set (MeteoSwiss, 2011b) to support the analysis of parameter seasonality. As for precipitation, the catchment-scale average temperature data set is obtained by averaging the daily time series of all pixels.

Before estimating rainfall frequency (λ_p) and average rainfall depth on rain days (α), the catchment-scale precipitation time series are pre-processed to remove losses from interception. This step requires information about land use. Of the retained 25 case study catchments, 22 are part of what is called "hydrological study areas" and have an associated extended data set, including land use (Aschwanden, 1996). For the other catchments (i.e., the Areuse, Rhône-Gletsch and Venoge), land use is obtained from the Swiss land use database (Federal Statistical Office, 2001). Details about the land use estimation are available in the Supplement).

4 Results

4.1 Discharge regimes and parameter seasonality

To gain further insights into the hydrological processes underlying the different regimes, Fig. 4 shows the within-year variability of the model parameters obtained by estimating the parameters in forward mode for moving and overlapping 90-day windows. The precipitation parameters α and λ_p do not show strong seasonal patterns, except for a few catchments such as the Goldach River (Fig. 4a). For snow and glacier regimes, the frequency of discharge-producing events, λ, increases strongly at the beginning of spring (Fig. 4b and c), which indicates the release of water from snowmelt or ice melt.

The inverse of the linear recession coefficient $\tau = k^{-1}$ shows a coherent annual cycle for all catchments, independent of the underlying discharge regime (Fig. 5). This seasonal pattern with consistently low τ values during summer for all catchments clearly justifies the choice of a common summer season (June, July, August) for all regimes. The amplitude of the annual cycle (the difference between high and low τ values) is stronger for snow or glacier regimes, which reflects the fact that in these regimes, parts of the catchment are effectively dormant during the winter (Schaefli et al., 2013).

4.2 Linear model

All estimated parameters for both forward and inverse estimations are summarized in Table 2, together with the values of the performance indicator c^{KS}. It can be noted that for 11 catchments (i.e., Rein da Sumvitg, Dischmabach, Alpbach, Grosstalbach, Rhône à Gletsch, Massa, Verzasca, Riale di Calneggia, Krummbach, Poschiavino and Ova da Cluozza), λ exceeds λ_p, contradicting the original description of the model (Botter et al., 2007b), which states that the discharge-producing frequency λ is smaller than the precipitation frequency λ_p. Such an exceedance of λ over λ_p should only happen in catchments or seasons with an additional source of water (in addition to rainfall), which in the present case is snowmelt or ice melt.

The exceedance of λ over λ_p increases with mean catchment elevation (Fig. 6), the limit of $\lambda = \lambda_p$ being at around 1500 m a.s.l. This important result is further discussed in Sect. 5.

The cdfs obtained from all estimated parameters are presented in Fig. 7 for the three example case studies. For the catchment with rainfall-driven discharges (GOL), it can be seen that the probabilities of occurrence of low flows are largely overestimated with forward estimation (Fig. 7a). This is a typical indication that the recession timescale is underestimated. The model values even exceed the envelopes that represent the natural variability of the discharges. In the presence of snow, the linear model in forward estimation mode tends to underestimate low flows, with satisfactory results for some cases, such as the Dischmabach (Fig. 7b).

Overall, there is a strong increasing trend of the linear model performance with mean catchment elevation (Fig. 9a). Despite this, the results of the linear model are not satisfactory for the forward estimation method for any of the regimes.

The inverse estimation of the model parameters improves the results significantly, but the c^{KS} performance indicator shows relatively high values and the curves are visually not accurate, especially for pluvial regimes. This suggests that the model with a linear discharge decay is overall not suitable for the studied catchments.

4.3 Nonlinear models

The results obtained from inverse parameter estimation for the nonlinear model are very good (Fig. 8, Table 2), and the nonlinear model outperforms the linear model for all catchments, both in terms of the KS performance and in terms of the Akaike criterion (Table 2). The relative model performance increase (as measured by r^{AIC}) shows furthermore a strong inverse trend with mean catchment elevation (Fig. 10), which results from the increasing performance of the linear model with increasing elevation (Fig. 9b).

It is noteworthy that the two catchments for which the performance increase in the nonlinear model over the linear models exceeds 20 % are the two catchments that have a strongly karst-influenced regime (Scheulte at Vicques and Venoge at Ecublens).

As for the linear model, the forward estimation mode gives worse results than the inverse estimation mode. For some catchments (i.e., Murg-Wängi, Gürbe, Sense, Ilfis, and Grosstalbach), the forward estimation mode gives nevertheless very good results with c^{KS} below 0.1. In general, for the catchments where the discrepancies between modeled and observed cdfs are due to an underestimation of τ, the nonlin-

Table 2. Parameter values and performance indicators for all the catchments for summer with linear model and forward estimation, summer linear model and inverse estimation, summer nonlinear model and forward estimation, winter nonlinear model and inverse estimation and winter linear model and forward estimation. $\overline{Q}$ stands for the mean observed discharge, P_s the mean total precipitation during summer, $\overline{T_s}$ the mean temperature during summer, I for interception depth and c^{KS} for the Kolmogorov–Smirnov distance. The indices stand for f, forward estimation; i, inverse estimation; l, linear model; n, nonlinear model.

Name	$\overline{Q}$ (mm d^{-1})	P_s (mm)	$\overline{T_s}$ (°C)	α (mm d^{-1})	λ_p mm	I (1/d)	λ (mm)	k_f (1/d)	c^{KS}_{lf}	k_i (1/d)	c^{KS}_{li}	c^{AIC}_{li}
Rein da Sumvitg – Sumvitg, Encardens	13.8	532	5.62	12.4	0.410	1.83	1.115	0.201	0.179	0.383	0.075	21 550
Dischmabach – Davos, Kriegsmatte	7.4	378	6.49	8.2	0.377	2.29	0.906	0.136	0.065	0.163	0.048	22 300
Goldach – Goldach, Bleiche	2.5	513	15.15	11.0	0.376	3.13	0.224	0.370	0.187	0.236	0.130	13 494
Necker – Mogelsberg, Aachsäge	3.3	600	14.22	12.2	0.393	3.30	0.273	0.435	0.183	0.275	0.125	16 467
Sitter – Appenzell	5.4	648	12.30	12.5	0.433	3.06	0.427	0.393	0.109	0.308	0.108	25 067
Murg – Wängi	1.7	432	16.07	9.6	0.348	3.13	0.174	0.282	0.293	0.105	0.120	13 636
Scheulte – Vicques	1.5	388	15.10	9.1	0.312	3.46	0.162	0.264	0.274	0.133	0.158	5262
Gürbe – Belp, Mülimatt	2.1	450	15.15	9.9	0.355	3.06	0.210	0.271	0.266	0.096	0.095	15 070
Rotenbach – Plaffeien, Schwyberg	4.3	616	13.29	14.0	0.378	3.16	0.309	0.550	0.202	0.339	0.161	22 856
Sense – Thörishaus, Sensematt	2.2	483	13.98	10.7	0.356	3.22	0.208	0.344	0.275	0.127	0.105	16 401
Areuse – Boudry	1.7	383	13.10	8.8	0.316	3.37	0.191	0.261	0.214	0.132	0.120	14 013
Ilfis – Langnau	2.7	567	13.79	12.4	0.373	3.40	0.220	0.362	0.287	0.149	0.123	8210
Sellenbodenbach – Neuenkirch	2.0	431	16.86	9.7	0.357	2.99	0.207	0.381	0.165	0.285	0.161	6617
Alpbach – Erstfeld, Bodenberg	16.5	457	7.29	8.9	0.477	1.28	1.858	0.171	0.081	0.276	0.014	30 444
Grosstalbach – Isenthal	6.0	598	8.97	11.8	0.444	2.35	0.504	0.195	0.128	0.106	0.053	22 256
Alp – Einsiedeln	4.7	687	13.03	14.1	0.415	3.40	0.335	0.521	0.204	0.318	0.144	9763
Rhone – Gletsch	17.1	473	3.58	9.0	0.505	1.00	1.905	0.092	0.197	0.419	0.064	32 412
Massa – Blatten bei Naters	17.1	739	3.48	13.9	0.533	1.00	1.228	0.130	0.112	0.272	0.049	32 418
Venoge – Ecublens, Les Bois	0.7	298	17.39	7.9	0.268	3.14	0.090	0.194	0.355	0.056	0.124	3737
Melera – Melera (Valle Morobbia)	3.1	562	12.64	18.1	0.273	3.87	0.174	0.142	0.176	0.079	0.096	2918
Verzasca – Lavertezzo, Campiòi	6.0	581	12.03	17.9	0.313	3.00	0.333	0.287	0.127	0.294	0.125	11 649
Riale di Calneggia – Cavergno, Pontit	8.9	482	9.96	13.5	0.332	2.04	0.655	0.173	0.192	0.352	0.071	25 838
Krummbach – Klusmatten	6.0	317	9.30	9.2	0.294	2.35	0.656	0.117	0.253	0.297	0.102	8673
Poschiavino – La Rösa	5.4	424	7.83	11.1	0.323	2.49	0.490	0.125	0.162	0.199	0.087	19 679
Ova da Cluozza – Zernez	5.2	329	6.58	8.4	0.342	1.77	0.619	0.215	0.047	0.192	0.058	21 954

Name	k_{nf}	a_f	c^{KS}_{nf}	k_{ni}	a_i	c^{KS}_{ni}	c^{AIC}_{ni}
Rein da Sumvitg – Sumvitg, Encardens	0.029	1.46	0.296	0.110	1.52	0.057	21 387
Dischmabach – Davos, Kriegsmatte	0.013	1.73	0.201	0.031	1.86	0.022	21 972
Goldach – Goldach, Bleiche	0.145	1.50	0.126	0.174	1.81	0.023	11 990
Necker – Mogelsberg, Aachsäge	0.125	1.63	0.107	0.156	1.81	0.023	15 015
Sitter – Appenzell	0.066	1.69	0.179	0.115	1.76	0.029	23 888
Murg – Wängi	0.099	1.70	0.062	0.081	1.98	0.019	11 978
Scheulte – Vicques	0.099	1.72	0.106	0.117	2.20	0.027	3978
Gürbe – Belp, Mülimatt	0.068	1.76	0.036	0.063	1.76	0.023	14 108
Rotenbach – Plaffeien, Schwyberg	0.080	1.81	0.218	0.154	1.87	0.043	20 753
Sense – Thörishaus, Sensematt	0.084	1.85	0.010	0.082	1.86	0.009	15 069
Areuse – Boudry	0.078	1.85	0.106	0.116	1.77	0.032	12 785
Ilfis – Langnau	0.068	1.96	0.042	0.069	2.04	0.025	7303
Sellenbodenbach – Neuenkirch	0.184	1.38	0.181	0.271	1.49	0.077	5776
Alpbach – Erstfeld, Bodenberg	0.057	1.17	0.168	0.156	1.21	0.020	30 420
Grosstalbach – Isenthal	0.017	1.88	0.070	0.025	1.86	0.016	21 768
Alp – Einsiedeln	0.089	1.76	0.160	0.110	1.97	0.036	8870
Rhone – Gletsch	0.107	0.87	0.216	0.897	0.70	0.043	32 234
Massa – Blatten bei Naters	0.052	1.13	0.181	0.585	0.70	0.034	32274
Venoge – Ecublens, Les Bois	0.119	1.65	0.103	0.104	2.00	0.030	2706
Melera – Melera (Valle Morobbia)	0.054	0.92	0.161	0.031	1.94	0.057	2702
Verzasca – Lavertezzo, Campiòi	0.041	1.70	0.261	0.081	1.94	0.030	10 738
Riale di Calneggia – Cavergno, Pontit	0.014	1.79	0.336	0.077	1.79	0.039	24 958
Krummbach – Klusmatten	0.032	1.37	0.354	0.064	1.99	0.060	8345
Poschiavino – La Rösa	0.014	1.75	0.318	0.042	2.05	0.035	18 837
Ova da Cluozza – Zernez	0.030	1.70	0.180	0.083	1.58	0.034	21 673

ear model yields a significant improvement. For catchments where the recession timescale is overestimated with the linear model, the nonlinear model in forward mode leads to a performance decrease.

5 Discussion

Our results show that analytical modeling framework for streamflow distributions proposed by Botter et al. (2007c) performs well for the 25 Swiss catchments across all studied discharge regimes. A detailed comparison between the performance of the linear and the nonlinear models considering the optimized parameters obtained from the inverse approach shows that the results for the nonlinear model are always better than for the linear model. This underlines that the nonlinear recession is better suited for the hydrological conditions of all studied catchments, which is in line with previous results (Ceola et al., 2010; Basso et al., 2015).

In forward estimation mode, the linear model outperforms the nonlinear model for catchments with summer high flows; the nonlinear model outperforms the linear model for catchments with rainfall-driven regimes (i.e., summer low flows). This results from the fact that for regimes with summer high flow, the linear model overestimates the recession timescale (resulting in a underestimation of the discharge variance). For regimes with summer low flow, the linear model in exchange underestimates the recession timescale. Given that the nonlinear model yields longer recessions, the nonlinear model shows accordingly a better performance for regimes with summer low flow.

The comparison between the forward and inverse estimation methods shows a clear underestimation of k_n for most of the catchments, which was already discussed by Dralle et al. (2015) and which is in line with previous work that tried to improve the results of the model in forward estimation mode, for the linear and the nonlinear formulation (Ceola et al., 2010; Basso et al., 2015). There is clearly a need to further improve the methods to estimate the recession parameters. Our results pinpoint that a key here might be the detailed investigation of recession analysis methods along elevational gradients and related hydrologic regimes.

Overall, good model performance in many different catchments with different regimes indicates that the modeling framework is suitable for the prediction of FDCs in Switzerland. A more detailed temporal model validation (e.g., with a split sample test; Klemeš, 1986) is not possible for this framework since the model parameters are obtained directly from observed data for each time period (i.e., they vary from period to period). The obtained model performances are comparable to the results obtained in previous studies, e.g., in the work of Ceola et al. (2010). For different case studies in Italy and the US, they obtained c^{KS} values varying between 0.030 and 0.409 for the nonlinear model using different methods of forward estimation, and c^{KS} values between 0.021 to 0.051 for inverse estimation. For the linear model, Ceola et al. (2010) obtained c^{KS} values between 0.054 and 0.567. Basso et al. (2015) and Doulatyari et al. (2017) studied some case studies that are included in the present paper (Sitter at Appenzell and Murg at Wängi).

Recomputing their results with their model parameters yields slightly different c^{KS} values for the nonlinear model for the Sitter (0.12 compared to our 0.19) and for the Murg (0.05 to 0.06 compared to our 0.06). These differences are small and can be explained by different data periods and by the methodological choices in the calculation of parameters.

The most remarkable result of the presented analysis is the fact that the modeling framework is applicable in its original formulation to catchments where summer flow is influenced by snow processes. The additional source of water from snow or ice melt is accommodated by increasing the frequency λ of discharge-producing events. This is in line with a common assumption in catchment-scale precipitation–runoff modeling (e.g., Schaefli et al., 2005), which is that runoff from snowmelt can be modeled with exactly the same functional relationships as for rainfall, by simply feeding so-called equivalent precipitation (sum of rainfall and simulated snowmelt) into the runoff-generation module.

Furthermore, the increase in the discharge-producing frequency to account for snow or ice melt is also coherent with the original description of the analytic modeling framework, which incorporates losses as a decrease in the discharge-producing frequency. This type of behavior can be identified in previous studies. For the Sitter at Appenzell, Basso et al. (2015) obtained λ values that are close to the precipitation frequency λ_p during spring; for the Thur at Jonschwil they obtained $\lambda = \lambda_p$ for spring. Both catchments have a mean elevation above 1000 m a.s.l., which suggests the presence of snow processes. Later on, Doulatyari et al. (2017) discussed that snow accumulation and melt could be affecting the streamflow pdf estimation for the Sitter at Appenzell, without, however, exploring the issue further.

As can be seen in Fig. 6, the switch from $\lambda < \lambda_p$ to λ_p to $\lambda > \lambda_p$ is located at around 1500 m a.s.l. This corresponds to a relatively low mean catchment elevation; for this mean elevation, it can a priori not be assumed that significant snowmelt continues throughout the summer. In fact, for most snow-influenced catchments, the majority of snowmelt happens during spring. Summer flows are nevertheless directly influenced by spring snowmelt since the summer discharge results from a continuous release of melt water stored in the catchment during the preceding snowmelt period. For high elevation catchments, the exceedance of λ over λ_p is directly related to significant snowmelt and ice melt inputs throughout the summer.

It should be kept in mind here that for the present study, λ is estimated directly from the relationship between discharge and precipitation (see Sect. 2.2 and Eq. 7). The question of how to estimate this parameter directly from catchment characteristics based on long-term snow cover statistics and data on glacier cover remains to be answered in future work.

Besides the important result that the model is applicable to snow-influenced catchments, additional insights can be obtained from the highlighted model performance trends with mean catchment elevation (Figs. 9 and 10). These performance trends are explained by the evolution of the regimes with mean catchment elevation, from rainfall-dominated (pluvial) regimes with summer low flow to snowfall-influenced (nival and glacier) regimes with summer high flow. This result suggests that mean catchment elevation is a good proxy for regime shifts, despite the fact that many other catchment characteristics vary strongly across the set of studied catchments (area, hypsometric curve, land use, etc.). Given the strong link between mean catchment elevation, mean catchment air temperature and snow accumulation, this opens interesting perspectives for parameter regionalization.

6 Conclusions

This application of the analytic framework of Botter et al. (2007c) to estimate summer streamflow probability distributions for 25 Swiss catchments shows that this framework performs well without any further methodological adjustments across a wide range of discharge regimes, including rainfall-driven regimes with summer low flows, but also regimes with snowmelt- and glacier-melt-influenced summer high flows. Given that the original framework was developed for purely rainfall-driven regimes, this result is unexpected. For snow-influenced catchments, the model has been shown here to accommodate the additional source of water from snowmelt by a relative increase in the discharge-producing frequency, which is coherent with the underlying analytic framework.

The detailed comparison between the performance of the linear and the nonlinear model formulation shows that the description of Swiss summer flows strongly benefits from using a nonlinear storage–discharge relationship, in particular for catchments with summer low flow and for the karst catchments. In general, the linear model performance increases for increasing total summer flows or, equivalently, for catchments with higher mean elevation. Future work will focus on improving the model parameter estimation directly from observed data (without parameter optimization), which is a precondition for parameter regionalization. Better insights into the physical drivers of the different parameters will also open new potential for extending the model framework to all four seasons for all Swiss streamflow regimes.

Competing interests. The authors declare that they have no conflict of interest.

Acknowledgements. The work of the first author is funded by the Portuguese Science and Technology Foundation (FCT), grant no. PD/BD/52663/2014. The work of Bettina Schaefli is funded by the Swiss National Science Foundation (SNSF), grant no. PP00P2_157611. We thank the editor, Fabrizio Fenicia, and two anonymous reviewers for their detailed comments on an earlier version of this manuscript.

Edited by: Fabrizio Fenicia

References

Addor, N. and Fischer, E. M.: The influence of natural variability and interpolation errors on bias characterization in RCM simulations, J. Geophys. Res.-Atmos., 120, 10180–10195, https://doi.org/10.1002/2014JD022824, 2015.

Aschwanden, A.: Caractéristiques physiographiques des bassins de recherches hydrologiques en Suisse, Berne, Service hydrologique et géologique national, Communications hydrologiques, 23, 1996.

Basso, S., Schirmer, M., and Botter, G.: On the emergence of heavy-tailed streamflow distributions, Adv. Water Res., 82, 98–105, https://doi.org/10.1016/j.advwatres.2015.04.013, 2015.

Bernet, D. B., Prasuhn, V., and Weingartner, R.: Surface water floods in Switzerland: what insurance claim records tell us about the damage in space and time, Nat. Hazards Earth Syst. Sci., 17, 1659–1682, https://doi.org/10.5194/nhess-17-1659-2017, 2017.

Biswal, B. and Marani, M.: Geomorphological origin of recession curves, Geophys. Res. Lett., 37, 24, https://doi.org/10.1029/2010gl045415, 2010.

Biswal, B. and Marani, M.: "Universal" recession curves and their geomorphological interpretation, Adv. Water Res., 65, 34–42, https://doi.org/10.1016/j.advwatres.2014.01.004, 2014.

Blanc, P. and Schädler, B.: Water in Switzerland – an overview, Tech. rep., available at: http://www.naturalsciences.ch/topics/water/ (in English; last access: 18 August 2015), 2013.

Botter, G., Peratoner, F., Porporato, A., Rodriguez-Iturbe, I., and Rinaldo, A.: Signatures of large-scale soil moisture dynamics on streamflow statistics across U.S. climate regimes, Water Resour. Res., 43, https://doi.org/10.1029/2007wr006162, 2007a.

Botter, G., Porporato, A., Daly, E., Rodriguez-Iturbe, I., and Rinaldo, A.: Probabilistic characterization of base flows in river basins: Roles of soil, vegetation, and geomorphology, Water Resour. Res., 43, https://doi.org/10.1029/2006wr005397, 2007b.

Botter, G., Porporato, A., Rodriguez-Iturbe, I., and Rinaldo, A.: Basin-scale soil moisture dynamics and the probabilistic characterization of carrier hydrologic flows: Slow, leaching-prone components of the hydrologic response, Water Resour. Res., 43, w02417, https://doi.org/10.1029/2006WR005043, 2007c.

Botter, G., Porporato, A., Rodriguez-Iturbe, I., and Rinaldo, A.: Nonlinear storage-discharge relations and catchment streamflow regimes, Water Resour. Res., 45, https://doi.org/10.1029/2008wr007658, 2009.

Botter, G., Basso, S., Rodriguez-Iturbe, I., and Rinaldo, A.: Resilience of river flow regimes, P. Natl. Acad. Sci. USA, 110, 12925–12930, https://doi.org/10.1073/pnas.1311920110, 2013.

Brutsaert, W. and Nieber, J. L.: Regionalized drought flow hydrographs from a mature glaciated plateau, Water Resour. Res., 13, 637–643, https://doi.org/10.1029/wr013i003p00637, 1977.

Burnham, K. P. and Anderson, D. R. (Eds.): Model Selection and Multimodel Inference, Springer New York, https://doi.org/10.1007/b97636, 2004.

Castellarin, A., Botter, G., Hughes, D. A., Liu, S., Ouarda, T. B. M. J., Parajka, J., Post, D. A., Sivapalan, M., Spence, C., Viglione, A., and Vogel, R. M.: Prediction of flow duration curves in ungauged basins, Runoff prediction in ungauged basins: Synthesis across processes, places and scales, Cambridge University Press, 135–162, 2013.

Ceola, S., Botter, G., Bertuzzo, E., Porporato, A., Rodriguez-Iturbe, I., and Rinaldo, A.: Comparative study of ecohydrological streamflow probability distributions, Water Resour. Res., 46, https://doi.org/10.1029/2010wr009102, 2010.

Doulatyari, B., Betterle, A., Basso, S., Biswal, B., Schirmer, M., and Botter, G.: Predicting streamflow distributions and flow duration curves from landscape and climate, Adv. Water Res., 83, 285–298, https://doi.org/10.1016/j.advwatres.2015.06.013, 2015.

Doulatyari, B., Betterle, A., Radny, D., Celegon, E. A., Fanton, P., Schirmer, M., and Botter, G.: Patterns of streamflow regimes along the river network: The case of the Thur river, Environ. Modell. Softw., 93, 42–58, https://doi.org/10.1016/j.envsoft.2017.03.002, 2017.

Dralle, D., Karst, N., and Thompson, S. E.: a, b careful: The challenge of scale invariance for comparative analyses in power law models of the streamflow recession, Geophys. Res. Lett., 42, 9285–9293, 2015.

Swiss Federal Office for Statistics: Geostat – Version 1997, Swiss spatial land use statistics data-base (Arealstatistik der Schweiz 1979/85 und Arealstatistik der Schweiz 1992/97), Neuchâtel, Switzerland, 35 p, 2001.

Federal Office for the Environment: Biogeographical regions of Switzerland (CH), available at: https://opendata.swiss/en/dataset/biogeographische-regionen-der-schweiz-ch (last access: 15 December 2017), 2004.

FOEN: Hydrological data and forecasts, Federal Office for the Environment (FOEN), Bern, Switzerland, available at: https://www.hydrodaten.admin.ch/en/stations-and-data.html, last access: 15 December 2017.

Gabella, M., Speirs, P., Hamann, U., Germann, U., and Berne, A.: Measurement of Precipitation in the Alps Using Dual-Polarization C-Band Ground-Based Radars, the GPM Spaceborne Ku-Band Radar, and Rain Gauges, Remote Sens., 9, 1147, https://doi.org/10.3390/rs9111147, 2017.

Gerrits, A. M. J.: The role of interception in the hydrological cycle, TU Delft, Delft University of Technology, 2010.

Helbling, A.: Topographical catchments of the FOEN gauging stations. Geodataset EZG hydrometrische Stationen, Tech. rep., available at http://www.bafu.admin.ch/wasser/13462/13496/15010/index.html?lang=en, last access: 11 May 2017, 2016.

Klemeš, V.: Operational testing of hydrological simulation models, Hydrol. Sci. J., 31, 13–24, https://doi.org/10.1080/02626668609491024, 1986.

Laio, F., Baldassarre, G. D., and Montanari, A.: Model selection techniques for the frequency analysis of hydrological extremes, Water Resour. Res., 45, https://doi.org/10.1029/2007wr006666, 2009.

MeteoSwiss, F. O. o. M. a. C.: Documentation of MeteoSwiss Grid-Data Products: Daily mean, minimum and maximum temperature, MeteoSwiss, Zürich, 4, 2011a.

MeteoSwiss, F. O. o. M. a. C.: Documentation of MeteoSwiss Grid-Data Products: Daily Precipitation (final analysis): RhiresD, MeteoSwiss, Zürich, 4, 2011b.

Müller, M. F. and Thompson, S. E.: Comparing statistical and process-based flow duration curve models in ungauged basins and changing rain regimes, Hydrol. Earth Syst. Sci., 20, 669–683, https://doi.org/10.5194/hess-20-669-2016, 2016.

Müller, M. F., Dralle, D. N., and Thompson, S. E.: Analytical model for flow duration curves in seasonally dry climates, Water Resour. Res., 50, 5510–5531, https://doi.org/10.1002/2014wr015301, 2014.

Mutzner, R., Bertuzzo, E., Tarolli, P., Weijs, S. V., Nicotina, L., Ceola, S., Tomasic, N., Rodriguez-Iturbe, I., Parlange, M. B., and Rinaldo, A.: Geomorphic signatures on Brutsaert base flow recession analysis, Water Resour. Res., 49, 5462–5472, https://doi.org/10.1002/wrcr.20417, 2013.

Naghettini, M.: Preliminary Analysis of Hydrologic Data, in: Fundamentals of Statistical Hydrology, 21–56, Springer International Publishing, https://doi.org/10.1007/978-3-319-43561-9_2, 2016.

Paschalis, A., Fatichi, S., Molnar, P., Rimkus, S., and Burlando, P.: On the effects of small scale space-time variability of rainfall on basin flood response, J. Hydrol., 514, 313–327, https://doi.org/10.1016/j.jhydrol.2014.04.014, 2014.

Rodriguez-Iturbe, I., Porporato, A., Ridolfi, L., Isham, V., and Coxi, D. R.: Probabilistic modelling of water balance at a point: the role of climate, soil and vegetation, P. Roy. Soc. A-Math. Phy., 455, 3789–3805, https://doi.org/10.1098/rspa.1999.0477, 1999.

Schaefli, B., Hingray, B., Niggli, M., and Musy, A.: A conceptual glacio-hydrological model for high mountainous catchments, Hydrol. Earth Syst. Sci., 9, 95–109, https://doi.org/10.5194/hess-9-95-2005, 2005.

Schaefli, B., Rinaldo, A., and Botter, G.: Analytic probability distributions for snow-dominated streamflow, Water Resour. Res., 49, 2701–2713, https://doi.org/10.1002/wrcr.20234, 2013.

Searcy, J. K.: Flow-duration curves, US Government Printing Office, 1959.

SwissTopo: DHM25- The digital height model of Switzerland DHM25, Product information, Wabern, Bern, 15, 2005.

Vogel, R. M. and Fennessey, N. M.: Flow-Duration Curves I: New Interpretation and Confidence Intervals, J. Water Res. Plann. Manage., 120, 485–504, https://doi.org/10.1061/(asce)0733-9496(1994)120:4(485), 1994.

Vogel, R. M. and Fennessey, N. M.: Flow duration curves II: a review of the applications in water resources planning, J. Am. Water Res. Assoc., 31, 1029–1039, https://doi.org/10.1111/j.1752-1688.1995.tb03419.x, 1995.

Weingartner, R. and Aschwanden, H.: Discharge regime–the basis for the estimation of average flows, Hydrological Atlas of Switzerland, Plate, 5, 26, 1992.

3

Regional analysis of parameter sensitivity for simulation of streamflow and hydrological fingerprints

Simon Höllering[1], **Jan Wienhöfer**[1], **Jürgen Ihringer**[1], **Luis Samaniego**[2], **and Erwin Zehe**[1]

[1]Karlsruhe Institute of Technology (KIT), Karlsruhe, Germany
[2]Helmholtz Centre for Environmental Research, Leipzig, Germany

Correspondence: Simon Höllering (simon.hoellering@kit.edu)

Abstract. Diagnostics of hydrological models are pivotal for a better understanding of catchment functioning, and the analysis of dominating model parameters plays a key role for region-specific calibration or parameter transfer. A major challenge in the analysis of parameter sensitivity is the assessment of both temporal and spatial differences of parameter influences on simulated streamflow response. We present a methodological approach for global sensitivity analysis of hydrological models. The multilevel approach is geared towards complementary forms of streamflow response targets, and combines sensitivity analysis directed to hydrological fingerprints, i.e. temporally independent and temporally aggregated characteristics of streamflow (INDPAS), with the conventional analysis of the temporal dynamics of parameter sensitivity (TEDPAS).

The approach was tested in 14 mesoscale headwater catchments of the Ruhr River in western Germany using simulations with the spatially distributed hydrological model mHM. The multilevel analysis with diverse response characteristics allowed us to pinpoint parameter sensitivity patterns much more clearly as compared to using TEDPAS alone. It was not only possible to identify two dominating parameters, for soil moisture dynamics and evapotranspiration, but we could also disentangle the role of these and other parameters with reference to different streamflow characteristics. The combination of TEDPAS and INDPAS further allowed us to detect regional differences in parameter sensitivity and in simulated hydrological functioning, despite the rather small differences in the hydroclimatic and topographic setting of the Ruhr headwaters.

1 Introduction

1.1 Analysis of parameter influences

The role of hydrological model parameters has been studied for a long time. The ill-posed nature of problems in hydrological modelling led to the awareness that parameter sets are not uniquely identifiable (Beven, 1993) and to the related branches of uncertainty assessment (e.g. Gupta et al., 1998) and automated parameter estimation (e.g. Hogue et al., 2000). Both are closely related to the sensitivity of model results to parameter variations. While a number of topics are often subsumed under sensitivity analysis, underlying objectives and methodological approaches can substantially differ from case to case (van Griensven et al., 2006; Saltelli et al., 2008; Zajac, 2010; Razavi and Gupta, 2015). Local and global strategies of sensitivity analysis have been shown to be helpful at different stages of the modelling process (McCuen, 1973; Hamby, 1994; Sieber and Uhlenbrook, 2005; Razavi and Gupta, 2015). Analogous to the number of different objectives and methods to assess parameter sensitivity, the results are subject to different forms of interpretation (Razavi and Gupta, 2015). The way that the outcome of sensitivity analysis is evaluated and illustrated can strongly affect the conclusions that are drawn. In this regard, results of sensitivity analysis can widely differ if varying objective functions are considered for the evaluation of parameter influences (Demaria et al., 2007; Wagener et al., 2009); for a comprehensive overview see Reusser et al. (2011).

Time-integrated sensitivity measures (van Griensven et al., 2006; Sudheer et al., 2011; Nossent and Bauwens, 2012) alone allow little more than rough estimates about the overall importance of parameters. Contrarily, McCuen (1973) early pointed out that parameter sensitivity should be analysed in a time-dependent context, as hydrological systems are subject to temporally dynamic processes. Guse et al. (2016b) argued that the study of temporal variations in sensitivity is essential to learn about the relation between dominant parameters and governing processes under changing hydrological conditions. The characterisation of temporal dynamics of parameter sensitivity (TEDPAS) has been accomplished in diverse ways (Cloke et al., 2008; Cibin et al., 2010; Reusser et al., 2011; Reusser and Zehe, 2011; Herman et al., 2013; Sanadhya et al., 2013; Guse et al., 2014; Pfannerstill et al., 2015; Pianosi and Wagener, 2016). The choice of the temporal resolution is an important factor which clearly influences the way parameters are identified and how inferences on related processes are made (Tang et al., 2007; Massmann and Holzmann, 2012; O'Loughlin et al., 2013). Necessarily, the timescale of sensitivity analysis is selected in accordance with the objective of the study and the dynamics of the system under investigation. The importance of parameters temporally varies as short periods of high flow alternate with longer periods of low flow (Massmann et al., 2014).

When model calibration and verification come into play, analysis of parameter sensitivity provides valuable information on the importance of each input factor in regard to simulated model output. On this basis, it can be decided for each parameter whether its value should be determined exactly, or if it could even be completely excluded, fixed at predetermined values (Reusser et al., 2011). Preferably, sensitivity analysis minimises the necessary number of parameters as hydrological models are often subject to overparameterisation (Beven, 2001; Kirchner, 2006; van Werkhoven et al., 2009; Samaniego et al., 2010b).

A common goal of sensitivity-guided studies dealing with an identification of dominant processes is the achievement of a suitable representation of real-world hydrological processes by understanding the reasons for model defectiveness. If non-sensitive parameters are detected, an indication of model structural deficits (Kirchner, 2006; Gupta et al., 2012), or a lack of the adequate model response target data might be given. Sensitivity analysis is, not just recently, considered as a helpful diagnostic tool to identify structural and performance deficits of hydrological models (McCuen, 1973; Sieber and Uhlenbrook, 2005; Yilmaz et al., 2008; Kavetski and Clark, 2010; Guse et al., 2014; Pfannerstill et al., 2015). Reusser and Zehe (2011) showed that a combined analysis of the temporally varying parameter dominance (sensitivity analysis) and model performance (error analysis) can effectively detect structural inadequacies of model components for a specific landscape.

1.2 Fingerprint-based sensitivity analysis

The characterisation of catchment functioning and the underlying hydrological processes can be addressed in various ways, at multiple scales and levels of complexity. Fingerprint metrics (hereinafter also referred to as fingerprints) are signatures of dynamic catchment response that change on different temporal and spatial scales (Sivapalan, 2005; Wagener et al., 2007; Winsemius et al., 2009).

In hydrological modelling, multiple fingerprint metrics have been adopted to enhance model evaluation beyond the minimisation of streamflow residuals. Fingerprints of catchment functioning may be classified into measures based on single-value (statistical) streamflow indices, and those based on characteristic curves, e.g. (cumulative) frequency curves, regime curves, or double mass curves. Examples for the two kinds of fingerprint metrics are the runoff ratio and the flow duration curve, respectively. Representatives of both categories can be selected to describe single components of streamflow regimes, namely the magnitude, frequency of occurrence, duration, timing, and flashiness of flow events (Poff et al., 1997; Olden and Poff, 2003), or of the general hydrological variability at different spatial and temporal scales.

In a comprehensive analysis of catchment functioning in order to understand dominant processes, the use of a single criterion is often not sufficient. Therefore, hydrological fingerprints have been jointly used as multivariate objectives to estimate the parameters of hydrological models (Shamir et al., 2005a, b; Pokhrel et al., 2008; Castiglioni et al., 2010) or to assess model performance and evaluate model structures (Farmer et al., 2003; Gupta et al., 2008; Yilmaz et al., 2008; Clark et al., 2011; Euser et al., 2013; Vrugt and Sadegh, 2013).

Sensitivity analysis related to streamflow characteristics was formerly mostly applied prior to model evaluation (e.g. Atkinson et al., 2003). For sensitivity analysis, different options have been selected as hydrological target variables. Sensitivity analysis to assess the influence of parameters can be directed to (i) simulated streamflow, (ii) different objective functions (e.g. van Werkhoven et al., 2009; Wagener et al., 2009; Herman et al., 2013; Sanadhya et al., 2013), (iii) simulated hydrological processes (e.g. Massmann and Holzmann, 2015; Pfannerstill et al., 2015; Guse et al., 2016a), or (iv) different hydrological fingerprints (this study). Previous studies applied fingerprint metrics but based the analysis of parameter sensitivity on only a few aspects of streamflow (e.g. limb densities; Shamir et al., 2005a) or on single (statistical) streamflow indices of different aggregation timescales (Shamir et al., 2005b).

In our view, multivariate sensitivity analysis geared towards fingerprint metrics as response targets has not received adequate consideration for model diagnostics. Especially in terms of joint fingerprints, using both single-value indices and characteristic curves along independent variables, the full potential for process-oriented model diagnostics has not been exploited. Some progress has been made by Guse et al. (2016b), who combined TEDPAS for different temporal resolutions with segments of the flow duration curve (FDC) to identify parameters and related processes that dominate at variable streamflow magnitudes of two distinct streamflow regimes.

1.3 Objectives, research questions and approach

The main objectives of this study are to analyse the parameter sensitivity of a mesoscale hydrological model for the simulation of streamflow response and hydrological fingerprints at a set of headwater catchments of the Ruhr in Germany. The approach extends TEDPAS along two avenues: the first is to investigate TEDPAS results in more detail to derive parameter sensitivities in different hydrological conditions; the second is to direct the analysis to other, temporally independent characteristics of streamflow response (INDPAS).

With this approach we explore the following three research questions:

- Which sensitive parameters can be identified with regard to specific hydrological response characteristics?
- How does parameter sensitivity change with different hydrological objectives (response targets) applied in global sensitivity analysis?
- How does parameter sensitivity change among different catchments with slightly distinct physiographic and hydroclimatic conditions?

The methodological approach combines streamflow hydrographs and fingerprint metrics as response targets for the analysis of first-order partial parameter sensitivity. The analysis rests on a state-of-the-art distributed hydrological model and is structured in the following steps:

- combining the application of a hydrological model with global sensitivity analysis to generate an ensemble of parameter sets;
- deriving fingerprint metrics (single-value indices and characteristic curves) from simulated streamflow time series;
- analysing parameter sensitivity to temporally resolved dynamics of streamflow response (TEDPAS);
- analysing parameter sensitivity to both temporally aggregated (single-value indices) and temporally independent (characteristic curves) characteristics of streamflow (INDPAS);
- assessing differences in parameter sensitivity between the two different methodological approaches (TEDPAS and INDPAS), and between the analysed headwaters.

In the study we will thus complement sensitivity analysis based on TEDPAS with fingerprint metrics of streamflow response (INDPAS), which include both temporally aggregated single-valued indices and temporally independent characteristic curves. In cases where characteristic curves (e.g. the FDC) are used, changes in parameter sensitivity will be analysed for changes in the independent variable (e.g. streamflow exceedance probability). We focus the study on the headwaters of the Ruhr catchment in western Germany based on available data sets.

From this we expect to pinpoint dominant parameters related to individual process components and to ease the interpretation of parameter sensitivity detached from the variability of timescales. Bearing in mind the complexity of the evaluation of spatially and temporally distributed model responses, our multilevel approach aims at providing further insight into the dominance of model parameters and related streamflow response processes.

2 Methods and models

First we detail the fingerprint metrics used to characterise streamflow response (Sect. 2.1). We implemented the Fourier amplitude sensitivity test (FAST; Sect. 2.2) to analyse the mesoscale hydrologic model (mHM, Sect. 2.3) in the Ruhr headwater catchments. Focusing on eight global mHM parameters (Sect. 2.4), we employed two different forms (TEDPAS and INDPAS) of simulated streamflow response (Sect. 2.5) in the sensitivity analysis. Finally, we introduce the catchment of the Ruhr River and the headwaters which were selected for this study (Sect. 2.6.1), and specify the data used for the analysis (Sect. 2.6.2).

2.1 Fingerprint metrics

Fingerprint metrics are often used in hydrology for characterising the hydrological response of catchments (Olden and Poff, 2003; Yadav et al., 2007; Yilmaz et al., 2008; Winsemius et al., 2009). The fingerprint metrics used in this study included single-value indices and the flow duration curve as an example for catchment-characteristic curves. These fingerprints were derived from model results and precipitation data (see Sect. 2.6.2), respectively.

We chose eight indices reflecting different aspects of the integral and long-term hydrological functioning of catchments in a single, time-aggregated number (Table 1). These fingerprints characterise the overall water balance (runoff ratio, RR), the variability of streamflow (coefficient of variation, CV), the frequency of flow events (high pulse count, HPC), the change rate of streamflow (slope of flow duration curve between 33 and 66 %, SLFDC), the streamflow during high-flow (high flow discharge, HFD) and low-flow (baseflow index, BFI) conditions, the streamflow recession behaviour (recession time constant, RTC), and the autocorrelation structure of streamflow (autocorrelation time, ACT), respectively. In this study, the slope of the flow duration curve (SLFDC) is the only single-value fingerprint that could

Table 1. Temporally aggregated single-value fingerprint metrics derived from FAST-mHM simulated streamflow and observed precipitation time series, serving as model response targets for sensitivity analysis (INDPAS).

Response characteristic	Fingerprint metric	Abbreviation	Unit	Derivation
Water balance	Runoff ratio	RR	(–)	Q_{Total} / P_{Total}
Streamflow variability	Coefficient of variation	CV	(–)	σ / μ
Frequency of flow events	High pulse count	HPC	(yr^{-1})	(number of time steps $Q > 3 \times Q_{\text{mean}}$)/ years
Rate of change in streamflow	Slope of flow duration curve	SLFDC	(%)	Slope of FDC between 33 and 66 % Q exceedance
High-flow conditions	High flow discharge	HFD	(–)	$Q_{\text{5th percentile}}$ / Q_{median}
Low-flow conditions	Baseflow index	BFI	(–)	Q_{Baseflow} / Q_{Total}
Streamflow recession	Recession time constant	RTC	(d)	Mdn (time required for Q to reach $1/e \times Q_{\text{Peak}}$)
Streamflow autocorrelation structure	Autocorrelation time	ACT	(d)	Lag time required for AC function to decrease below 0.5

Q: streamflow; P: precipitation; σ: standard deviation; μ: mean; Mdn: median; AC: autocorrelation.

not be directly determined from streamflow hydrographs. Instead, the FDC was used as a basis for its derivation. The eight single-value fingerprints were implemented as model response targets for sensitivity analysis (Sect. 2.5.2).

As an example for more complex characteristics than single-valued indices, we also used entire flow duration curves as model response targets for sensitivity analysis (Sect. 2.5.2).

2.2 Fourier amplitude sensitivity test (FAST)

FAST is a partial variance-based method to determine first-order sensitivities of parameter changes on the outcome of monotonic and non-monotonic numeric models (Cukier et al., 1973; Schaibly and Shuler, 1973; Cukier et al., 1975). The general idea of FAST is (a) to vary parameters of interest with independent frequencies along a predefined number of model runs, and (b) to perform a Fourier analysis of the simulated target variable across the ensemble of model runs to obtain a power spectrum. In the case of TEDPAS, the spectrum is calculated for each simulation time step. The variance σ_i^2 that is explained by a parameter i is determined by normalising the corresponding power with the total power in the spectrum, which corresponds to the total variance σ_{tot}^2 within the model ensemble. The sensitivity to model output of parameter i is then calculated as the partial variance, which is the ratio $\sigma_i^2/\sigma_{\text{tot}}^2$. Parameter interactions, i.e. higher-order sensitivity, are not detected by this method. For more details on FAST the reader is referred to Reusser et al. (2011).

FAST was originally applied to study parametric model sensitivities of chemical reaction systems. In recent decades, the method has been used and evaluated in a variety of fields such as hydrogeology (Fontaine et al., 1992), atmospheric sciences (Rodríguez-Camino and Avissar, 1998), geologic nuclear waste disposal modelling (Lu and Mohanty, 2001), food safety risk assessment (Frey and Patil, 2002), or ecologic forestry (Song et al., 2013). A number of studies treat the application of FAST in hydrological modelling (Reusser et al., 2011; Reusser and Zehe, 2011; Sanadhya et al., 2013; Guse et al., 2014; Pfannerstill et al., 2015; Guse et al., 2016a, b).

FAST is a highly efficient computational method that requires significantly fewer model runs to yield similar results for parameter sensitivity than other approaches (Saltelli and Bolado, 1998; Reusser et al., 2011). The number of model runs (hence parameter sets) in FAST is determined by the number of analysed model parameters. This means that always the same number of model runs is required for a given number of parameters, independent of model, catchment or type of parameter.

2.3 Mesoscale hydrologic model (mHM)

The mHM (Kumar et al., 2010; Samaniego et al., 2010b) accounts for diverse processes of the hydrological cycle: canopy interception, evapotranspiration, snow, soil moisture dynamics, overland flow, infiltration, interflow, subsurface storage, groundwater recharge, baseflow, discharge attenuation, as well as flood routing. The mHM is conceptualised on the basis of grid cells, and has been applied to a wide range of mesoscale river catchments (10^1–10^4 km^2; Kumar et al., 2010; Samaniego et al., 2010a, 2011; Cuntz et al., 2015; Rakovec et al., 2016). Gridded information is implemented in mHM at three levels: morphology (level 0), hydrology (level 1), meteorology (level 2), with $l_0 \ll l_1 \leq l_2$ denoting the relative sizes of the grid cells at the respective data level (Kumar et al., 2010).

The parameterisation of mHM is based on a simultaneous regionalisation technique called multiscale parameter regionalisation to account for the physiographic sub-grid and hydrological process variability (Samaniego et al., 2010b; Kumar et al., 2013). Hydrological process parameters at level

1 are derived from physiographic characteristics at level 0 using (pedo-)transfer functions with coefficients (in the following referred to as global mHM parameters). Hence, mHM is calibrated indirectly, by altering the 52 level-0 parameters of the transfer functions instead of the hydrological level-1 parameters. This procedure not only reduces the problem of overparameterisation and the dependence on specific hydrological scales (Beven, 2001) but also reduces the amount of time that is needed for grid-wise calibration (Samaniego et al., 2010b).

2.4 Model setup for sensitivity analysis

To facilitate the selection of the most sensitive parameters, we first carried out a preliminary FAST analysis at the local scale for gauge Wenholthausen (WEN; Fig. 1) including all 52 global mHM parameters to reveal parameter sensitivities to streamflow simulations. For this initial analysis, 21 803 model runs were conducted and the streamflow hydrographs were analysed with FAST. We found 14 parameters with a maximal sensitivity value of more than 0.01 (1 %). When inspecting the model equations we identified correlations between these parameters, which led to the removal of six parameters from this set.

The eight uncorrelated parameters (Table 2) were used for the regional sensitivity analysis in the 14 headwater catchments. All other mHM parameters were kept fixed on calibrated values found via global automatic optimisation using the dynamically dimensioned search algorithm (Tolson and Shoemaker, 2007) at WEN for the period 2002 to 2006. The value ranges for the parameters were selected from mHM literature (Samaniego et al., 2014), partly extended based on the results from the preliminary analysis. For eight parameters, the FAST method requires 243 model runs based on different parameter combinations originating from variation with independent frequencies inside the parameter ranges (Fig. 2). The same 243 combinations of mHM parameter sets were used for streamflow simulations in each of the 14 catchments. Differences between catchments in terms of hydroclimatic forcing and physiographic attributes were included in the model by the locally specific meteorological and morphological input on data levels l_2 and l_0.

The hydrological model level l_1 and the meteorological l_2 of mHM were set to a spatial resolution of 1 km, whereas for level 0 with the physiographic catchment data (morphology), a finer resolution of $l_0 = 200$ m was selected as an adequate spatial discretisation. Model simulations were conducted for each of the 14 headwater catchments (see Sect. 2.6.1) with a daily time step for the 10-year period of 1997 to 2006.

2.5 Sensitivity analysis

We analysed the parameter sensitivity in different forms to be able to evaluate the dominance of parameters and to potentially detect local differences among the headwaters related to various aspects of streamflow response in a more specific way. We used simulated streamflow hydrographs (TEDPAS; Sect. 2.5.1) and both temporally aggregated and temporally independent fingerprint metrics of simulated streamflow response (INDPAS; Sect. 2.5.2) as model response targets for the sensitivity analyses.

2.5.1 TEDPAS – temporal dynamics and sensitivity duration

Using simulated hydrographs with FAST provided daily time series of partial parameter sensitivities for each headwater catchment for the simulation period 1997–2006. These temporal dynamics of parameter sensitivity (TEDPAS; Reusser et al., 2011) were analysed and compared for the Ruhr headwater catchments (Sect. 3.1).

We also calculated sensitivity duration curves (SDCs) for each parameter, which we defined in analogy to other well-known cumulative frequency curves like the FDC. Each SDC is specific for one of the eight parameters, for one catchment and for the period (1997–2006) in which sensitivity analysis is performed. SDCs were developed for each catchment by arranging the daily sensitivity values from FAST by magnitude in ascending order and by plotting them as a line against the percentage of time during which the sensitivity equalled or exceeded the specified values. Sensitivities were normalised by the highest sensitivity value found for each parameter among all headwaters. These curves reveal whether a parameter is consistently (non-)sensitive or if its importance changes during the simulation period (Sect. 3.2).

2.5.2 INDPAS – parameter sensitivity to fingerprint metrics

For each catchment we calculated eight single-valued fingerprint metrics (Sect. 2.1 and Table 1) from each of the 243 simulated streamflow hydrographs. Using these fingerprint metrics as target variables for FAST yielded the partial sensitivities of the model parameters with regard to each fingerprint (Sect. 3.3.1).

In a similar way, for each headwater catchment, 243 flow duration curves were derived from the simulated streamflow time series and analysed with FAST. This yielded parameter sensitivities along the axis of streamflow exceedance probability as an independent variable, revealing which parameters dominate streamflow simulations during high-, intermediate- or low-flow conditions. As a supplementary step, the parameters showing the highest sensitivity for a given streamflow exceedance probability were extracted, revealing patterns of dominant parameters across the spectrum of streamflow in each headwater catchment (Sect. 3.3.2).

Table 2. Eight global mHM parameters (dimensionless): function, value range for FAST, range and arithmetic mean of sensitivity values (TEDPAS) across 14 Ruhr headwater catchments (1997–2006).

Parameter	Process	Description	Value range (–)	Sensitivity range (–)	Sensitivity mean (–)
DegdayForest	snow	Determination of degree daily factor and maximum degree-day factor	0–4	0–0.74	0.012
ThetaSconst	soil moisture	Estimation of water content at saturation of soil (constant part)	0.65–0.95	0–0.68	0.018
Ksconst	soil moisture	Estimation of saturated vertical hydraulic conductivity	−1.9–0.0	0–0.77	0.392
InfilShapeFactor	soil moisture	Determination of numerical index of rooting distribution	1–4	0–0.29	0.017
AspectcorrPET	meteo correction	Account for aspect dependent correction of PET	0.70–1.50	0–0.78	0.138
ExpslowInterflow	interflow	Determination of exponent for the interflow reservoir	0.05–0.3	0–0.22	0.062
RechargeCoeff	percolation	Determination of percolation coefficient	0–70	0–0.38	0.057
GeoParam	baseflow	Determination of baseflow recession parameter	0–1000	0–0.47	0.012

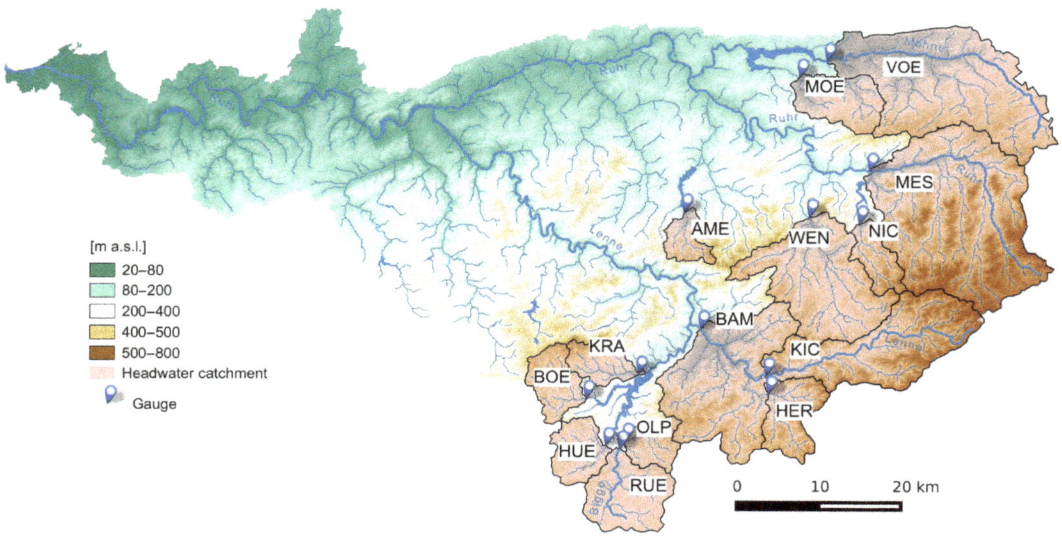

Figure 1. The Ruhr catchment with altitudinal zones, river network and 14 gauged headwater catchments: Amecke (AME), Bamenohl (BAM), Börlinghausen (BOE), Herrntrop (HER), Hüppcherhammer (HUE), Kickenbach (KIC), Kraghammer (KRA), Meschede1 (MES), Möhnesee-Neuhaus (MOE), Nichtinghausen (NIC), Olpe (OLP), Rüblinghausen (RUE), Völlinghausen (VOE), and Wenholthausen (WEN).

2.6 Study area and data

2.6.1 The Ruhr headwater catchments

The Ruhr (Fig. 1) has a catchment area of 4485 km^2, and originates from a spring at about 670 m a.s.l. on the northern slope of the Ruhrkopf (842 m a.s.l.). The Ruhr joins the Rhine at Duisburg-Ruhrort (20 m a.s.l.) after 219 km. The landscape characteristics of the catchment range from densely wooded and scarcely populated lower mountain ranges in the Sauerland to widely sealed urban areas in the river valleys and in the western part close to the mouth. The area belongs to the geology and geography of the Rhenish Slate Mountains to the east of the Rhine (Brudy-Zippelius, 2003). The average discharge at the confluence with the Rhine is about 80.5 m^3 s^{-1} (Bode et al., 2003). With a total of eight dams and five reservoirs the Ruhr and its tributaries form a complex hydrological system. The total stored water surface area of about 35 km^2 equates to about 480 million m^3 of water retained behind damming structures (Ruhrverband, 2011). An intensive use of water resources (e.g. reservoirs,

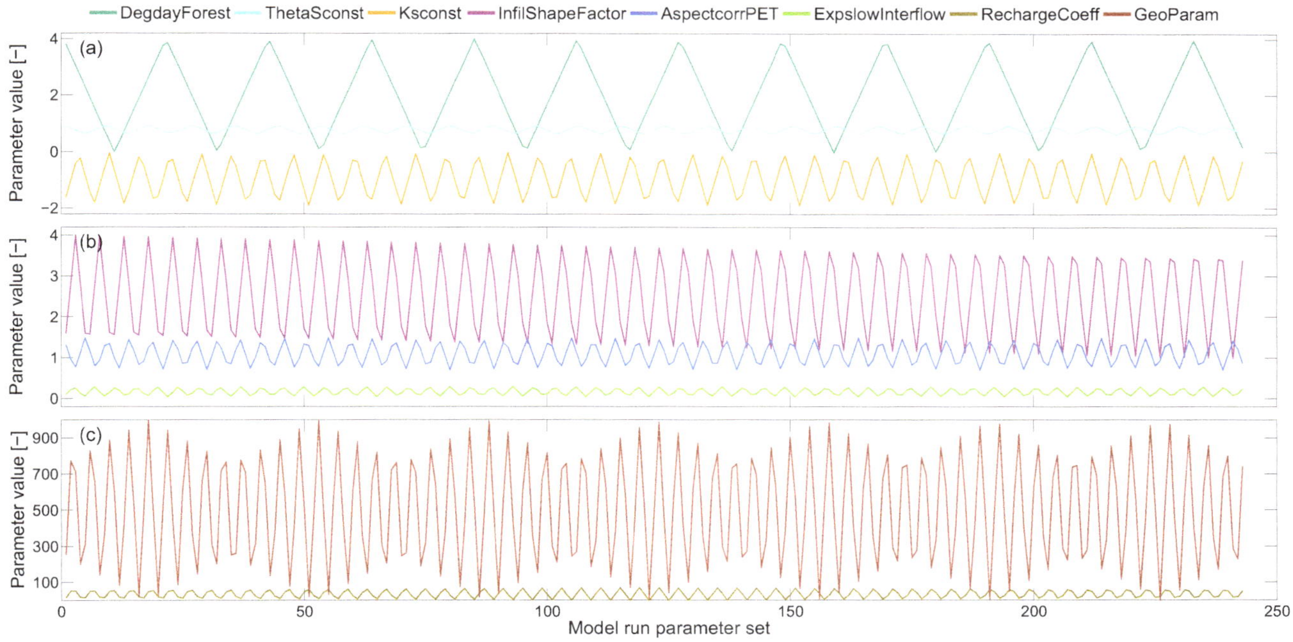

Figure 2. (a–c) Variations of the eight selected global mHM parameters for 243 model runs with independent frequencies according to the FAST sampling plotted as connected curves (see also Table 2).

barrages, withdrawals, inlets) supplies almost 5 million people with drinking and processing water along the Ruhr, within its catchment and to adjacent watersheds.

Our investigations concentrate on 14 headwater catchments (Fig. 1) of the Ruhr River and its tributaries (e.g. Bigge, Lenne, and Möhne), where the hydrological regimes are much less affected by water management measures. The headwaters are situated in the eastern, rural part of the Ruhr basin with higher altitudes, and cover an area of 1742 km^2 in total. Individual catchment sizes range from 28.7 km^2 at gauge Amecke (AME) to 453.1 km^2 at gauge Bamenohl (BAM). Average catchment slopes vary between 10.8 % (Rüblinghausen, RUE) and 26.1 % (Kickenbach, KIC). The dominant form of land cover is forest (39.7–87.3 %) followed by pasture (0.8–47.5 %), cropland (7.6–43.9 %) plus a few predominantly dispersed settlements (0.0–13.2 %; Table 3). The climatic conditions are humid warm-temperate (Göppert et al., 1998) with warm summers and moderate winters. Annual mean temperature ranges between 8.45 and 5.45 °C at the lower and higher altitudes in the study area, respectively. Annual precipitation ranges from 1025 mm in the northeast to 1425 mm in the southwest (1997–2006; Table 3).

2.6.2 Data

Different kinds of observation data were used to set up and calibrate the hydrological model, to perform simulations for sensitivity analysis, and to derive the fingerprint metrics and a set of physiographic catchment descriptors.

Meteorological input data were daily values for precipitation (HYRAS; Rauthe et al., 2013), temperature (HYRAS; Frick et al., 2014), and potential evapotranspiration (AMBAV; Löpmeier, 1994), all at a spatial resolution of 1 km^2. Streamflow observations were available for all headwater catchments from 2002 to 2006. Spatial physiographic data were a digital elevation model (50 m × 50 m), CORINE land cover data (100 m × 100 m; European Environment Agency, 2009), a soil map (1 : 200 000; Bundesanstalt für Geowissenschaften und Rohstoffe, 2015a), and a geological map (1 : 1 000 000; Bundesanstalt für Geowissenschaften und Rohstoffe, 2015b).

A set of 14 descriptors to characterise the hydroclimatic and physiographic setting of the headwaters and to capture characteristics that might jointly control relevant hydrological functions as defined by Black (1997) has been compiled in Table 3. Each descriptor in Table 3 was assigned to one of five main classes of catchment characteristics, i.e. climate (1), landform (2), topography (3), land cover (4), and soil (5), as proposed by Yadav et al. (2007). The choice of climate and physiographic descriptors originates from correlation analysis of catchment descriptors within each category (Yadav et al., 2007), multivariate statistical analysis techniques (Di Prinzio et al., 2011), regionalisation models (Plate et al., 1988), GIS-based analysis of the digital elevation model and, in the case of the baseflow index (BFI), from comparison of methods for baseflow separation (Duband et al., 1993). Table 3 includes the BFI as an intermediate form between physiographic descriptors of soil hydrological

Table 3. Physiographic and climate descriptors characterising the topographic and hydroclimatic (1997–2006) setting of 14 Ruhr headwater catchments.

Characteristic	Descriptor	Abbreviation	Unit	Headwater													
				NIC	VOE	MOE	AME	BOE	RUE	HUE	OLP	KRA	KIC	HER	BAM	WEN	MES
(1) Climate	Wetness index	P/PET	(–)	2.3	2.1	2.1	2.4	2.5	2.2	2.3	2.2	2.2	2.4	2.3	2.3	2.2	2.5
	Annual precipitation	PMEAN	(mm)	1170	1025	1062	1240	1425	1286	1334	1232	1263	1261	1278	1237	1156	1192
(2) Landform	Area	A	(km^2)	37	293	66	29	48	86	47	35	38	187	61	453	184	426
	Longest drainage path	LDP	(km)	13	44	16	10	13	19	9	10	13	36	13	52	21	39
(3) Topography	Slope	SLOPE	(%)	22.6	10.9	11.2	17.3	14.9	10.8	13.2	17.5	16.0	26.1	21.1	22.5	16.2	20.9
	Weighted slope	I_g	(%)	1.7	0.5	1.1	2.1	0.9	0.5	0.8	1.2	0.9	0.7	1.6	0.5	0.6	0.7
	Elevation range	ELR	(m)	430	426	316	361	297	394	204	274	391	533	415	425	440	585
	Aspect	ASP	(°)	195	189	187	189	167	175	170	197	159	201	199	188	186	190
	Flow accumulation	FACC	(km^{-1})	0.7	0.3	0.6	0.7	0.6	0.5	0.5	0.7	0.9	0.5	0.5	0.2	0.3	0.2
(4) Land cover	Forest	FOR	(%)	55	50	87	51	54	43	40	62	52	72	75	67	41	60
	Urban area	URB	(%)	0	7	1	3	2	13	5	7	4	4	3	6	3	5
	Pasture	PAST	(%)	21	13	2	20	8	26	48	6	1	15	6	10	16	10
	Cropland	CROP	(%)	24	29	10	26	36	18	8	25	44	9	17	17	40	24
(5) Soil	Baseflow index	BFI	(–)	0.39	0.34	0.33	0.38	0.37	0.38	0.39	0.39	0.42	0.35	0.36	0.34	0.34	0.31

characteristics and temporally aggregated fingerprint metrics introduced in Sect. 2.1.

3 Results

3.1 Temporal dynamics of parameter sensitivity (TEDPAS)

TEDPAS analysis for the 14 headwaters in the period of 1997–2006 showed a strong temporal dependence of the fraction of the total variance explainable by first-order sensitivities for hydrograph simulation. The sum of all eight parameter sensitivity values per time step ranged between 0.26 and 0.87, while the average sum of the eight sensitivity values per time step was 0.71. The spread between the maximal and minimal sum per time step was found to be smaller in the southwestern (e.g. Rüblinghausen RUE, 0.48) than in the northeastern (e.g. Völlinghausen VOE, 0.61) headwaters.

Minimal and maximal (sensitivity range) and the average (sensitivity mean) sensitivity values of the eight parameters, summarised across all headwaters (Table 2), give a first impression that the soil moisture parameter Ksconst generally exhibited the highest influence (sensitivity mean 0.392), while AspectcorrPET showed the largest range (sensitivity range 0–0.78). The interflow parameter ExpslowInterflow had the smallest sensitivity range (0–0.22), whereas parameters for snow (DegdayForest) and baseflow (GeoParam) had the overall lowest sensitivity mean values of 0.012. Across all headwaters, the parameters listed in terms of their average sensitivity to streamflow simulations are (in descending order): Ksconst, AspectcorrPET, ExpslowInterflow, RechargeCoeff, ThetaSconst, InfilShapeFactor, GeoParam, and DegdayForest.

TEDPAS did not reveal many differences between the headwaters. For instance, Ksconst consistently had a highly dynamic course of sensitivity with frequently high values (Fig. 3a and b). Nevertheless, some of the parameters showed differences between the headwaters, for example for DegdayForest (January–March; Fig. 3a and b) and AspectcorrPET (November–April; Fig. 3c and d). AspectcorrPET allows us to include the aspect of slopes, controlling insulation, in evapotranspiration estimations, while DegdayForest is a parameter related to snow dynamics in forested areas.

The example of these two parameters also illustrates the seasonality in sensitivity dynamics. AspectcorrPET showed highest sensitivity in the summer period from April to August, when evapotranspiration processes dominate and streamflow dynamics are low (Fig. 3c and d, g and h). During that period, the parameter showed an alternating course of sensitivity compared to Ksconst (Fig. 3b and d) with local maxima connected to (simulated) streamflow peaks (Fig. 3h).

Higher sensitivities of DegdayForest were found for periods (e.g. February) when snow processes (accumulation and melting) can occur. This was predominantly observed in catchments at higher altitudes – for example, for the headwater VOE (up to 630 m a.s.l.) rather than for RUE (450 m a.s.l.; Fig. 3a and b). Also, VOE (50 %) exhibits a higher percentage of forest cover (FOR) than RUE (43 %; Table 3). A similar distinction between summer and winter patterns was found for InfilShapeFactor, although at lower sensitivity levels (Fig. 3c and d). For the rest of the parameters either no seasonal patterns (e.g. RechargeCoeff; Fig. 3e and f) could be seen or only very low sensitivity values were found (e.g. ThetaSconst; Fig. 3a and b).

The ensembles of simulated streamflow compared reasonably well with the observed hydrographs (Fig. 3g and h), although the simulation ensemble underestimated some high-flow periods.

3.2 Sensitivity duration

SDCs for the 14 headwaters revealed distinct influences of the eight parameters on streamflow simulations (Fig. 4). Different sensitivity characteristics were identifiable among

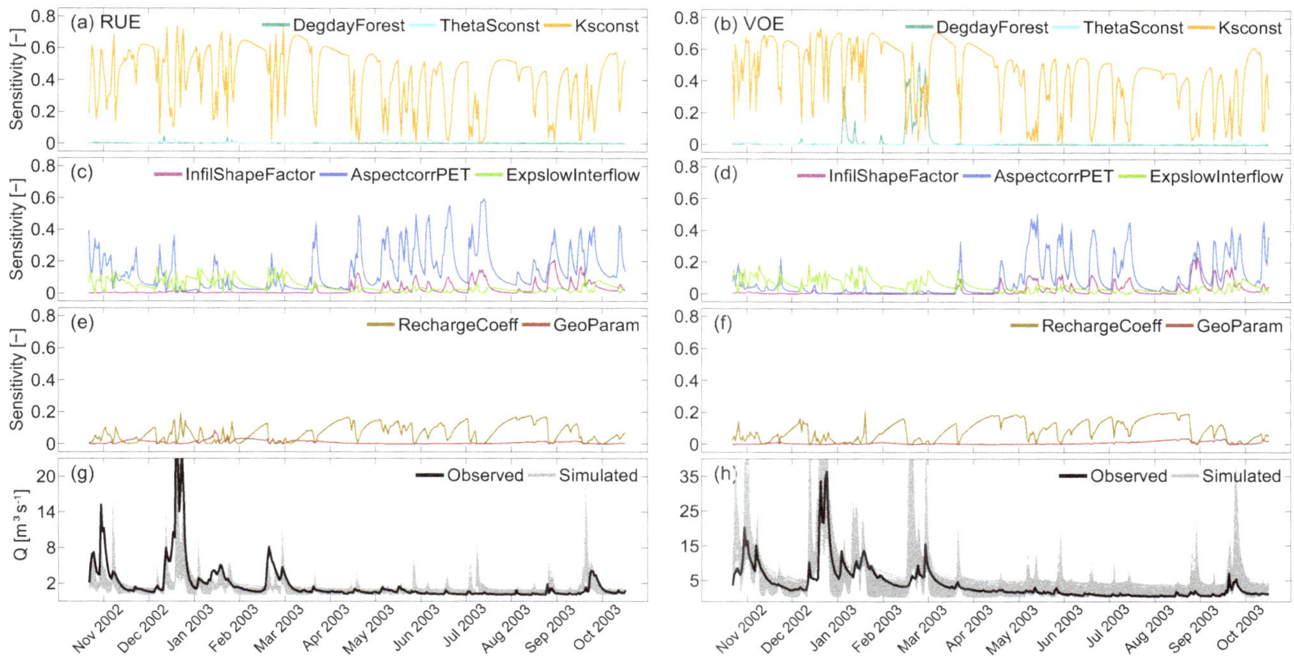

Figure 3. Time-dependent FAST sensitivities (TEDPAS) of eight global mHM parameters for two headwater gauges RUE **(a, c, e)** and VOE **(b, d, f)**, and observed and FAST-mHM simulated streamflow ensembles at gauges RUE **(g)** and VOE **(h)**. The results are shown for the hydrological year 2003 consisting of a wet season (November 2002–March 2003) and a long dry period from April 2003 to September 2003. Please note the different axis scaling for streamflow for the two headwaters **(g, h)**.

the parameters, with either very low (e.g. DegdayForest; Fig. 4a and ThetaSconst; Fig. 4b), intermediate (RechargeCoeff; Fig. 4g) or high (Ksconst; Fig. 4c) influence with respect to sensitivity exceedance probability.

Some of the parameters showed a regional variation of SDCs. Four of eight parameters, i.e. Ksconst (Fig. 4c), InfilShapeFactor (Fig. 4d), AspectcorrPET (Fig. 4e), and ExpslowInterflow (Fig. 4f), revealed certain differences among the headwaters. The SDCs of the two most influential parameters Ksconst (Fig. 4c) and AspectcorrPET (Fig. 4e) showed a systematic spread for the different headwaters, with the curve of gauge RUE plotting at the lower (Fig. 4c) and upper (Fig. 4e) margins of the group of headwaters, respectively. For InfilShapeFactor (Fig. 4d), the headwater Möhnesee-Neuhaus (MOE), and for ExpslowInterflow (Fig. 4f) both RUE and MOE deviated from the other headwaters and showed lower SDC values.

In the case of AspectcorrPET (Fig. 4e), the SDCs were sorted from the southwestern (e.g. RUE) to the northeastern (e.g. MES) headwaters (Fig. 1). In the southwestern headwaters (e.g. RUE) the slopes are gentler with lower relief energy than further northeast, where valleys are more deeply incised (e.g. NIC, SLOPE and ELR; Table 3). The slopes in the southwestern headwaters are on average facing southeast, compared to the more southwest-facing slopes in the northern and eastern Ruhr headwaters (ASP; Table 3). Besides showing a different aspect, the southwestern headwater RUE also has the highest proportion of urban areas (URB, 13 %; Table 3). Both factors influence the estimation of evapotranspiration in mHM and hence streamflow simulations.

The SDCs of the most sensitive parameter Ksconst (Fig. 4c) showed concave curvatures, in contrast to the other parameters which had convex SDCs. The SDCs of the largest headwater Bamenohl (BAM) fell in between the other catchments, showing a kind of transitional behaviour of sensitivity duration (Fig. 4), except for ExpslowInterflow.

3.3 Parameter sensitivity to fingerprints (INDPAS)

3.3.1 Single-value indices

Similar patterns of parameter sensitivities to single-value fingerprints were consistently found across all 14 headwaters. Figure 5 shows the matrix representations for four representative headwaters (RUE, VOE, HER, and WEN). All of them comprise eight rows for the parameters and eight columns for the fingerprint metrics. Sensitivity to a specific fingerprint is arranged column-wise.

As in the TEDPAS analysis, Ksconst was by far the most sensitive parameter for the simulation of five of the fingerprints (CV, HPC, HFD, BFI, RTC) in all 14 headwater catchments. The parameters ExpslowInterflow and RechargeCoeff were identified as the second and the third most sensitive parameter in these cases. In terms of the fingerprint RR, in con-

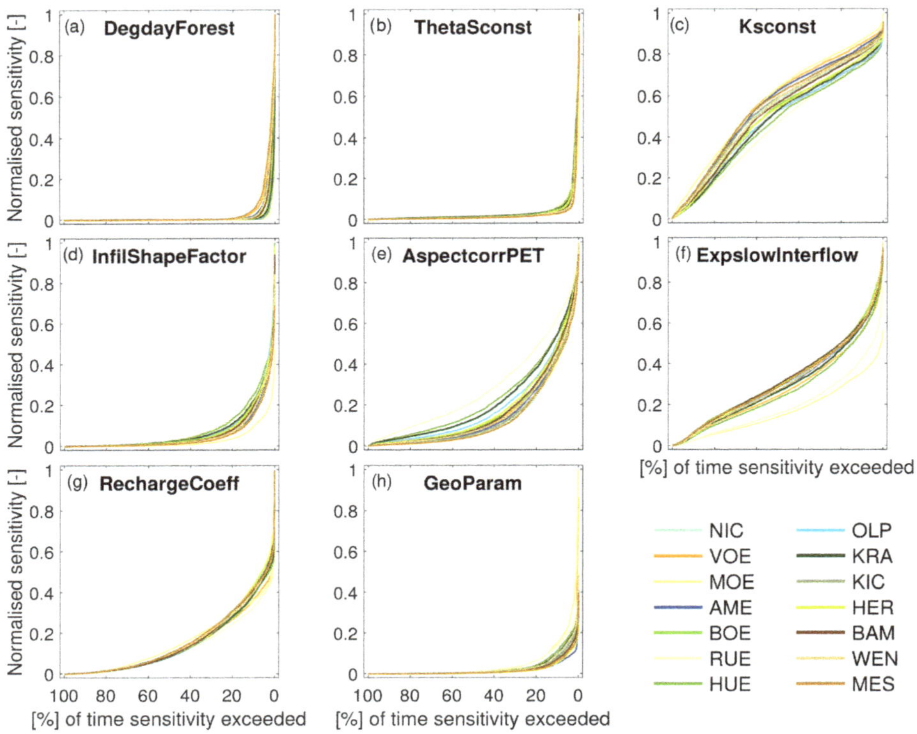

Figure 4. Sensitivity duration curves (SDCs) for eight global mHM parameters of 14 Ruhr headwater catchments (1997–2006): DegdayForest **(a)**, ThetaSconst **(b)**, Ksconst **(c)**, InfilShapeFactor **(d)**, AspectcorrPET **(e)**, ExpslowInterflow **(f)**, RechargeCoeff **(g)**, and GeoParam **(h)**. SDCs are shown normalised by the highest sensitivity value for each parameter among the headwaters. The eight parameters in terms of their average importance to streamflow simulations (TEDPAS) among all headwaters listed in descending order: Ksconst, AspectcorrPET, ExpslowInterflow, RechargeCoeff, ThetaSconst, InfilShapeFactor, GeoParam, and DegdayForest.

trast, AspectcorrPET was the most sensitive parameter, while others, including Ksconst, showed almost no sensitivity to the simulation of the overall water balance. The parameters RechargeCoeff and Ksconst had similarly highest importance for the simulation of the fingerprint SLFDC (slope of the flow duration curve). Other parameter–fingerprint combinations revealed parameters with very low sensitivity values, e.g. DegdayForest or GeoParam, which showed very low sensitivities to all of the eight fingerprint metrics (Fig. 5).

Only minor differences in these patterns occurred between the catchments, and these related to small deviations in absolute sensitivity values or in the order of the second and third rank, e.g. for the fingerprint ACT (Fig. 5).

3.3.2 Flow duration curve

Using FDCs as model response targets revealed parameter sensitivities to different streamflow magnitudes. Again, a high proportion of similarities among the headwaters was found. The highest influence was alternately exerted by the parameters Ksconst and AspectcorrPET (Fig. 6a–d); their courses of parameter sensitivity were highly anticorrelated (mean correlation across all headwaters $r = -0.975$). The soil moisture parameter Ksconst clearly dominated the very high flows (0–10 % of time Q is exceeded) and the entire mid- and low-flow sections (40–100 %); moderate high flows between 10 and 40 % were most affected by changes in the evapotranspiration parameter AspectcorrPET. These changes in the dominating parameter are additionally illustrated in Fig. 6 by a catchment-specific strip showing the pattern of parametric dominance along the FDC, which showed only slight differences in the lengths of the intermittent parts (AspectcorrPET) between the headwaters.

The other parameters reached overall lower sensitivity levels. The patterns were again similar for all headwaters, with minor differences regarding the absolute sensitivity values and the order of importance in the third and higher ranks. The parameters RechargeCoeff and ExpslowInterflow revealed a bimodal sensitivity distribution. RechargeCoeff showed a first peak between 0 and 20 %, and a steady increase from 40 % to its maximum sensitivity value at very low flows with 100 % of streamflow exceedance, which was a sensitivity value of about 0.25 in the case of gauge HER (Fig. 6c). ExpslowInterflow had its highest parametric influence at very high flows (0–15 %), and at moderate- to low-flow magnitudes. The curves of InfilShapeFactor alternated with ExpslowInterflow along the FDCs (Fig. 6a–d), while the rest of the parameters did not show notable sensitivity values.

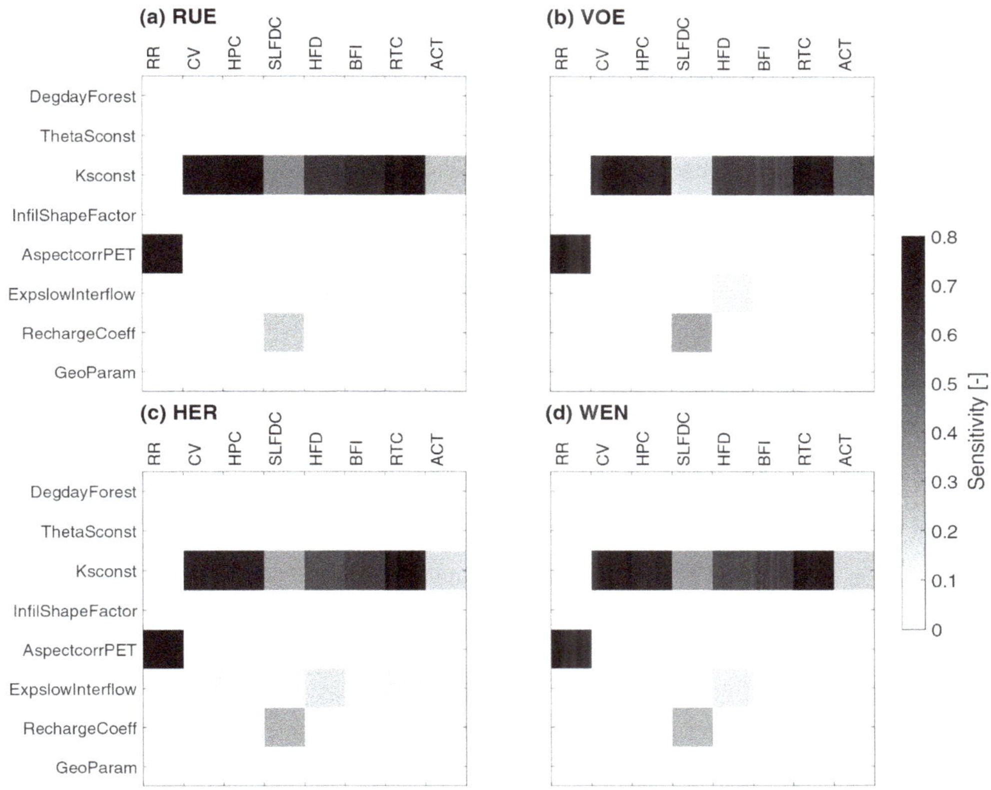

Figure 5. Sensitivity of eight global mHM parameters to eight single-value fingerprint metrics (RR, CV, HPC, SLFDC, HFD, BFI, RTC, ACT) in four Ruhr headwater catchments: RUE **(a)**, VOE **(b)**, HER **(c)**, WEN **(d)**. The simulation period was 1997 to 2006.

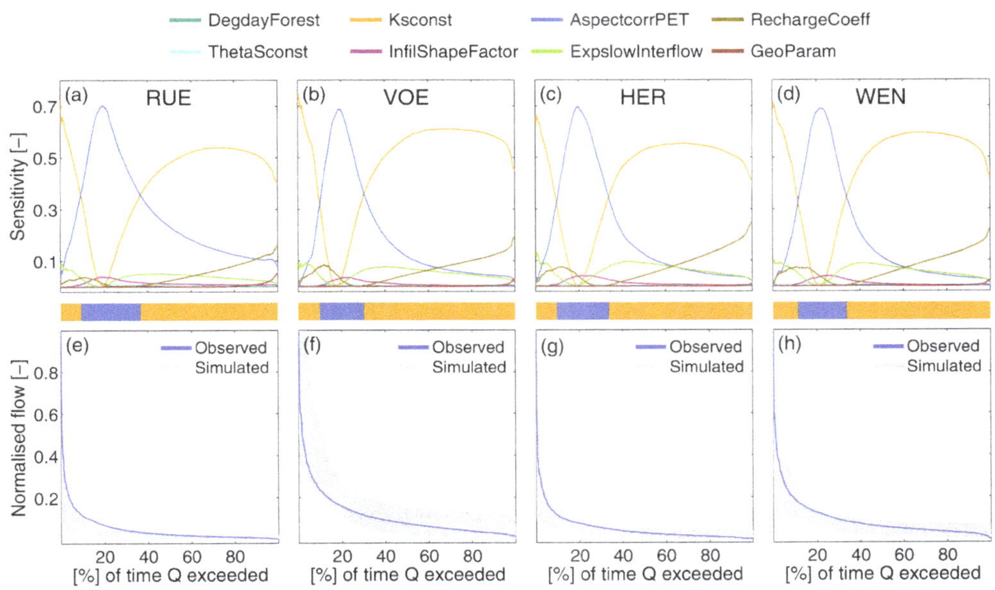

Figure 6. (a–d) Sensitivity of eight global mHM parameters along streamflow exceedance probability (flow duration curve, FDC) for four Ruhr headwater catchments RUE, VOE, HER, and WEN. Catchment-specific strips show the parameters with the highest sensitivity along the FDC. **(e–h)** Observed flow duration curves and the corresponding ensembles of each 243 simulated FDCs for the period 2002 to 2006, normalised by the highest streamflow value observed at each of the four headwater gauges.

Interestingly, the ensembles of normalised FDCs showed distinct differences between the catchments (Fig. 6e–h), although the sensitivity dynamics were similar, and the same 243 parameter variations from FAST were used for each headwater. The largest spread of FDCs was found for the northeastern headwater VOE (Fig. 6f) with the largest catchment size among the four shown headwaters. The smaller headwaters WEN, HER, and RUE (Fig. 1) showed a decreasing spread of the FDCs from northeast to southwest (Fig. 6h, g, e). Additionally, the widths of the simulation envelopes changed for different streamflow magnitudes. All ensembles of FDCs showed a constriction point located at about 20 % of streamflow exceedance (Fig. 6e–h), which is the same point where Ksconst and AspectcorrPET showed lowest and highest sensitivity values, respectively (Fig. 6a–d). The ensembles of simulated FDCs encompassed the observed FDC in most cases (e.g. VOE, HER, WEN; Fig. 6f–h). In some cases the observed FDC was outside the simulated range (e.g. RUE; Fig. 6e).

4 Discussion

4.1 Parameter sensitivities from TEDPAS and INDPAS

The combination of TEDPAS and INDPAS created a detailed sensitivity pattern for the response characteristics of the hydrological model mHM. Overall, the soil moisture dynamics parameter Ksconst and the evapotranspiration parameter AspectcorrPET were found most relevant for the simulation of the streamflow response of 14 Ruhr headwaters.

The TEDPAS analysis confirmed, as expected, a seasonality of sensitivity for parameters controlling snow (DegdayForest) or evapotranspiration (AspectcorrPET) processes. More interestingly, TEDPAS also showed an alternating dominance of Ksconst and AspectcorrPET during the temporal course of the simulation. Using flow duration curves as response targets (INDPAS) clarified that this was related to different streamflow magnitudes. The soil moisture dynamics (Ksconst) dominated at both high and low flows, which may be attributed to the dual role of Ksconst in parameterising both storage (field capacity) and transmission (conductivity) of soil water in mHM. During intermediate-flow conditions, evapotranspiration (AspectcorrPET) governed the streamflow simulations, probably because of the high influence of evapotranspiration on the shallow soil storage and thus the system state of the catchment. This high sensitivity of an evapotranspiration parameter during intermediate-flow conditions coincides with the findings of former studies (e.g. Guse et al., 2014, 2016b).

The specific influence of AspectcorrPET was mainly on the water balance (fingerprint RR) in all of the headwaters. Ksconst was the most sensitive parameter for most of the other single-value fingerprints, encompassing those quantifying low-flow (BFI) as well as high-flow conditions or the flashiness of hydrological response (CV, HPC, HFD, RTC). Only for one single-value fingerprint (SLFDC, related to the rate of change in streamflow) another parameter (RechargeCoeff) was found to be as sensitive as Ksconst. The moderate relevance of the groundwater-related RechargeCoeff increased during low-flow periods, as illustrated by INDPAS using FDCs and TEDPAS.

A temporally resolved sensitivity makes it difficult to reveal clear patterns of dominant parameters when dealing with long time periods. Guse et al. (2016b) also recognised that parameter sensitivity by TEDPAS based on the streamflow hydrograph should be analysed on different temporal aggregation levels and should be related to different streamflow magnitudes for a detailed assessment of dominant model parameters and temporal process dynamics. While their methodological approach was purely based on aggregation and reordering of TEDPAS sensitivity and streamflow time series, we added additional value with INDPAS aiming at multiple response targets including FDCs. The consideration of flow duration curves enabled analysing streamflow free of autocorrelation and time dependence. The FDC as a model response target for sensitivity analysis provided information on parameter sensitivity along the independent variable of streamflow exceedance probability. In contrast, for classical hydrograph inspection, which is the basis of TEDPAS, time is the independent variable. INDPAS along FDCs allowed us to draw conclusions about parametric influences at specific streamflow magnitudes.

Regardless of the chosen model response target, in the case of 14 Ruhr headwaters only one or a very small group of parameters were identified as relevant for streamflow response. In this context, Herman et al. (2013) showed that the long-term water balance is dominated by only very few parameters, irrespective of the hydrological conditions and of the model. Cuntz et al. (2015) performed a global Sobol sensitivity analysis on the hydrologic model mHM. For three distinct humid and arid European catchments this analysis always resulted in about 20 informative parameters, though the dominant parameter sets were composed very differently. Their criteria to select the sensitive parameters was substantially different from our approach, which renders a direct comparison between the studies difficult. The different number of dominant parameters might also be due to correlated mHM parameters which we sorted out before sensitivity analysis. In contrast, Cuntz et al. (2015) considered the degree of correlation between mHM parameters as rather minor to be interfering with parameter identification.

Especially the application of fingerprints as model response targets for sensitivity analysis revealed further details of parametric dominance that can help in more directed model applications. Dominant parameters can be additionally adjusted pinpointed to fingerprint metrics in a subsequent calibration step, while insensitive parameters can be disregarded for model calibration directed to the respective response target (van Werkhoven et al., 2009). This reduces the parameter space and need for constraining parameter values and thus facilitates model calibration. A missing sensitivity signal to a fingerprint (e.g. RechargeCoeff to RR) can reveal that the chosen response target might not be relevant to further constrain the parameter identification in a certain catchment.

The spread of the simulated response ensembles allows us to judge whether a fingerprint metric is a reliable response target for sensitivity analysis or whether different fingerprints, e.g. the master recession curve or the double mass curve, should be considered instead. For example, the hydrographs and FDCs showed significant spread that also differed between the catchments (Figs. 3 and 6). In contrast, the spread in simulated values for the Autocorrelation Time (ACT) was small for all catchments. Accordingly, INDPAS analysis directed to ACT also showed only moderate sensitivities for the set of eight parameters (Fig. 5). Our preselection of eight parameters can potentially lead to the elimination of other storage parameters that might be most sensitive to ACT. Thus, one might conclude that parameter selection based on INDPAS would result in a different choice in the set of the most sensitive parameters. In the case of ACT this is not very likely, since storage parameters (e.g. ExpslowInterflow and GeoParam) were still included. Instead, the precipitation time series has a large impact on the autocorrelation structure of streamflow; the ACT metric is thus less informative than others that depend less on the hydroclimatic boundary conditions.

The resulting partial variances for each fingerprint are comparable as they portray the relative influence of the parameters on the variation of the target, regardless of the specific values of the targets. In order to take into account the impact of the spread of the simulation results on the parameter sensitivities, a weighting factor for partial parameter sensitivities might be helpful. An impact-weighted INDPAS might then be used along with catchment class-specific response fingerprints to select relevant parameters for the specific hydrological conditions. In Fig. 6 we normalised the FDCs by the maximum value of each time series. For the visual comparison of sites this is a necessary step, but it might lead to a different form of appearance, including the spread of the simulation ensemble. If absolute fingerprint values are replaced by normalised quantities (Samaniego et al., 2010a; He et al., 2011), dimensions should be considered explicitly when determining sensitivity weighting factors.

Our findings using different model response targets confirm the necessity of a multivariate sensitivity analysis. This was similarly recognised by Wagener et al. (2009), who applied three standard error metrics as objective functions for sensitivity analysis. Their results for parameter sensitivity were found to change spatially when the objective function was replaced. Razavi and Gupta (2015) similarly pointed out that even conflicting conclusions could be drawn if different properties of the model response were applied in sensitivity analysis. To avoid misinterpretation of sensitivity results we propose that the selection of specific fingerprint metrics should be determined by the purpose of the modelling; for instance, sensitivity to fingerprint metrics for peak flow is suitable if flood prediction is the focus. Redundancy is not problematic if several similar metrics for a specific streamflow characteristic are selected, e.g. HPC, CV, and HFD for high flows. A multivariate analysis with metrics of several, even partly similar streamflow characteristics (frequency of high flow, magnitude of high flows etc.) is rather helpful to ensure complete parameter identification for different catchments. Aggregated and temporally independent fingerprints like the FDC proved to be especially applicable.

4.2 Regional differences in parameter sensitivity

Although the most sensitive parameters and the corresponding sensitivity patterns of streamflow response were found to be similar for the 14 investigated Ruhr headwater catchments, the analysis with TEDPAS and INDPAS revealed certain regional differences.

Especially the analysis of sensitivity duration curves derived from TEDPAS revealed regional differences of parameter sensitivity between headwaters. For half of the eight selected parameters we found regional differences in SDCs (Sect. 3.2 and Fig. 4). The most sensitive parameters exhibited the largest spread of SDCs (e.g. Ksconst and AspectCorrPET; Fig. 4c and e), and their SDCs were systematically ordered according to the geographical location (southwest–northeast) and the physiographic setting (ASP, URB; Table 3). Some catchments deviate from the general pattern in SDCs for evapotranspiration and interflow parameters (e.g. RUE; Fig. 4e and f). In these cases, the specific combination of catchment characteristics (degree of soil sealing, topographic gradients, land cover) might have led to different processes in streamflow simulations. For the catchment of RUE, the smallest slope value among the headwaters in conjunction with the highest percentage of urban area (SLOPE, URB; Table 3) can explain the deviation from the general pattern of SDCs for the two parameters. SDCs thus provided a convenient means to identify regionally different sensitivity characteristics for each of the analysed parameters.

The results from the INDPAS analysis directed to single-valued fingerprints also showed some differences between the headwaters. In particular, the sensitivity patterns of the second and third ranked parameters were found to differ, although not with the same systematic ordering as for the SDCs. The patterns of the most sensitive parameters along streamflow exceedance probability (catchment-specific strips in Fig. 6) provided visually condensed diagnostic information for different streamflow magnitudes, but showed only minor differences between the catchments.

Together this shows that even the small physiographic gradients in the Ruhr headwater catchments can cause differences in parameter sensitivity to streamflow response characteristics. This finding is partly contrary to those of Guse et al.

(2014), who reported almost no differences of parameter sensitivities among different subcatchments of the Treene in northern Germany in a similar analysis. This was explained by the absence of a pronounced heterogeneity in their study area.

Given that the same parameter sets are applied to all headwater catchments, any regional differences in parameter sensitivity originate from differences either in the hydroclimatic or in the physiographic setting. In the case of the Ruhr headwaters, the local hydroclimatic and physiographic differences (Table 3) seem to be sufficient to be discriminated by the hydrological model structure in the form of a different variation in streamflow response. Due to their geographical proximity, the 14 Ruhr headwaters are generally similar in terms of the long-term hydroclimate (Wetness Index) and the physiography, e.g. in their soil hydrological characteristics (BFI). They show, however, some differences in annual precipitation amount, topographic gradient, and land cover. The average annual precipitation in the simulation period between 1997 and 2006 reveals a hydroclimatic gradient with lower to higher precipitation rates from northeast to southwest (PMEAN), similar to the geographical ordering of the SDCs in Fig. 4. Song et al. (2013) also attributed local differences of parameter sensitivity to the spatial distribution of meteorologic forcing. Demaria et al. (2007) similarly concluded that parameter sensitivity was more strongly determined by climate gradients than by changes in soil properties in their Monte Carlo-based sensitivity study. Under different hydrological conditions, regional sensitivity patterns or the number of parameters which influence streamflow simulations might be different from the present example (Cuntz et al., 2015).

As parameter sensitivity is a prerequisite for parameter identifiability, even slight differences in sensitivity reveal information about how identifiability can change among different catchments. Scale-dependent limitations have to be kept in mind to avoid a levelling out of the explanatory value of a physiographic descriptor (Blöschl and Sivapalan, 1995), possibly resulting in an intermediate course of sensitivity duration as seen for the largest headwater of Bamenohl (BAM; Sect. 3.2 and Fig. 4). van Griensven et al. (2006) remarked that local differences indicate that results of global sensitivity analysis for one catchment cannot be directly applied to other, even nearby locations, but may be used as reasonable estimates within the same catchment category. Locations with intermediate-sensitivity characteristics (e.g. BAM) could at least serve as a starting point for parameter transfer to closely located ungauged sites. As the local differences between the Ruhr headwaters are not very large, the most sensitive parameters found for WEN in the first step of the analysis with all model parameters were also dominant in the other subcatchments, which was corroborated by the TEDPAS analysis with eight selected parameters on all subcatchments. Any local differences in parameter sensitivity revealed by the analysis of sensitivity duration or INDPAS could then be handled during individual model calibration for each catchment.

5 Conclusions

We used FAST for a global sensitivity analysis of the hydrological model mHM in 14 headwater catchments of the Ruhr River in Germany. Our multilevel approach not only reveals the dominating parameters for streamflow simulation but also pinpoints the influences of the analysed parameters to diverse aspects of hydrological response processes. Especially the application of several hydrological fingerprints as response targets allows for detailed model diagnostics. Comparison of streamflow response characteristics and analysing along the range of streamflow magnitudes shows how the parametric dominances and the most influential parameters can change with streamflow conditions, for example the complementary sensitivity of soil moisture dynamics and evapotranspiration in our case. The combination of TEDPAS and INDPAS also provides a means to unveil the slight differences in catchment-specific patterns between the closely located headwaters. The general similarity in the sensitivity patterns indicates, however, that a parameter transfer to other catchments might be possible, provided that the interplay of catchment structure and local hydroclimate has evolved in a similar way.

The results provide in-depth diagnostics on the model and its parameters, which can support future improvements of model structures, and facilitate case-specific model calibration in a reduced parameter space. The methodological approach of the multilevel sensitivity analysis with fingerprints as response targets may be generalised to any hydrological model or kind of catchment. The findings of this study motivate to include further catchments as regional end-members within different physiographic and climate settings to evaluate how parameter identifiability and the simulated hydrological functioning regionally changes among distant catchments. The consideration of multivariate response variables in diagnostics of hydrological models is beneficial for a complete identification, not only of parameter influences on simulated hydrological functioning but also of the regional relevance for model calibration.

Data availability. The streamflow data and the digital elevation model used in this study are not publicly accessible, and were provided by the Ruhrverband in Essen.

The further spatial physiographic data can be downloaded from the web: land cover data (https://land.copernicus.eu/pan-european/corine-land-cover/clc-2006?tab=download); soil data (https://produktcenter.bgr.de/terraCatalog/OpenSearch.do?search=3E80DA1A-A9A7-45A3-9CC7-79796FE9ABA4&type=/Query/OpenSearch.do); geological data (https://produktcenter.bgr.de/terraCatalog/DetailResult.do?fileIdentifier=1C60DDA9-EF73-47B9-9ED7-FCD22B3226C1).

Meteorological input data are available from Deutscher Wetterdienst. Data for precipitation (HYRAS; Rauthe et al., 2013) and temperature (HYRAS; Frick et al., 2014) can be obtained via https://www.dwd.de/DE/leistungen/hyras/hyras.html. The database for potential evapotranspiration (AMBAV; Löpmeier, 1994) can be accessed via ftp://ftp-cdc.dwd.de/pub/CDC/grids_germany/daily/evapo_p.

Competing interests. The authors declare that they have no conflict of interest.

Acknowledgements. The streamflow data and the digital elevation model used in this study were provided by the Ruhrverband in Essen. We thank Linda Bergmann, who supported the technical implementation of the FAST analysis. We are grateful to Fanny Sarrazin, Sibylle Haßler, Björn Guse, Shervan Gharari, and one anonymous reviewer for their critical comments on this and an earlier version of the manuscript, which greatly helped to improve the paper. We acknowledge support by Deutsche Forschungsgemeinschaft and Open Access Publishing Fund of Karlsruhe Institute of Technology.

The article processing charges for this open-access publication were covered by a Research Centre of the Helmholtz Association.

Edited by: Ross Woods

References

Atkinson, S. E., Sivapalan, M., Woods, R. A., and Viney, N. R.: Dominant physical controls on hourly flow predictions and the role of spatial variability: Mahurangi catchment, New Zealand, Adv. Water Res., 26, 219–235, https://doi.org/10.1016/S0309-1708(02)00183-5, 2003.

Beven, K.: Prophecy, reality and uncertainty in distributed hydrological modelling, Adv. Water Res., 16, 41–51, https://doi.org/10.1016/0309-1708(93)90028-E, 1993.

Beven, K.: How far can we go in distributed hydrological modelling?, Hydrol. Earth Syst. Sci., 5, 1–12, https://doi.org/10.5194/hess-5-1-2001, 2001.

Black, P.: Watershed functions, J. Am. Water Resour. Assoc., 33, 1–11, 1997.

Blöschl, G. and Sivapalan, M.: Scale issues in hydrological modelling: A review, Hydrol. Process., 9, 251–290, https://doi.org/10.1002/hyp.3360090305, 1995.

Bode, H., Evers, P., and Albrecht, D. R.: Integrated water resources management in the Ruhr River Basin, Germany, Water Sci. Technol., 47, 81–86, 2003.

Brudy-Zippelius, T.: Wassermengenbewirtschaftung im Einzugsgebiet der Ruhr : Simulation und Echtzeitbetrieb, PhD thesis, Universität Karlsruhe (TH), 159 pp., 2003.

Bundesanstalt für Geowissenschaften und Rohstoffe: Bodenübersichtskarte der Bundesrepublik Deutschland 1:200.000 (BÜK200), 2015a.

Bundesanstalt für Geowissenschaften und Rohstoffe: Geologische Karte der Bundesrepublik Deutschland 1:1000000 (GK1000), 2015b.

Blöschl, G. and Sivapalan, M.: Scale issues in hydrological modelling: A review, Hydrol. Process., 9, 251–290, https://doi.org/10.1002/hyp.3360090305, 1995.

Bode, H., Evers, P., and Albrecht, D. R.: Integrated water resources management in the Ruhr River Basin, Germany, Water Sci. Technol., 47, 81–86, 2003.

Brudy-Zippelius, T.: Wassermengenbewirtschaftung im Einzugsgebiet der Ruhr : Simulation und Echtzeitbetrieb, PhD thesis, Universität Karlsruhe (TH), 159 pp., 2003.

Bundesanstalt für Geowissenschaften und Rohstoffe: Bodenübersichtskarte der Bundesrepublik Deutschland 1:200.000 (BÜK200), 2015a.

Bundesanstalt für Geowissenschaften und Rohstoffe: Geologische Karte der Bundesrepublik Deutschland 1:1000000 (GK1000), 2015b.

Castiglioni, S., Lombardi, L., Toth, E., Castellarin, A., and Montanari, A.: Calibration of rainfall-runoff models in ungauged basins: A regional maximum likelihood approach, Adv. Water Res., 33, 1235–1242, https://doi.org/10.1016/j.advwatres.2010.04.009, 2010.

Cibin, R., Sudheer, K. P., and Chaubey, I.: Sensitivity and identifiability of stream flow generation parameters of the SWAT model, Hydrol. Process., 24, 1133–1148, https://doi.org/10.1002/hyp.7568, 2010.

Clark, M. P., McMillan, H. K., Collins, D. B. G., Kavetski, D., and Woods, R. A.: Hydrological field data from a modeller's perspective: Part 2: Process-based evaluation of model hypotheses, Hydrol. Process., 25, 523–543, https://doi.org/10.1002/hyp.7902, 2011.

Cloke, H. L., Pappenberger, F., and Renaud, J.-P.: Multi-Method Global Sensitivity Analysis (MMGSA) for modelling floodplain hydrological processes, Hydrol. Process., 22, 1660–1674, https://doi.org/10.1002/hyp.6734, 2008.

Cukier, R. I., Fortuin, C. M., Shuler, K. E., Petschek, A. G., and Schaibly, J. H.: Study of the sensitivity of coupled reaction systems to uncertainties in rate coefficients. I Theory, J. Chem. Phys., 59, 3873, https://doi.org/10.1063/1.1680571, 1973.

Cukier, R. I., Schaibly, J. H., and Shuler, K. E.: Study of the sensitivity of coupled reaction systems to uncertainties in rate coefficients. III. Analysis of the approximations Theory, J. Chem. Phys., 63, 1140, https://doi.org/10.1063/1.431440, 1975.

Cuntz, M., Mai, J., Zink, M., Thober, S., Kumar, R., Schäfer, D., Schrön, M., Craven, J., Rakovec, O., Spieler, D., Prykhodko, V., Dalmasso, G., Musuuza, J., Langenberg, B., Attinger, S., and Samaniego, L.: Computationally inexpensive identification of noninformative model parameters by sequential screening, Water Resour. Res., 51, 6417–6441, https://doi.org/10.1002/2015WR016907, 2015.

Demaria, E. M., Nijssen, B., and Wagener, T.: Monte Carlo sensitivity analysis of land surface parameters using the Variable Infiltration Capacity model, J. Geophys. Res., 112, 1–15, https://doi.org/10.1029/2006JD007534, 2007.

Di Prinzio, M., Castellarin, A., and Toth, E.: Data-driven catchment classification: application to the pub problem, Hydrol. Earth Syst. Sci., 15, 1921–1935, https://doi.org/10.5194/hess-15-1921-2011, 2011.

Duband, D., Obled, C., and Rodriguez, J.: Unit hydrograph revisited: an alternate iterative approach to UH and effective precipitation identification, J. Hydrol., 150, 115–149, https://doi.org/10.1016/0022-1694(93)90158-6, 1993.

European Environment Agency: CORINE Land Cover (CLC2006), 2009.

Euser, T., Winsemius, H. C., Hrachowitz, M., Fenicia, F., Uhlenbrook, S., and Savenije, H. H. G.: A framework to assess the realism of model structures using hydrological signatures, Hydrol. Earth Syst. Sci., 17, 1893–1912, https://doi.org/10.5194/hess-17-1893-2013, 2013.

Farmer, D., Sivapalan, M., and Jothityangkoon, C.: Climate, soil, and vegetation controls upon the variability of water balance in temperate and semiarid landscapes: Downward approach to water balance analysis, Water Resour. Res., 39, 1035, https://doi.org/10.1029/2001wr000328, 2003.

Fontaine, D. D., Havens, P. L., Blau, G. E., and Tillotson, M.: Groundwater Risk Modeling for Pesticides, Weed Technol., 6, 716–724, 1992.

Frey, H. C. and Patil, S. R.: Identification and review of sensitivity analysis methods, Computat. Studies, 22, 553–578, https://doi.org/10.1111/0272-4332.00039, 2002.

Frick, C., Steiner, H., Mazurkiewicz, A., Riediger, U., Rauthe, M., Reich, T., and Gratzki, A.: Central European high-resolution gridded daily data sets (HYRAS): Mean temperature and relative humidity, Meteorol. Z., 23, 15–32, https://doi.org/10.1127/0941-2948/2014/0560, 2014.

Göppert, H., Ihringer, J., Plate, E. J., and Morgenschweis, G.: Flood forecast model for improved reservoir management in the Lenne River catchment, Germany, Hydrolog. Sci. J., 43, 215–242, https://doi.org/10.1080/02626669809492119, 1998.

Gupta, H. V., Sorooshian, S., and Yapo, P. O.: Toward improved calibration of hydrologic models: Multiple and noncommensurable measures of information, Water Resour. Res., 34, 751–763, https://doi.org/10.1029/97WR03495, 1998.

Gupta, H. V., Wagener, T., and Liu, Y.: Reconciling theory with observations: elements of a diagnostic approach to model evaluation, Hydrol. Process., 22, 3802–3813, https://doi.org/10.1002/hyp.6989, 2008.

Gupta, H. V., Clark, M. P., Vrugt, J. A., Abramowitz, G., and Ye, M.: Towards a comprehensive assessment of model structural adequacy, Water Resour. Res., 48, 1–16, https://doi.org/10.1029/2011WR011044, 2012.

Guse, B., Reusser, D. E., and Fohrer, N.: How to improve the representation of hydrological processes in SWAT for a lowland catchment – temporal analysis of parameter sensitivity and model performance, Hydrol. Process., 28, 2651–2670, https://doi.org/10.1002/hyp.9777, 2014.

Guse, B., Pfannerstill, M., Gafurov, A., Fohrer, N., and Gupta, H.: Demasking the integrated information of discharge: Advancing sensitivity analysis to consider different hydrological components and their rates of change, Water Resour. Res., 52, 8724–8743, https://doi.org/10.1002/2016WR018894, 2016a.

Guse, B., Pfannerstill, M., Strauch, M., Reusser, D. E., Lüdtke, S., Volk, M., Gupta, H., and Fohrer, N.: On characterizing the temporal dominance patterns of model parameters and processes, Hydrol. Process., 30, 2255–2270, https://doi.org/10.1002/hyp.10764, 2016b.

Hamby, D. M.: A review of techniques for parameter sensitivity analysis of environmental models, Environ. Monit. Assess., 32, 135–154, https://doi.org/10.1007/BF00547132, 1994.

He, Y., Bárdossy, A., and Zehe, E.: A catchment classification scheme using local variance reduction method, J. Hydrol., 411, 140–154, https://doi.org/10.1016/j.jhydrol.2011.09.042, 2011.

Herman, J. D., Reed, P. M., and Wagener, T.: Time-varying sensitivity analysis clarifies the effects of watershed model formulation on model behavior, Water Resour. Res., 49, 1400–1414, https://doi.org/10.1002/wrcr.20124, 2013.

Hogue, T. S., Sorooshian, S., Gupta, H., Holz, A., and Braatz, D.: A Multistep Automatic Calibration Scheme for River Forecasting Models, J. Hydrometeorol., 1, 524–542, https://doi.org/10.1175/1525-7541(2000)001<0524:AMACSF>2.0.CO;2, 2000.

Kavetski, D. and Clark, M. P.: Ancient numerical daemons of conceptual hydrological modeling: 2. Impact of time stepping schemes on model analysis and prediction, Water Resour. Res., 46, 1–27, https://doi.org/10.1029/2009WR008896, 2010.

Kirchner, J. W.: Getting the right answers for the right reasons: Linking measurements, analyses, and models to advance the science of hydrology, Water Resour. Res., 42, 1–5, https://doi.org/10.1029/2005WR004362, 2006.

Kumar, R., Samaniego, L., and Attinger, S.: The effects of spatial discretization and model parameterization on the prediction of extreme runoff characteristics, J. Hydrol., 392, 54–69, https://doi.org/10.1016/j.jhydrol.2010.07.047, 2010.

Kumar, R., Samaniego, L., and Attinger, S.: Implications of distributed hydrologic model parameterization on water fluxes at multiple scales and locations, Water Resour. Res., 49, 360–379, https://doi.org/10.1029/2012WR012195, 2013.

Löpmeier, F.-J.: Berechnung der Bodenfeuchte und Verdunstung mittels agrarmeteorologischer Modelle, Zeitschrift f. Bewässerungswirtschaft, 29, 157–167, 1994.

Lu, Y. C. and Mohanty, S.: Sensitivity Analysis of a complex, proposed geological waste disposal system using the Fourier Amplitude Sensitivity Test Method, Reliab. Eng. Syst. Safe., 72, 275–291, https://doi.org/10.1016/s0951-8320(01)00020-5, 2001.

Massmann, C. and Holzmann, H.: Analysis of the behavior of a rainfall-runoff model using three global sensitivity analysis methods evaluated at different temporal scales, J. Hydrol., 475, 97–110, https://doi.org/10.1016/j.jhydrol.2012.09.026, 2012.

Massmann, C. and Holzmann, H.: Analysing the Sub-processes of a Conceptual Rainfall-Runoff Model Using Information About the Parameter Sensitivity and Variance, Environ. Monit. Assess., 20, 41–53, https://doi.org/10.1007/s10666-014-9414-6, 2015.

Massmann, C., Wagener, T., and Holzmann, H.: A new approach to visualizing time-varying sensitivity indices for environmental model diagnostics across evaluation time-scales, Environ. Modell. Softw., 51, 190–194, https://doi.org/10.1016/j.envsoft.2013.09.033, 2014.

McCuen, R. H.: The role of sensitivity analysis in hydrologic modeling, J. Hydrol., 18, 37–53, https://doi.org/10.1016/0022-1694(73)90024-3, 1973.

Nossent, J. and Bauwens, W.: Multi-variable sensitivity and identifiability analysis for a complex environmental model in view of integrated water quantity and water quality modeling, Water Sci. Technol., 65, 539–549, https://doi.org/10.2166/wst.2012.884, 2012.

Olden, J. D. and Poff, N. L.: Redundancy and the choice of hydrologic indices for characterizing streamflow regimes, River Res. Appl., 19, 101–121, https://doi.org/10.1002/rra.700, 2003.

O'Loughlin, F., Bruen, M., and Wagener, T.: Parameter sensitivity of a watershed-scale flood forecasting model as a function of modelling time-step, Hydrol. Res., 44, 334–350, https://doi.org/10.2166/nh.2012.157, 2013.

Pfannerstill, M., Guse, B., Reusser, D., and Fohrer, N.: Process verification of a hydrological model using a temporal parameter sensitivity analysis, Hydrol. Earth Syst. Sci., 19, 4365–4376, https://doi.org/10.5194/hess-19-4365-2015, 2015.

Pianosi, F. and Wagener, T.: Understanding the time-varying importance of different uncertainty sources in hydrological modelling using global sensitivity analysis, Hydrol. Process., 30, 3991–4003, https://doi.org/10.1002/hyp.10968, 2016.

Plate, E. J., Ihringer, J., and Lutz, W.: Operational models for flood calculations, J. Hydrol., 100, 489–506, 1988.

Poff, N. L., Allan, J. D., Bain, M. B., Karr, J. R., Prestegaard, K. L., Richter, B. D., Sparks, R. E., and Stromberg, J. C.: The Natural Flow Regime: A paradigm for river conservation and restoration N, BioScience, 47, 769–784, https://doi.org/10.2307/1313099, 1997.

Pokhrel, P., Yilmaz, K. K., and Gupta, H. V.: Multiple-criteria calibration of a distributed watershed model using spatial regularization and response signatures, J. Hydrol., 418–419, 49–60, https://doi.org/10.1016/j.jhydrol.2008.12.004, 2008.

Rakovec, O., Kumar, R., Attinger, S., and Samaniego, L.: Improving the realism of hydrologic model functioning through multivariate parameter estimation, Water Resour. Res., 52, 613–615, https://doi.org/10.1002/2016WR019430, 2016.

Rauthe, M., Steiner, H., Riediger, U., Mazurkiewicz, A., and Gratzki, A.: A Central European precipitation climatology – Part I: Generation and validation of a high-resolution gridded daily data set (HYRAS), Meteorol. Z., 22, 235–256, https://doi.org/10.1127/0941-2948/2013/0436, 2013.

Razavi, S. and Gupta, H. V.: What do we mean by sensitivity analysis? The need for comprehensive characterization of "global" sensitivity in Earth and Environmental systems models, Water Resour. Res., 51, 3070–3092, https://doi.org/10.1002/2014WR016527, 2015.

Reusser, D. E. and Zehe, E.: Inferring model structural deficits by analyzing temporal dynamics of model performance and parameter sensitivity, Water Resour. Res., 47, W07550, https://doi.org/10.1029/2010WR009946, 2011.

Reusser, D. E., Buytaert, W., and Zehe, E.: Temporal dynamics of model parameter sensitivity for computationally expensive models with the Fourier amplitude sensitivity test, Water Resour. Res., 47, W07551, https://doi.org/10.1029/2010WR009947, 2011.

Rodríguez-Camino, E. and Avissar, R.: Comparison of three land-surface schemes with the Fourier amplitude sensitivity test (FAST), Tellus A, 50, 313–332, https://doi.org/10.3402/tellusa.v50i3.14529, 1998.

Ruhrverband: Ruhrwassermenge 2011, Tech. rep., Ruhrverband, Essen, 2011.

Saltelli, A. and Bolado, R.: An alternative way to compute Fourier amplitude sensitivity test (FAST), Comput. Stat. Data An., 26, 445–460, https://doi.org/10.1016/S0167-9473(97)00043-1, 1998.

Saltelli, A., Ratto, M., Andres, T., Campolongo, F., Cariboni, J., and Saisana, M.: Global Sensitivity Analysis. The Primer, John Wiley & Sons, Ltd., 2008.

Samaniego, L., Bárdossy, A., and Kumar, R.: Streamflow prediction in ungauged catchments using copula-based dissimilarity measures, Water Resour. Res., 46, W02506, https://doi.org/10.1029/2008WR007695, 2010a.

Samaniego, L., Kumar, R., and Attinger, S.: Multiscale parameter regionalization of a grid-based hydrologic model at the mesoscale, Water Resour. Res., 46, 1–25, https://doi.org/10.1029/2008WR007327, 2010b.

Samaniego, L., Kumar, R., and Jackisch, C.: Predictions in a data-sparse region using a regionalized grid-based hydrologic model driven by remotely sensed data, Hydrol. Res., 42, 338–355, https://doi.org/10.2166/nh.2011.156, 2011.

Samaniego, L., Cuntz, M., Craven, J., Dalmasso, G., Kumar, R., Mai, J., Musuuza, J., Prykhodko, V., Schneider, C., Spieler, D., Thober, S., and Zink, M.: multiscale Hydrologic Model mHM-Documentation for version 5.1., Tech. rep., Helmholtz Centre for Environmental Research – UFZ, 2014.

Sanadhya, P., Gironás, J., and Arabi, M.: Global sensitivity analysis of hydrologic processes in major snow-dominated mountainous river basins in Colorado, Hydrol. Process., 28, 3404–3418, https://doi.org/10.1002/hyp.9896, 2013.

Schaibly, J. H. and Shuler, K. E.: Study of the sensitivity of coupled reaction systems to uncertainties in rate coefficients. II Applications, J. Chem. Phys., 59, W07551, https://doi.org/10.1063/1.1680572, 1973.

Shamir, E., Imam, B., Gupta, H. V., and Sorooshian, S.: Application of temporal streamflow descriptors in hydrologic model parameter estimation, Water Resour. Res., 41, 1–16, https://doi.org/10.1029/2004WR003409, 2005a.

Shamir, E., Imam, B., Morin, E., Gupta, H. V., and Sorooshian, S.: The role of hydrograph indices in parameter estimation of rainfall-runoff models, Hydrol. Process., 19, 2187–2207, https://doi.org/10.1002/hyp.5676, 2005b.

Sieber, A. and Uhlenbrook, S.: Sensitivity analyses of a distributed catchment model to verify the model structure, J. Hydrol., 310, 216–235, https://doi.org/10.1016/j.jhydrol.2005.01.004, 2005.

Sivapalan, M.: Pattern, Process and Function: Elements of a Unified Theory of Hydrology at the Catchment Scale, in: Encyclopedia of Hydrological Sciences, APRIL 2006, John Wiley & Sons, Ltd., 193–219, https://doi.org/10.1002/0470848944, 2005.

Song, X., Bryan, B. A., Almeida, A. C., Paul, K. I., Zhao, G., and Ren, Y.: Time-dependent sensitivity of a process-based ecological model, Ecol. Modell., 265, 114–123, https://doi.org/10.1016/j.ecolmodel.2013.06.013, 2013.

Sudheer, K. P., Lakshmi, G., and Chaubey, I.: Application of a pseudo simulator to evaluate the sensitivity of parameters in complex watershed models, Environ. Modell. Softw., 26, 135–143, https://doi.org/10.1016/j.envsoft.2010.07.007, 2011.

Tang, Y., Reed, P., Wagener, T., and van Werkhoven, K.: Comparing sensitivity analysis methods to advance lumped watershed model identification and evaluation, Hydrol. Earth Syst. Sci., 11, 793–817, https://doi.org/10.5194/hess-11-793-2007, 2007.

Tolson, B. A. and Shoemaker, C. A.: Dynamically dimensioned search algorithm for computationally efficient watershed model calibration, Water Resour. Res., 43, 1–16, https://doi.org/10.1029/2005WR004723, 2007.

van Griensven, A., Meixner, T., Grunwald, S., Bishop, T., Diluzio, M., and Srinivasan, R.: A global sensitivity analysis tool for the parameters of multi-variable catchment models, J. Hydrol., 324, 10–23, https://doi.org/10.1016/j.jhydrol.2005.09.008, 2006.

van Werkhoven, K., Wagener, T., Reed, P., and Tang, Y.: Sensitivity-guided reduction of parametric dimensionality for multi-objective calibration of watershed models, Adv. Water Res., 32, 1154–1169, https://doi.org/10.1016/j.advwatres.2009.03.002, 2009.

Vrugt, J. A. and Sadegh, M.: Toward diagnostic model calibration and evaluation: Approximate Bayesian computation, Water Resour. Res., 49, 4335–4345, https://doi.org/10.1002/wrcr.20354, 2013.

Wagener, T., Sivapalan, M., Troch, P., and Woods, R.: Catchment Classification and Hydrologic Similarity, Geography Compass, 1, 1–31, https://doi.org/10.1111/j.1749-8198.2007.00039.x, 2007.

Wagener, T., van Werkhoven, K., Reed, P., and Tang, Y.: Multiobjective sensitivity analysis to understand the information content in streamflow observations for distributed watershed modeling, Water Resour. Res., 45, 1–5, https://doi.org/10.1029/2008WR007347, 2009.

Winsemius, H. C., Schaefli, B., Montanari, A., and Savenije, H. H. G.: On the calibration of hydrological models in ungauged basins: A framework for integrating hard and soft hydrological information, Water Resour. Res., 45, W12422, https://doi.org/10.1029/2009WR007706, 2009.

Yadav, M., Wagener, T., and Gupta, H.: Regionalization of constraints on expected watershed response behavior for improved predictions in ungauged basins, Adv. Water Res., 30, 1756–1774, https://doi.org/10.1016/j.advwatres.2007.01.005, 2007.

Yilmaz, K. K., Gupta, H. V., and Wagener, T.: A process-based diagnostic approach to model evaluation: Application to the NWS distributed hydrologic model, Water Resour. Res., 44, 1–18, https://doi.org/10.1029/2007WR006716, 2008.

Zajac, Z. B.: Global sensitivity and uncertainty analysis of spatially distributed watershed models, PhD thesis, University of Florida, 195 pp., 2010.

The river absorption capacity determination as a tool to evaluate state of surface water

Paweł Wilk, Paulina Orlińska-Woźniak, and Joanna Gębala

Section of Modelling Water Quality Surface – Institute of Meteorology and Water Management, National Research Institute, 01-673 Warsaw, 61 Podleśna St., Poland

Correspondence: Paweł Wilk (pawel.wilk@imgw.pl)

Abstract. In order to complete a thorough and systematic assessment of water quality, it is useful to measure the absorption capacity of a river. Absorption capacity is understood as a pollution load introduced into river water that will not cause permanent and irreversible changes in the aquatic ecosystem and will not cause a change in the classification of water quality in the river profile. In order to implement the method, the Macromodel DNS/SWAT basin for the Middle Warta pilot (central Poland) was used to simulate nutrient loads. This enabled detailed analysis of water quality in each water body and the assessment of the size of the absorption capacity parameter, which allows the determination of how much pollution can be added to the river without compromising its quality class. Positive values of the calculated absorption capacity parameter mean that it is assumed that the ecosystem is adjusted in such a way that it can eliminate pollution loads through a number of self-purification processes. Negative values indicate that the load limit has been exceeded, and too much pollution has been introduced into the ecosystem for it to be able to deal with through the processes of self-purification. Absorption capacity thus enables the connection of environmental standards of water quality and water quality management plans in order to meet these standards.

1 Introduction

The Water Framework Directive (WFD) (Directive 2000/60/EC) implemented in the European Union treats water as a common good and obligates the sustainable management of its resources, among other things, by preventing degradation, improving the state of water resources, and protecting aquatic and water-dependent ecosystems (Orlińska-Wozniak et al., 2013; Boeuf and Fritsh, 2016). The WFD determines the search for new and more effective solutions for the state of the aquatic environment and, consequently, the assessment of the effectiveness of action plans in areas deemed polluted. What is important here is that surface water pollution, in most cases, is not an irreversible state, so pollutant input to surface water will not always lead to irreversible contamination. Water has the ability to self-purify (Dubnyak and Timchenko, 2000; Gorecki, 2007; Jancarkova et al., 1997; Karrasch et al., 2006; Jarosiewicz and Dalszewska, 2008; McColl, 1974; Vagnetti, 2003; Zagorc-Končan and Somen, 1999; Zalewski, 2003).

Generally, self-purification is a biochemical transformation of pollutants (containing mainly organic compounds) into simpler forms, often inorganic, that have microorganism activity; this is done at the expense of the consumption of oxygen from the air and water and through sedimentation processes that cause the physical elimination of water pollution (Jarosiewicz, 2007; Kowalewski, 2009; Elósegui et al., 1995; Vagnetti, 2003). The most important factors influencing a river's self-purification ability include, among others, topography, soil type, vegetation, the hydraulic characteristics of the river, the retention time of water in the catchment, biodiversity and temperature (Spellman and Drinan, 2001; Schulz et al., 2003; Vaikasas and Dumbrauskas, 2010; Marsili-Libelli and Giusti, 2008; Popek, 2011; Van der Lee et al., 2004). To evaluate indirectly the possibility of a river performing self-purification – through the comparison of actual loads in the river to limit loads – the river absorption capacity (RAC) parameter is useful. The establishment of limits aims to determine a threshold below which there is a possibility of a river's self-purification.

It should be noted that beyond a certain critical level of pollutants it may be impossible for an aquatic ecosystem to return to its original condition (Kowalkowski, 2009; Nixon, 2009). This was confirmed, inter alia, by Duarte et al. (2009) in the article "Return to Neverland...", which referred to the process of eutrophication. This work was motivated by attempts to improve water quality aimed at restoring ecosystems to their original conditions. Observations of many ecosystems were made and, despite the elimination of sources of pollution, they did not return to their original state, even after more than 30 years. Owing to this, the authors concluded that above a certain critical value of pollution, an ecosystem is not able to return to its original state. Understanding an ecosystem's response to multiple shifting baselines is essential in setting reliable targets for restoration efforts.

The most advanced work related to the determination of the so-called assimilative capacity of the river, which has similar assumptions to the river absorption capacity, is led by the US Environmental Protection Agency (EPA) through the implementation of the Clean Water Act (Federal Water Pollution Control Act, 2002). Institute of Meteorology and Water Management (IMGW) National Research Institute (PIB) started studies on river the utilization of the absorption capacity parameter in water management in 2011 by proposing the calculation of absorption capacity on the basis of the advanced Macromodel Discharge–Nutrient–Sea (DNS), which has the ability to determine of the amount of point and nonpoint sources of pollution in a given river section.

This paper presents a method of assessing the surface water state by determining the river absorption capacity parameter. Absorption capacity is defined as the maximum load that can be input into a river without exceeding limit load and changing the water quality state class or, when the absorption capacity is negative, the load that should be removed to accomplish limit loads. The method assumes the use of mathematical modelling. The Macromodel DNS, developed at IMGW-PIB, was used (Ostojski, 2012). It allows, inter alia, for daily flow simulations, as well as average daily loads of selected pollutants, which is essential for the determination of river retention. The simulation of these loads allows for the calculation of river absorption capacity.

The proposed method of evaluating the status of surface water by using the river absorption capacity parameter is an alternative to the ratio commonly used by the EPA, defined as the assimilative capacity of the river, and which is called the total maximum daily load (TMDL) (Bulsathaporn et al., 2013; Magley and Joyner, 2008; Mohlar, 2011; Radcliffe et al., 2009; Reckhow and Wostl, 2001).

This is defined as the maximum amount of a pollutant that can occur in a waterbody, designates the necessary reductions to one or more pollutant sources, allocates a planning tool and points out a potential starting point for restoration or protection activities with the ultimate goal of attaining or maintaining water quality standards (EPA, 2016). Procedures for determining the TMDL were first defined in 1992 and have been repeatedly revised and updated since then. This method is based on the identification of point and nonpoint pollution and then the calculation of their sum for the river sections previously recognized as endangered by excessive pollution.

$$\mathrm{TMDL} = \Sigma \mathrm{WLAs} + \Sigma \mathrm{LAs} + \mathrm{MOS}$$

WLAs are the point source loads, LAs are the nonpoint source loads, and MOS is the margin of safety.

Knowledge of a river's absorption capacity enables the implementation of action plans aimed at preventing the degradation of water quality, and consequently that of aquatic and water-dependent ecosystems, which occurs as a result of human activity, that is, "anthropopressure". Currently there is no universal methodology to determine the absorption capacity of a river. In the few publications on the subject, only general equations are available (Chmielowski and Jarząbek, 2008; Monka, 2005; Tyszewski et al., 2008). The general RAC equations in this publication for calculating this parameter are sufficient for their purpose. However, when a more in-depth analysis of both the calculations and the results is required, knowledge of the entire RAC methodology is required, taking into account the pollutant load used for the calculation and a selected flow characteristic. There is also no information on the natural background of pollution identified as actual natural load (ANL), which should be taken into account when analysing the RAC parameter.

The aim of the study was to develop a method for the evaluation of the physico-chemical surface water state through the definition of the absorption capacity of a river. For this purpose, the Macromodel DNS/SWAT was used, which was calibrated, verified and validated for the Middle Warta pilot catchment. The model was prepared to take into account the individual processes of pollutant transformation and transport, from the moment pollutants enter the environment until their discharge into the sea, and to allow the determination of nutrient loads at estuary profiles of water bodies. Using data from the simulation, the method of calculating the absorption capacity of the river was implemented. The absorption capacity is understood as pollution load entered into a section of river that will not cause permanent and irreversible changes in the aquatic ecosystem and will not change the classification of water quality in the profile (Chmielowski and Jarząbek, 2008). Afterwards the results of absorption capacity at river profiles were evaluated and the consequences of positive and negative values of absorption capacity for the analysed area were analysed, linking the reference of water quality environmental standards with water quality management plans in order to meet these standards.

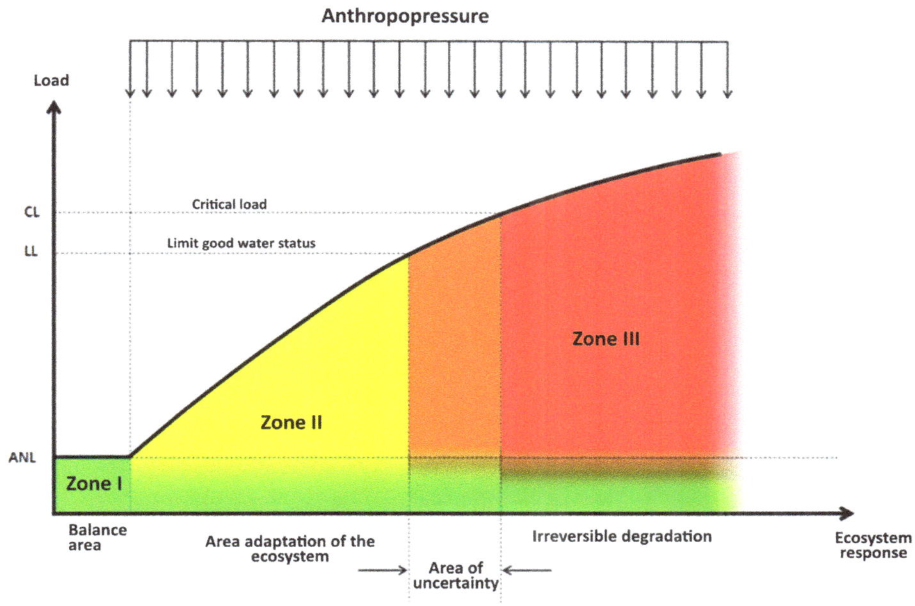

Figure 1. The impact of anthropopressure on the ecosystem surface water at the profile.

2 Materials and methods

2.1 Absorption capacity

The phenomenon of absorption capacity is described by the schematic impact of pollutant discharges originating from human activity on the ecosystem of the basin presented in Fig. 1. When a natural river devoid of anthropopressure has only natural background pollution (Brodie et al., 2009; Helsinki Commission, 2004; Henriksson and Milijokonsulter, 2007), then we talk about the state of balance in the river (zone I), while the maximum pollution load in the river can be called ANL. This is a very rare situation and in Europe there are virtually no basins without human pressure. Some such basins are in the northern parts of Scandinavia and Russia (Helsinki Commission, 2004). Therefore, the only way to determine the natural background is to use a mathematical model which allows the creation of a scenario in which the entire human pressure on a catchment is removed. At the moment that anthropogenic discharge appears in a catchment, an ecosystem changes and enters a state of adaptation (zone II). This means that the pollution introduced to the basin will interfere with the balance of the ecosystem, but not cause permanent changes in it. An ecosystem adapts to pollution's "elimination" through a series of processes collectively called the self-purification of the river. Generally, it is assumed that after the removal of a pollution source, an ecosystem will return to a state of equilibrium (or close to it). In river ecosystems in zone II, retention can be up to 30 % of the total amount of nitrogen (Dziopak, 2007; Neverova-Dziopak, 2009). Therefore, the pollution discharged into a stream at any point will not be equal to the load of these pollutants registered at the control profile located below this discharge. Moreover, if there is too much pollution introduced into the ecosystem, the critical load (CL) can be exceeded, which will cause irreversible changes in the ecosystem. This causes an "overload" of the ecosystem (zone III). CL cannot be defined as the critical size of load in practice, as it it is very difficult to determine. Of course, it is possible to try to estimate the critical load value for selected catchment fragments, but this will always be approximate. The critical load can also be defined as the limit beyond which a return to good status is no longer possible. The amount of pollutants in a river is then so great that the previous self-cleaning processes are permanently impaired and no longer function. To prevent irreversible consequences of exceeding the CL, the concept of limit load (LL) is introduced, and its value is defined as the limit of good water status set by administrative decisions in accordance with the WFD and the regulation of the Minister of the Environment (Regulation of the Minister of the Environment, 2011). Limit load is the maximum load of a selected pollutant, which may be in a selected river section of an analysed river, which has been classified in class II of clean water (good water status). This load was calculated on the basis of the limit concentration (LC), which is determined in Poland by the ordinance of the Minister of the Environment, and also the selected characteristic flow (CF). A good water status is defined as waters that meet, for most water quality indicators, the requirements for surface waters used to supply the population with water for consumption. As well as the values of biological water quality, indicators show little indication of anthropogenic impact.

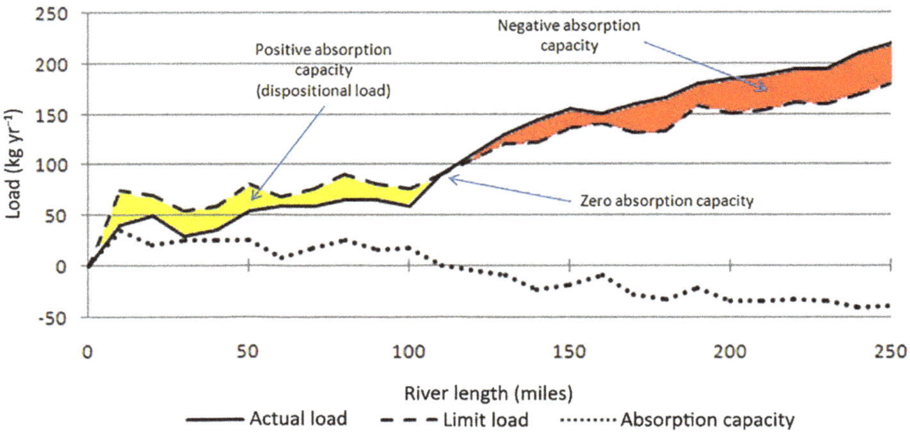

Figure 2. Possible cases of absorption capacity values for selected section of the river.

Actual natural load (background pollution) is determined based on the concentration of pollutants in the river without the anthropopressure impact (ANC) and characteristic flow (CF), which is the extreme value of water levels observed in the analysed period (Eq. 1).

$$\mathrm{ANL} = \mathrm{ANC} \cdot \mathrm{CF} \tag{1}$$

CL is difficult to determine in practice, mainly because there is an area of uncertainty between the limit load value and the critical load value (Fig. 1). It is impossible to determine the boundaries of the area. Incorrect determination of CL can lead to erroneous conclusions, and consequently actions taken on this basis could result in irreversible changes in an ecosystem. Therefore, it is preferred that the LL is determined on the basis of LC of good water status (Regulation of the Minister of the Environment, 2011) and the CF. LL should be lower than the CL. It is the difference between LL and CL that defines whether we are dealing with a positive or negative RAC parameter (Fig. 2). In addition, the LL along with the CL define the so-called area of uncertainty (Fig. 1). The size of this area will be different for each water body, and its precise definition is very difficult because it is practically impossible to determine the critical load value beyond which a catchment will be permanently and irreversibly polluted.

Absorption capacity of the river is the difference between two loads: the first of these is the limit load calculated on the basis of a limit concentration determined in Poland for different types of water by the Regulation of the Minister of the Environment (Regulation of the Minister of the Environment, 2011); the second is the actual load calculated based on the actual concentration at a selected river profile. When calculating both mentioned loads, the selected characteristic flow is used. Absorption capacity of the river is calculated for each pollutant separately and should consider all potential sources of pollution (both point and nonpoint sources). The results of absorption capacity are obtained for selected river profiles.

River absorption capacity for a selected control profile is described by the following equation:

$$\mathrm{RAC} = \mathrm{LL} - \mathrm{AL}, \tag{2}$$

where LL is the limit load for selected pollutant ($10^3\,\mathrm{kg\,yr^{-1}}$), and AL is the actual load for selected pollutant ($10^3\,\mathrm{kg\,yr^{-1}}$).

The actual load at a control profile is described by the following equation:

$$\mathrm{CL} = \mathrm{AC} \cdot \mathrm{CF}, \tag{3}$$

where AC is the actual concentration of selected pollutant ($\mathrm{mg\,L^{-1}}$), and CF is the characteristic flow ($\mathrm{m^3\,s^{-1}}$).

While the limit load at a control profile is described by the following equation:

$$\mathrm{LL} = \mathrm{LC} \cdot \mathrm{CF}. \tag{4}$$

LC is the limit concentration of the selected pollutant ($\mathrm{mg\,L^{-1}}$).

In the research, the mean low flow (SNQ), which is the arithmetic mean of the lowest yearly flows, was chosen as a characteristic flow.

$$Q_{\mathrm{SNQ}} = \mathrm{SNQ} = \frac{\sum_{i=1}^{n} NQ_i}{n}, \tag{5}$$

where n is the number of elements in the analysed set.

2.2 Absorption capacity cases

Analysis of the relation between the LL and actual load (AL) of a selected pollutant at a control profile shows that RAC of a river section can have positive values, negative values and, theoretically, values equal to zero. These cases are shown in Fig. 2.

Values of RAC are as follows.

- Positive – means that AL at the profile is lower than LL of selected pollutant:

$$AL < LL. \tag{6}$$

Thus, there is a possibility of additional load input into the river section equal to the difference between the AL and LL without causing an exceedance of environmental limits;

- Equal to zero – means that the AL and LL at the profile are equal:

$$AL = LL. \tag{7}$$

This is a hypothetical situation in which there is no possibility of the input of any load of pollutant, although there is still no necessity to make radical steps to purify the river water. In practice, the absorption equal to zero is difficult to determine.

- Negative – means that at the profile the AL is greater than LL of a selected pollutant:

$$AL > LL. \tag{8}$$

Therefore, there is an exceedance of the limit value for environmental standards for a good water state at the profile and there is an obligation to take action aimed at reducing pollutant loads discharged into the river. If the critical load is not exceeded, in the case of negative absorption capacity, the reduction of pollutant loads should cause a spontaneous return of the ecosystem to its original state, understood as a state before a change caused by anthropopressure.

2.3 Macromodel DNS/SWAT

The appropriate amount of monitoring data covering a sufficiently long period of time is pivotal to the analysis of the state of surface water pollution. When monitoring data are limited, which is a common situation, it becomes essential to use supplemental tools as mathematical models. They provide an opportunity not only for complementing spatial and temporal resolution data, but also allow analysis to be carried out on, inter alia, processes responsible for the self-purification of the river. Utilization of the correctly chosen and adapted mathematical model for the determination absorption capacity enables extensive knowledge about the state of surface water to be obtained and the selected scenarios of action programmes to be simulated to improve water quality on a selected river section.

The Macromodel DNS (Discharge–Nutrient–Sea) was designed at The Institute of Meteorology and Water Management – National Research Institute (Poland) for the analysis of processes taking place in a catchment, such as water and matter cycles (Ostojski, 2012). The Macromodel is

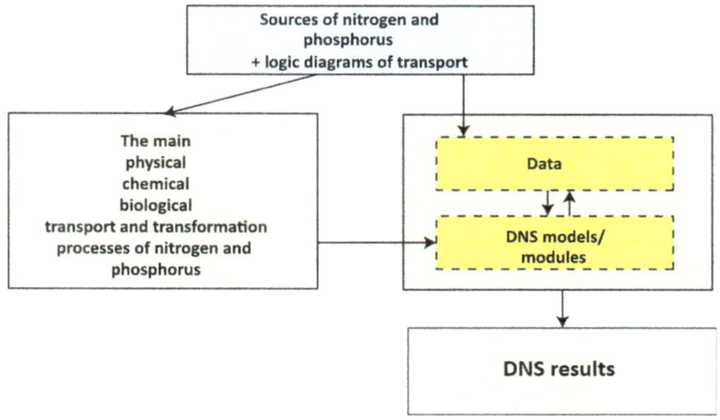

Figure 3. Macromodel DNS scheme (source: Ostojski, 2012).

a unified tool combining existing and verified mathematical models and equations of hydrological transport process units. It allows the simulation of the long-term impact of land use on water quality and the impact of pollutant discharges to surface waters. It is a merger of data processing modules, data replenishment modules, water quantity models and water quality models (Figs. 3 and 4). The Macromodel DNS defines the AL description at a control profile as follows:

$$AL = L_{POINT} + L_{NON} + L_{INF} + L_{DEP} + AL_{-1} - R. \tag{9}$$

L_{POINT} is the sum of load discharged from point sources, L_{NON} is the sum of load discharged from nonpoint sources, L_{INF} is the sum of load discharged from infiltration, L_{DEP} is the sum of load discharged from atmospheric deposition and AL_{-1} is the load flowing from upper river profile, R is the section retention.

The values of the individual components of Eq. (9) were obtained by using the Macromodel DNS in conjunction with the SWAT module. The possibility of introducing detailed catchment data allows for modelling data on both point and nonpoint pollutants (L_{POINT}, L_{NON}), pollutant loads entering the river along with infiltration waters (L_{INF}) and retention (R). Data on atmospheric deposition (L_{DEP}) were obtained from the Norwegian Meteorological Institute, whose mathematical models cover the area of Poland.

SWAT (Soil and Water Assessment Tool) (Neitsch et al., 2004, 2005) can be one of the modules of the Macromodel DNS. SWAT is a continuous long-term yield model. SWAT is a physically based model where processes associated with water and nutrient cycles are directly modelled by internal algorithms to describe the relationship between input and output variables. Physical processes are simulated within hydrologic response units (HRUs). HRUs are lumped land areas within a sub-basin that are comprised of unique land cover, soil and management combinations. To accurately predict the movement of pesticides, sediment or nutrients, firstly the hydrologic cycle is simulated. The simulation is divided into two major phases – a land phase which controls the amount of water (and nutrients) loading to the main channel, and

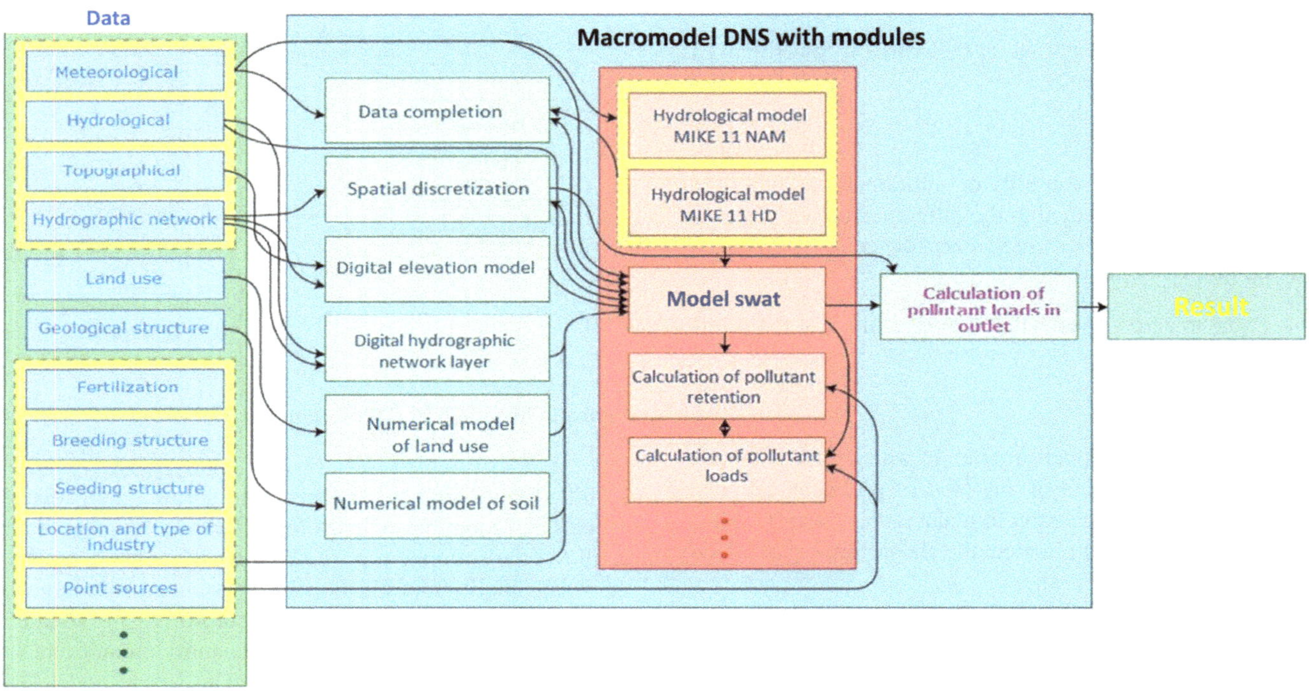

Figure 4. Macromodel DNS/SWAT schematic diagram (source: Ostojski, 2012).

a routing phase which is the movement of water (and nutrients) through the channel network of a watershed to the outlet (Neitsch et al., 2011). The Macromodel DNS containing the SWAT model as an integral module was called the Macromodel DNS/SWAT (Fig. 4).

With the use of the Macromodel DNS/SWAT, all the elements form a homogenous, numerical catchment model that enables the analysis of different scenarios of catchment exploitation in different meteorological and hydrologic conditions. The Macromodel DNS/SWAT can be used to analyse the loads of nutrients at any selected control point (Gębala et al., 2014; Ostojski, 2012).

2.3.1 Research area

The methodology proposed in the article for calculating absorption capacity of a river has been implemented on the example of a fragment of the catchment of the Warta (the Middle Warta). The Warta is the third longest river in Poland. The selected basin has an area of 6039 km^2, which represents approximately 11 % of the entire Warta catchment. The study area was divided into water bodies that are the basic unit of water management in the EU. Numbers were assigned from 1 to 70 to each water body, and water bodies located in the main stream received numbers from 56 to 63. As the beginning of the basin, the Nowa Wies Podgórna profile was selected, and at the end the profile Oborniki was chosen (Fig. 5). The analysed part of the catchment is characterized by a significant amount of area exposed to nitrogen pollutants of an agricultural origin. The area is characterized by a high proportion of nitrate-vulnerable zones (NVZs) – areas particularly vulnerable to nitrogen pollution from agricultural sources (Directive 91/676/EEC). The main soil types on the selected catchment area are light and very light soils. The major sources of pollution are constant and seasonal discharges of domestic, economic and industrial sewage from cities located near the river as well as surface runoff from agricultural areas. In the basin area is the largest metropolitan area of Warta catchment, Poznań, which is the fifth most populous city in Poland and has a very rapidly growing suburban area. Monitoring research which has been conducted for many years on the water state of the Warta indicates that the quality of its waters is strongly differentiated into individual river sections, and pollution flowing into the river can locally affect the process of eutrophication, among other aspects.

2.3.2 Data

For the pivotal river catchments and for the use of the Macromodel DNS/SWAT, the input data were prepared, namely digital elevation model (DEM), hydrology map, soil map, land use map, data concerning a wastewater treatment plant, the daily meteorological and hydrological data, and the amount of fertilizers. The gathered data were developed in

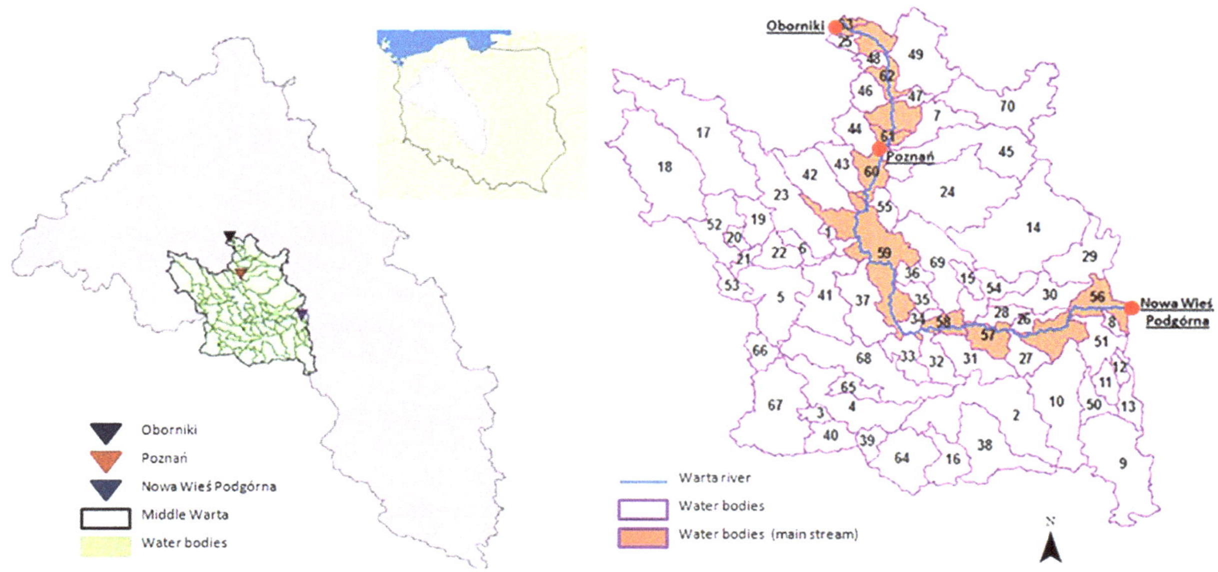

Figure 5. Location of catchments in Poland the Middle Warta catchment with an indication of profiles for calibration and verification (Poznań) and validation (Oborniki).

the form of a database required by the model (Abbaspour, 2008; Srinivasan, 2006, 2011).

The DEM remains the national, central geodesic and cartographic resource and is created on the basis of aerial photographs within a flat and rectangular system of coefficients. The Map of Hydrographical Divisions of Poland (MPHP, 2009) is the basis for the information system of water management. The map containing the details of river networks and water bodies within the boundaries of the analysed catchments on a scale of 1 : 50 000 was used. Data concerning wastewater treatment plants located in the area of the analysed catchments were obtained from the National Water Management Authority in Poland. The data contained detailed information, including the geographic coordinates of a given wastewater treatment plant, the amount of public wastewater treated within a year in thousands of cubic metres per year, total suspended solids ($mg\,L^{-1}$), total nitrogen ($mg\,L^{-1}$) and total phosphorus ($mg\,L^{-1}$). Meteorological input data had a daily time step and included solar radiation, wind speed, precipitation, relative moisture, and maximum and minimum temperatures. Soil maps on a scale of 1 : 100 000 with the soil types very light, light, average and heavy (Table 1) were obtained from the Institute of Soil Science and Plant Cultivation National Research Institute (IUNG, 2009).

Land use maps of the Middle Warta catchment were created based on the CORINE Land Cover information system (Bossard, 2000; CORINE, 2009), which divides land use into five classes, each attributed to the relevant abbreviations that can be accepted and read by the model (Table 2).

Table 1. Soil class with the percentage of participation in the catchment area (source: IUNG, 2009).

Soil class	Catchment Middle Warta
Very light	32.9
Light	30.6
Average	33.9
Heavy	2.4

Table 2. Land use types with the percentage of participation in the catchment area (source: CORINE, 2009).

Land use types	Catchment Middle Warta
Artificial surfaces	6.17
Agricultural areas	72.82
Forests	20.04
Wetland areas	0.1
Water bodies	0.85
Meadows	–

Input data used to calculate phosphorus loads from manure and mineral fertilizers were obtained from the Polish Local Database (BDL) and included information regarding livestock and the surface area of arable lands in hectares at the provincial level. The average dose of nitrate fertilizers was $158.5\,kg\,N\,ha^{-1}$ and phosphate fertilizers $47.08\,kg\,P\,ha^{-1}$. The Middle Warta River catchment has been divided into 70 sub-basins in line with the boundaries of water bodies, which are the basic unit of water management in Poland, according to Directive 2000/60/EC.

2.3.3 Sensitivity analysis and calibration

Sensitivity analysis demonstrates the impact that change to an individual input parameter has on the model response and can be performed using a number of different methods. The method in the ArcSWAT Interface combines the Latin hypercube (LH) and one-factor-at-a-time (OFAT) sampling. During sensitivity analysis, the SWAT runs $(p+1)\cdot m$ times, where p is the number of parameters being evaluated and m is the number of LH loops. For each loop, a set of parameter values is selected such that a unique area of the parameter space is sampled. That set of parameter values is used to run a baseline simulation for that unique area. Then, using OAT sampling, a parameter is randomly selected, and its value is changed from the previous simulation by a user-defined percentage. SWAT is run on the new parameter set, and then a different parameter is randomly selected and varied. After all the parameters have been varied, the LH algorithm locates a new sampling area by changing all the parameters.

In further work, sensitivity analysis of the parameters in the model was carried out. The main purpose of applying sensitivity analysis is to define a set of parameters with the highest sensitivity, meaning those which have the greatest impact on the parameters affecting flow and phosphorus load in the analysed profile of the river. The parameters were developed for ranges typical for Polish conditions.

After conducting the sensitivity analysis, the next stage of study was the model calibration. Model calibration was performed through an iterative value selection process of a single parameter of the model, in order to achieve the greatest possible modelling accuracy in regard to observational data. The estimation of model parameters, in the assumed conditions, in order to achieve the highest convergence of the simulation and observation results, was carried out with the OAT method (one-at-a-time), a repeated iterative loop. The values of parameters received during the sensitivity analysis (Table 4) were successively changed in ranges with a high probability of occurrence in a given area. These values were based on expertise gained from analysis and consulting in the field of hydrology as well as the sources and dynamics of phosphorus change in surface waters in the area of the pilot catchment. It was recognized that such a calibration method enables the fitting of the appropriate model to real conditions, especially for general phosphorus, for which automatic calibration is problematic due to the small amount of observational data. To evaluate model matching with observation in subsequent iterations of the loop, three statistical measures, R2 (coefficient of determination), percent bias (PBIAS) and the Nash–Sutcliffe efficiency (NSE), were used (Moriasi and Arnold, 2007; Ostojski et al., 2016).

The calibration of the flow, total nitrogen and total phosphorus loads was carried out for the data derived from multi-year analyses from 1 January 2003 to 31 December 2007, with the daily time step for Poznań – Most Rocha measuring point, located at 241.76 km (150.22 miles). The verification of the models was conducted from 1 January 2008 to 31 December 2009 with the daily time step for Poznań – Most Rocha measuring point. As for validation, this was conducted from 1 January 2003 to 31 December 2006 with a daily time step for the Oborniki measuring point located at 205.2 km (127.5 miles). In the mentioned periods, the full range of daily data for flow was available, as was 3 to 7 % of data for both total nitrogen loads and total phosphorus loads (Table 3). The flow rate data were derived from the Hydrology System conducted by the IMGW. Total nitrogen and total phosphorus data were from the SEM database.

The monitoring data we used to calibrate, validate and validate the mathematical model come from the years 2003–2009. It was decided that calibration and validation of the mathematical model for this period would be made, as IMGW was conducting its own parallel SEM monitoring of the Middle Warta then. Therefore, for this period of time, we have the largest possible database to better match modelling results to observations (especially for general nitrogen). The monitoring measures currently underway have confirmed that the values of general nitrogen concentrations in the Middle Warta basin remain at a similar level. In a revised version of the article, we will describe the data so that it is clear to every reader why this time period has been used. For each of these processes, robust statistics were used to calculate the winsorized robust statistical measures (Ostojski et al., 2016).

3 The modelling results

For the pilot catchment model Middle Warta and Rega (within the functionality of SWAT, which is in this case a DNS Macromodel module), sensitivity analysis of parameters associated with the flow, total nitrogen and total phosphorus was conducted according to the description in Sect. 3.3.3. The results of this sensitivity analysis are presented in Table 4. There are 14 parameters presented which are most sensitive and associated with the flow in the control point. For total nitrogen loads, from a range of parameters that may be manipulated during the calibration of the model, four parameters obtained the highest sensitivity and seven parameters for total phosphorus loads. The parameters selected during the sensitivity analysis were used during the model calibration.

Charts showing the matching of modelling results and monitoring results for the period of calibration and verification are shown in Fig. 6, with their validation in Fig. 7.

To describe the results of the calibration, verification and validation, three statistical measures were used: the coefficient of determination (R^2), the PBIAS and NSE (Alansi et al., 2009; Bosch et al., 2011; Chu et al., 2004; Pai et al.,

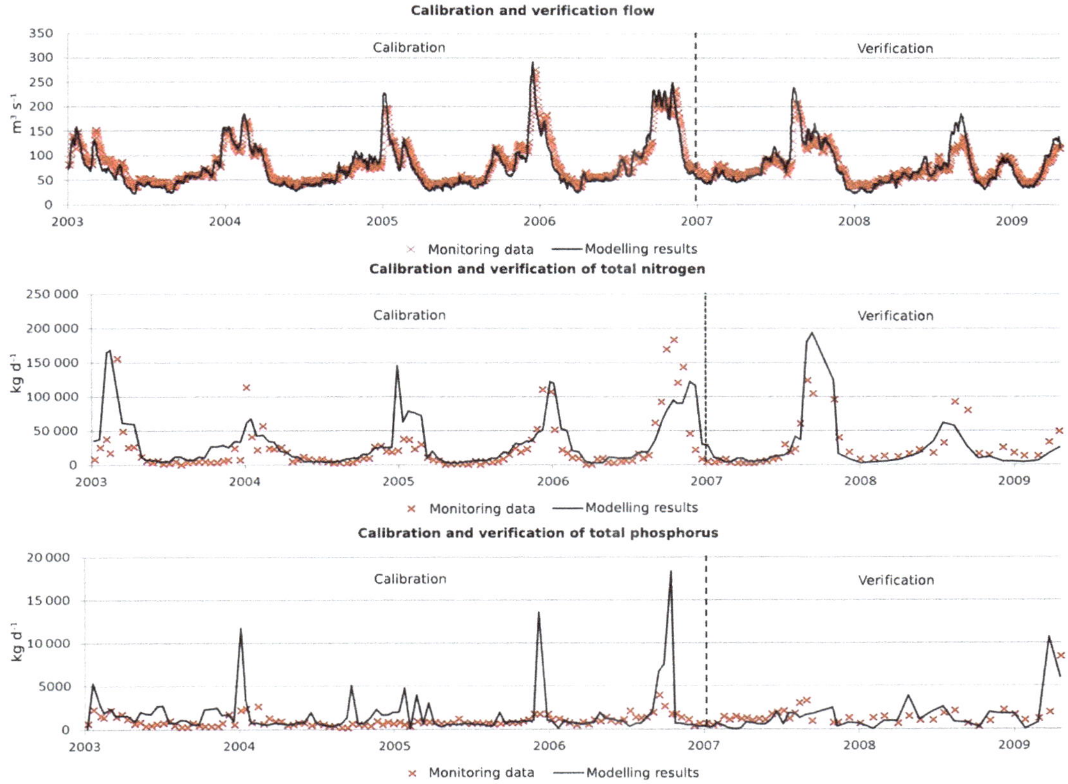

Figure 6. Matching of modelling results and monitoring results for the period of calibration and verification.

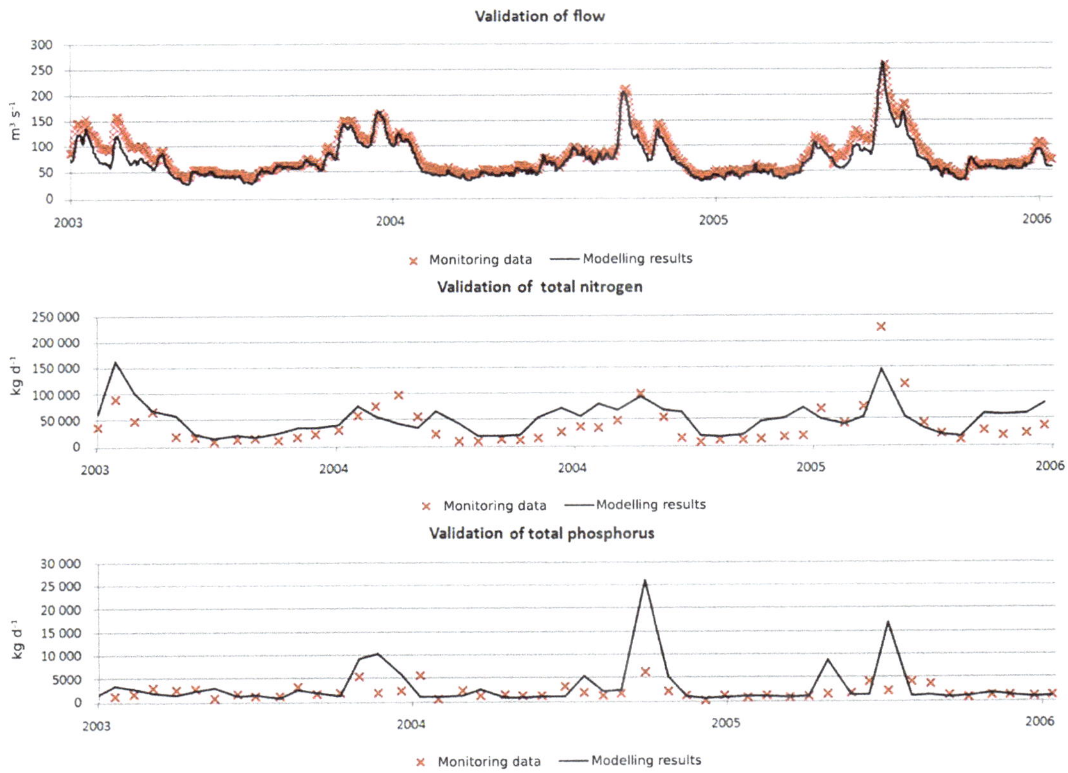

Figure 7. Matching of modelling results and monitoring results for the period of validation.

Table 3. The amount of data available from the State Environmental Monitoring and IMGW for selected catchments. TDQ: total data quantity.

Process	Total data	Flow		Total N		Total P	
	Quantity (TDQ)	Data quantity	Percentage of TDQ	Data quantity	Percentage of TDQ	Data quantity	Percentage of TDQ
Calibration	1826	1826	100 %	120	7 %	120	7 %
Verification	731	731	100 %	25	3 %	25	3 %
Validation	1461	1461	100 %	48	3 %	48	3 %

Table 4. The most sensitive parameters obtained from the sensitivity analysis in SWAT model for the analysed catchments (source: Gębala, 2015; Wilk, 2015).

Parameter	Parameter description
	Flow parameters
ALPHA_BF	Baseflow alpha factor (days)
CANMX	Maximum canopy storage (mm H_2O)
CH_K(1)	Effective hydraulic conductivity in tributary channel alluvium ($mm\,h^{-1}$)
CH_K(2)	Effective hydraulic conductivity in main channel alluvium ($mm\,h^{-1}$)
CN2	Initial SCS runoff curve number for moisture condition II
EPCO	Plant uptake compensation factor
ESCO	Soil evaporation compensation factor
GWQMN	Threshold depth of water in the shallow aquifer required for return flow to occur (mm H_2O)
GW_REVAP	Groundwater "revap" coefficient
RCHRG_DP	Deep aquifer percolation factor
SOL_ALB	Moist soil albedo
SOL_K	Saturated hydraulic conductivity ($mm\,h^{-1}$)
SURLAG	Surface runoff lag coefficient
TIMP	Snow pack temperature lag factor
	Nitrogen and phosphorus parameters
ERORGP	Phosphorus enrichment ratio for loading with sediment
PHOSKD	Phosphorus soil partitioning coefficient ($10\,m^3\,Mg^{-1}$)
PPERCO	Phosphorus percolation coefficient ($10\,m^3\,Mg^{-1}$)
PSP	Phosphorus availability index
P_UPDIS	Phosphorus uptake distribution parameter
SOL_ORGN	Initial organic N concentration in the soil layer ($mg\,N\,kg^{-1}$ soil)
SOL_ORGP	Initial organic P concentration in the soil layer ($mg\,P\,kg^{-1}$ soil)
NPERCO	Nitrogen percolation coefficient ($10\,m^3\,Mg^{-1}$)
SOL_NO3	Initial NO3 concentration in the soil layer ($mg\,N\,kg^{-1}$ soil)
CMN	Rate factor for humus mineralization of active organic nutrients (N and P)

2011; Rathjens and Oppelt, 2012). The results are presented in Table 5. An assessment of the modelling conducted using robust statistics and winsorized L estimators were used (Ostojski et al., 2016).

In the case of calibration, verification and validation of flow, all statistical measures obtained "very good" or "good" (Alansi et al., 2009; Chiang et al., 2012; Krause, 2005; Moriasi et al., 2007) results of model fit. For calibration and verification of total nitrogen, "very good" or "good" fit results of the model were obtained. Only in the case of the validation for the coefficient of determination R^2 and the NSE coefficient were the achieved results "satisfactory". As expected, the biggest difficulty during calibration, verification and validation was with total phosphorus loads – this was due to, among other reasons, the high volatility of daily and seasonal concentrations of this parameter in the environment and the limited amount of monitoring data. The coefficients of determination R^2 at the stage of calibration and verification of the received values were "satisfactory" and "unsatisfactory" respectively. PBIAS in all cases obtained values classifying it as "very good", as opposed to the NSE values, which were in all cases "unsatisfactory". Due to the lack of methodol-

Table 5. The results of the calibration, verification and validation for flow, total nitrogen and total phosphorus.

Parameter	Flow			Total nitrogen			Total phosphorus		
Coefficient/phases	R^2	PBIAS	NSE	R^2	PBIAS	NSE	R^2	PBIAS	NSE
Calibration	0.93	6.07	0.91	0.65	0.14	0.57	0.41	−0.36	−2.49
Verification	0.92	−0.84	0.81	0.81	0.14	0.56	0.01	0.36	−1.05
Validation	0.94	14.51	0.85	0.47	−0.58	0.06	0.65	−0.22	−5.91

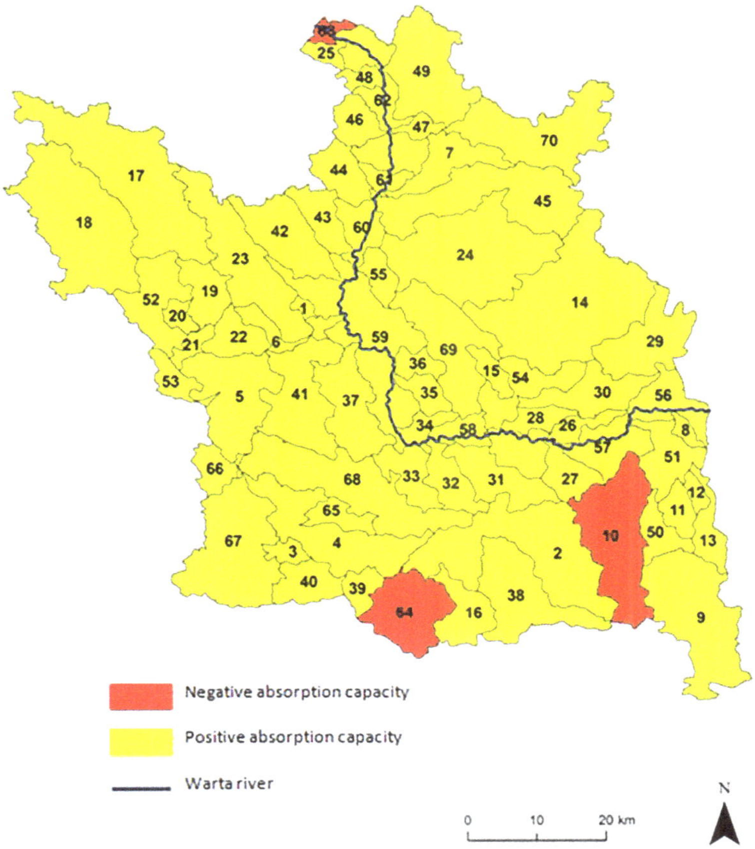

Figure 8. Actual absorption capacity for each water body of the Middle Warta for total nitrogen based on characteristic flow SNQ.

ogy enabling a joint assessment of these three coefficients, R^2 was prioritized and the results of total phosphorous calibration were evaluated as satisfactory.

3.1 Absorption capacity results

After the calibration, verification and validation processes of the SWAT module, daily loads of nitrogen and total phosphorus for the selected period of time were obtained. This database has enabled the calculation of the precise absorption capacity at closing profiles of all 70 water bodies located in the analysed basin. Absorption capacity was calculated for total nitrogen and total phosphorus where, for the characteristic flow, average low-flow QSNQ was used.

3.1.1 Total nitrogen

Absorption capacity values of total nitrogen for individual water bodies in the vast majority (67 water bodies) obtained positive values (Fig. 8). The highest positive values of absorption capacity were observed at closing profiles of the individual water bodies located in the main stream of the Middle Warta sub-catchments 56 to 60 (values of up to 3500 $t\,yr^{-1}$). A clear drop in absorption capacity on the main stream was noted from the closing profile of 61 water bodies where absorption reached 880 $t\,yr^{-1}$, that is, on water bodies located directly behind the city of Poznań. The lowest values, that are still positive, of absorption capacity were at the closing profiles of small streams characterized by low flows.

Table 6. Total nitrogen absorption capacity for each profile of Middle Warta water bodies based on characteristic flow SNQ. The negative values are in boldface.

Water body number	$t\,yr^{-1}$	Water body number	$t\,yr^{-1}$	Water body number	$t\,yr^{-1}$
1	2.4	25	4.7	49	7.3
2	85.4	26	6	50	3.1
3	4.1	27	2.7	51	3.3
4	344.3	28	7.1	52	32.3
5	909.1	29	42.8	53	20.6
6	1065	30	59.8	54	21.4
7	142.4	31	22.8	55	32.2
8	1.3	32	16.1	56	2981
9	44.4	33	28.6	57	2203.4
10	**−54.6**	34	2.8	58	2009.5
11	2.1	35	6.7	59	2345.3
12	1.2	36	13.4	60	3041.4
13	3.6	37	71.9	61	879.8
14	135.6	38	18.4	62	588.3
15	9.8	39	0.1	63	**−883.3**
16	4.8	40	0.4	64	**−1.9**
17	274.3	41	6.1	65	0.9
18	95.6	42	3.9	66	2.4
19	67.9	43	11	67	2.4
20	14.3	44	14.2	68	13.9
21	62	45	19.5	69	7.1
22	17.3	46	0.1	70	7.4
23	137.5	47	0.2	–	–
24	285.1	48	0.2	–	–

Negative absorption capacity values for total nitrogen only occurred at three water bodies (10, 63 and 64). Two of them (10 and 64) are small reaches with low flows which should limit the amount of total nitrogen by about 55 and $2\,t\,yr^{-1}$, respectively. Basin 63 is the last section of the main stream of the analysed catchment where absorption capacity obtained the value of $880\,t\,yr^{-1}$.

The results of absorption capacity of total nitrogen load based on a characteristic flow of QSNQ are shown in Fig. 8. A summary of the results for all analysed closing profiles of water bodies are presented in Table 6.

3.1.2 Total phosphorus

The values of absorption capacity for most closing profiles of water bodies (58) for total phosphorus were positive, as shown in Fig. 9. The highest values were obtained for water bodies located between Nowa Wieś Podgórna and the city of Poznań – these were up to $130\,t\,yr^{-1}$. There is a clearly visible negative impact of the city of Poznań on the absorption capacity of profiles located on the main stream below the agglomeration (60, 61, 62, 63). There, negative values of absorption capacity were up to $-1500\,t\,yr^{-1}$. Other water bodies that obtained negative values of absorption are numbers 10, 27, 38, 43, 64, 69, 50 and 51. Most of these are located in the southern part of the analysed catchment area. Graphic results of the absorption capacity of total phosphorus load, based on a characteristic flow of QSNQ, are shown in Fig. 9. A summary of the results for all analysed closing profiles of water bodies are presented in Table 7.

Table 7. Total phosphorous absorption capacity for each profile of Middle Warta water bodies based on characteristic flow SNQ. The negative values are in boldface.

Water body number	$t\,yr^{-1}$	Water body number	$t\,yr^{-1}$	Water body number	$t\,yr^{-1}$
1	0.098	25	0.189	49	0.292
2	0.983	26	0.241	50	**−0.016**
3	0.165	27	**−0.215**	51	**−3.218**
4	5.304	28	0.283	52	1.136
5	16.128	29	0.843	53	0.669
6	8.014	30	1.315	54	0.444
7	3.476	31	0.913	55	0.663
8	0.05	32	0.644	56	129.257
9	0.484	33	1.146	57	123.965
10	**−15.519**	34	0.113	58	104.055
11	0.082	35	0.269	59	79.82
12	0.049	36	0.538	60	**−151.807**
13	0.145	37	2.881	61	**−592.318**
14	5.01	38	**−1.234**	62	**−1069.137**
15	0.393	39	0.005	63	**−1485.195**
16	0.193	40	0.014	64	**−1.008**
17	10.239	41	0.095	65	0.035
18	3.692	42	0.124	66	0.096
19	1.285	43	**−0.484**	67	0.096
20	0.572	44	0.568	68	0.519
21	0.736	45	0.782	69	**−0.035**
22	0.693	46	0.004	70	0.15
23	3.096	47	0.007	–	–
24	8.823	48	0.01	–	–

4 Summary

The Macromodel DNS/SWAT, developed at the Institute of Meteorology and Water Management (Poland), is an effective tool for the determination of river absorption capacity. The utilization of mathematical modelling for the calculation of absorption capacity allowed the precise determination of total nitrogen and total phosphorous overloads at each river profile, according to limit values, as well as the determination of the disposal capacity of the reach when the absorption capacity has positive values.

Research results show, simply but precisely, the water state at the estuarine profiles of selected basins. The method assumes the performance of the analysis at closing profiles of water bodies. For the analysed pilot catchment, 95 % of estuarine profiles of water bodies have a positive absorption capacity for the load of total nitrogen and 80 % for a load of total phosphorus, which means that in these areas the ecosystem is able to adapt to the "elimination" of these pollutants

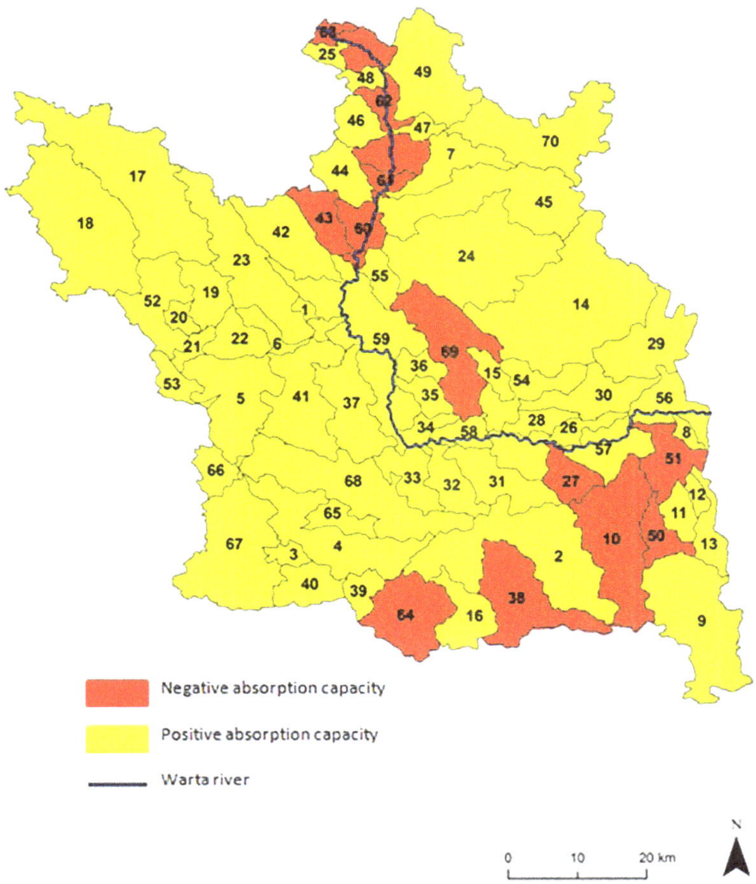

Figure 9. Actual absorption capacity for each water body of the Middle Warta for total phosphorous based on characteristic flow SNQ.

by a series of processes generally referred to as the self-purification of the river. In such cases, it is assumed that after the removal of pollution sources, an ecosystem returns to a state of equilibrium, or close to it (Fig. 1). The highest values of absorption capacity for both total nitrogen and total phosphorus were observed at profiles of water bodies located within the main stream, to the north of the city of Poznań (water bodies 56 to 59). This area is characterized by low population density and low industrialization, although there are a large number of large-scale farms. It should also be taken into account that the main reach of the Warta River is characterized by a high flow (average 195 $m^3 s^{-1}$), which has a direct impact on the dilution of contaminants in the water mass. It may thus be concluded that major streams characterized by low time variability of flow are more resistant to contamination, and higher values of absorption can be expected there. Negative absorption capacity for total nitrogen was noted for three water bodies (10, 63, 64), which means that limit load was exceeded there. Two profiles belong to two water bodies located in the southern part of the basin (10, 64). These basins are characterized by very low flows (they sometimes dry up in the summer) and industrial plants, and numerous illegal sewage discharges are located in their territory. A small reservoir characterized by a high concentration of nitrates is also located in water body 10. The third case of negative absorption capacity for total nitrogen is the last control profile of the main stream – estuarine profile 63 of the pilot Middle Warta catchment. There is a problem here related to the rapid expansion of the city of Oborniki and the insufficient development of its sewage system; there is also intensive agriculture in this area. A large number of leaking septic tanks are responsible for a significant load of nutrients in surface waters in the area.

For total phosphorous, 12 water bodies were obtained with negative values of absorption capacity (10, 27, 38, 43, 50, 51, 60, 61, 62, 63, 64, 69). Half of them are located in the southern part of the study area and are water bodies of small streams characterized by low flows; their territories feature large-scale farms, industrial plants, sewage treatment plants and, as has already been mentioned, numerous illegal discharges of wastewater (water bodies numbers 10, 27, 38, 50, 51, 64). Negative values of absorption capacity for total phosphorus were also noted at closing profiles of water bodies on the main reach located south of the city of Poznań (water bodies numbers 60, 61, 62, 63), which proves that high loads of pollutants are discharged from the area of this agglomeration to the waters of the Warta. The results of the proposed method coincide with previous field research conducted by the Voivodeship Inspectorate for Environmental Protection.

5 Discussion

Data on the analysed Middle Warta catchment obtained with the use of the calibrated, verified and validated Macromodel DNS/SWAT were used to calculate the absorption capacity of the river. The results of general phosphorus calibration using the Macromodel DNS/SWAT are worse than for general nitrogen. This is due to the high daily and seasonal volatility of this element in river waters and the relatively small amount of monitoring data needed to calibrate, verify and validate the model. In spite of this, results of the statistical measures identified as "unacceptable" were obtained mainly for the NSEs, so the authors decided to use the obtained model data for general phosphorus. The obtained RAC results for general phosphorus largely coincide with the actual state occurring in the basin; the water bodies that have been negatively affected by the RAC parameter are located in, among other areas, the main watercourse south of the city of Poznań, which has a negative impact on the waters of the Warta. The remaining water bodies, which had negative RAC values for total phosphorus, are small watercourses in the southern part of the analysed basin, characterized by low flow rates and the location of sewage treatment sites and other points of wastewater discharges. Discussion of the results in the article will be extended with information on this subject. The use of mathematical modelling was intended to simulate daily loads of pollutants at any selected river profile and then to classify the sources of pollution occurring in the basin. The absorption capacity determined on this basis for individual water bodies enables detailed knowledge of the condition of the aquatic environment and the possibility of its adaptation to pollution to be obtained. This is key information for assessing whether the intended water management has a significant impact on the state of a water body. The obtained information concerns the location of polluted areas as well as areas not currently at risk of pollution and where reserves of absorption capacity occurred; this is important during the development of management plans for water in basins, as well as in other contexts.

The article presents the possible response of the river ecosystem to the pollutant load conveyed to it (Fig. 1). In addition to the three zones marked I, II and III, there is one more area of uncertainty. A river's width defines the difference between CL and LL, but only the LL value is known in this case. As previously described, it is not possible to determine the CL value beyond which it will no longer return to good water status. The actual volume of CL loads is influenced by a number of processes occurring in surface waters; these vary with the intensity of the river flow and the seasons. It is not known how wide the zone, defined as the area of uncertainty, is. Knowledge of the existence of the area of uncertainty and CL is important in various contexts, including for governmental authorities dealing with the definition of LL which should be designed to guarantee the protection of the aquatic environment.

The knowledge of ANL is important information for the proper determination of pollutant loads in rivers. Much of the biogenic compounds in surface waters are found there as a result of human activity, but nutrients are also leached from the soil under natural conditions. Estimating ANL is difficult because in Europe there are no longer virtually any anthropogenic drainage basins that could become a benchmark for determining ANL. The Macromodel DNS/SWAT can be used to help create scenarios for the use of catchments. This allows you to "remove" the anthropopressure from the analysed basin and estimate the amount of naturally occurring pollutants. This will be the subject of further research.

In this study, river absorption capacity was calculated for 70 estuarine profiles of individual water bodies for the pilot catchment. This enabled the identification of those water bodies which need urgent action aimed at reducing the amount of nutrients entering the surface water from point and nonpoint sources. However, there were areas identified where acceptable limits of pollution are not exceeded and, moreover, it was possible to accurately determine the pollutant load, which, if necessary, can even lead to the river not changing its class of quality state. In both cases exploitation of the Macromodel DNS/SWAT does not have to be limited to only determining the amount of pollutant loads but can also allow the assessment of the impact of planned activities on the catchment. In the case of the described studies a division of the basin into water bodies was used, but if necessary it is possible to split pilot catchment into any selected basin areas.

During planning of the research described in the paper, similar methods used in other regions of the world were analysed. Primarily, the TMDL ratio, described in Sect. 1, were analysed. However, this parameter is based only on the identification of pollution sources in the catchment without the utilization of limit loads. In Poland, limit loads are placed in acts of national law. As well as this, developing TMDLs for nutrients is also difficult because nitrogen and phosphorus can come from any number of sources – e.g. a significant amount of nitrogen can come from agricultural or atmospheric sources. Therefore, the proposed method of determining the absorption capacity is especially suitable for nutrients due to the fact that it is based largely on the Macromodel DNS/SWAT adapted to the modelling of nutrients, and this refers to the load limit of pollutants for a specific class of water quality, as described in this article. Both the parameter TMDL and river absorption capacity are parameters related to the determination of the assimilative capacity of the river;

however, the calculation of these parameters is based on different assumptions.

For both TMDL and absorption capacity one of the most important elements for calculations is the proper selection of characteristic flow. If the absorption capacity of the river is a parameter on the basis of which decisions of water quality improvement could be made, the choice of characteristic flow becomes crucial. Depending on the country, different characteristic flows are used for environmental calculations. More and more often in publications (Dyson et al., 2003; European Commission, 2015) the question is being raised concerning the use of environmental flows (taking into account the meaning of medium and high flows in maintaining a good state of river ecosystems). However, the determination of environmental flows requires field and computation research of hydraulic and ecological characteristics. At the current stage of knowledge and research related to Polish catchments, this type of analysis for all estuarine profiles of water bodies is impossible to implement. This is one of the reasons why in Poland, similarly to in the United States, commonly used characteristic flows are those emphasizing only low flows (SNQ, 7Q10). The 7Q10 parameter (the lowest flow of a 7-day period for a decade) statistically has a predisposition for often achieving zero values; this is a clear disadvantage of using 7Q10 in environmental analysis. It also requires a 10-year measurement series. For the purposes of research conducted on the possibility of using the absorption capacity as a parameter to control the quality of water in the catchment, the SNQ flow was chosen as adequate. The proposed designation of absorption capacity includes water quality standards in accordance with the law acts, which also recommend the use of SNQs to calculate the environmental calculation. In addition, for basins with a negative absorption, the use of flow lower than SNQ results in lowering the load that should be removed from the river in order to achieve environmental standards. However, the use of either SNQ or 7Q10 is a hydrological approach characterized by the simplicity of calculations and the possibility of its utilization at uncontrolled reaches due to the revised method of handling this type of statistic between the profiles. It is assumed that, in the case of research described in the paper, hydrological conditions reflect biological needs.

6 Conclusion

1. To assess the quality of surface water, it is useful to use the concept of a river's absorption capacity. This is understood as the maximum load of a pollutant which can still enter the river without exceeding the limit load and, consequently, without changing its quality class; or, when the load limit is exceeded, this is the load that must be removed from the water in order meet environmental standards. Knowledge of the river absorption capacity enables the planning of actions that prevent the degradation of water quality in the river and, consequently, damage to aquatic and water-dependent ecosystems that occurs due to anthropogenic pressure.

2. Calculations of absorption capacity of a river segment are based on pollutant loads from the nearest profiles; however, it is not equal to a load introduced directly into the surface water on that segment due to the self-purification processes of the river water on the way from the introduction of the pollutant into the river to the control profile. The dynamics of self-purification processes affect absorption capacity.

3. Positive values of absorption capacity mean that an ecosystem adapts to the elimination of introduced pollution loads through a number of self-purification processes. It is assumed that pollution that is already introduced does not cause permanent change and after the elimination of the source of the pollution, the ecosystem will return to equilibrium, or close to a state of balance. Positive values of absorption capacity allow the disposable load of pollutant in the catchment to be determined.

4. Negative values of absorption capacity indicate areas where anthropogenic activities – agricultural or municipal – are responsible for excessive pollution of nutrients. This means that the limit load has been exceeded, and too much pollution has been introduced to the ecosystem for the possibility of the river's self-purification. There is an amount of load that must be removed from the river segment to attain water quality standards; however, this is not tantamount to a complete return to the initial state of the ecosystem, especially for nutrients, and this may affect many processes in the catchments. It is also unknown whether the critical load has been exceeded or not. Nevertheless, in this case, the excessive pollution should be reduced and methods should be implemented to revitalize the ecosystem.

5. The Macromodel DNS/SWAT can be an effective tool for the analysis of the absorption capacity of a river segment, as was shown with the example of the Middle Warta basin for average daily loads of nitrogen and total phosphorus.

6. Information on the river absorption capacity can be a good basis for the creation of action scenarios which could help in determining the impact of the realization of land development plans or investment planning on surface water and aquatic and water-dependent ecosystems.

7. For a more detailed analysis of water quality for selected catchments, hydrological units smaller than water bodies should be considered.

8. To summarize, the developed absorption capacity parameter is a valuable element of the assessment of water quality. It allows the connection of water quality according to reference values of environmental standards and water quality management plans in order to meet these standards. This parameter is designated by detailed mathematical calculations in combination with precise modelling techniques. This enables the consideration within the modelling of many hydrological, geological and soil conditions; weather and water quality parameters; and then the implementation of the obtained results during mathematical calculations. Absorption capacity of a river segment explains the relationship between the load of pollutants in the river, limit values of environmental standards and water quality assessment, along with identifying sources of pollution and possibilities for improving water quality and the state of aquatic and water-dependent ecosystems.
9. The analysis of absorption capacity using QSNQ as a characteristic flow has proven to be efficient and useful. However, at the stage of practical implementation of this parameter in order to allow in-depth assessment of nutrient pollution (e.g. in terms of eutrophication), the utilization of characteristic flow based on low flows may not be sufficient. Therefore, it is necessary to continue research on the utilization of environmental flows which additionally takes into account medium and high flow, as well as the number of elements affecting the hydrological regime, such as the size, time of occurrence, duration and frequency of floods, and low waters.

Author contributions. PW worked on the development of a method for calculating the river absorption capacity using the Macromodel DNS/SWAT as well as on the calibration, verification and validation of the model. POW and JG prepared the data necessary to build the Macromodel DNS/SWAT. MO worked on the development of the Macromodel DNS.

Competing interests. The authors declare that they have no conflict of interest.

Special issue statement. This article is part of the special issue “Coupled terrestrial-aquatic approaches to watershed-scale water resource sustainability”. It is not associated with a conference.

Edited by: Zhenyao Shen

References

Abbaspour, K. C.: SWAT-CUP2: SWAT Calibration and Uncertainty Programs – A User Manual, Department of Systems Analysis, Integrated Assessment and Modelling (SIAM), Eawag, Swiss Federal Institute of Aquatic Science and Technology, Switzerland, 2008.

Alansi, A. W., Amin, M. S. M., Abdul Halim, G., Shafri, H. Z. M., and Aimrun, W.: Validation of SWAT model for stream flow simulation and forecasting in Upper Bernam humid tropical river basin, Malaysia, Hydrol. Earth Syst. Sci. Discuss., https://doi.org/10.5194/hessd-6-7581-2009, 2009.

Arnold, J. G., Kiniry, J. R., Srinivasan, R., Williams, J. R., Haney, E. B., and Neitsch, S. L.: Soil and Water Assessment Tool input/output file documentation: Version 2009, Texas Water Resources Institute Technical Report No. 365, 2011.

Boeuf, B. and Fritsch, O.: Studying the implementation of the Water Framework Directive in Europe: a meta-analysis of 89 journal articles, Ecol. Soc., 21, 19, https://doi.org/10.5751/ES-08411-210219, 2016.

Bosch, N. S., Allan, J. D., Dolan, D. M., Han, H., and Richards, R. P.: Application of the soil and water assessment tool for six watersheds of Lake Erie: model parameterization and calibration, J. Great Lakes Res., 37, 263–271, 2011.

Bossard, M., Feranec, J., and Otahel, J.: CORINE Land Cover Technical Guide, EEA Copenhagen, Denmark, 2000.

Brodie, E., Waterhouse, J., Lewis, E., Bainbridge, T., and Johnson, J.: Current loads of priority pollutants discharged from Great Barrier Reef Catchments to the Great Barrier Reef. ACTFR Report Number 09/02, Australian Centre for Tropical FreshWater Research, Townsville, 2009.

Bulsathaporn, A., Bhaktikul, K., Arunlertaree, Ch., Sueadee, W., and Suttigarn, A.: The Application of Mathematical Models for an Environmental Flow Assessment and Total Maximum Daily Load (TMDL) of Prachinburi-Bangpakong River, Thai National Committee on Irrigation Drainage, World Water, Mahidol University, Thailand, 2013.

Chiang, L. C., Chaubey, I., Hong, N. M., Lin, Y. P., and Huang, T.: Implementation of BMP strategies for adaptation to climate change and land use change in a pasture-dominated watershed, Int. J. Env. Res. Pub. He., 9, 3654–3684, 2012.

Chmielowski, W. Z. and Jarząbek, A.: Ćwiczenia i projekty z przedmiotu gospodarka wodna: pomoc dydaktyczna, Politechnika Krakowska, Kraków, 2008.

Chu, T. W., Shirmohammadi, A., Montas, H., and Sadeghi, A.: Evaluation of the SWAT model's sediment and nutrient components in the Piedmont physiographic region of Maryland, T. ASAE, 47, 1523–1538, 2004.

CORIN: CORINE LAND COVER, available at: http://www.eea.europa.eu/themes/landuse/interactive/clc-download, last access: 12 July 2013.

Directive: Nitrate, Council Directive 91/676/EEC of 12 December 1991 concerning the protection of waters against pollution caused by nitrates from agricultural sources, Official Journal L, 375, 8 pp., 1991.

Duarte, C. M., Conley, D. J., Carstensen, J., and Sánchez-Camacho, M.: Return to Neverland: shifting baselines affect eutrophication restoration targets, Estuar. Coast., 32, 29–36, 2009.

Dubnyak, S. and Timchenko, V.: Ecological role of hydrodynamic processes in the Dnieper reservoirs, Ecol. Eng., 16, 181–188, 2000.

Dyson, M., Bergkamp, G., and Scanlon, J.: Flow, International Union for Conservation of Nature and Natural Resources, IUCN, Gland, Switzerland, 2003.

Elósegui, A., Arana, X., Basaguren, A., and Pozo, J.: Self-purification processes along a medium-sized stream, Environ. Manage., 19, 931–939, 1995.

EPA – Environmental Protection Agency: Implementing Clean Water Act Section 303(d): Impaired Waters and Total Maximum Daily Loads (TMDLs), available at: https://www.epa.gov/tmdl, last access: 20 September 2016.

European Commission: Ecological flows in the implementation of the Water Framework Directive, Technical Report, Policy Summary, CIS Guidance Document No. 31, 2015.

Federal Water Pollution Control Act: As Amended Through P. L. 107–303, US Congress, Washington, DC, 2002.

Gębala, J., Orlińska-Woźniak, P., and Wilk, P.: Zastosowanie modelu SWAT w metodykach identyfikacji źródełzanieczyszczeń związkami biogennymi w zlewniach. Wybrane przykłady systemów wspomagania decyzji i modelowania w gospodarce wodnej, IMGW, Warszawa, 2014.

Gębala, J.: Method for assessing the impact of agricultural anthropopressure on surface water quality, PhD thesis, IMGW-PIB, 2015 (in Polish).

Górecki, K.: Zmiany stężenia i ładunku składników biogennych w wodzie rzeki warty między przekrojami Oborniki i Skwierzyna, Acta Scientiarum Polonorum. Formation Circumiectus, 6, 29–42, 2007.

Helsinki Commission: The Fourth Baltic Sea Pollution Load Compilation (PLC-4), Baltic Marine Environment Protection Commission, Helsinki, Baltic Sea Environment Proceedings No. 93, 2004.

Henriksson, A. and Miljökonsulter, A.: Actions against phosphorus losses from agriculture in the countries surrounding the Baltic Sea, Sveriges lantbruksuniversitet, Uppsala, 2007.

IUNG: Mapy kategorii glebowych Ministerstwa Rolnictwa i Rozwoju Wsi – Instytut Uprawy Nawożenia i Gleboznawstwa PIB, available at: http://www.susza.iung.pulawy.pl/index.html?str=mapkat, last access: 20 July 2012, 2009.

Jancarkova, I., Larsen, T. A., and Gujer, W.: Distribution of nitrifying bacteria in a shallow stream, Water Sci. Technol., 36, 161–166, 1997.

Jarosiewicz, A.: Proces samooczyszczania w ekosystemach rzecznych, Słupskie Prace Biologiczne, 4, 27–41, 2007.

Jarosiewicz, A. and Dalszewska, K.: Dynamika składników biogenicznych w rzece Słupi-ocena zdolności samooczyszczania rzeki, Słupskie prace biologiczne, 5, 63–73, 2008.

Karrasch, B., Parra, O., Cid, H., Mehrens, M., Pacheco, P., Urrutia, R., and Zaror, C.: Effects of pulp and paper mill effluents on the microplankton and microbial self-purification capabilities of the Biobio River, Chile, Sci. Total Environ., 359, 194–208, 2006.

Kowalewski, Z.: Metody oceny stanu troficznego wód powierzchniowych. IV Krakowska Konferencja Młodych Uczonych, Akademia Górniczo-Hutnicza im., Kraków, Stanisława Staszica w Krakowie, 343–351, 2009.

Kowalkowski, T.: Classification of nutrient emission sources in the Vistula River system, Environ. Pollut., 157, 1867–1872, 2009.

Krause, P., Boyle, D. P., and Bäse, F.: Comparison of different efficiency criteria for hydrological model assessment, Adv. Geosci., 5, 89–97, https://doi.org/10.5194/adgeo-5-89-2005, 2005.

Magley, W. and Joyner, D.: Total Maximum Daily Load for Nutrients for the Lower St. Johns River, Bureau of Watershed Management, Florida Department of Environmental Protection, 2008.

Marsili-Libelli, S. and Giusti, E: Water quality modelling for small river basins, Environ. Modell. Softw., 23, 451–463, 2008.

McColl, R. H. S.: Self-purification of small freshwater streams: phosphate, nitrate, and ammonia removal, New Zeal. J. Mar. Fresh., 8, 375–388, 1974.

Mohlar, R.: Penobscot River, Phosphorus Waste Load Allocation, Maine Department of Environmental Protection (DEP), Maine, USA, 2011.

Mońka, B.: Planowanie w gospodarowaniu wodami w Polsce przed i po wejściu do Unii Europejskiej, Regionalny Zarząd Gospodarki Wodnej we Wrocławiu, Wrocław, 2005.

Moriasi, D. N., Arnold, J. G., Van Liew, M. W., Bingner, R. L., Harmel, R. D., and Veith, T. L.: Model evaluation guidelines for systematic quantification of accuracy in watershed simulations, T. ASABE, 50, 885–900, 2007.

Neitsch, S. L., Arnold, J. G., Kiniry, J. R., and Williams, J. R.: Soil and water assessment tool theoretical documentation version 2009, Texas Water Resources Institute, USA, 2011.

Neverova-Dziopak, E.: New approach to estimation of permissible discharge of biogenic matter into water bodies (example of the Neva estuary of the Baltic Sea), Environ. Prot. Eng., 33, 121–128, 2007.

Neverova-Dziopak, E.: Problems of ecological monitoring of surface waters, Environ. Prot. Eng., 35, 81–91, 2009.

Nixon, S. W.: Eutrophication and the macroscope, Hydrobiologia, 629, 5–19, 2009.

Orlińska-Woźniak, P., Wilk, P., and Gębala, J.: Water availability in reference to water needs in Poland, Meteorology Hydrology and Water Management, Research and Operational Applications, IMGW, Warsaw 2013.

Ostojski, M. S.: Modelowanie procesów odprowadzania do Bałtyku związków biogennych: na przykładzie azotu i fosforu ogólnego, Wydawnictwo Naukowe PWN, Warszawa, 2012.

Ostojski, M. S., Gębala, J., Orlińska-Woźniak, P., and Wilk, P.: Implementation of robust statistics in the calibration, verification and validation step of model evaluation to better reflect processes concerning total phosphorus load occurring in the catchment, Ecol. Model., 332, 83–93, 2016.

Pai, N., Saraswat, D., and Daniels, M.: Identifying priority subwatersheds in the Illinois river drainage area in Arkansas watershed using a distributed modeling approach, T. ASABE, 54, 2181–2196, 2011.

Popek, Z.: Analiza możliwości zwiększania retencji na obszarach zurbanizowanych w dorzeczu Wisły Środkowej–stan wiedzy i dalsze kierunki działań. Program bezpieczeństwa powodziowego w dorzeczu Wisły Środkowej, Warszawa, 2011.

Radcliffe, D. E., Freer, J., and Schoumans, O.: Diffuse phosphorus models in the United States and Europe: their usages, scales, and uncertainties, J. Environ. Qual., 38, 1956–1967, 2009.

Rathjens, H. and Oppelt, N.: SWAT model calibration of a grid-based setup, Adv. Geosci., 32, 55–61, https://doi.org/10.5194/adgeo-32-55-2012, 2012.

Reckhow and Pahl-Wostl, C.: Directive, W. F. Water Policy in the USA and the EU KH Case Study: Total Maximum Daily Load and Water Framework Directive, Ecol. Econ., 64, 484–495, 2001.

Regulation of the Minister of the Environment: W sprawie sposobu klasyfikacji stanu jednolitych części wód powierzchniowych oraz środowiskowych norm jakości dla substancji priorytetowych, Dz. U. Nr 257, Poz. 1545, Warsaw, 2011.

Schulz, M., Kozerski, H. P., Pluntke, T., and Rinke, K: The influence of macrophytes on sedimentation and nutrient retention in the lower River Spree (Germany), Water Res., 37, 569–578, 2003.

Spellman, F. R. and Drinan, J.: Stream Ecology and Self Purification: An Introduction, CRC Press, https://doi.org/10.1016/S0043-1354(02)00276-2, 2001.

Srinivasan, R.: Soil and Water Assessment Tool, Introductory Manual – Teaching Materials, Texas A&M University, Texas, USA, 2011.

Srinivasan, R., Hadley, J., Uhlenbrook, S., Van Griensven, A., Holvoet, K., and Bauwens, W.: European SWAT summer school 2006, UNESCO – IHE, Institute for Water Education, Delft, Netherlands, 2006.

Tyszewski, S., Herbich, P., Indyk, W., Jarząbek, A., Pusłowska-Tyszewska, D., and Rutkowski, M.: Metodyka opracowywania warunków korzystania z wód regionu wodnego oraz warunków korzystania z wód zlewni, PRO WODA, Warszawa, 2008.

Vagnetti, R., Miana, P., Fabris, M., and Pavoni, B.: Self-purification ability of a resurgence stream, Chemosphere, 52, 1781–1795, 2003.

Vaikasas, S. and Dumbrauskas, A.: Self-purification process and retention of nitrogen in floodplains of River Nemunas, Hydrol. Res., 41, 338–345, 2010.

Van der Lee, G. E. M., Olde Venterink, H., and Asselman, N. E. M.: Nutrient retention in floodplains of the Rhine distributaries in the Netherlands, River Res. Appl., 20, 315–325, 2004.

WFD (Water Framework Directive): Directive 2000/60/EC of the European Parliament and of the Council Establishing a Framework for the Community Action in the Field of Water Policy, The European Parliament and the Council of the European Union, Brussels, Belgium, 2000.

Wilk, P. (2015) The method of calculating river absorption capacity (RAC) as a tool to assess the physicochemical state of surface flowing waters, PhD thesis, IMGW-PIB (in Polish).

Zagorc-Končan, J. and Šömen, J.: A simple test for monitoring biodegradable industrial pollution in a receiving stream, Water Sci. Technol., 39, 221–224, 1999.

Zalewski, M.: Ecohydrology – Integrative tool for achieving good ecological status of freshwater ecosystems, International Centre for Ecology of the Polish Academy of Sciences, Warsaw/Lodz Department of Applied Ecology University of Lodz, available at: http://www.inbo-news.org/IMG/pdf/EH_Krakow.pdf (last access: July 2016), 2003.

5

Topography significantly influencing low flows in snow-dominated watersheds

Qiang Li[1], **Xiaohua Wei**[1], **Xin Yang**[2,3,4], **Krysta Giles-Hansen**[1], **Mingfang Zhang**[5], **and Wenfei Liu**[6]

[1]Department of Earth, Environmental and Geographic Sciences, University of British Columbia (Okanagan campus), 1177 Research Road, Kelowna, British Columbia, V1V 1V7, Canada
[2]Key Laboratory of Virtual Geographic Environment, Nanjing Normal University, Ministry of Education, Nanjing, 210023, China
[3]State Key Laboratory Cultivation Base of Geographical Environment Evolution, Nanjing, 210023, China
[4]Jiangsu Center for Collaborative Innovation in Geographical Information Resources Development and Application, Nanjing, 210023, China
[5]School of Resources and Environment, University of Electronic Science and Technology of China, Chengdu, 611731, China
[6]Institute of Ecology and Environmental Science, Nanchang Institute of Technology, Nanchang, 330099, China

Correspondence: Xin Yang (xxinyang@163.com)

Abstract. Watershed topography plays an important role in determining the spatial heterogeneity of ecological, geomorphological, and hydrological processes. Few studies have quantified the role of topography in various flow variables. In this study, 28 watersheds with snow-dominated hydrological regimes were selected with daily flow records from 1989 to 1996. These watersheds are located in the Southern Interior of British Columbia, Canada, and range in size from 2.6 to 1780 km^2. For each watershed, 22 topographic indices (TIs) were derived, including those commonly used in hydrology and other environmental fields. Flow variables include annual mean flow (Q_{mean}), $Q_{10\,\%}$, $Q_{25\,\%}$, $Q_{50\,\%}$, $Q_{75\,\%}$, $Q_{90\,\%}$, and annual minimum flow (Q_{min}), where $Q_{x\,\%}$ is defined as the daily flow that occurred each year at a given percentage (x). Factor analysis (FA) was first adopted to exclude some redundant or repetitive TIs. Then, multiple linear regression models were employed to quantify the relative contributions of TIs to each flow variable in each year. Our results show that topography plays a more important role in low flows (flow magnitudes $\leq Q_{75\,\%}$) than high flows. However, the effects of TIs on different flow magnitudes are not consistent. Our analysis also determined five significant TIs: perimeter, slope length factor, surface area, openness, and terrain characterization index. These can be used to compare watersheds when low flow assessments are conducted, specifically in snow-dominated regions with the watershed size less than several thousand square kilometres.

1 Introduction

Topography plays a critical role in geomorphological, biological, and hydrological processes (Moore et al., 1991; Quinn et al., 1995). Many topographic indices (TIs) have been derived to describe the spatial patterns of a landscape (Yokoyama et al., 2002), locate spatial patterns of species (Jenness, 2004), and simulate spatial soil moisture (Park et al., 2001). In hydrology, hydrological responses are forced by climatic inputs (e.g. precipitation) but are controlled by topography and other factors such as land use and land cover (Beven and Kirkby, 1979; Hewlett and Hibbert, 1967). In describing the role of topography in hydrology, numerous TIs have been developed and applied to help understand hydrological processes and to explain the variation between watersheds (Moore et al., 1991). Although the importance of topography in controlling various flow magnitudes has been widely recognized (Price, 2011), the quantitative relationship between specific TIs and various flow variables is not well understood.

TIs can be categorized into two groups, namely, primary and secondary (or compounded) indices (Moore et al., 1988). Primary indices (e.g. slope, elevation, and aspect) are normally directly calculated from a digital elevation model (DEM), while secondary indices are the combination of primary indices that are used to explain the role of topography in geomorphological, biological, and hydrological processes. For instance, the topographic wetness index (TWI) is defined as ln (α/tanβ), where α is the upslope contributing area per unit contour length and β is the slope. TWI is a required primary input for TOPMODEL and other hydrological applications (Beven, 1995; Beven and Kirkby, 1979; Quinn et al., 1995). Hydrological studies mainly focus on primary TIs and a few secondary TIs; however, these have limited explanatory power as they largely fail to explain the variation in hydrological processes. Several TIs (e.g. terrain characterization index, topographic openness) have been widely adopted in geomorphology and biology, but are seldom used in hydrological studies. A thorough examination of existing TIs is needed to identify those that best account for hydrological variations between watersheds.

Studies of watershed topography on hydrological processes often include topics such as specific discharge (Karlsen et al., 2016), spatial baseflow distribution (Shope, 2016), transit time (McGuire et al., 2005; McGuire and McDonnell, 2006), and hydrological connectivity (Jencso and McGlynn, 2011). These studies were often based on a short period of data (< 5 years), limiting our ability to draw general conclusions on how topography affects hydrological processes. Moreover, hydrological responses are compounded by the spatially diverse effects of climate, vegetation, soil, and topography (Li et al., 2017; Wei et al., 2018; Zhang et al., 2017). For example, several hydrological models have been applied to test the effects of spatial distribution of a hydrological variable (e.g. specific discharge, soil moisture, or groundwater recharge) (Erickson et al., 2005; Gómez-Plaza et al., 2001; Li et al., 2014). However, the effects of topography alone on hydrology are not usually addressed in those studies. Finally, understanding how topography influences hydrology has significant implications for sustainable management of aquatic ecosystems (Zhang et al., 2016). Therefore, the major objectives of this study were (1) to examine the role of topography in various flow magnitudes in 28 selected watersheds with snow-dominated hydrological regimes in the Southern Interior of British Columbia, Canada; and (2) to identify the most important topographic indices that can be used to compare variations in flow magnitudes between watersheds under similar climatic conditions.

2 Data

2.1 Study watersheds

In this study, 28 watersheds were selected ranging in size from 2.6 to 1780 km^2 (Fig. 1 and Table S1 in the Supplement). The watersheds are located between 51° N, 122° W and 49° N, 118° W in the Southern Interior of British Columbia, Canada, where hydrological regimes are snow-dominated. In this region, the Pacific Decadal Oscillation shifted from a cool to a warm phase around 1977 (Fleming et al., 2007; Wei and Zhang, 2010), resulting in more precipitation and lower temperatures, and consequently affecting hydrological regimes. In addition, an extensive mountain pine beetle infestation caused large-scale forest cover change from 2003 onwards. To avoid the uncertainties associated with these perturbations and to maximize the sample size, the period of 1989–1996 was selected, during which daily flow records of selected stations are complete. In addition, we further confirm that vegetation changes (using the leaf area index (LAI) as a proxy) did not significantly alter annual mean flow during this period (see the Supplement, Sect. S2).

Annual mean temperature (T) and precipitation (P) of the study watersheds were calculated from the ClimateBC dataset (Wang et al., 2006). ClimateBC is a stand-alone program that extracts and downscales PRISM (Daly et al., 2008) monthly climate normal data and calculates seasonal and annual climate variables for specific locations based on latitude, longitude, and elevation. Annual P, T, and potential evapotranspiration (PET) were determined at a spatial resolution of 500 m and averaged for each watershed. PET was calculated using the Hargreaves method (Liu et al., 2016; Wang et al., 2006; Zhang and Wei, 2014). The average mean annual P and PET of all 28 watersheds were 813 ± 205 and 586 ± 58 mm for 1989–1996, respectively (Figs. 1 and S1–S4).

2.2 Topographic indices

Based on availability and representation of TIs in literature, 22 topographic indices (TIs) were derived using a gridded DEM at a spatial resolution of 25 m (Table 1). The DEM, geospatial streamflow networks, and lakes and wetland coverage were obtained from GeoBC (Government of British Columbia, available at http://www2.gov.bc.ca/gov/content/data/about-data-management/geobc/geobc-products). All data were transformed to the same projected coordinate system prior to the calculations of TIs. Calculation of the TIs was made in ArcGIS 10.4.1 (ERSI®) and SAGA GIS 2.1.2. Detailed information on the calculation and interpretation of TIs can be found in the references listed in Table 1.

Table 1. Topographic indices and descriptions.

No.	Abbreviation	TI	Description and references
1	**UCA**	Upslope contributing area	UCA is the area that can potentially produce runoff to a given location (Erskine et al., 2006).
2	**DDG**	Downslope distance gradient	DDG is a hydrologic measure of the impact of the local slope characteristics on a hydraulic gradient. Values are low on concave slope profiles and high on convex slope profiles (Hjerdt et al., 2004).
3	DDGD	Downslope distance gradient difference	DDGD describes the difference between DDG and local or neighbour gradients (Hjerdt et al., 2004).
4	FLD	Downstream flow length	This describes the downslope distance of a pixel along the flow path to the outlet of a watershed (Greenlee, 1987).
5	ME	Median elevation	ME refers to the median elevation among all DEM pixels in a watershed.
6	**Relief**	Relief	Relief describes the difference between the highest and lowest elevations within a local analysis window. An 11×11 grid cell window is used in this paper.
7	Roughness	Roughness	Roughness is calculated as 1/cos(slope) of each DEM pixel.
8	**Slope**	Slope degree	Slope refers to the slope degree of each DEM pixel (Burrough et al., 2015).
9	**LS**	Slope length factor	LS is a combined factor of slope length and slope gradient. It represents the ratio of soil loss per unit area on a site to the corresponding loss from a 22.1 m long experimental plot with a 9 % slope (Desmet and Govers, 1996).
10	**SCA**, also known as $\mathbf{A_s}$	Specific contributing area	The SCA is the upslope contributing area per unit length of contour (Quinn et al., 1991).
11	STRD	Stream density	The STRD is the ratio of the sum of all stream length to watershed area.
12	**TCI**	Terrain characterization index	$\text{TCI} = \text{Cs}^*\log_{10}(\text{SCA})$, where Cs is the surface curvature index; the higher positive TCI values reflect higher aggradation of soil materials at a certain point along the hillslope (Park and van de Giesen, 2004).
13	**TRI**	Terrain ruggedness index	TRI expresses the degrees of difference in elevation among adjacent cells (Riley, 1999). It calculates the sum changes between a grid cell and its eight neighbour grid cells. Higher values indicate more ruggedness of a watershed.
14	TPI	Topographic position index	$\text{TPI} \approx 0$ indicates flat area. $\text{TPI} > 0$ tends towards ridge tops and hilltops. $\text{TPI} < 0$ tends towards the valley and canyon bottoms (Jenness, 2006). A 9×9 grid cell window is used in this paper.
15	TWI	Topographic wetness index	$\text{TWI} = \ln(\text{SCA}/\tan(\text{slope}))$; this shows the spatial distribution of zones of surface saturation and soil water content (Ambroise et al., 1996; Quinn et al., 1995).
16	Wetland	Wetland coverage	This describes the percentage wetland area to the total watershed area.
17	Length	Length of main river	This refers to the total length of mainstream.
18	Roundness	Roundness coefficient	This is the ratio of watershed area to the area of a circle with the same perimeter. A lower value indicates a longer and narrow watershed.
19	**Openness**	Positive topographic openness	Openness describes the degree of dominance or enclosure of a location on an irregular surface. Values are high for convex and low for concave forms, respectively (Yokoyama et al., 2002).
20	**SA**	Surface area	SA describes the land area of each DEM, which may provide a better estimation of the surface roughness than the planimetric area (Jenness, 2004). A lower value indicates a more gentle topography.
21	**Perimeter**	Perimeter of a watershed	The perimeter of a watershed is utilized.
22	Total	Total curvature	The standard curvature combines profile and planform curvatures (Moore et al., 1991).

Note that bold TIs are selected by the factor analysis test.

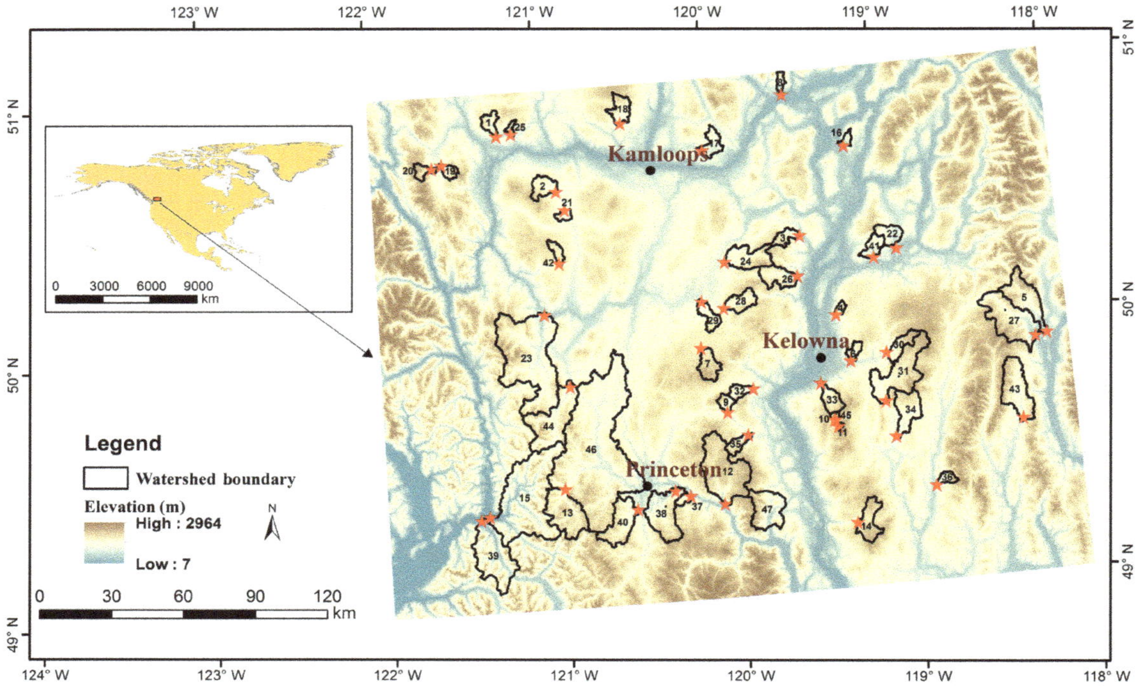

Figure 1. Locations and elevations of the 28 study watersheds, and their hydrometric stations (red star symbols).

3 Methods

3.1 Definitions of the selected flow variables

Annual mean flow (Q_{mean}) and other flow variables generated by the annual flow duration curve were used in this study. The selected flow variables include $Q_{10\,\%}$, $Q_{25\,\%}$, $Q_{50\,\%}$, $Q_{75\,\%}$, $Q_{90\,\%}$, and annual minimum flow ($Q_{\min}$) (mm day^{-1}). They are defined as the daily flow that occurred each year at the given percentage. For example, $Q_{90\,\%}$ is the flow magnitude at 90 % of the time in a year (Cheng et al., 2012). To account for the confounding effects of climate, each annual flow variable was standardized with annual (P) and expressed as Q_X/P.

3.2 Factor analysis

Because some initially selected TIs may be highly related, the first step was to introduce factor analysis (FA) to reduce the number of TIs while still retaining important topographic information. FA can be interpreted in a similar manner as principal component analysis. The major difference between the two approaches is that FA not only considers the total variance but also makes the distinction between common and unique variance (Lyon et al., 2012). As TIs were calculated in a region with similar topography, the average TIs at the watershed level are not varied dramatically between watersheds (McGuire and McDonnell, 2006; Price, 2011). Therefore, to ensure better differentiation, the standard deviations of TIs of a watershed were used for the FA test. It should be noted that the flow variables were not included in the FA test.

Three criteria were used in the FA procedure to exclude redundant TIs: the Kaiser–Meyer–Olkin (KMO) test, Bartlett's test, and anti-image correlation. KMO is a measure of sampling adequacy which tests whether partial correlations among variables are small enough to ensure the validity of the FA test. Bartlett's test of sphericity assesses the level of correlation between the variables in the FA to determine if the combination of variables is suitable for such analysis (Lyon et al., 2012). The diagonals of the anti-image correlation matrix are a measure of sampling adequacy of specific TIs, which ensures that TIs are adequate for the FA. If a TI makes the FA indefinite, namely KMO <0.7, Bartlett's test $P>0.05$, and the diagonals of anti-image correlation <0.7, then this TI is excluded from further consideration. With this iterative approach, the group of TIs with the largest KMO, Bartlett's test $P<0.05$, and the diagonals of anti-image correlation >0.7 are determined as the final group of TIs. In this study, FA tests were conducted in the IBM® SPSS® Statistics Version 22.

3.3 Relative contributions of each TI to flow variables

The nonparametric Kendall's tau correlation examined the statistical correlations between flow variables and the FA-selected TIs in the 28 study watersheds. If a significant correlation is detected, it indicates high topographic control on that flow variable. Multiple linear regression (MLR) models were then built for each year between 1989 and 1996 for each flow variable (see Sect. 5 in the Supplement for details). The purposes of the MLR models were (1) to further exclude those TIs that were insignificantly related to flow variables, and (2) to quantify the relative contributions of the selected TIs to each flow variable in each year. In the MLR models, each flow variable was treated as a dependent variable, while all FA-selected TIs were regarded as independent variables. To exclude insignificant TIs to flow variables, all the 11 selected TIs by the FA were initially included in the MLR model. The ANOVA (analysis of variance) test was then adopted to identify the statistical significance between TIs and each flow variable in each year. If one TI was insignificant ($P > 0.05$), then it was removed from the model. The ANOVA test was then re-run for the rest of the TIs to ensure that all significant TIs were selected. By this trial and error process, the final models with only significant TIs to flow variables were determined. The detailed procedure can also be found in Li et al. (2014). Then, the *R* package "relaimpo" was used to quantify the relative contributions of the selected TIs to each flow variable (Gromping, 2006). In particular, the relative contributions of each dependent variable to R^2 were calculated for each model. In this way, the contributions of the significant TIs to flow variables were derived for each MLR model.

To quantify the role of each TI in regulating each flow variable, we defined a contribution index (CI), which can be expressed as $\mathrm{CI} = O \times C$, where O is the number of the selected TIs appearing in the final MLR models (the maximum number is eight because each flow variable is studied for 8 years), and C is the average relative contribution of each TI to the specific flow variable . Therefore, a higher CI indicates a higher influence of that TI on a flow variable. The lumped CI of each TI to each flow variable is treated as the total contribution of the TI to all flow variables. Finally, the TIs with the CI values that are higher than the average of the lumped CI of all TIs were determined as the final set of TIs. As such, TIs with higher contributions were selected for the flow variables.

4 Results

4.1 Factor analysis

A subset of 11 TIs was selected from the initial 22 calculated TIs using the FA procedure. The KMO test (0.853), Bartlett's test ($P < 0.001$), and the diagonals of anti-image correlation (> 0.7) on the 11 TIs further confirm that our selected TIs are adequate to represent topographical characteristics of a watershed in our study region. In the FA analysis, the first and second factors explained 80.9 and 11.7 % of total variance of TIs in the selected watersheds, respectively (Fig. 2). TIs included in the first factor are DDG, LS, openness, relief, slope, SA, TCI, and TRI, while those in the second factor

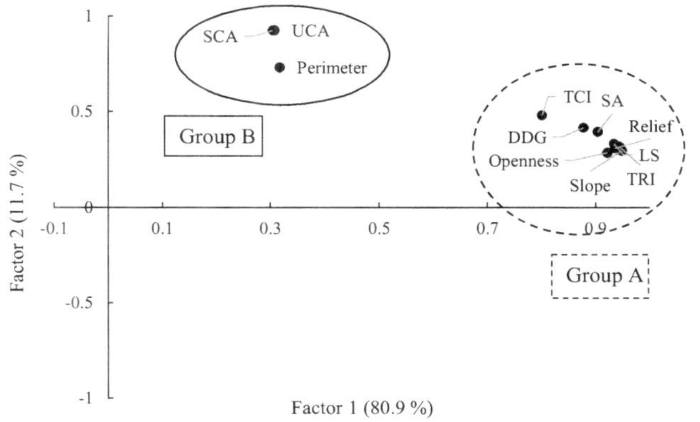

Figure 2. Factor analysis of topographic indices (TIs) among 28 watersheds. The first and second factors explained 80.9 and 11.7 % of the total variance, respectively.

include SCA, perimeter, and UCA. Based on the definition of each TI, we further conclude that the first factor represents watershed roughness or complexity, while the second factor describes the watershed size. Therefore, the selected 11 TIs were subsequently classified into two groups representing complexity (Group A) and area (Group B) (Fig. 2).

4.2 Relative contributions of TIs to flow variables

The nonparametric Kendall's tau test revealed significant correlations between the TIs and each flow variable from 1989 to 1996 (Tables S2 to S8). The number of significant TIs in each year increased from 1 to 11 with decreasing flow magnitudes. A larger number of the TIs were correlated to low flow variables ($Q_{75\,\%}$, $Q_{90\,\%}$, and Q_{min}), indicating that topography plays a more pronounced role in regulating lower flows. Here, flows lower than or equal to $Q_{75\,\%}$ are defined as the low flows. Thus, MLR models were only carried out for $Q_{75\,\%}$, $Q_{90\,\%}$, and Q_{min}.

The regression models between flow variables and the selected TIs were all significant ($P < 0.01$) with R^2 values ~ 0.5, indicating that selected TIs can be used to explain the variations of flow variables (Table S9). The details of MLR models are listed in the Supplement. The TIs that were included in each model are shown in Fig. 3 and Tables S10–S13. The relative contributions of each TI to $Q_{75\,\%}$, $Q_{90\,\%}$, and Q_{min} showed large variations between years (Fig. 3a to c, and Tables S10–S12). Figure 3 revealed that each TI influences flow variables differently (e.g. SA played a prominent role in $Q_{90\,\%}$ but did not significantly contribute to the variation of Q_{min}, while openness had the opposite role). This means that each TI cannot be used to explain variation in the same way for each flow variable. Based on the lumped CI values, the relative contribution of the perimeter TI was the highest in Group B as well as among all TIs (Fig. 3d). In Group A, the TIs above the average of the lumped CI were LS, SA, openness, and TCI receiving high contributions to low flows. Therefore, we conclude that the five above-mentioned TIs are significant topographic indices influencing low flow variables, which can be used to assess and compare low flows in any watershed studies where there are similarities in watershed size and climate.

5 Discussion

In this study, our results show that a limited number of TIs are significantly related to the Q_{mean}, $Q_{10\,\%}$, and $Q_{25\,\%}$, suggesting that topography plays a limited role in the variations of annual mean flow and high flows. This study area is characterized by snow-dominated hydrological regimes, with high flows (e.g. $Q_{25\,\%}$ or greater) coming predominantly from the snowmelt process in early March to late May (e.g. Fig. S6). Snowmelt processes are significantly related to elevation and climate in our study region (Winkler et al., 2005). As the study watersheds in the Southern Interior of British Columbia, Canada, have similar elevation ranges and climate variability, it is not surprising that only a limited number of TIs were significantly related to high or mean flows. In contrast, more TIs were significantly associated with low flows, suggesting that topography plays a more important role in low flows than high flows in the study region. Low flows often occur in the later summer (late August) and winter (October to February) (e.g. Fig. S6) and are mainly driven by groundwater discharges and small amounts of precipitation. A watershed with more complexities of topography would likely have a higher water retention ability due to longer flow paths and residence time and consequently promote more groundwater recharge and higher low flows (Price, 2011). Our Kendall's tau correlation tests uncover a positive relationship between the selected TIs and low flow variables (Tables S6–S8), indicating that the rougher or more complex that a watershed is, the higher the low flow yields are. Therefore, we conclude that topography plays a more important role in low flows than in high flows in the study region.

Five TIs including perimeter, LS, SA, openness, and TCI were identified as the major contributors to flow variables in this study. As far as we know, no studies have quantified topographic controls on various flow magnitudes. Nevertheless, the relationship between topography and the mean transit time (McGuire and McDonnell, 2006), temporal specific discharge (Karlsen et al., 2016), and hydrological connectivity (Jencso and McGlynn, 2011) have been investigated. There is no doubt that topography is one of the major contributors to hydrological variations (Price, 2011; Smakhtin, 2001). Although these studies pinpointed specific TIs and their interactions with hydrological responses, only a limited number of TIs were quantitatively assessed. In contrast, a total number of 22 TIs were calculated for 28 watersheds in this study. The much higher number of TIs that were initially included, along with the filtering methods applied, allowed us to select more suitable and significant TIs. Through this study design, we expect that the five selected TIs can effectively be used to support assessment or comparisons of low flows between watersheds in the study region. It should also be noted that we only selected the first five TIs that had substantially higher contributions than the other calculated TIs. The rest of the hydrological-related TIs showed only a minor ability to explain flow variations.

Among the five selected TIs, the perimeter, a primary TI, is commonly used in scientific studies to describe the characteristics of watershed topography. Our study further proves that it has a large influence on low flow variables. However, four secondary TIs (LS, SA, openness, and TCI) are mainly used in geomorphology to characterize ruggedness or roughness of landscapes and to identify topographic functioning of ecosystems. For examples, TCI has been used to map soil organic matter concentration (Zeng et al., 2016). Park et al. (2001) revealed that TCI is a better TI to predict soil depth than TWI, plan curvature, and profile curvature (see definitions in Table 1). LS is one of the key inputs to the Universal Soil Loss Equation being used to quantify soil erosion hazards (Desmet and Govers, 1996). SA was used to estimate animal species and habitat (Jenness, 2004) and map the spatial patterns of a floodplain (Scown et al., 2015). Openness was initially adopted to identify the boundary of different geological units and can be used to identify surface convexities and concavities, which is better than the commonly used profile and plan curvature (Yokoyama et al., 2002). In this study, the five selected TIs were initially filtered by the FA test, indicating that each selected TI has uniqueness in describing watershed topographic characteristics and outperformed the other tested TIs in describing variation in flow variables in our study region. Therefore, we expect that the five selected TIs can be applied to support hydrological analysis and modelling.

To our surprise, some commonly used TIs in hydrology, such as slope, median elevation, upslope contribution area, and wetland areas are not included in the FA list as the topographic information contained in those primary TIs also exists in some secondary TIs. Secondary TIs have the advantage of describing the hydrology-related landscapes in fuller detail. For example, slope is directly included in calculations of the TWI (secondary) and DDG (secondary). UCA (primary TI) is included in TWI and TCI (secondary). It is also worth mentioning that some secondary TIs played critical roles in determining the spatial heterogeneity of ecological and geomorphological processes, but their roles in hydrological processes in our study region were not demonstrated. These TIs were, therefore, not selected in this study. For example, the TWI has long been used as a key input variable for TOPMODEL and is an indicator of soil moisture (Beven, 1995). Our study identified that TWI and wetland area were not significantly related to flow variables, indicating that these factors played a limited role in the selected flow variables in our region. This may be because the wa-

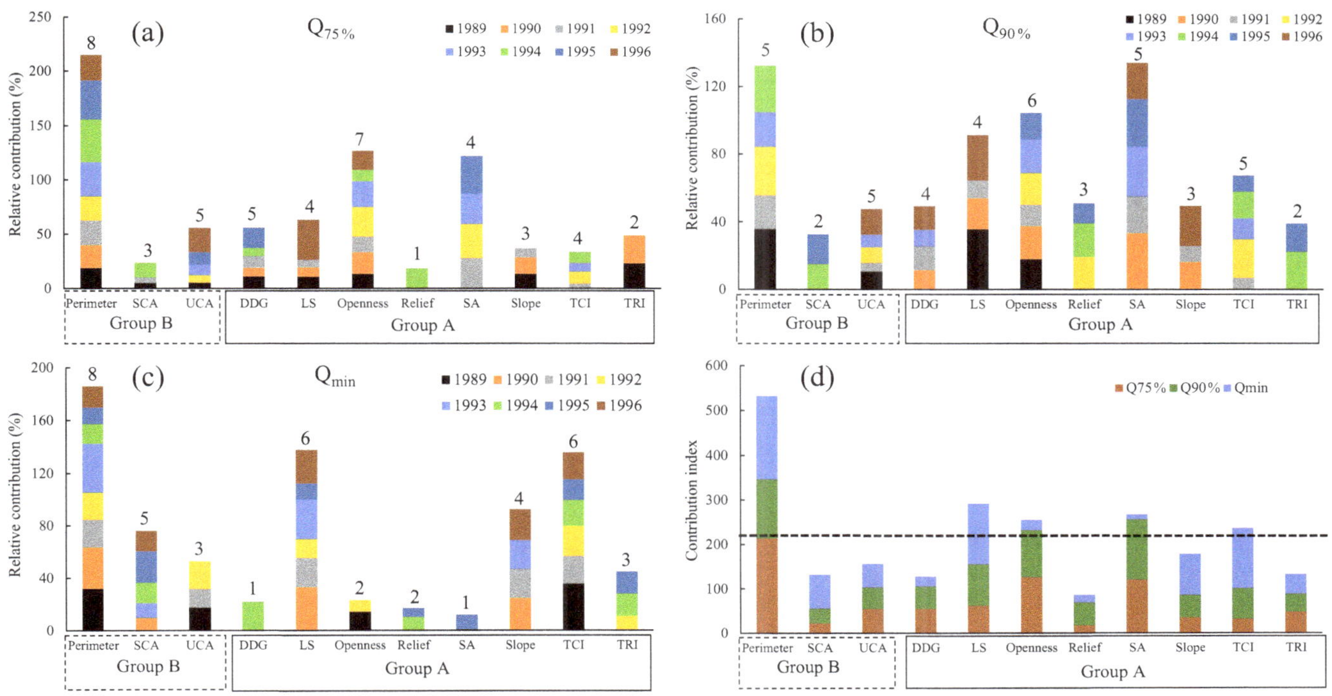

Figure 3. Relative contributions of each topographic index (TI) to $Q_{75\%}$ **(a)**, $Q_{90\%}$ **(b)**, and $Q_{\min}$ **(c)** from 1989 to 1996. Note that the numbers above the bars indicate the number of years when the given TIs were included in the regression models. **(d)** Contribution index (CI) of the 11 topographic indices (TIs) selected by factor analysis (FA) to $Q_{75\%}$, $Q_{90\%}$, and $Q_{\min}$ respectively. The dashed line is the average of the lumped CI.

tersheds in this area need to overcome a soil moisture storage threshold prior to releasing water (Karlsen et al., 2016). In summary, our selected five TIs significantly represent low flow characteristics of the watersheds in the Southern Interior of British Columbia, Canada, which is characterized by a snow-dominated hydrological regime.

In our study, FA was first introduced to exclude reductant TIs. The MLR models were further employed to exclude the insignificant TIs to various flow magnitudes. There is a concern whether the TIs excluded by FA were significantly related to flow magnitudes but not included in the MLR models. To address this concern, we re-ran a subset of analyses by conducting the MLR models for $Q_{\min}$ in 1989, 1990, and 1991 as examples (Sect. 6 in the Supplement). The results showed that the MLR models would include several redundant TIs (e.g. wetland coverage, roundness, stream length) if the FA test were not conducted (Table S14). The Kendall's tau correlation tests uncovered that they were not significantly correlated with low flows (Table S8). This further demonstrates that some redundant and nonsignificant TIs would be introduced if the FA test were not initially conducted. In addition, our research framework was developed based on a large number of watersheds with long-term hydrological data, all in a similar climatic region that is snow-dominated. As far as we know, there are no such studies conducted in either snow- or rain-dominated regions. Our research methodology or framework can be applied in rain-dominated regions to assess how topography controls flow in any regions where there are sufficient long-term monitoring data.

There are several uncertainties in our study. Firstly, hydrological responses are the resultant effects of climate, soil, vegetation, topography, and geology (Price, 2011; Smakhtin, 2001). In this study, the LAI, which represented the variations of vegetation cover in different watersheds, was included in our analysis in order to minimize the effects of different levels of vegetation coverage in the study watersheds. However, it was excluded by the FA test, confirming that the differences in vegetation cover and their effects were minor; therefore our selected watersheds are comparable in terms of forest coverage. In addition, the climate could be a confounding factor affecting our comparisons. In this study, annual flow variables were standardized by the annual precipitation to minimize the effects of climate on flow variables. In this way, the effects of climate variability were considered to some extent, but not in full detail. For examples, in extreme dry or wet years (e.g. 1994 and 1996 are the driest and wettest years in our study period, respectively) (Figs. S2–S5), hardly any of the TIs were significantly correlated to $Q_{90\%}$ (Table S7), but almost all were correlated to $Q_{\min}$ (Table S8). This is because that the flows at $Q_{90\%}$ in 1994 and 1996 in the selected watersheds occurred in the winter (October–

February) (e.g. Fig. S6), while the flows at $Q_{\min}$ in the majority of watersheds were in the late summer (August and September). The Kendall's tau correlation tests also confirm that topography played a minor role in $Q_{90\,\%}$ (Table S7), but a significant role in $Q_{\min}$ (Table S8). Thus, hydrological responses are mainly controlled by the combined effects of climate and topography. Secondly, LS and TCI are commonly used indicators of soil erodibility and were included in this study to capture the influence of soil conditions. For example, Park et al. (2001) showed that the areas with high values of TCI have high rates of soil erosion. Therefore, the variation of soil properties was considered to some degree by the inclusion of these TIs, but not in full detail. In fact, it is impossible to derive accurate soil properties over a large area. Thirdly, potential impacts of the DEM resolution on the calculation of TIs were not considered. However, Panagos et al. (2015) indicated that the resolution of 25 m is adequate for the calculation of slope factor at the European scale. Similarly, DEM resolutions were not found to significantly affect the calculation of DDG (Hjerdt et al., 2004). Thus, considering the same DEM data with the resolution of 25 m, we assume that the error caused by the DEM resolution would also be minor.

6 Conclusions

This study concludes that topography plays a significant role in low flows, while its role in high flows is limited. A total number of five topographic indices, including perimeter, LS (slope length factor), SA (surface area), openness, and TCI (topographic characteristic index), were identified with significant contributions to low flow variables. It is recommended that these five above-mentioned TIs can be used to assess the magnitude of low flows in a study region which is characterized by a snow-dominated hydrological regime with watershed sizes up to several thousand square kilometres. Our research methodology can be applied to other regions for investigating how topography controls flow magnitudes.

Competing interests. The authors declare that they have no conflict of interest.

Acknowledgements. We thank Environment Canada and GeoBC for their data on streamflow and DEM. The authors would also like to thank the editor (Thom Bogaard) and two reviewers (Petra Hulsman and Niclas Hjerdt) for their constructive comments on the earlier version of the paper. The research funding for supporting this project was provided by the Natural Sciences and Engineering Research Council of Canada (RGPIN-2015-06032), the National Natural Science Foundation of China (numbers 41771415 and 31770759), and the Priority Academic Program Development of Jiangsu Higher Education Institutions (number: 164320H116).

Edited by: Thom Bogaard

References

Ambroise, B., Beven, K., and Freer, J.: Toward a generalization of the TOPMODEL concepts: Topographic indices of hydrological similarity, Water Resour. Res., 32, 2135–2145, 1996.

Beven, K.: Linking Parameters across Scales – Subgrid Parameterizations and Scale-Dependent Hydrological Models, Hydrol Process, 9, 507–525, https://doi.org/10.1002/hyp.3360090504, 1995.

Beven, K. and Kirkby, M. J.: A physically based, variable contributing area model of basin hydrology/Un modèle à base physique de zone d'appel variable de l'hydrologie du bassin versant, Hydrolog. Sci. J., 24, 43–69, 1979.

Burrough, P. A., McDonnell, R., McDonnell, R. A., and Lloyd, C. D.: Principles of geographical information systems, Oxford University Press, 2015.

Cheng, L., Yaeger, M., Viglione, A., Coopersmith, E., Ye, S., and Sivapalan, M.: Exploring the physical controls of regional patterns of flow duration curves – Part 1: Insights from statistical analyses, Hydrol. Earth Syst. Sci., 16, 4435–4446, https://doi.org/10.5194/hess-16-4435-2012, 2012.

Daly, C., Halbleib, M., Smith, J. I., Gibson, W. P., Doggett, M. K., Taylor, G. H., Curtis, J., and Pasteris, P. P.: Physiographically sensitive mapping of climatological temperature and precipitation across the conterminous United States, Int. J. Climatol., 28, 2031–2064, 10.1002/joc.1688, 2008.

Desmet, P. J. J. and Govers, G.: A GIS procedure for automatically calculating the USLE LS factor on topographically complex landscape units, J. Soil Water Conserv., 51, 427–433, 1996.

Erickson, T. A., Williams, M. W., and Winstral, A.: Persistence of topographic controls on the spatial distribution of snow in rugged mountain terrain, Colorado, United States, Water Resour. Res., 41, W04014, https://doi.org/10.1029/2003wr002973, 2005.

Erskine, R. H., Green, T. R., Ramirez, J. A., and MacDonald, L. H.: Comparison of grid-based algorithms for computing upslope contributing area, Water Resour. Res., 42, W09416, https://doi.org/10.1029/2005WR004648, 2006.

Fleming, S. W., Whitfield, P. H., Moore, R. D., and Quilty, E. J.: Regime-dependent streamflow sensitivities to Pacific climate modes cross the Georgia-Puget transboundary ecoregion, Hydrol. Process., 21, 3264–3287, https://doi.org/10.1002/hyp.6544, 2007.

Gómez-Plaza, A., Martınez-Mena, M., Albaladejo, J., and Castillo, V.: Factors regulating spatial distribution of soil water content in small semiarid catchments, J. Hydrol., 253, 211–226, 2001.

Greenlee, D. D.: Raster and vector processing for scanned linework, Photogramm. Eng. Rem. S., 53, 1383–1387, 1987.

Gromping, U.: Relative importance for linear regression in R: The package relaimpo, J. Stat. Softw., 17, 27, https://doi.org/10.18637/jss.v017.i01, 2006.

Hewlett, J. D. and Hibbert, A. R.: Factors affecting the response of small watersheds to precipitation in humid areas, Forest Hydrology, 1, 275–290, 1967.

Hjerdt, K. N., McDonnell, J. J., Seibert, J., and Rodhe, A.: A new topographic index to quantify downslope controls on local drainage, Water Resour. Res., 40, W0560, https://doi.org/10.1029/2004wr003130, 2004.

Jencso, K. G. and McGlynn, B. L.: Hierarchical controls on runoff generation: Topographically driven hydrologic connectivity, geology, and vegetation, Water Resour. Res., 47, W11527 https://doi.org/10.1029/2011wr010666, 2011.

Jenness, J.: Topographic Position Index (tpi_jen. avx) extension for ArcView 3. x, v. 1.3 a. Jenness Enterprises, available at: http://www.jennessent.com/arcview/tpi.htm (last access: 16 March 2018), 2006.

Jenness, J. S.: Calculating landscape surface area from digital elevation models, Wildlife Soc. B., 32, 829–839, https://doi.org/10.2193/0091-7648(2004)032[0829:Clsafd]2.0.Co;2, 2004.

Karlsen, R. H., Grabs, T., Bishop, K., Buffam, I., Laudon, H., and Seibert, J.: Landscape controls on spatiotemporal discharge variability in a boreal catchment, Water Resour. Res., 52, 6541–6556, https://doi.org/10.1002/2016WR019186, 2016.

Li, Q., Qi, J. Y., Xing, Z. S., Li, S., Jiang, Y. F., Danielescu, S., Zhu, H. Y., Wei, X. H., and Meng, F. R.: An approach for assessing impact of land use and biophysical conditions across landscape on recharge rate and nitrogen loading of groundwater, Agr. Ecosyst. Environ., 196, 114–124, https://doi.org/10.1016/j.agee.2014.06.028, 2014.

Li, Q., Wei, X., Zhang, M., Liu, W., Fan, H., Zhou, G., Giles-Hansen, K., Liu, S., and Wang, Y.: Forest cover change and water yield in large forested watersheds: A global synthetic assessment, Ecohydrology, 10, e1838, https://doi.org/10.1002/eco.1838, 2017.

Liu, W., Wei, X., Li, Q., Fan, H., Duan, H., Wu, J., Giles-Hansen, K., and Zhang, H.: Hydrological recovery in two large forested watersheds of southeastern China: the importance of watershed properties in determining hydrological responses to reforestation, Hydrol. Earth Syst. Sci., 20, 4747–4756, https://doi.org/10.5194/hess-20-4747-2016, 2016.

Lyon, S. W., Nathanson, M., Spans, A., Grabs, T., Laudon, H., Temnerud, J., Bishop, K. H., and Seibert, J.: Specific discharge variability in a boreal landscape, Water Resour. Res., 48, W08506, https://doi.org/10.1029/2011wr011073, 2012.

McGuire, K. J. and McDonnell, J. J.: A review and evaluation of catchment transit time modeling, J. Hydrol., 330, 543–563, 10.1016/j.jhydrol.2006.04.020, 2006.

McGuire, K. J., McDonnell, J. J., Weiler, M., Kendall, C., McGlynn, B. L., Welker, J. M., and Seibert, J.: The role of topography on catchment-scale water residence time, Water Resour. Res., 41, W05002, https://doi.org/10.1029/2004wr003657, 2005.

Moore, I., Burch, G., and Mackenzie, D.: Topographic effects on the distribution of surface soil water and the location of ephemeral gullies, Trans. ASAE, 31, 1098–1107, 1988.

Moore, I. D., Grayson, R. B., and Ladson, A. R.: Digital Terrain Modeling – a Review of Hydrological, Geomorphological, and Biological Applications, Hydrol. Process., 5, 3–30, https://doi.org/10.1002/hyp.3360050103, 1991.

Panagos, P., Borrelli, P., and Meusburger, K.: A new European slope length and steepness factor (LS-Factor) for modeling soil erosion by water, Geosciences, 5, 117–126, 2015.

Park, S. J. and van de Giesen, N.: Soil-landscape delineation to define spatial sampling domains for hillslope hydrology, J. Hydrol., 295, 28–46, https://doi.org/10.1016/j.jhydrol.2004.02.022, 2004.

Park, S. J., McSweeney, K., and Lowery, B.: Identification of the spatial distribution of soils using a process-based terrain characterization, Geoderma, 103, 249–272, https://doi.org/10.1016/S0016-7061(01)00042-8, 2001.

Price, K.: Effects of watershed topography, soils, land use, and climate on baseflow hydrology in humid regions: A review, Prog. Phys. Geog., 35, 465–492, 2011.

Quinn, P., Beven, K., Chevallier, P., and Planchon, O.: The Prediction of Hillslope Flow Paths for Distributed Hydrological Modeling Using Digital Terrain Models, Hydrol. Process., 5, 59–79, https://doi.org/10.1002/hyp.3360050106, 1991.

Quinn, P. F., Beven, K. J., and Lamb, R.: The Ln(a/Tan-Beta) Index – How to Calculate It and How to Use It within the Topmodel Framework, Hydrol. Process., 9, 161–182, https://doi.org/10.1002/hyp.3360090204, 1995.

Riley, S. J.: Index That Quantifies Topographic Heterogeneity, Intermountain Journal of Sciences, 5, 23–27, 1999.

Scown, M. W., Thoms, M. C., and De Jager, N. R.: Measuring floodplain spatial patterns using continuous surface metrics at multiple scales, Geomorphology, 245, 87–101, 10.1016/j.geomorph.2015.05.026, 2015.

Shope, C. L.: Disentangling event-scale hydrologic flow partitioning in mountains of the Korean Peninsula under extreme precipitation, J. Hydrol., 538, 399–415, https://doi.org/10.1016/j.jhydrol.2016.04.050, 2016.

Smakhtin, V. U.: Low flow hydrology: a review, J. Hydrol., 240, 147–186, https://doi.org/10.1016/S0022-1694(00)00340-1, 2001.

Wang, T., Hamann, A., Spittlehouse, D. L., and Aitken, S. N.: Development of scale-free climate data for western Canada for use in resource management, Int. J. Climatol., 26, 383–397, https://doi.org/10.1002/Joc.1247, 2006.

Wei, X., Li, Q., Zhang, M., Giles-Hansen, K., Liu, W., Fan, H., Wang, Y., Zhou, G., Piao, S., and Liu, S.: Vegetation cover – another dominant factor in determining global water resources in forested regions, Glob. Change Biol., 24, 786–795, https://doi.org/10.1111/gcb.13983, 2018.

Wei, X. H. and Zhang, M. F.: Quantifying streamflow change caused by forest disturbance at a large spatial scale: A single watershed study, Water Resour. Res., 46, W12525, https://doi.org/10.1029/2010wr009250, 2010.

Winkler, R. D., Spittlehouse, D. L., and Golding, D. L.: Measured differences in snow accumulation and melt among clearcut, juvenile, and mature forests in southern British Columbia, Hydrol. Process., 19, 51–62, https://doi.org/10.1002/hyp.5757, 2005.

Yokoyama, R., Shirasawa, M., and Pike, R. J.: Visualizing topography by openness: a new application of image processing to digital elevation models, Photogramm. Eng. Rem. S., 68, 257–266, 2002.

Zeng, C. Y., Yang, L., Zhu, A. X., Rossiter, D. G., Liu, J., Liu, J. Z., Qin, C. Z., and Wang, D. S.: Mapping soil organic matter concentration at different scales using a mixed geographically weighted regression method, Geoderma, 281, 69–82, https://doi.org/10.1016/j.geoderma.2016.06.033, 2016.

Zhang, M. and Wei, X.: Contrasted hydrological responses to forest harvesting in two large neighbouring watersheds in snow hydrology dominant environment: implications for forest management and future forest hydrology studies, Hydrol. Process., 28, 6183–6195, https://doi.org/10.1002/hyp.10107, 2014.

Zhang, M., Wei, X., and Li, Q.: A quantitative assessment on the response of flow regimes to cumulative forest disturbances in large snow-dominated watersheds in the interior of British Columbia, Canada, Ecohydrology, 9, 843–859, https://doi.org/10.1002/eco.1687, 2016.

Zhang, M. F., Liu, N., Harper, R., Li, Q., Liu, K., Wei, X. H., Ning, D. Y., Hou, Y. P., and Liu, S. R.: A global review on hydrological responses to forest change across multiple spatial scales: Importance of scale, climate, forest type and hydrological regime, J. Hydrol., 546, 44–59, https://doi.org/10.1016/j.jhydrol.2016.12.040, 2017.

6

Rapid surface-water volume estimations in beaver ponds

Daniel J. Karran[1], Cherie J. Westbrook[1], Joseph M. Wheaton[2], Carol A. Johnston[3], and Angela Bedard-Haughn[4]

[1]Department of Geography and Planning, Centre for Hydrology, University of Saskatchewan, Saskatoon, SK S7N 5C8, Canada
[2]Department of Watershed Sciences, Utah State University, Logan, UT 84322-5210, USA
[3]Department of Natural Resource Management, South Dakota State University, Brookings, SD 57007, USA
[4]Department of Soil Science, University of Saskatchewan, Saskatoon, SK S7N 5A8, Canada

Correspondence to: Daniel J. Karran (daniel.karran@usask.ca)

Abstract. Beaver ponds are surface-water features that are transient through space and time. Such qualities complicate the inclusion of beaver ponds in local and regional water balances, and in hydrological models, as reliable estimates of surface-water storage are difficult to acquire without time- and labour-intensive topographic surveys. A simpler approach to overcome this challenge is needed, given the abundance of the beaver ponds in North America, Eurasia, and southern South America. We investigated whether simple morphometric characteristics derived from readily available aerial imagery or quickly measured field attributes of beaver ponds can be used to approximate surface-water storage among the range of environmental settings in which beaver ponds are found. Studied were a total of 40 beaver ponds from four different sites in North and South America. The simplified volume–area–depth (V–A–h) approach, originally developed for prairie potholes, was tested. With only two measurements of pond depth and corresponding surface area, this method estimated surface-water storage in beaver ponds within 5 % on average. Beaver pond morphometry was characterized by a median basin coefficient of 0.91, and dam length and pond surface area were strongly correlated with beaver pond storage capacity, regardless of geographic setting. These attributes provide a means for coarsely estimating surface-water storage capacity in beaver ponds. Overall, this research demonstrates that reliable estimates of surface-water storage in beaver ponds only requires simple measurements derived from aerial imagery and/or brief visits to the field. Future research efforts should be directed at incorporating these simple methods into both broader beaver-related tools and catchment-scale hydrological models.

1 Introduction

The volume of water stored at the surface of wetlands, ponds, and lakes (as a function of stage) is of great concern to those responsible for assessing risks and balancing water supplies to societal demands. Arriving at reliable estimates of such storage is difficult without some knowledge of the feature's morphometry, i.e. information that is often time consuming and impractical to acquire, especially when the features are numerous and transient through space and time (Milly et al., 2008). This is particularly true for beaver ponds owing to their cyclic creation and abandonment.

Beaver dams and their associated ponds are ubiquitous in streams and wetlands in the Northern Hemisphere and southern South America (Whitfield et al., 2015). Beaver dam densities have been reported to exceed 40 dams per kilometre (Macfarlane et al., 2017), making them one of the most frequent obstructions to flowing water (Naiman et al., 1986; Pollock et al., 2003). Beaver dams increase the open-water area within watersheds (Hood and Bayley, 2008) and ponds bring numerous ecosystem benefits (Johnston, 2012), but beaver ponds can also be viewed as burdensome or even dangerous from an anthropomorphic perspective (Butler and Malanson, 2005; Green and Westbrook, 2009). Such concerns, whether positive or negative, generally centre around the pond's capacity to store water and sediment, highlighting the need for quick and accurate surface-water storage estimation methods.

Numerous hydrological investigations have sought to estimate surface-water storage in other types of wetlands (Trigg et al., 2014; Xu and Singh, 2004). For hydrological modellers, an ideal approach is one that overcomes the need for often time-intensive topographic surveys and that is more practical for use in models at varying scales and locations. Previous studies have set about this by defining statistical relationships between surface area and volume for wetlands of specific physiographic regions (Gleason et al., 2007; Hubbard, 1982; Lane and D'Amico, 2010; Wiens, 2001). Such approaches have been found useful for modelling entire watersheds (Gleason et al., 2007), but limited for estimating storage in individual wetlands because depth and basin morphometry (i.e. surface area, volume, depth) are not considered (Huang et al., 2011; Lane and D'Amico, 2010; Wiens, 2001). Brooks and Hayashi (2002) presented an equation that includes depth and basin morphometry, but to use it, basin morphometry must be predefined and no such information yet exists for beaver ponds.

Another approach, the simplified volume–area–depth (V–A–h) method (Hayashi and van der Kamp, 2000), accounts for depth and calculates basin morphometry for each individual wetland. Requiring only two measurements of depth and surface area, it has been shown to provide reliable estimates of surface-water storage in the pothole wetlands of the North American prairies for which it was designed (Minke et al., 2010). Prairie potholes are depressional wetlands that have fairly regular shapes, i.e. concave profiles with smooth slopes. Beaver ponds, by contrast, typically encompass a bathymetry that is far more complex because their size and shape is controlled by the dimensions of the dam and the land surface that becomes flooded upon dam establishment (Johnston and Naiman, 1987). Whether statistical or analytical approaches can reliably estimate water storage in beaver ponds has yet to be determined. Our goal was thus to explore tools useful for estimating surface-water storage in beaver ponds. We studied beaver ponds across much of their habitat range and (i) evaluated the utility of the simplified V–A–h method in estimating surface-water storage, (ii) evaluated correlations between surface-water storage and beaver pond morphometry, and (iii) described beaver pond morphometry in relation to surface-water storage capacity.

2 Methods

2.1 The simplified V–A–h method

The simplified V–A–h method is based on a simple power equation (Hayashi and van der Kamp, 2000), where the area of a pond (A), at a given height above the pond bottom (h), is described as

$$A = s\left(\frac{h}{h_0}\right)^{2/p}, \quad (1)$$

where h_0 is the unit height of the water surface (e.g. 1 m for SI units), s is a scaling coefficient that represents the area of a circle (m^2) with a radius that corresponds to h_0, and p is a dimensionless morphometry coefficient that represents the shape of the bathymetric curve (i.e. the area–depth relationship of the pond). The volume of the pond is then determined by integrating all the area profiles below h to give

$$V(h) = \int_0^h s\left(\frac{h^*}{h_0}\right)^{2/p} \mathrm{d}h^* = \left(\frac{s}{1+2/p}\right)\left(\frac{h^{1+2/p}}{h_0^{2p}}\right). \quad (2)$$

Using Eqs. (1) and (2) requires parameterizing the s and p coefficients. The simplified V–A–h method arrives at these values by rearranging Eq. (1) to give (Minke et al., 2010)

$$s = A_1\left(\frac{h_1}{h_2}\right)^{-2/p}, \quad (3)$$

and

$$p = 2\left(\frac{\log(h_1/h_2)}{\log(A_1/A_2)}\right), \quad (4)$$

where A_1 and A_2 are surface areas of the pond at corresponding depths of h_1 and h_2, respectively, and $h_1 < h_2$. With only two measurements of area and depth in time, Eqs. (3) and (4) can be used to calculate s and p coefficients that are then reinserted into Eqs. (1) and (2) to define the entire area–depth and volume–depth relationship of the pond.

2.2 Beaver pond morphometry

2.2.1 Metrics for surface-water volume estimations

A beaver pond's capacity to store surface water is defined simply by its bathymetry, and can be directly calculated if an accurate topographic survey is available. The problem here relates to how well we can approximate that volume given some simple measures of the dam and pond dimensions. To discover if metrics exist, a series of morphometric variables were generated in addition to the p coefficient described in Eq. (1). They include the maximum dam height (h_{max}) defined as the difference in elevation (m) between the dam crest and the lowest point in the pond, the maximum surface area (m^2) of the pond (A_{max}) at h_{max}, and the length (m) of the dam (D_{len}) measured along its crest. Regression analysis was then used to determine if any of the variables are correlated to the maximum volume of the pond (V_{max}).

2.2.2 Morphometric analysis

Understanding the underlying mechanics of the simplified V–A–h method and how morphometry relates to a pond's capacity to store water requires a deeper analysis of the bathymetric curve. The bathymetric curve is equivalent to the hypsometric curve defined by Strahler (1952) as the ground surface area of a land mass with respect to elevation. To compare curves for ponds of different size and relief, it is necessary

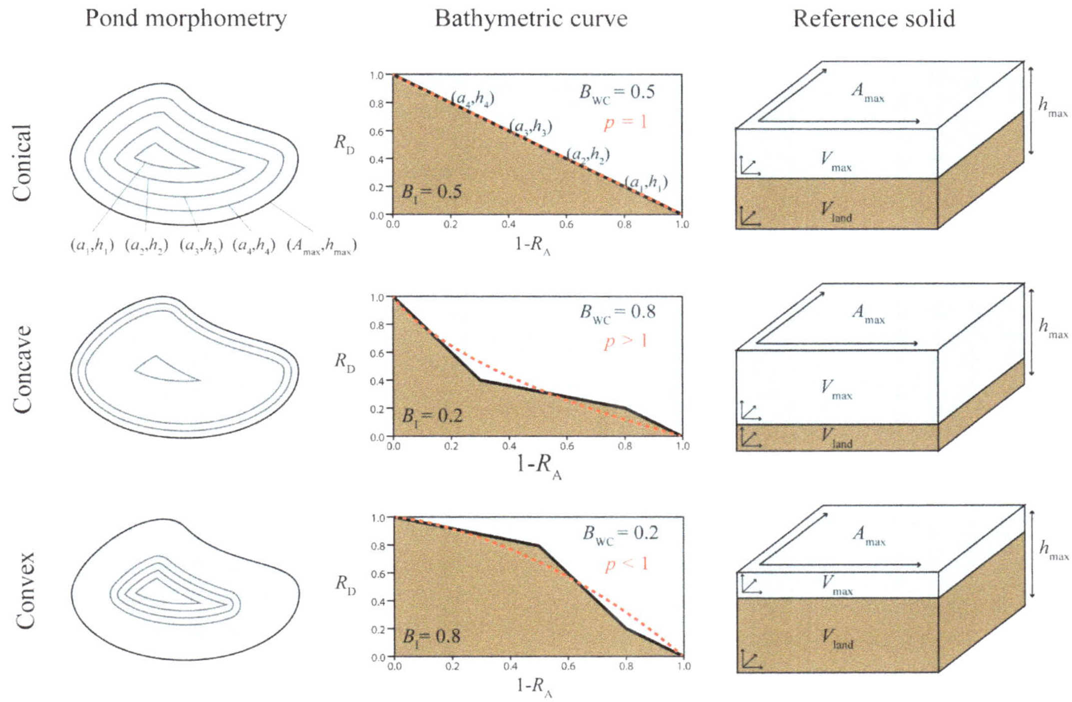

Figure 1. Perceptual diagram of the relationship between morphometric variables. The area (a) at a given stage of the pond (h) is a point on the bathymetric curve (thick black line), where R_A is the relative area and R_D is the relative depth. The bathymetric integral (B_I) is the integration of everything below the bathymetric curve and the pond's capacity to store water (B_WC) is the integration of everything above the bathymetric curve. The morphometry (p) coefficient represents the shape of the bathymetric curve in the power function equation (red-dashed line; Eq. 7). The reference solid is the box created by multiplying the maximum height of the dam (h_max) by the maximum surface area created by the pond (A_max), and is entirely comprised of land (V_land) and/or water (V_max) proportional to B_I and B_WC.

to express the variables as relative depth (R_D) and relative area (R_A) as

$$R_\mathrm{D} = \frac{h}{h_\mathrm{max}}, \tag{5}$$

and

$$R_\mathrm{A} = \frac{a}{A_\mathrm{max}}, \tag{6}$$

where h is the stage (m) elevation of the pond and a is the corresponding surface area (m^2) at any given h. For ease of visual interpretation, we express the bathymetric curve as R_D vs. $1 - R_\mathrm{A}$ (Fig. 1). Power functions described by Eq. (1) can then be fit to a bathymetric curve with the following equation:

$$R_\mathrm{D} = (1 - R_\mathrm{A})^{p/2}, \tag{7}$$

where the p coefficient here is equal to the p coefficient in Eq. (1). This allows for a visual aid in the analysis of error by superimposing estimated curves produced via either Eq. (1) or Eq. (4) to the pond's actual bathymetric curve. It also eliminates issues of scale between different ponds so that bathymetric curves can be visually compared to one another.

From the relative bathymetric curve, it is possible to compute the bathymetric integral (B_I), a modified form of the hypsometric integral defined as the measure of land mass volume with respect the entire reference solid created by the maximum dimensions of the pond (Fig. 1; Strahler, 1952):

$$B_\mathrm{I} = \frac{V_\mathrm{land}}{h_\mathrm{max} A_\mathrm{max}} = \int_0^1 R_\mathrm{A} \mathrm{d} R_\mathrm{D}. \tag{8}$$

Equation (10) produces values between 0 and 1, with 1 representing a reference solid entirely composed of land mass. Using the B_I, we introduce a new metric that represents the pond's bathymetric capacity to store water (B_WC). Since the total volume of the reference solid is comprised of either land or water, the B_WC, relative to the reference solid, is expressed as

$$B_\mathrm{WC} = 1 - B_\mathrm{I} = \frac{V_\mathrm{max}}{h_\mathrm{max} A_\mathrm{max}}. \tag{9}$$

The B_I and B_WC are quantitative measurements of the pond's morphometry and capacity to store water, respectively. The value in using these metrics is that they facilitate the comparison of surface-water storage capacity among beaver ponds and other wetland types.

Finally, we described the shape of the beaver pond surface using a dimensionless shape index (S_I), which is essentially the ratio of the pond perimeter to the circumference of a circle with the same area (Hutchinson, 1957):

$$S_I = \frac{P}{2\sqrt{\pi A_{max}}}, \tag{10}$$

where P is the perimeter of the pond (m). Ponds with $S_I = 1$ have shapes that are perfectly circular, whereas ponds with $S_I > 1$ are increasingly complex. Pond shape is an important metric as much of the interaction between surface water and groundwater happens at the shoreline (Shaw and Prepas, 1990). We chose S_I as it is easy to interpret and enables a relative comparison between the shapes of beaver ponds and other types of wetlands (Minke et al., 2010).

2.3 V–A–h models for surface-water storage estimation in beaver ponds

Three versions of the power function model described by Eq. (1) were tested in this study. They are referred to as the full, simplified, and optimized models. The simplified model is the actual test of the simplified V–A–h method and the other two models were included to aid in the analysis of this approach.

The full model is a power function fitted to the complete data set of each pond's bathymetry (i.e. empirical fit). We arrive at values for s and p by fitting a simple power function, $y = ax^b$, to the pond's bathymetric curve, and assume $a = s$ and $b = 2/p$ in accordance with Eq. (1). Non-linear least-squares regression was used to determine the best fit; the ability of this model to make accurate area and volume estimates depends on its "goodness of fit" to the data set. Analysis of the full model was included to (i) identify the p coefficient that best describes each beaver pond's morphometry and (ii) assess the overall suitability of power functions to describe beaver pond bathymetry.

The simplified model is a power function using s and p coefficients created from the same two relative measurements of depth (i.e. h_1 and h_2 as a percentage of h_{max}) in each pond. Minke et al. (2010) evaluated the simplified V–A–h method by applying it to two scenarios: a dry one where h_1 and h_2 are taken at 0.1 m and 25 % of h_{max}, and a wet one where h_1 and h_2 are taken at 50 and 75 % of h_{max}. They found that estimation errors were lowest using the wet scenario; therefore, we chose this scenario to simulate the application of the simplified V–A–h method as it may be practically used in the field.

The optimized model differs from the simplified model through parameterizing coefficients via the optimum combination of h_1 and h_2 for each pond. This required calculating s and p coefficients at every possible combination of h_1 and h_2 along the bathymetric curve (note that h_1 and h_2 are expressed as a percentage of h_{max} from 1 to 100; therefore, the total number of combinations where $h_1 < h_2$ is 5000 for each pond). Each set of s and p coefficients was then reinserted into Eqs. (1) and (2) to estimate area and volume, respectively, and the set that produced the least combined area and volume error was selected as the optimum. The optimum model was included in this study to discover how best to use the simplified V–A–h method with regards to differences in pond morphometry.

Error for all three models was evaluated using root mean square error (E_{RMS}), defined as

$$E_{RMS} = \sqrt{\frac{1}{m}\sum_{i=1}^{m}(D_{ACT} - D_{EST})^2}, \tag{11}$$

where m is the number of data points, D_{ACT} is the point on the actual bathymetric curve calculated from the pond itself, and D_{EST} is the point on the estimated bathymetric curve derived from the s and p coefficients at a given combination of h_1 and h_2. Finally, to allow for coherent comparisons of error among the different beaver ponds, the magnitude of error, referred to as A_{ERR} (%) for area and V_{ERR} (%) for volume, was calculated by dividing the E_{RMS} by the actual area and volume of the pond at 80 % of h_{max}. This particular depth was chosen to avoid inconsistencies in error magnitudes that arise when the evaluation depth is set too close to the minimum and maximum (Minke et al., 2010).

2.4 Test sites

Forty beaver ponds were selected for this study and simulated in digital elevation models (DEMs). Our sample design captured the range of structures built by beaver along streams with mineral and organic substrates in both mountainous and lowland terrain. Beaver ponds were thus analyzed from multiple locations where bathymetric data existed, which included Kananaskis Provincial Park, Alberta, Canada; Escondido, Tierra del Fuego, Argentina; the Logan River watershed, Utah, USA; and Voyageurs National Park, Minnesota, USA. Details of the location, terrain, number of ponds, survey methods, and survey resolution for each site are provided in Table 1.

2.5 DEM creation and manipulation for variable calculations

Sites selected for this study were former beaver ponds that had drained sufficiently to either reveal pond bottom bathymetry or allow it to be surveyed. Beaver ponds extracted from lidar, when available, were fully drained with visible relic dams, whereas some ponds surveyed by total station and real-time kinetic geographical positioning system (rtkGPS) often were still full with water up to their crest elevations, but not enough to impede point collection by wading. DEMs that relied on total station and rtkGPS surveys were created with Surfer®v10 (Golden Software, Colorado) using ordinary kriging. The beaver ponds were then isolated

Table 1. Site locations, characteristics, and details of topographic pond surveys. "*n*" is the number of ponds studied at each site.

Site	Latitude and longitude (degree, ′)	*n*	Soil substrate type	Terrain	Survey method	DEM resolution (m)
Kananaskis Provincial Park, AB, Canada	51°3.553′ N, 114°52.009′ W	10	Organic	Mountainous	rtkGPS	1
Escondido, Tierra del Fuego, Argentina	54°36.908′ S, 67°44.540′ W	3	Organic	Mountainous	rtkGPS	1
Logan River Watershed, UT, USA	41°50.327′ N, 111°33.668′ W	14	Mineral	Mountainous	Total station	0.1
	41°49.568′ N, 111°34.516′ W	2	Mineral	Mountainous	Total station	0.1
	41°48.868′ N, 111°35.553′ W	5	Mineral	Mountainous	Total station	0.1
Voyageurs National Park, MN, USA	48°32.773′ N 93°4.328′ W	1	Organic	Lowland	Lidar	1
	48°27.975′ N 92°53.864′ W	3	Mineral	Lowland	Lidar	1
	48°30.405′ N 92°40.331′ W	1	Mineral	Lowland	Lidar	1
	48°31.262′ N 92°52.794′ W	1	Mineral	Lowland	Lidar	1

from the unneeded areas of the DEM by extracting all the points in the raster below and upstream of the dam crest (i.e. h_{max}). This was done in ArcGIS v10.2 (ESRI, 2015) as was the calculation of the morphometric variables. The V–h relationship, as well as bathymetric curve of each pond, was calculated at 5 cm increments using a script written in Python™ that utilizes the "volume" feature of ArcGIS Toolbox. The V–h relationship and bathymetric curve of each pond were the primary inputs for the three models, which were built and run in RStudio (RStudio Team, 2015). Finally, the bathymetric curve for each pond was established using linear interpolation to create 100 points, i.e. 1–100 % of h_{max}.

3 Results

3.1 Beaver pond morphometry

Pond morphometric characteristics are provided in Table 2 and examples of the DEMs from each location are provided in Fig. 2. The 40 ponds well represented the various types of beaver ponds that are created in riverine and wetland habitats (Baker and Hill, 2003), with maximum dam heights (h_{max}) ranging from 0.25 to 2 m and dam lengths (D_{len}) spanning 3–308 m, with medians of 0.83 and 40 m, respectively. Pond volumes (V_{max}) ranged between 1 and 9001 m^3 and showed strong power correlations to D_{len}, h_{max}, and A_{max} (Fig. 3). Among the ponds, there was considerable variability in shape as S_I values ranged from 1.5 to 5.3 (mean = 2.6). No strong correlations (i.e. $-0.10 > R^2 < 0.10$) were found between S_I and the other morphometric variables used in this study (i.e. p, B_I, B_{WC}, D_{len}, h_{max}).

The p coefficients for the beaver ponds followed a log-normal distribution, and ranged from 0.45 to 2.58 (median of 0.91) (Fig. 4). Of the 40 beaver ponds analyzed, 70 % (28) had p coefficients that were < 1, indicating that beaver ponds tend to have convex bathymetries. Most beaver ponds tended to be more convex than they are concave, given the shape of the bathymetric curves (Fig. 5) and the range of B_I (0.45–0.85; median of 0.69). In all but one case, V_{land} was greater than 50 % of the total volume of space, indicating that most beaver ponds are shallow, which limits the volume of surface water they can store. This phenomenon is well described by the strong exponential relationship between the p coefficient ($R^2 = 0.96$) and B_I and B_{WC} (Fig. 6). Soil substrate type (Table 1; organic vs. mineral) did not affect the value of the p coefficient, as evidenced by a t test ($P = 0.97$).

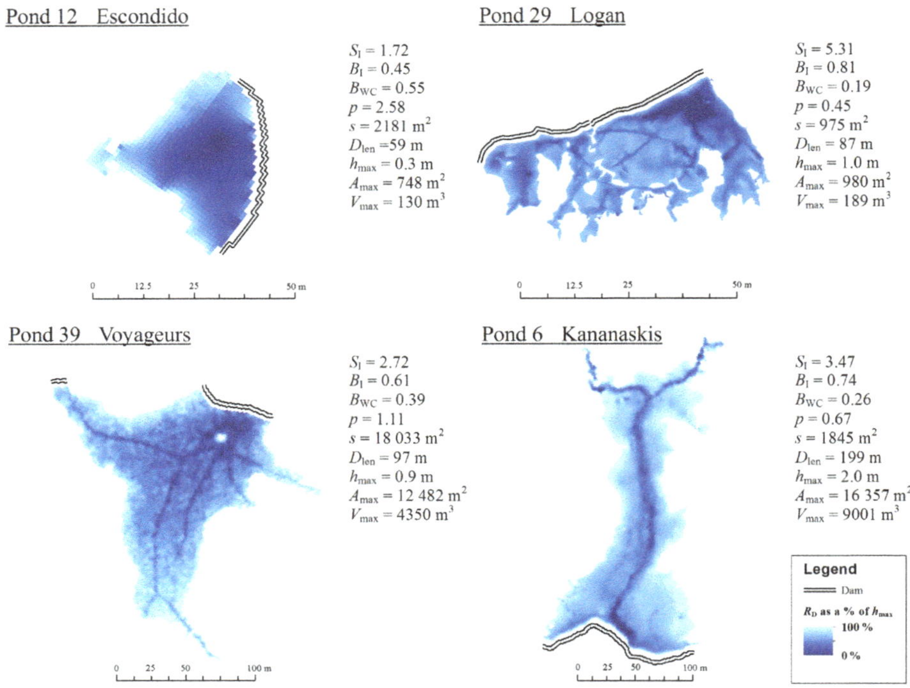

Figure 2. Four examples of detrended beaver pond DEMs used for this study, one from each study area (S_I = shape index, B_I = bathymetric integral, B_{WC} = bathymetric water capacity, p = morphometry coefficient (full model), s = scaling coefficient, D_{len} = dam length, h_{max} = maximum height of the dam, A_{max} = maximum surface area of the pond, and V_{max} = maximum volume of the pond).

3.2 surface-water storage estimations

The full model had the least A_{ERR}, and the optimized model had the least V_{ERR} (Fig. 7; Table 3). The highest A_{ERR} and V_{ERR} was associated with simplified model estimates, which also produced the greatest variability of error among the different ponds. With regards to study locations, full V_{ERR} ranked as Escondido < Voyageurs < Logan < Kananaskis, whereas full A_{ERR} ranked Logan < Escondido < Kananaskis < Voyageurs. Overall, the beaver ponds in Kananaskis proved most difficult to model (i.e. highest V_{ERR} and A_{ERR} overall); however, mean error for the full model remained below 5 % for both area and volume estimates.

Compared to the full model (Fig. 7), the simplified model had higher V_{ERR} in 65 % of cases (26 ponds) and higher A_{ERR} in 98 % of cases (39 ponds), whereas the optimized model had lower V_{ERR} in 100 % of cases but slightly (< 1 %) higher A_{ERR} in 100 % of cases. The optimum p coefficients for volume tended to be slightly different than the optimum p coefficients for area, which are the coefficients derived from the empirical fit of the Full model. The optimum model proved useful for revealing the two points on the bathymetric curve that can be used to obtain the optimum p coefficient for volume estimates. Pond 7 had the largest A_{ERR} and V_{ERR} (Fig. 7), and therefore was selected for more detailed study (Fig. 8). The optimum points were found at the approximate location of where the empirical fit intersects with the bathymetric curve. Thus, using the optimum points in Eq. (4) computes a p coefficient that is closest to the same coefficient generated by the curve fitted by non-linear least-squares regression. The points used by the simplified model for Pond 7 fall on segments of the bathymetric curve that are farther away in distance from the empirical fit; hence, the p coefficient generated by these points creates a curve that is farther away from the bathymetric curve, which ultimately leads to a less accurate estimate of volume.

In a number of ponds, the empirical fit nearly overlapped the entire bathymetric curve, and in such cases, there were many combinations of h_1 and h_2 that produced reasonable estimates of volume. For example, Pond 10 had the lowest full A_{ERR} and V_{ERR} of all the beaver ponds. In this case, there were 1899 combinations of h_1 and h_2 that produced estimates with total error below 5 %, and the distance between the points ranged from 1 to 84 % of h_{max}. Overall, the error was not sensitive to distance between h_1 and $h_2$if the points were on or near the full fitted curve. That said, the average minimum and maximum for h_1 (for all the optimum combinations for each pond) was 18–74 %, and for h_2 it was 42–98 %.

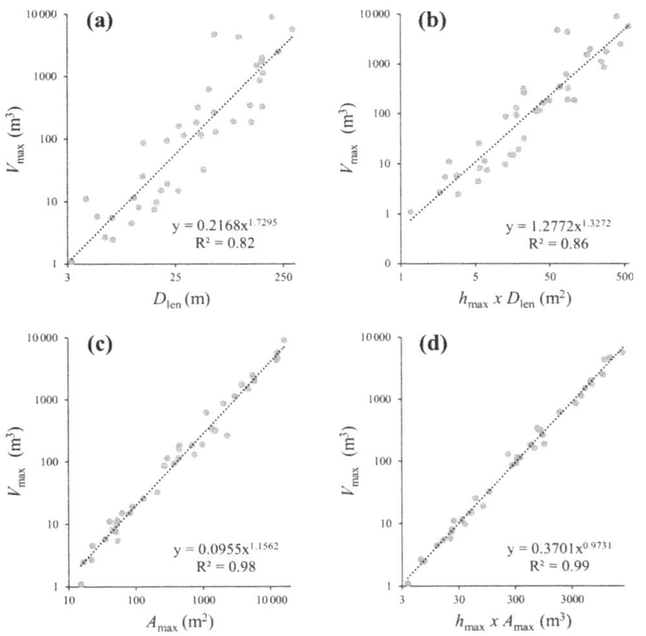

Figure 3. Power regression relationships between the maximum volume of the beaver ponds (V_{max}) and **(a)** the length of the beaver dams (D_{len}), **(b)** the product of the maximum depth of the ponds (h_{max}) and the length of the beaver dams, **(c)** the maximum surface area (A_{max}) of the ponds, and **(d)** the product of the maximum surface area and maximum depth of the pond.

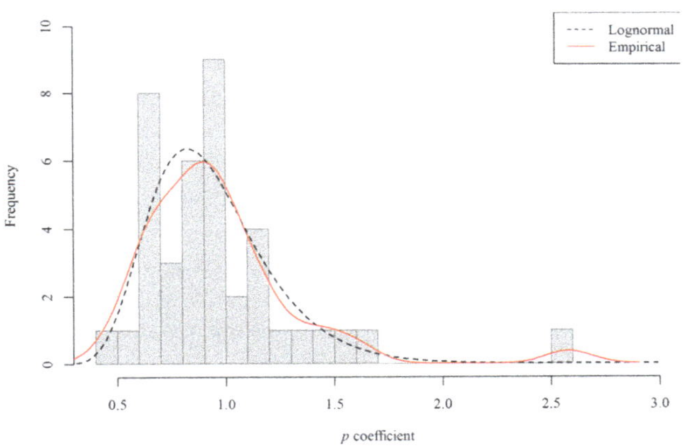

Figure 4. Distribution of morphometry (p) coefficients (full model) for all beaver ponds sampled ($n = 40$).

4 Discussion

The simplified V–A–h method estimated surface-water storage in the beaver ponds with high accuracy. Also, strong statistical relationships were found between surface-water storage capacity in beaver ponds and the dimensions of the dam and pond. The beaver ponds studied have a convex shape that permits less water storage than do other open-water wetland types. surface-water storage estimates can be made in beaver

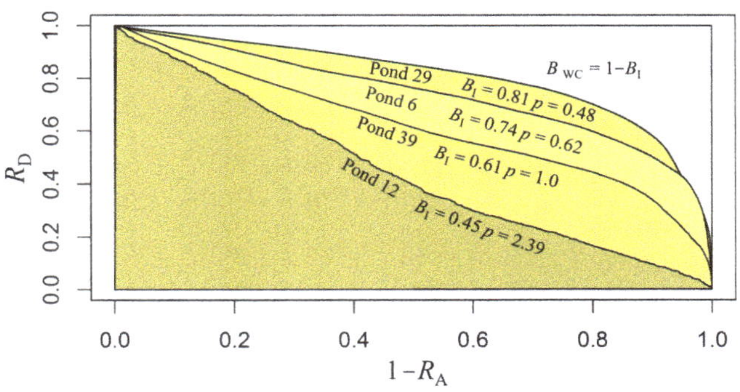

Figure 5. Bathymetric curves for ponds shown in Fig. 1. R_D is relative depth, R_A is relative area, B_I is the bathymetric integral, B_{WC} is the bathymetric water capacity, and p is the optimum morphometry coefficient.

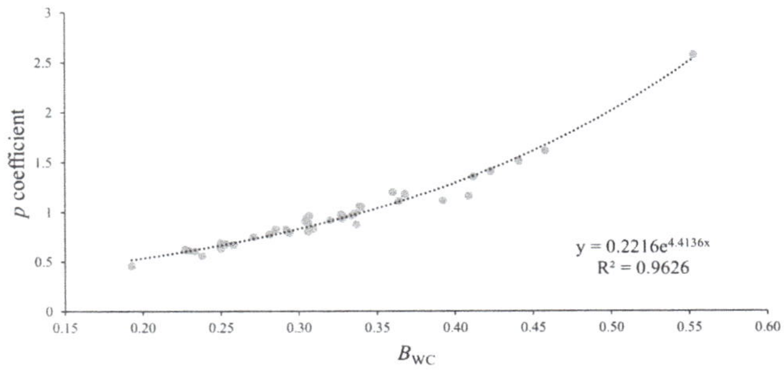

Figure 6. Relationship between the morphometry (p) coefficient (full model) and the bathymetric water capacity (B_{WC}).

ponds without the need for topographic surveys if pond morphology is used instead.

4.1 V–A–h model performance in beaver ponds

The low full A_{ERR} and V_{ERR} overall indicates that beaver pond morphometry is adequately described by power functions. This is because the bathymetric curve proved resilient to fluctuations in "elevation" inherent to the impounded land surface. Also, the dams, intricate canals and holes that beavers create in the areas they inhabit (Hood and Larson, 2015) do not warp the shape of the bathymetric curve enough that a power function becomes inappropriate to sufficiently describe it. However, it appears that volume estimations are more resilient to aberrations in the bathymetric curve than are area estimates. The power functions in the full model are fitted to pond bathymetry. When the power curve moves up and down, A_{ERR} will increase, but sometimes the V_{ERR} can decrease because volume is the integration of everything above the bathymetric curve. When the curve moves slightly up or down from the empirical fit, irregularities on the bathymetric curve are captured, which improves volume estimations at the sacrifice of area estimations. This explains why the optimum p coefficients for volume are different than they are for area. It also explains why, in many cases, the

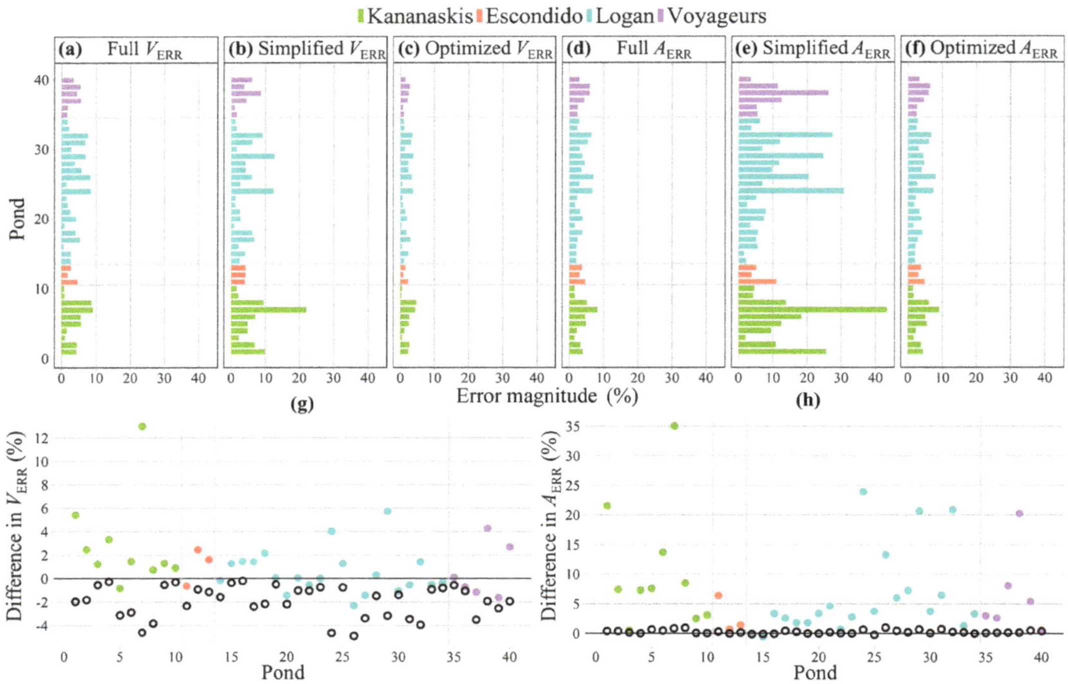

Figure 7. Volume (V_{ERR}) and area error (A_{ERR}) from each beaver pond using the three different approaches **(a–f)**. Plots on the bottom show the difference in volume **(g)** and area **(h)** error of the simplified (solid circles) and optimized (open circles) models relative to the full model (the full model is represented by the solid black line at zero on the *y* axis). Bars and solid circles are colour coded by location as per the legend at the top of the figure.

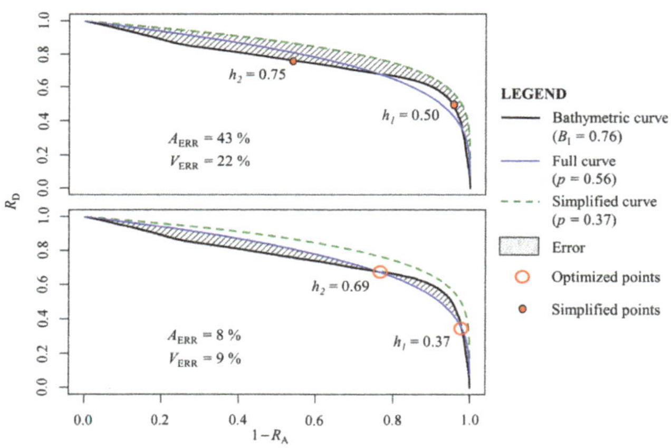

Figure 8. Comparison of the bathymetric curve for Pond 7 with the full and simplified curve. The top shows the area (A_{ERR}) and volume (V_{ERR}) error associated with the simplified curve that was calculated using simplified depths h_1 and h_2 and the bottom shows the error associated with the full curve and the optimum location for depths h_1 and h_2. R_D is relative depth, R_A is relative area, B_I is the bathymetric integral, and p is the morphometry coefficient.

simplified model had V_{ERR} that was less than 10 %, while A_{ERR} was greater than 25 %. Without a complete set of pond bathymetry, it is unlikely that users of the simplified V–A–h method would be able to discern the optimum points for h_1 and h_2; however, as long the chosen values for h_1 and h_2 are selected within the range identified here (i.e. 18–74 % of h_{max} for h_1 and 42–98 % of h_{max} for h_2), fairly accurate estimates of surface-water storage should be expected. Overall, the simplified model performed reasonably, exceeding 10 % V_{ERR} in only three cases. Given that the simplified V–A–h method appears to work well across the broad range of beaver pond bathymetry reported here, and across a wide range of prairie potholes (e.g. Minke et al., 2010), it should be a robust enough approach to be used other open-water wetlands.

4.2 Beaver pond morphometry and surface-water storage capacity

Our results show that p coefficients in beaver ponds are lower overall than those reported in prairie wetlands (Hayashi and van der Kamp, 2000) and those reported in forest pools in New England (Brooks and Hayashi, 2002). Because of the strong exponential relationship between p coefficients and B_{WC}, we can conclude that beaver ponds typically store less water. For example, the prairie potholes studied by Hayashi and van der Kamp (2000) had a median p coefficient of 3.22. Using Fig. 6, this p coefficient is equivalent to a B_{WC} of 0.61, which is almost double the median beaver pond B_{WC}equivalent of 0.32. The most likely explanation for this is the ontogeny of beaver ponds compared to other open water wetland types. Beaver ponds occur via inunda-

Table 2. Pond morphometric characteristics, including the full model morphometry (p) and scaling (s) coefficients, shape index (S_I), bathymetric integral (B_I), bathymetric water capacity (B_{WC}), length of the dam (D_{len}), and maximum depth (h_{max}), area (A_{max}), and volume (V_{max}) of the ponds.

Location	Pond no.	S_I	B_I	B_{WC}	p	s (m^2)	D_{len} (m)	h_{max} (m)	A_{max} (m^2)	V_{max} (m^3)
Kananaskis	1	2.05	0.75	0.25	0.69	889	164	1.50	2974	1135
	2	2.37	0.77	0.23	0.61	356	152	1.75	2006	867
	3	2.57	0.69	0.31	0.97	959	127	0.85	686	186
	4	1.79	0.77	0.23	0.61	123	27	1.50	446	163
	5	3.71	0.77	0.23	0.62	705	226	1.95	5496	2503
	6	3.47	0.74	0.26	0.67	1845	199	2.00	16357	9001
	7	1.76	0.76	0.24	0.56	1334	308	1.85	12912	5734
	8	2.55	0.75	0.25	0.63	701	159	1.80	3787	1757
	9	1.71	0.68	0.32	0.92	290	39	1.25	448	184
	10	1.51	0.66	0.34	1.05	247	30	1.10	297	113
Escondido	11	2.32	0.59	0.41	1.16	5352	162	0.55	1528	325
	12	1.72	0.45	0.55	2.58	2181	59	0.30	748	130
	13	1.99	0.54	0.46	1.61	3223	124	0.55	1342	344
Logan	14	2.19	0.66	0.34	1.06	438	7	0.30	54	6
	15	2.03	0.72	0.29	0.83	464	3	0.25	15	1
	16	1.89	0.56	0.44	1.51	87	4	0.60	41	11
	17	2.63	0.75	0.25	0.67	112	17	0.75	52	10
	18	2.14	0.70	0.30	0.91	91	19	0.80	63	15
	19	2.17	0.67	0.33	0.97	138	10	0.65	53	11
	20	1.95	0.67	0.33	0.94	352	16	0.45	50	8
	21	2.47	0.64	0.36	1.11	179	11	0.50	45	8
	22	2.70	0.67	0.33	0.98	96	7	0.45	17	2
	23	1.90	0.64	0.36	1.20	56	10	0.55	23	5
	24	1.97	0.69	0.31	0.80	430	27	0.60	82	15
	25	2.37	0.59	0.41	1.36	154	6	0.30	22	3
	26	2.83	0.75	0.25	0.68	124	21	0.90	90	19
	27	2.79	0.73	0.27	0.75	114	5	0.60	36	6
	28	1.73	0.67	0.33	0.96	278	13	1.00	265	87
	29	4.32	0.81	0.19	0.45	975	87	1.00	980	189
	30	3.43	0.71	0.29	0.79	620	21	0.85	374	94
	31	5.31	0.69	0.31	0.90	551	43	0.85	432	115
	32	2.61	0.69	0.31	0.83	1647	46	0.50	210	32
	33	2.59	0.66	0.34	0.99	409	51	1.65	1123	621
	34	2.40	0.58	0.42	1.41	470	12	0.45	130	26
Voyageurs	35	4.65	0.71	0.29	0.83	3683	144	1.10	4725	1517
	36	3.82	0.70	0.31	0.94	4539	161	1.10	5928	1999
	37	3.54	0.72	0.28	0.78	36105	57	0.40	2297	264
	38	2.52	0.66	0.34	0.88	11836	58	1.10	12985	4740
	39	2.72	0.61	0.39	1.11	18033	97	0.90	12482	4350
	40	2.78	0.63	0.37	1.18	4867	41	0.55	1504	316

tion of an existing channel and adjacent riparian area surface, whereas prairie potholes are bowl shaped geomorphic depressions created by the deposition of glacial till (Richardson et al., 1994). These different origins are reflected in the shape of the bathymetric curves, and they also explain the strong statistical relationships between surface-water storage capacity and the dimensions of the dam and pond. The stream channel in Fig. 3, for example, is represented on the far-right side of the bathymetric curve. Beaver ponds built on deeper and narrower stream channels tend to have lower p coefficients than ponds built on wider, fewer constrained channels. This happens because there is a rapid expansion of surface area inundated as the dam exceeds the height and width of the stream channel; a phenomenon that is well described by the "power" relationships between D_{len}, h_{max}, A_{max} and V_{max}. Pond 12 is a good example of this; the p coefficient was high-

Table 3. V–A–h model performance comparisons based on the mean (± standard deviation) volume (V_{ERR}) and area (A_{ERR}) error magnitude. "n" is the number of ponds studied at each site.

Site	n	Full		Simplified		Optimized	
		V_{ERR} (%)	A_{ERR} (%)	V_{ERR} (%)	A_{ERR} (%)	V_{ERR} (%)	A_{ERR} (%)
Kananaskis	10	4.3 ± 3.1	3.8 ± 2.1	7.2 ± 6.0	14.6 ± 12.3	2.3 ± 1.6	4.2 ± 2.5
Escondido	3	3.1 ± 1.4	3.8 ± 0.7	4.3 ± 2.5	6.7 ± 3.8	1.6 ± 0.7	4.0 ± 0.9
Logan	21	4.0 ± 2.6	3.6 ± 1.7	4.6 ± 3.5	9.9 ± 8.5	2.0 ± 1.2	3.9 ± 1.9
Voyageurs	6	3.8 ± 1.8	4.1 ± 1.6	3.8 ± 1.8	4.1 ± 1.6	1.9 ± 0.9	4.4 ± 1.7
All ponds	40	4.0 ± 2.5	3.8 ± 1.7	5.2 ± 4.1	11.0 ± 9.4	2.1 ± 1.2	4.0 ± 1.9

est (2.58) and a distant outlier compared to the other ponds. The uniqueness of this site is that the beaver built a small dam (0.3 m) with excavated peat and impounded groundwater seepage rather than damming channel flows. Even though the dam is relatively small, it has a large B_{WC} (0.55) relative to the other ponds because the dam is entirely dedicated to impounding a mostly flat land surface. In contrast, Pond 6, which was also built in a peatland, has a much lower B_{WC} (0.26) because most of the dam height (2 m) is dedicated to impounding water in an incised stream channel. An advantage of using the B_{WC} metric over pond volumes is that it allows for a comparison of surface-water storage capability in a way that is independent of pond size and shape.

4.3 Tools for surface-water storage estimation in beaver ponds

There are a variety of ways our results can be used to estimate surface-water storage in beaver ponds under different data availability scenarios. In situations where only aerial or remotely sensed imagery is available (i.e. world wide), dam length and pond area can be approximated and used in the generalized power regression relationships presented in Fig. 3. This is a quick and easy way to incorporate beaver pond surface-water storage capacity into land use planning decisions and watershed-scale hydrological models. However, this approach is not suitable for detailed study in individual beaver ponds as it does not account for pond morphometry (Huang et al., 2011; Wiens, 2001). Including dam height should improve estimates. Measuring dam height in the field is quick and straight forward, but it can also be reasonably approximated with remotely sensed imagery alone using spectral-depth correlation methods (e.g. Passalacqua et al., 2015). If dam heights are available, we recommend using our median p coefficient (0.91) for beaver ponds in the equation presented by Brooks and Hayashi (2002):

$$V_{max} = \frac{A_{max} \times h_{max}}{1 + 2/p}. \quad (12)$$

This equation is a modified form of Eq. (2) used to estimate surface-water storage capacity. It is easily incorporated into spatially distributed hydrological models. Fang et al. (2010) had success in using this approach, albeit for prairie potholes, in their Cold Regions Hydrological Model.

With a moderate amount of data, the simplified V–A–h method offers an alternative that produces surface-water storage estimates with minimal error. The advantage of this method over the others is that it is robust, it is customized to each pond's basin morphometry, and it calculates a coefficient of scale (i.e. s coefficient) for use in estimating surface-water storage across the range of pond stages, unlike the generalized power regression models and Eq. (12), which are limited to estimates of V_{max}. Combined with a few field visits and something as simple as automated water level observations, the simplified V–A–h method can be a powerful tool. But, it also has practical application in relatively data rich environments. For example, many lidar data sets are collected when beaver ponds are not fully drained. If the beaver pond is not entirely full, the measurements for A_2 and h_2 can be measured within the vertical distance between the crest of the dam and the surface of the water, thus allowing for an appropriate p coefficient to be derived. Furthermore, the simplified V–A–h method is increasingly practical with the advent of new technologies. For example, structure from motion software facilitates the creation of high resolution DEMs from ordinary photographs (Javernick et al., 2014). Theoretically, with both tools, one field visit to collect a few pictures and depths measurements should be all that is needed to make reliable estimates of wetland surface-water storage.

4.4 Implications of study results

The results of our study provide some simple tools that enable surface-water storage in beaver ponds to be estimated without the need for topographic surveys. This allows environmental managers to better assess the risks and benefits associated with beaver ponds that appear on landscapes, and allows for the easy inclusion of the surface-water storage component of beaver ponds into hydrological models at various scales. This study also demonstrates that beaver pond morphometry is different than other types of wetlands, which requires consideration. For example, based on this analysis we might expect beaver ponds to reach their capacity faster during rainfall events, while impounding larger surface areas than depressional wetlands. Although we show that some beaver ponds store less surface-water than other wetland types, their relevance to local and regional water balances should not be underestimated. Beaver population recovery, post fur trade, has led to the creation of between 9494 and 42 236 km^2 of new beaver ponds globally (Whitfield et al., 2015). Using the estimates of Whitfield et al. (2015) and our median p coefficient (0.91) and median dam height (0.83 m) in Eq. (12), we crudely estimate that between 2.5 and 11 km^3 of water are stored in beaver ponds.

5 Conclusions

The primary goal of this study was to test the utility of readily applicable tools for estimating surface-water storage in beaver ponds. We examined whether the simplified V–A–h method was appropriate for this purpose and described beaver pond morphology to explore its relationship to surface-water storage capacity. A number of valuable insights were revealed. The simplified V–A–h method proved to be a simple and effective tool as it was able to estimate beaver pond surface-water storage with an average volume error of 5 %. The median basin coefficient for beaver ponds was 0.91, suggesting that they tend to have a convex basin morphometry, and that they typically store less water than other wetlands studied in the same way. Pond capacity was strongly correlated to the dimensions of the dam and surface area of the pond, further cementing the idea that beaver ponds exhibit characteristic traits in pond morphometry that make reliable estimates of surface-water storage possible without the need for topographic surveys. Future research efforts should be directed at applying these simple methods more remotely, and incorporating them into both broader beaver-related planning tools and catchment-scale hydrological models.

Competing interests. The authors declare that they have no conflict of interest.

Acknowledgements. We thank Carlie Elliott, Kirsten Allen, Larisa Barber, Kenny DeMeurichy, Ryan Lokteff, and Konrad Hafen for field assistance. We also thank the staff and students of the Centre for Hydrology, particularly Sebastian Krogh, Chris Marsh, and Phillip Harder for their assistance with computer programming. The University of Calgary Biogeoscience Institute provided accommodation and access to their facilities. The project was funded through a Natural Sciences and Engineering Research Council (NSERC) of Canada Doctoral Graduate scholarship, an NSERC Discovery Grant (RGPIN 32837-20), a Global Institute for Water Security grant, a United States Department of Agriculture Grant (NIFA Hatch Project SD00H505-15), and a National Science Foundation Grant (DEB-1349240). The Utah field data collection was funded by the United States Forest Service (cooperative agreement 09-CS-1300001885391 awarded to Utah State University, award 100030) and facilitated by USFS partner Brett Roper. Research in Argentina was conducted under SDSyA Resolution no. 414/2013 and facilitated by Drs. Christopher Anderson and Alejandro Valenzuela. S. Archfield, S. Markstrom, and one anonymous reviewer helped improve the quality of this paper through their constructive critiques.

Edited by: S. Archfield

References

Baker, B. W. and Hill, E. P.: Beaver (Castor canadensis), in: Wild Mammals of North America: Biology, Management, and Conservation, edited by: Feldhamer, G., Thompson, B., and Chapman, J., The John Hopkins University Press, Baltimore, Maryland, 288–310, 2003.

Brooks, R. T. and Hayashi, M.: Depth-area-volume and hydroperiod relationships of ephemeral (vernal) forest pools in southern New England, Wetlands, 22, 247–255, doi:10.1672/0277-5212(2002)022[0247:DAVAHR]2.0.CO;2, 2002.

Butler, D. R. and Malanson, G. P.: The geomorphic influences of beaver dams and failures of beaver dams, Geomorphology, 71, 48–60, doi:10.1016/j.geomorph.2004.08.016, 2005.

ESRI: ArcGIS Desktop, available at: http://www.esri.com (last access: 20 June 2016), 2015.

Fang, X., Pomeroy, J. W., Westbrook, C. J., Guo, X., Minke, A. G., and Brown, T.: Prediction of snowmelt derived streamflow in a wetland dominated prairie basin, Hydrol. Earth Syst. Sci., 991–1006, doi:10.5194/hess-14-991-2010, 2010.

Gleason, R., Tangen, B., Laubhan, M. K., Kermes, K. E., and Euliss, N. H.: Estimating water storage capacity of existing and potentially restorable wetland depressions in a subbasin of the Red River of the North, US Geological Survey Open File Report 2007-1159, US Geological Survey, Virginia, USA, 2007.

Green, K. C. and Westbrook, C. J.: Changes in riparian area structure, channel hydraulics, and sediment yield following loss of beaver dams, BC, J. Ecosyst. Manage., 10, 68–79, 2009.

Hayashi, M. and van der Kamp, G.: Simple equations to represent the volume–area–depth relations of shallow wetlands in small topographic depressions, J. Hydrol., 237, 74–85, doi:10.1016/S0022-1694(00)00300-0, 2000.

Hood, G. A. and Bayley, S. E.: Beaver (Castor canadensis) mitigate the effects of climate on the area of open water in boreal wetlands in western Canada, Biol. Conserv., 141, 556–567, doi:10.1016/j.biocon.2007.12.003, 2008.

Hood, G. A. and Larson, D. G.: Ecological engineering and aquatic connectivity: a new perspective from beaver-modified wetlands, Freshwater Biol., 60, 198–208, doi:10.1111/fwb.12487, 2015.

Huang, S., Young, C., Feng, M., Heidemann, K., Cushing, M., Mushet, D. M., and Liu, S.: Demonstration of a conceptual model for using LiDAR to improve the estimation of floodwater mitigation potential of Prairie Pothole Region wetlands, J. Hydrol., 405, 417–426, doi:10.1016/j.jhydrol.2011.05.040, 2011.

Hubbard, D. E.: Surface area-volume relationship of wetland basins in the interior of the Coteau Des Prairies, Proc. S. Dakota Acad. Sci., 61, 154–158, 1982.

Hutchinson, G. E.: A Treatise on Limnology, in: Vol. 1, Geology, Physics and Chemistry, John Wiley and Sons, New York, 1957.

Javernick, L., Brasington, J., and Caruso, B.: Modeling the topography of shallow braided rivers using Structure-from-Motion photogrammetry, Geomorphology, 213, 166–182, doi:10.1016/j.geomorph.2014.01.006, 2014.

Johnston, C. A.: Beaver Wetlands, in: Wetland habitats of North America: Ecology and Conservation Concerns, edited by: Batzer, D. P. and Baldwin, A. H., University of California Press, Berkeley, 161–171, 2012.

Johnston, C. A. and Naiman, R. J.: Boundary dynamics at the aquatic-terrestrial interface: The influence of beaver and geomorphology, Landsc. Ecol., 1, 47–57, doi:10.1007/BF02275265, 1987.

Lane, C. R. and D'Amico, E.: Calculating the ecosystem service of water storage in isolated wetlands using LiDAR in north central Florida, USA, Wetlands, 30, 967–977, doi:10.1007/s13157-010-0085-z, 2010.

Macfarlane, W. W., Wheaton, J. M., Bouwes, N., Jensen, M. L., Gilbert, J. T., Hough-Snee, N., and Shivik, J. A.: Modeling the capacity of riverscapes to support beaver dams, Geomorphology, 277, 72–99, doi:10.1016/j.geomorph.2015.11.019, 2017.

Milly, P. C. D., Betancourt, J., Falkenmark, M., Hirsch, R. M., Kundzewicz, Z. W., Lettenmaier, D. P., and Stouffer, R. J.: Climate change. Stationarity is dead: whither water management?, Science, 319, 573–574, doi:10.1126/science.1151915, 2008.

Minke, A. G., Westbrook, C. J., and van der Kamp, G.: Simplified volume–area–depth method for estimating water storage of prairie potholes, Wetlands, 30, 541–551, doi:10.1007/s13157-010-0044-8, 2010.

Naiman, R. J., Melillo, J. M., and Hobbie, J. E.: Ecosystem alteration of boreal forest streams by beaver (Castor Canadensis), Ecology, 67, 1254–1269, 1986.

Passalacqua, P., Belmont, P., Staley, D. M., Simley, J. D., Arrowsmith, J. R., Bode, C. A., Crosby, C., DeLong, S. B., Glenn, N. F., Kelly, S. A., Lague, D., Sangireddy, H., Schaffrath, K., Tarboton, D. G., Wasklewicz, T., and Wheaton, J. M.: Analyzing high resolution topography for advancing the understanding of mass and energy transfer through landscapes: A review, Earth-Sci. Rev., 148, 174–193, doi:10.1016/j.earscirev.2015.05.012, 2015.

Pollock, M. M., Heim, M., and Werner, D.: Hydrologic and geomorphic effects of beaver dams and their influence on fishes, Am. Fish. Soc. Symp., 37, 213–233, 2003.

Richardson, J. L., Arndt, J. L., and Freeland, J.: Wetland soils of the prairie potholes, Adv. Agron., 52, 121–171, doi:10.1016/S0065-2113(08)60623-9, 1994.

RStudio Team: Integrated Development for R, available at: http://www.rstudio.com/ (last access: 20 June 2016), 2015.

Shaw, R. D. and Prepas, E. E.: Groundwater-lake interactions: II. Nearshore seepage patterns and the contribution of groundwater to lakes in central Alberta, J. Hydrol., 119, 121–136, 1990.

Strahler, A. N.: Hypsometric (area–altitude) analysis of erosional topography, Geol. Soc. Am. Bull., 63, 1117–1142, doi:10.1130/0016-7606(1952)63[1117:HAAOET]2.0.CO;2, 1952.

Trigg, M. A., Cook, P. G., and Brunner, P.: Groundwater fluxes in a shallow seasonal wetland pond: The effect of bathymetric uncertainty on predicted water and solute balances, J. Hydrol., 517, 901–912, doi:10.1016/j.jhydrol.2014.06.020, 2014.

Whitfield, C. J., Baulch, H. M., Chun, K. P., and Westbrook, C. J.: Beaver-mediated methane emission: The effects of population growth in Eurasia and the Americas, Ambio, 44, 7–15, doi:10.1007/s13280-014-0575-y, 2015.

Wiens, L. H.: A Surface Area–Volume Relationship for Prairie Wetlands in the Upper Assiniboine River Basin, Saskatchewan, Can. Water Resour. J., 26, 503–513, doi:10.4296/cwrj2604503, 2001.

Xu, C.-Y. and Singh, V. P.: Review on regional water resources assessment models under stationary and changing climate, Water Resour. Manage., 18, 591–612, doi:10.1007/s11269-004-9130-0, 2004.

7

Extrapolating regional probability of drying of headwater streams using discrete observations and gauging networks

Aurélien Beaufort, Nicolas Lamouroux, Hervé Pella, Thibault Datry, and Eric Sauquet

Irstea, UR RiverLy, centre de Lyon-Villeurbanne, 5 rue de la Doua CS 20244, 69625 Villeurbanne, France

Correspondence: Aurélien Beaufort (aurelien.beaufort@irstea.fr)

Abstract. Headwater streams represent a substantial proportion of river systems and many of them have intermittent flows due to their upstream position in the network. These intermittent rivers and ephemeral streams have recently seen a marked increase in interest, especially to assess the impact of drying on aquatic ecosystems. The objective of this paper is to quantify how discrete (in space and time) field observations of flow intermittence help to extrapolate over time the daily probability of drying (defined at the regional scale). Two empirical models based on linear or logistic regressions have been developed to predict the daily probability of intermittence at the regional scale across France. Explanatory variables were derived from available daily discharge and groundwater-level data of a dense gauging/piezometer network, and models were calibrated using discrete series of field observations of flow intermittence. The robustness of the models was tested using an independent, dense regional dataset of intermittence observations and observations of the year 2017 excluded from the calibration. The resulting models were used to extrapolate the daily regional probability of drying in France: (i) over the period 2011–2017 to identify the regions most affected by flow intermittence; (ii) over the period 1989–2017, using a reduced input dataset, to analyse temporal variability of flow intermittence at the national level. The two empirical regression models performed equally well between 2011 and 2017. The accuracy of predictions depended on the number of continuous gauging/piezometer stations and intermittence observations available to calibrate the regressions. Regions with the highest performance were located in sedimentary plains, where the monitoring network was dense and where the regional probability of drying was the highest. Conversely, the worst performances were obtained in mountainous regions. Finally, temporal projections (1989–2016) suggested the highest probabilities of intermittence (> 35 %) in 1989–1991, 2003 and 2005. A high density of intermittence observations improved the information provided by gauging stations and piezometers to extrapolate the temporal variability of intermittent rivers and ephemeral streams.

1 Introduction

Headwater streams represent a substantial proportion of river systems (Leopold et al., 1964; Nadeau and Rains, 2007; Benstead and Leigh, 2012). From an ecological point of view, headwater catchments are at the interface between terrestrial and aquatic ecosystems and they often harbour a unique biodiversity with a very high spatial turnover (Meyer et al., 2007; Clarke et al., 2008; Finn et al., 2011). Their contribution to the functioning of hydrographic networks is essential: sediment flows, inputs of particulate organic matter and nutrients, refugia/colonisation, sources of aquatic organisms (Meyer et al., 2007; Finn et al., 2011).

Headwater streams are generally naturally prone to flow intermittence, i.e. streams which stop flowing or dry up at some point in time and space, mainly due to their upstream position in the network and their high reactivity to natural or human disturbances (Benda et al., 2005; Datry et al., 2014b). These waterways which cease flow and/or dry are referred to as intermittent rivers and ephemeral streams (IRES). The geographic extent of IRES is poorly documented due to mapping limitations (digital elevation models, satellite images, aerial photos) and because of their size and their location (Leopold et al., 1994; Nadeau and Rains, 2007; Benstead and Leigh, 2012; Fritz et al., 2013). However, the proportion of

IRES in hydrological networks can be very large: for example, they represent 60 % of the length of rivers in the United States (Nadeau and Rains, 2007) and are considered to represent probably more than 50 % of the global hydrological network (Larned et al., 2010; Datry et al., 2014b). Considering only gauging stations with continuous records may lead to severe underestimation of their regional extent (Snelder et al., 2013; De Girolamo et al., 2015; Eng et al., 2016).

Recently, IRES have seen a marked increase in interest stimulated by the challenges of water management facing the global change context (water scarcity issues, climate change impact, etc.) (Acuña et al., 2014; Datry et al., 2016b). Studies have characterised the hydrological functioning of IRES (Gallart et al., 2012; Costigan et al., 2016; Sarremejane et al., 2017) to assess the effects of flow intermittence on aquatic ecosystems (Larned et al., 2010; Datry et al., 2016b; Leigh et al., 2016; Leigh and Datry, 2017). IRES have been altered due to human actions (abstraction, hill dams, low-water support, pollution, etc.) despite their high and unique biodiversity (Datry et al., 2014a; Garcia et al., 2017a). In addition, some perennial streams are becoming intermittent due to global change, water abstraction or river damming (Skoulikidis, 2009), and the extent of IRES may increase in the future (Döll and Schmied, 2012; Jaeger et al., 2014; Pumo et al., 2016; Garcia et al., 2017b; De Girolamo et al., 2017a).

A better hydrological understanding of IRES is now essential and improved management requires knowledge of both the spatial extent and arrangement of IRES within the river network (Boulton, 2014; Acuña et al., 2017). Efforts have been made to estimate the spatial distribution of IRES at the catchment scale (Skoulikidis et al., 2011; Datry et al., 2016a), at the regional scale (Gómez et al., 2005) and at the national scale (Snelder et al., 2013). In France, Snelder et al. (2013) suggested a classification of IRES regimes and spatialised their distribution. Based on an analysis of the continuous gauging network, they showed that the proportion of IRES accounted for 20 to 39 % of the hydrographic network. The accuracy of the obtained map is highly dependent on the density of the flow monitoring network. The installation of additional gauging stations is expensive and headwater systems may be difficult to monitor due to active geomorphology processes or to difficult access.

As a promising tool to advance the mapping of IRES, citizen science creates opportunities to overcome the lack of hydrological data, contributes to densifying the flow-state observation network (Turner and Richter, 2011; Buytaert et al., 2014; Datry et al., 2016b) and could be used for hydrological model calibration (van Meerveld et al., 2017). In France, Datry et al. (2016a) used such data to describe the spatio-temporal dynamics of aquatic and terrestrial habitats within five river catchments located in the western part of France. They showed that processes resulting in flow intermittence were complex at a fine scale and could vary substantially among nearby catchments. However, these data were only available in a few catchments, limiting any attempt to map large-scale patterns of flow intermittence in river networks. Since this first attempt, new sources of observational data have become available in France thanks to the ONDE network (Observatoire National des Etiages, https://onde.eaufrance.fr, last access: 22 May 2018). This unique network in Europe provides frequent discrete field observations (five inspections per year) of the flow intermittence across more than 3300 sites throughout France and located mostly in headwater areas.

However, discrete observations of intermittence do not provide any information on the persistence of dry conditions between two consecutive dates of observation. The rewetting–drying events could have significant impacts on communities whose survival is conditioned by the duration/frequency of drying. The duration of drying is of importance for ecologists, as one key driver for the composition and persistence of aquatic species (Vardakas et al., 2017; Kelso and Entrekin, 2018; Vadher et al., 2018). Temporal extrapolations of river flow regimes are thus necessary to summarise the different facets of flow intermittence at various timescales, from daily to inter-annual.

The main objective of this paper is to use discrete (in space and time) field observations of flow intermittence to extrapolate over time the daily probability of drying (averaged at the regional scale). We first carried out a quantitative analysis of the ONDE network data in order to characterise the information that they contribute in comparison with the data resulting from the conventional hydrological monitoring. Then, we developed two empirical models based on linear or logistic regressions to convert discontinuous series of flow intermittence observation from ONDE into continuous daily probability of drying, defined at the regional scale across France. Explanatory variables were derived from available continuous daily discharge and groundwater-level data of a dense gauging/piezometer network, and models were calibrated using the ONDE discrete observations. The robustness of the models was tested using (1) an independent, dense regional dataset of intermittence observations and (2) observations of the year 2017 excluded from the calibration. Finally, the resulting models were used to extrapolate the regional probability of drying in France: (i) over the period 2012–2017 to identify the regions most affected by flow intermittence; (ii) over the period 1989–2017, using a reduced input dataset, to analyse temporal variability of flow intermittence at the national level.

2 Material and methods

2.1 Study area

The study area is continental France and Corsica ($550\,000\,\text{km}^2$). France is located in a temperate zone characterised by a variety of climates due to the influences of the Atlantic Ocean, the Mediterranean Sea and mountain areas.

We defined regions as combinations of "level-2 Hydro-EcoRegions" (HER2) and classes of hydrological regimes (HR). Hydro-EcoRegion (HER) corresponds to a typology developed for river management in accordance with the European Water Framework Directive. The Hydro-EcoRegion classification includes 22 "level-1 Hydro-EcoRegions" (HER1) based on geology, topography and climate, and considered the primary determinants of the functioning of water ecosystems (Wasson et al., 2002). HER2 correspond to a finer classification accounting for stream size. HER2 have a mean drainage area of 5000 km^2 (between 100 and 27 000 km^2). The hydrological regime classes (HR) were identified by reference to the work carried out by Sauquet et al. (2008) where it was possible to distinguish rainfall-fed regimes, transition and snowmelt-fed river flow regimes. Overall, we used 280 regions (that is, HER2–HR combinations) with a mean drainage area of 1400 km^2 (between 4 and 20 000 km^2).

2.2 ONDE dataset discrete national flow-state observations

The ONDE network was set up in 2012 by the French Biodiversity Agency (AFB, formerly ONEMA) with the aim of constituting a perennial network recording summer low-flow levels and used to anticipate and manage water crisis during severe drought events (Nowak and Durozoi, 2012).

There are 3300 ONDE sites distributed throughout France (Fig. 1). ONDE sites are located on headwater streams with a Strahler order of strictly less than 5 and balanced across HER2 regions to take into account the representativeness of the hydrological contexts (Nowak and Durozoi, 2012). The ONDE network is stable over time. Observations have been made monthly (around the 25th) by trained AFB staff, between April and September, every year since 2012. One of the statuses is assigned at each observation among "visible flow", "no visible flow" and "dried out". Here, we consider two intermittency statuses: "Flowing" when there is visible flow across the channel ("visible flow") and "Drying" when the channel is entirely devoid of surface water ("dried out") or when there is still water in the river bed but without visible flow (disconnected pools, lentic systems) ("no visible flow"). The proportion of drying sites determined on the basis of the ONDE network for each HER2–HR combination is considered a good estimate of the daily regional probability of drying ($\mathrm{RPoD_{ONDE}}$) of streams with a Strahler order of less than 5. Observed values of $\mathrm{RPoD_{onde}}$ are calculated as follows:

$$\mathrm{RPoD_{ONDE}}(d) = \frac{(N_{\mathrm{drying}})_{\mathrm{HER2-HR}}}{(N_{\mathrm{flowing}} + N_{\mathrm{drying}})_{\mathrm{HER2-HR}}}, \quad (1)$$

where d denotes the observation date of the ONDE network, N_{drying} and N_{flowing} are the number of drying and of flowing

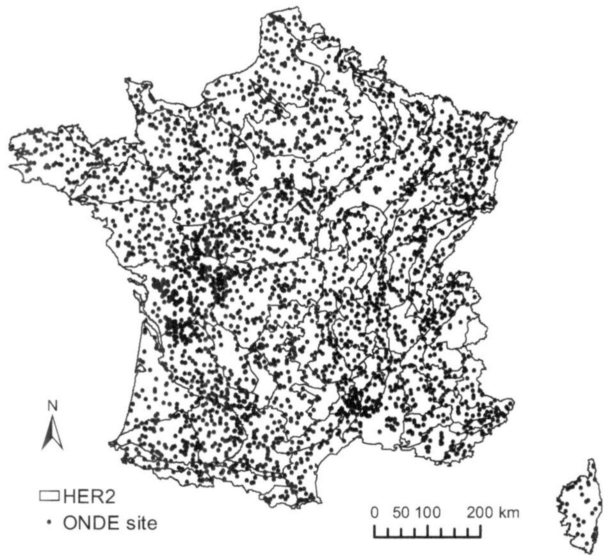

Figure 1. Location of the 3300 ONDE sites and partition into HER2.

statuses observed at ONDE sites located in the same HER2–HR combination at the observation date d, respectively.

Figure 2 illustrates the complementary nature of the ONDE network to the already existing HYDRO (http://www.hydro.eaufrance.fr, last access: 22 May 2018) French river flow monitoring network. The ONDE sites and a set of 1600 gauging stations available in the HYDRO database have been projected on the RHT (Theoretical Hydrographic Network; Pella et al., 2012) river network and the drainage area and the elevation have been estimated. A large part of ONDE sites are located on small headwater streams, with 70 % of the sites with a drainage area of less than 50 km^2, while most of the gauging stations record flows of catchments of medium size (between 100 and 500 km^2). Only four stations display a drainage area of more than 1000 km^2. The distributions of elevation of the two databases look similar. The ONDE sites are mostly located on rivers with an elevation below 200 m (75 % of sites). The ONDE sites are sparse at high elevations (95 sites located above 1000 m). This bias is likely due to access difficulties in mountainous areas.

2.3 POC dataset: a denser regional dataset used for independent validation

A spatially denser citizen science dataset of flow-state observations in western France (Poitou-Charente region) (http://atlas.observatoire-environnement.org, last access: 22 May 2018) has been used as a validation dataset to test the robustness of our models calibrated with the ONDE dataset. The POC monitoring (2011–2013) covered more than 4000 km of river length across 20 catchments. Each river was entirely surveyed every 1st and 15th of each month be-

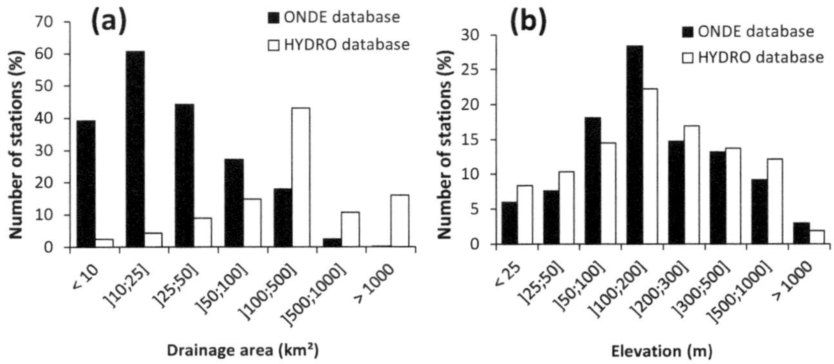

Figure 2. Distribution of the 3300 ONDE sites and of the 1600 gauging stations available in the HYDRO database against **(a)** drainage area and **(b)** elevation.

tween June and October, resulting in eight observations per year. Four intermittency statuses were available in the POC dataset (Datry et al., 2016a), but to allow comparisons with the ONDE network, we pooled the two "Flowing" and "Low Flow" POC statuses into a single "Flowing" status and the two "No flow" and "Dry" statuses into the "Drying" status. This dataset is available as maps with flow states assigned to the inspected streams. Values of RPoD at each POC observation date are calculated in the same way as $\mathrm{RPoD}_{\mathrm{ONDE}}$. Thus $\mathrm{RPoD}_{\mathrm{POC}}$ is given by the ratio between the number of drying statuses and the total number of observations at each inspected stream located in the same HER2–HR.

2.4 Explanatory discharge dataset

Two discharge datasets (continuous daily time series) were used as explanatory variables of discrete intermittence observations, with the objective of extrapolating the intermittence frequency over time. The two datasets included time series of daily discharge extracted from the French river flow monitoring network (HYDRO database, http://www.hydro.eaufrance.fr/, last access: 22 May 2018): (i) the 2011–2017 dataset with full records available between 1 January 2011 and 31 June 2017; (ii) the 1989–2017 dataset concerning a reduced number of gauging stations and providing daily discharges between 1 January 1989 and 31 June 2017. According to the hydrometric services in charge of the selected gauging stations, high quality of measurements was ensured and observed discharges were not or only slightly altered by human actions.

The 2011–2017 dataset was composed of 1600 gauging stations distributed across France. Each stream where a HYDRO gauging station is located has been defined as IRES or perennial. Several definitions of IRES can be found in the literature (Huxter and van Meerveld, 2012; Eng et al., 2016; Reynolds et al., 2015). In this study, we considered stations to be intermittent when five consecutive days with discharge less than $1\,\mathrm{L\,s^{-1}}$ have been observed during the period of record.

The 1989–2017 dataset consisted of 630 gauging stations selected with less than 5 % of missing data (continuous or not) during the period 1989–2017. This dataset was thereafter used to estimate the regional probability of drying before the creation of the ONDE network.

2.5 Explanatory groundwater-level dataset

Because groundwater resources influence stream intermittence, we used available time series of the daily groundwater level available in the ADES database (http://www.ades.eaufrance.fr, last access: 22 May 2018) at sites identified as involved in groundwater–surface water exchanges (Brugeron et al., 2012). Similarly to the discharge data, two sets of groundwater-level data with records available over the two periods 2011–2017 and 1989–2017 have been selected. The level of alteration of groundwater levels by water withdrawal is unknown because no information is available at this scale.

The 2011–2017 dataset was composed of 750 piezometers with daily groundwater-level data with less than 5 % of missing data (continuous or not). The selection of the 1989–2017 dataset was not easy because few groundwater-level measurements were available in the database before 2000. For example, only five piezometers met the tolerance limit on missing values considered for the 1989–2017 discharge dataset. In order to extend the dataset and because groundwater levels were less variable than stream discharges, the proportion of permitted gaps was fixed to 20 % between 1989 and 2017. This led us to select 150 piezometers. Thereafter, when the missing data period was less than 10 days, groundwater levels were reconstructed by linear interpolation in order to reduce the proportion of missing values to less than 5 % for the 150 piezometers selected.

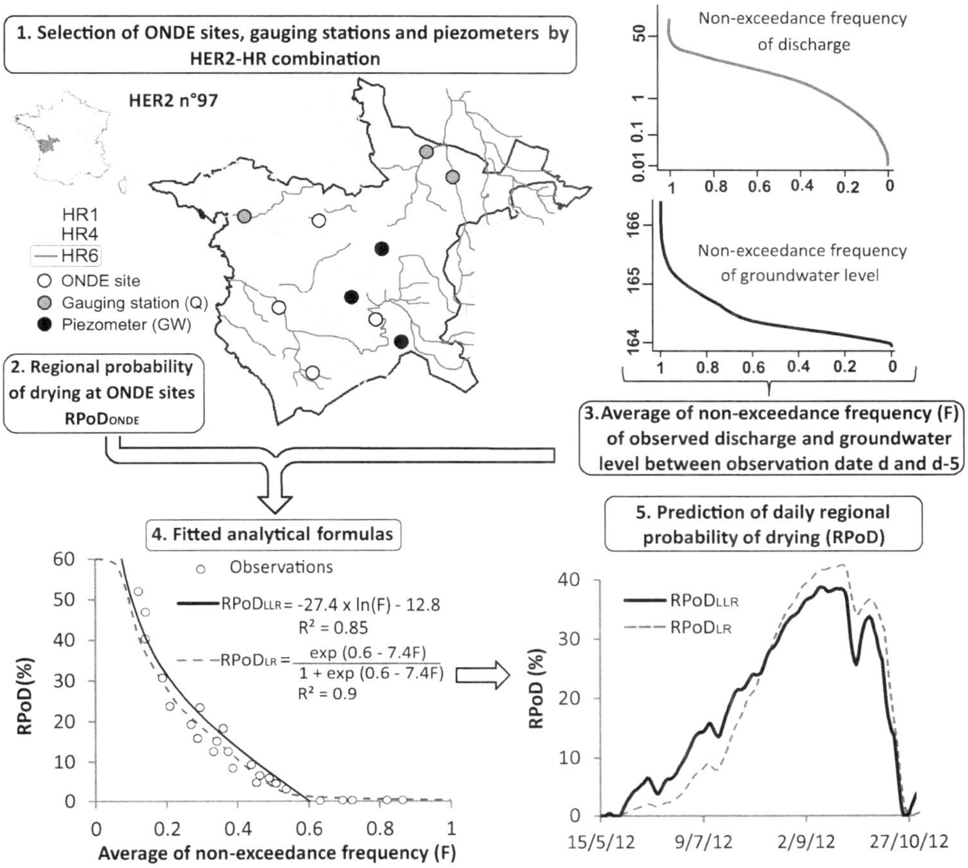

Figure 3. Strategy of parametric modelling (steps 1–4) developed to predict (step 5) the regional probability of drying (RPoD) by HER2–HR combination in France.

2.6 Statistical modelling of regional probability of drying

The parametric modelling strategy was based on five main steps (Fig. 3). The first step consisted in selecting all ONDE sites, gauging stations and piezometers located in the same HER2–HR combination. When the total number of gauging stations and piezometers was less than five for a HER2–HR combination, we merged the HER2–HR combination with a neighbouring one located in the same HER1. This was done for 20 of the 280 regions. The second step consisted in calculating the RPoD$_{\mathrm{ONDE}}$ for each observation date (5 year^{-1}) and for all selected ONDE sites. In a third step, a flow duration curve was determined for each selected HYDRO gauging station. The average non-exceedance frequency of the observed discharge at gauging stations was averaged for the date of observation (d) at ONDE sites and the 5 days preceding the observation. The lag of 6 days accounted for the fact that ONDE survey dates in a region could differ by 5 days, and accounted for the inertia of physical processes (e.g. storage capacity). The same operation was carried out with selected piezometers. Finally the hydrological conditions are described by the average (across stations) F of the non-exceedance frequencies of discharge (F_q) and groundwater levels (F_{gw}) with respect to the relative proportions of gauging stations and piezometers:

$$F(d) = \frac{\sum_{i=1}^{i=N_q} F_{q_i} + \sum_{j=1}^{j=N_{\mathrm{gw}}} F_{\mathrm{gw}_j}}{(N_q + N_{\mathrm{gw}})}, \tag{2}$$

where F_{q_i} denotes the average non-exceedance frequency of discharge at the gauging station i calculated between d and $d-5$; F_{gw_j} the average non-exceedance frequency of groundwater levels at the piezometer j calculated between d and $d-5$; N_q the number of gauging stations selected in a HER2–HR combination; and Ngw the number of selected piezometers selected in the HER2–HR combination. The fourth step consisted in estimating the RPoD$_{\mathrm{ONDE}}$ as a function of F. Two types of regression were fitted for each HER2–HR combination across France:

- A truncated logarithmic linear regression (LLR), with two parameters α_1 and β_1:

$$\mathrm{RPoD_{LLR}}(d) = \begin{cases} \min(1; \alpha_1 \times \ln(F(d)) + \beta_1) \text{ when } F < F0 \\ 0 \text{ when } F \geq F0. \end{cases} \tag{3}$$

$F0$ was fixed as the value of non-exceedance frequencies of discharge and groundwater levels at which no more drying status was observed across the ONDE network ($\mathrm{RPoD_{ONDE}} = 0$).

- A logistic regression (LR) with two parameters α_2 and β_2:

$$\mathrm{Logit}(\mathrm{RPoD_{LR}}(d)) = \ln\left(\frac{\mathrm{RPoD_{LR}}(d)}{1 - \mathrm{RPoD_{LR}}(d)}\right) = \alpha_2 \times F(d) + \beta_2. \tag{4}$$

LR is a multivariate analysis method well known for its relevance in binary classification issues (Lee, 2005). The $\mathrm{RPoD_{LR}}$ was then calculated as follows in Eq. (5):

$$\mathrm{RPoD_{LR}}(d) = \frac{\exp(\alpha_2 F(d) + \beta_2)}{1 + \exp(\alpha_2 F(d) + \beta_2)}. \tag{5}$$

Models were calibrated against observations available during the same period, 2012–2016, leaving out the year 2017 for an independent validation test. However, for the continuous temporal extrapolations (one over 2011–2017, the other over 1989–2017), two models were built with different piezometers and gauging stations selected as explanatory variables (see Sects. 2.4 and 2.5). Thus there are two sets of regression parameters specific to each dataset for both the LLR and LR models, leading to different prediction of RPoD.

Finally, in a fifth step, a daily regional probability of drying (RPoD) could be predicted for each HER2–HR combination with both models following analytical formulas (Eqs. 3 and 5).

2.7 Model robustness: validation using independent datasets

We used (1) the POC independent data and (2) the 2017 ONDE year to test the robustness of the LLR and LR model to predict the intermittence frequency (1) in space and (2) over time. Note that when predicting on the POC datasets, a new model was calibrated using only ONDE sites located out of POC streams.

For both datasets (POC and ONDE, 2017), the relative performance of the LLR and LR models was compared in multiple ways using both the 2011–2017 and 1989–2017 datasets. The performance of each model was evaluated by the Nash–Sutcliffe efficiency criterion (NSE) (Nash and Sutcliffe, 1970):

$$\mathrm{NSE} = 1 - \frac{\sum_{i=1}^{N} (\mathrm{RPoD_{ONDEi}} - \mathrm{RPoD_{pri}})^2}{\sum_{i=1}^{N} (\mathrm{RPoD_{ONDEi}} - \overline{\mathrm{RPoD_{ONDEi}}})^2}, \tag{6}$$

where $\mathrm{RPoD_{ONDEi}}$ is the average proportion of drying statuses over the ONDE sites located in the HER2–HR combination at the ith observation date, $\mathrm{RPoD_{pri}}$ is the predicted regional probability of drying at the ith observation date, $\overline{\mathrm{RPoD_{ONDEi}}}$ is the mean of $\mathrm{RPoD_{ONDEi}}$ over the period and N is the total number of observations in the ONDE network for each HER2–HR combination.

2.8 Model prediction

Both models have been calibrated over the period 2012–2016 and were then applied in a fifth step to predict the daily RPoD in France (Fig. 3). The RPoD was firstly predicted over the period 2012–2016 in order to identify the most affected regions by flow intermittence using the 2011–2017 datasets. The second application concerned the extrapolation of RPoD in France over a longer period using the 1989–2017 dataset to analyse the temporal variability of flow intermittence at the national level. It should be noted that model predictions only concern streams with a Strahler order of lower than 5 due to the ONDE site location.

3 Results

3.1 Quantitative analysis

3.1.1 Inter-annual intermittence according to the raw discrete ONDE network

A total of 1127 ONDE sites have recorded at least one drying event during the period 2012–2016 representing 35 % of the 3300 ONDE sites. From the ONDE database the probability of drying at the country scale was computed as the total number of drying statuses over France divided by the total number of ONDE observations available during statuses the same year (Fig. 4a). Between 2012 and 2016, the most critical year is 2012, with 15 % of drying statuses followed by 2016 (14 %) and 2015 (14 %) (Fig. 4a). The years 2013 and 2014 are less affected with only 6 % of drying statuses observed (Fig. 4a).

Drying events mainly occur between July and September, but the evolution of the month's proportion of drying can differ between years (Fig. 4b). In more detail, water levels in 2012 decrease in August when the proportion of drying is 27 %, and the situation lasts until the end of September with 25 % of drying (Fig. 4b). In 2013, the proportion of drying is

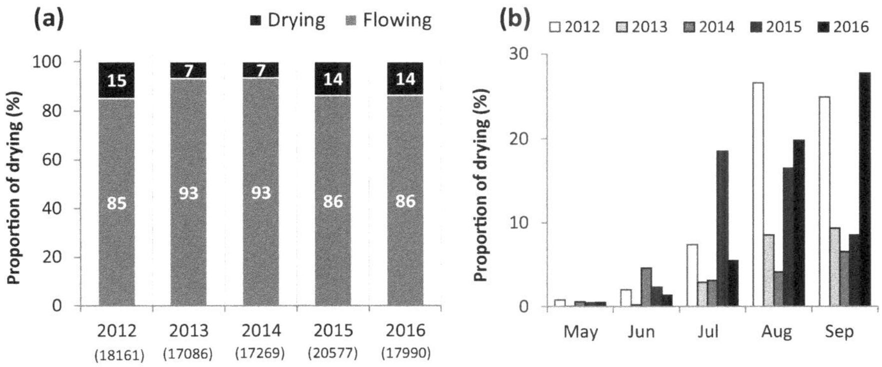

Figure 4. **(a)** Distribution of the yearly proportion of drying observed with the ONDE network with the total yearly number of ONDE observations written in brackets and **(b)** distribution of proportions of drying per year and per month.

lower than in 2012, but follows the same pattern with an increase at the end of July (3 %), reaching 9 % in August and in September. In 2014, the first peak of drying (5 %) is reached early in June. Then, the proportion of drying decreases in July (3 %) and increases slightly in August (4 %), reaching 7 % in September. In 2015, the critical period occurs at the end of July with 19 % of drying statuses, and the proportion of drying decreases slightly at the end of August (17 %) until it reaches 9 % in September. Finally, in 2016, the situation gradually deteriorates every month, reaching 20 % of drying statuses in August, and 28 % in September.

Between 2012 and 2016, a proportion of drying higher than 50 % is recorded on 93 ONDE sites and their spatial distribution is very patchy at the scale of France (black and dark grey dots, Fig. 5a). There are only 158 ONDE sites with at least one drying event every year and a variability of drying locations can be observed across years. The south-east of France is heavily affected by rivers drying, where the proportion of drying can exceed 75 % annually (black dots, Fig. 5b–f). The north-western part of France is less affected, although many ONDE sites show a proportion of drying observed above 50 % in 2014 and 2016 (Fig. 5d and f). North-eastern France is rather affected in 2012, 2014 and 2015, where several ONDE sites have more than 75 % of drying statuses (Fig. 5b, d and e). South-western France is particularly affected in 2012 and 2015 (Fig. 5b and e).

3.1.2 Comparison of flow intermittence between the raw ONDE and HYDRO datasets

The HYDRO dataset includes 90 gauging stations located on streams considered IRES, which represents only 5.6 % of the 1600 gauging stations against 35 % for ONDE sites. At the national scale, the number of IRES seems underrepresented in the south-western, central, and north-eastern parts of France and Corsica in comparison with sites experiencing drying in the ONDE network (Fig. 6).

Table 1. Annual statistics on flow intermittence calculated on HYDRO gauging stations between 1 May and 30 September.

	Stations with at least one drying event	Stations with drying > 50 %	Frequency of discharge < $1\,\mathrm{L\,s^{-1}}$
2012	79	19	32.7
2013	47	14	37.9
2014	54	15	32.9
2015	76	21	31.1
2016	71	19	28.6

The number of gauging stations with at least one drying event (discharge < $1\,\mathrm{L\,s^{-1}}$) observed between May and September varies between 79 in 2012 and 47 in 2014 (Table 1). The lowest numbers of gauging stations with drying events are observed in the years 2013 and 2014, while the highest numbers are related to the years 2012, 2015 and 2016. This finding is consistent with the analysis of the ONDE network (Fig. 5a, d). The frequency of drying, corresponding to the ratio between the number of dry days and the total number of days between 1 May and 30 September (153 days), in contrast, is quite constant over the years (~ 30 %). The number of gauging stations with drying events over more than 50 % of the time varies little between wet years (14 in 2013) and dry years (21 in 2015), unlike ONDE observations, suggesting a significant temporal variability in the frequency of drying between dry and wet years (Fig. 5).

3.2 Validation of the predicted regional probability of drying

3.2.1 Regression results

The LLR and LR models, calibrated over the period 2012–2016, perform well with the 2011–2017 dataset, with a mean NSE of 0.8 with the LR model against 0.7 with the LLR

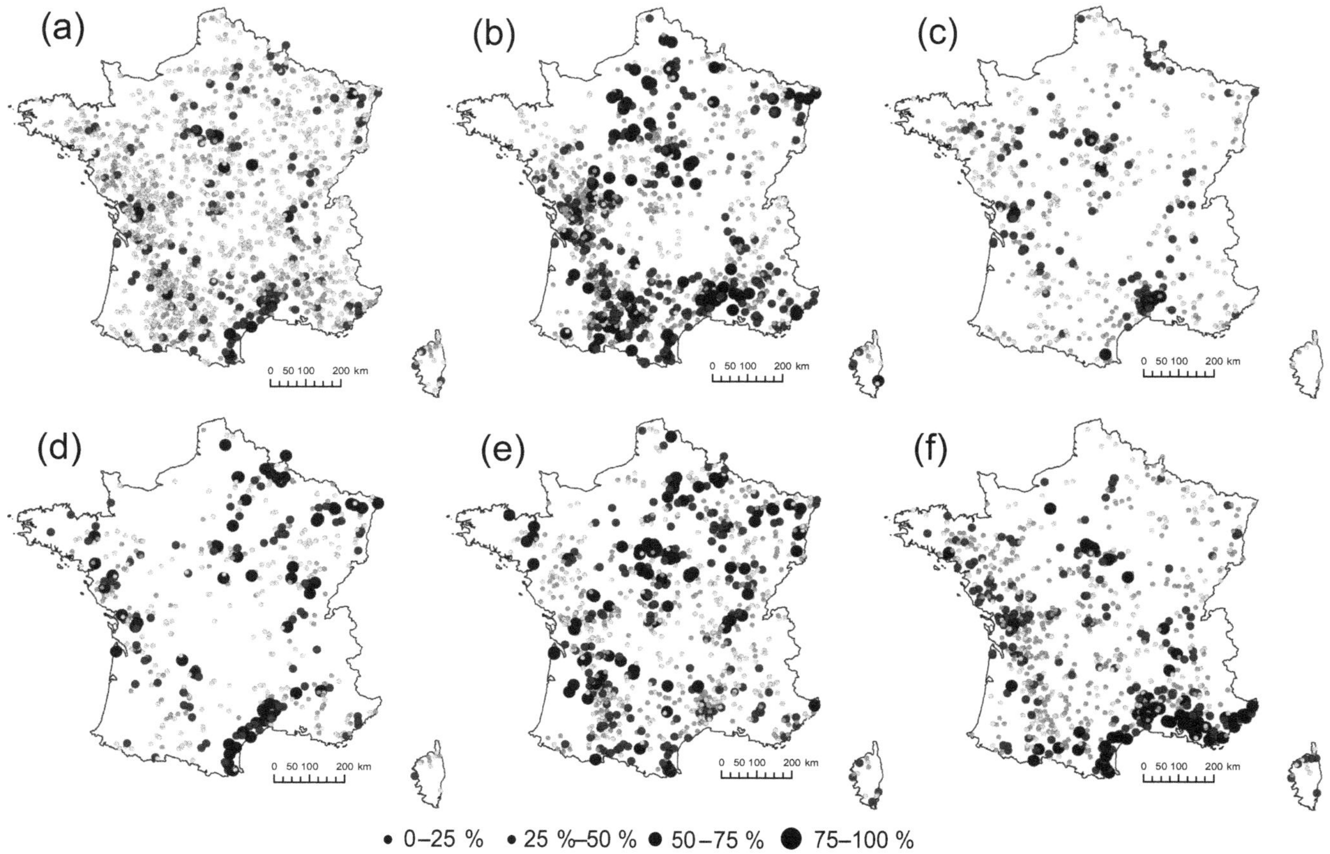

Figure 5. Distribution of the percentages of drying observed at ONDE sites for the years **(a)** 2012–2016, **(b)** 2012, **(c)** 2013, **(d)** 2014, **(e)** 2015 and **(f)** 2016.

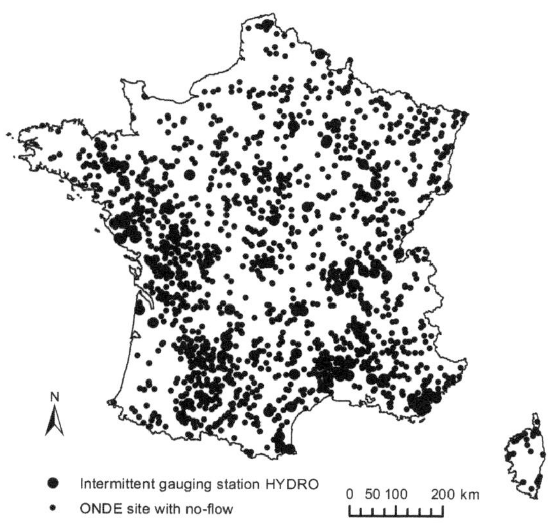

Figure 6. Map of ONDE sites and HYDRO gauging stations with at least one drying.

model (Fig. 7a and b). With the LR model, 50 % of the HER2–HR combinations obtain a NSE greater than 0.8, representing a coverage of 65 % of the French territory, while 33 % of HER2–HR combinations display a NSE higher than 0.8 (50 % of coverage of France) with the LLR model. Regions with the highest performances are located in sedimentary plains, in the south-east of France and in the Pyrenees Mountains. Conversely, the worst performances are obtained in the mountainous regions of the Alps as well as in the Massif Central. In these regions the size of the HER2 is rather small and the number of ONDE sites, gauging stations and piezometers per HER2–HR combination are certainly too few to derive reliable relations. Despite pooling, estimating RPoD remains impossible for nine HER2–HR combinations (4.5 % of coverage of France) because the number of ONDE sites, gauging stations and piezometer sites is insufficient (less than five) to perform the regression analysis.

The performance level is lower when the 1989–2017 dataset is used in models: the mean NSE with the LR and LLR models is 0.7 and 0.6, respectively (Fig. 7c and d).

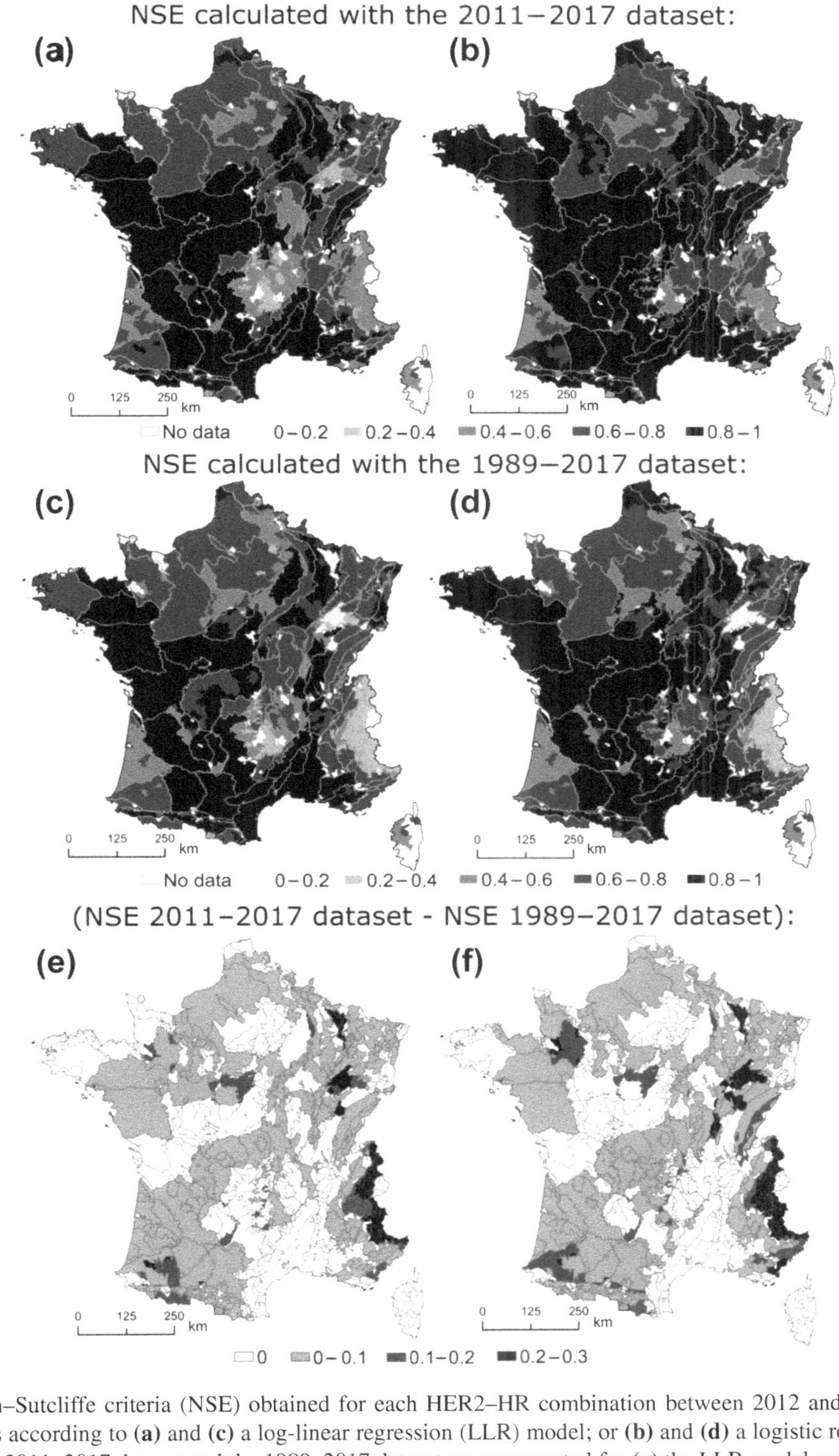

Figure 7. Map of Nash–Sutcliffe criteria (NSE) obtained for each HER2–HR combination between 2012 and 2016 with the 2011–2017 and 1989–2017 datasets according to **(a)** and **(c)** a log-linear regression (LLR) model; or **(b)** and **(d)** a logistic regression (LR) model. NSE differences between the 2011–2017 dataset and the 1989–2017 dataset are represented for **(e)** the LLR model and **(f)** the LR model.

The LR and LLR models lead to a similar performance range. However, the LR model outperforms the LLR model in terms of number of HER2–HR combinations, with NSE greater than 0.8 (Fig. 7c and d). The performance is sensitive to the dataset. As expected, the best results are obtained with the denser network. A decrease in NSE by more than 0.2 is identified for 5 % of the French territory when the 1989–2017 dataset is used (black areas; Fig. 7e and f). The regions with the most degraded values of NSE are small HER2–HR combinations located in eastern France (Fig. 7e and f).

The decrease in performance is mainly due to the difference in the number of gauging stations and piezometers between the two datasets (Fig. 8). The most degraded NSEs correspond to HER2–HR combinations where the number of gauging stations and piezometers considered in regressions is the most reduced, i.e. with a loss higher than 50 % of the stations (black and dark grey dots; Fig. 8a and b). However, the decrease in performance remains low (the difference in NSE is below 0.1 for 75 and 64 % of HER2–HR combinations with the LLR and LR models, respectively).

3.2.2 Comparison to the POC database

The observed proportion of drying $RPoD_{POC}$ is rather well simulated by both LLR and LR models with the 2011–2017 explanatory dataset (NSE > 0.7 except for the year 2011; Fig. 9). In addition, the models are able to capture small fluctuations of $RPoD_{POC}$ during the summer period. The best results during the year 2011 are obtained with the LLR model (black curve; Fig. 9) and the LR model overestimates $RPoD_{POC}$ by 3 % (dashed grey curve; Fig. 9). In 2012, the decline in water levels is more gradual than in 2011 and a marked peak is reached in September with 40 % of $RPoD_{POC}$ (Fig. 9). This pattern is well reproduced by both models with a good fit to all observation points (Fig. 9). The year 2013 is less affected by drying occurrence and the maximum $RPoD_{POC}$ does not exceed 20 % (Fig. 9). Curves of both models fit to observations well until the end of August. Note that the LR model is slightly closer to the observations around the peak in September compared to the LLR model. However, the LR model overestimates the $RPoD_{POC}$ at the end of September and in October.

When the 1989–2017 dataset is used for explanatory variables, the simulations of RPoD are weakly degraded with both models (Fig. 9d, e, f). However, the simulated pattern is similar to the observed one. The LLR model outperforms the LR model during the 3 years of validation with the 1989–2017 dataset (black curve; Fig. 9d, e, f).

3.2.3 Temporal patterns assessment of models between 2012 and 2017

During the calibration period, the LLR and LR models tend to better simulate the RPoD during dry years 2012 and 2016 (NSE = 0.8 with the LLR and LR models; Table 2) than during wet years (e.g. 2014 with NSE < 0.7). The NSEs are lower during the months of May and June when few drying events are observed, while NSEs are much better during the driest months of August and September.

During the validation year of 2017, both models obtain a similar performance over the year independent of datasets (NSE = 0.7).

Monthly NSEs in 2017 follow the same trend as monthly NSEs of the calibration period, with lower NSEs in May (NSEs < 0.4) and June (NSEs = 0.5) and higher NSEs in July, August and September (NSEs = 0.6) with both models independent of datasets. Figure 10 shows the dispersion between predicted RPoD and drying statuses observed at ONDE sites in the scatter plot during the validation year 2017 (Fig. 10a and b) in comparison with the year 2012, which obtains the better NSE during calibration period (Fig. 10c and d). The NSEs obtained in 2017 are 0.72 with the LLR model and 0.68 with the LR model against 0.83 and 0.81 in 2012, respectively. The performance is slightly lower in 2017 but remains acceptable with NSEs close to 0.7, and both models seem able to predict RPoD from the calibration period.

3.3 Application of regional models

3.3.1 Modelling of intermittencies severity between 2012 and 2016

Both models have been applied using the 2011–2017 dataset. Figure 11 displays the maximum number of consecutive days ($D_{RPoD > 20\,\%}$) with RPoD higher than 20 % simulated by both the LLR and LR models. The most affected regions are located in the south-east of France and in the sedimentary plains, which is consistent with the spatial pattern obtained from the ONDE observations (Fig. 5). The most impacted year followed the same hierarchy: the year 2012 is the most critical year, with 30 % of France displaying $D_{RPoD > 20\,\%}$ higher than 60 days followed by the year 2015 (20 % of France with $D_{RPoD > 20\,\%}$ > 60 days) and 2016 (15 % of France with $D_{RPoD > 20\,\%}$ > 60 days) (Fig. 11). The years 2013 and 2014 are weakly affected, with 5 % and 6 % of France with $D_{RPoD > 20\,\%}$ higher than 60 days, respectively.

The LR model tends to simulate shorter periods of drying, particularly in HER2–HR combinations located in south-eastern France in 2013 and 2014 (Fig. 11). However, there is an overall agreement between RPoD simulated by both models in terms of spatial and temporal extent of dry streams.

3.3.2 Reconstitution of historical regional probability of drying

The trend temporal patterns of RPoD predicted by the two models, considering the 1989–2017 dataset, look similar between 1989 and 2016, and the simulated RPoD fit well to $RPoD_{ONDE}$ (Fig. 12).

The proportion of drying is highly variable over the total simulation period, with alternating dry (1989 to 1991, 2003 to 2006, 2009 to 2012) and wet (1994 to 1995, 2000 to 2002, 2013 to 2014) phases. In spite of inter-annual variability, peaks of RPoD occur regularly between August and September, whether in dry years or wet years. This finding is consistent with the preeminence of rainfall-fed river flow regimes with low flows in summer, in France.

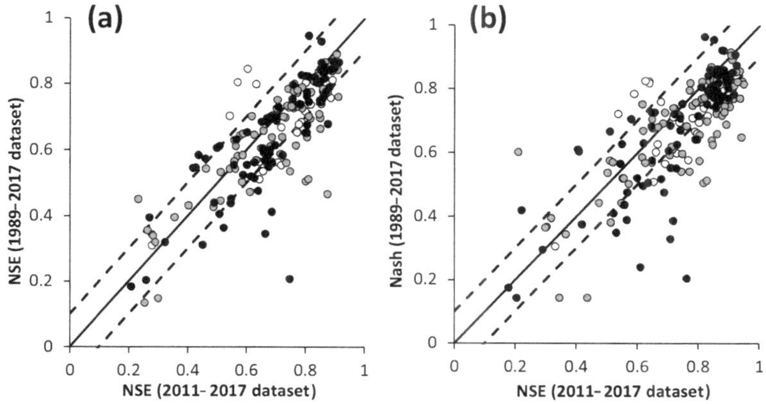

Figure 8. NSE calculated for each HER2–HR combination between 2012 and 2016 with the 1989–2017 dataset as a function of NSE calculated with the 2011–2017 dataset with, respectively, **(a)** the LLR model and **(b)** the LR model. The colours of the dots represent the proportion of gauging stations and piezometers lost between the 2011–2017 database and the 1989–2017 database: losses < 50 % (white); losses between 50 and 75 % (grey); losses > 75 % (black).

The highest values of RPoDs (above 35 % over France) are observed in 1989, 1990, 1991, 2003 and 2005 (black curve, Fig. 12a and b). The RPoDs simulated during these dry years are out of the range of the observed values over the calibration period (2012–2016). Estimations are thus uncertain. However, the high values of RPoD are consistent with observations reported in previous studies (e.g. Larue and Giret, 2004; Snelder et al., 2013; Caillouet et al., 2017). Conversely, the years less affected by drying are simulated in 1994, 2001 and 2014, with an average RPoD below 15 % throughout the year (black curve, Fig. 12a and b).

Results obtained with the LLR model are more contrasted in terms of extreme values than those obtained with the LR model (Fig. 12b).

4 Discussion

4.1 ONDE network complementarity with conventional flow monitoring network

The analysis of the ONDE observations shows that the proportion of rivers undergoing drying is significantly higher (35 %) than that observed with the conventional monitoring (HYDRO database, 8 %). This proportion, although related to a short period of records 2012 and 2016, is consistent with the percentage of 39 % of river segments classified as intermittent by Snelder et al. (2013). This analysis confirms the under-representation of IRES in the French HYDRO database, and probably others in other countries (flows are often uncontrolled in IRES). Without gauging stations located on headwaters, Snelder et al. (2013) were unable to predict IRES in eastern France (see Fig. 9, p. 2694). The high density of ONDE sites makes it possible to improve the detection of drying events and lead to better understand the spatial distribution of IRES located at the upstream extent of the hydrographic network. The ONDE network encompasses various hydrological conditions, which provides a more accurate assessment of inter-annual variability, differentiating between dry years (2012, 2015 and 2016) and wet years (2013, 2014) with clearly few drying occurrences.

The validation of the LR and LLR models against the spatially dense POC database also demonstrates the spatial representativeness of the ONDE network. Thanks to the qualitative information provided and to models such as the statistical models developed here, it is now possible to capture drying events at the regional scale.

The ONDE sites are located on small headwater streams which can be very reactive to external disturbances (rainfall deficit, change in air temperature, increase in water withdrawals, etc.) and by nature are more likely to be IRES. The gauging stations available in the HYDRO database are located on larger streams and their hydrologic response to changes in external factors (environmental or human) is slower and drying occurred with greater inertia under temperate climate. Their uneven distribution across France does not allow us to accurately characterise the inter-annual variability of drying development. Overall, the ONDE network provides very complementary information to conventional flow monitoring, leading to a better understanding of the processes of drying in upstream catchments.

4.2 Dependency on spatial gauging network density

The performance obtained with the LR and LLR models is slightly better with the 2011–2017 dataset (mean NSE = 0.75) than those obtained with the 1989–2017 dataset (mean NSE > 0.65), whose network is less dense. HER2–HR combinations are the most degraded where the number of monitoring stations is the most decreased between the two

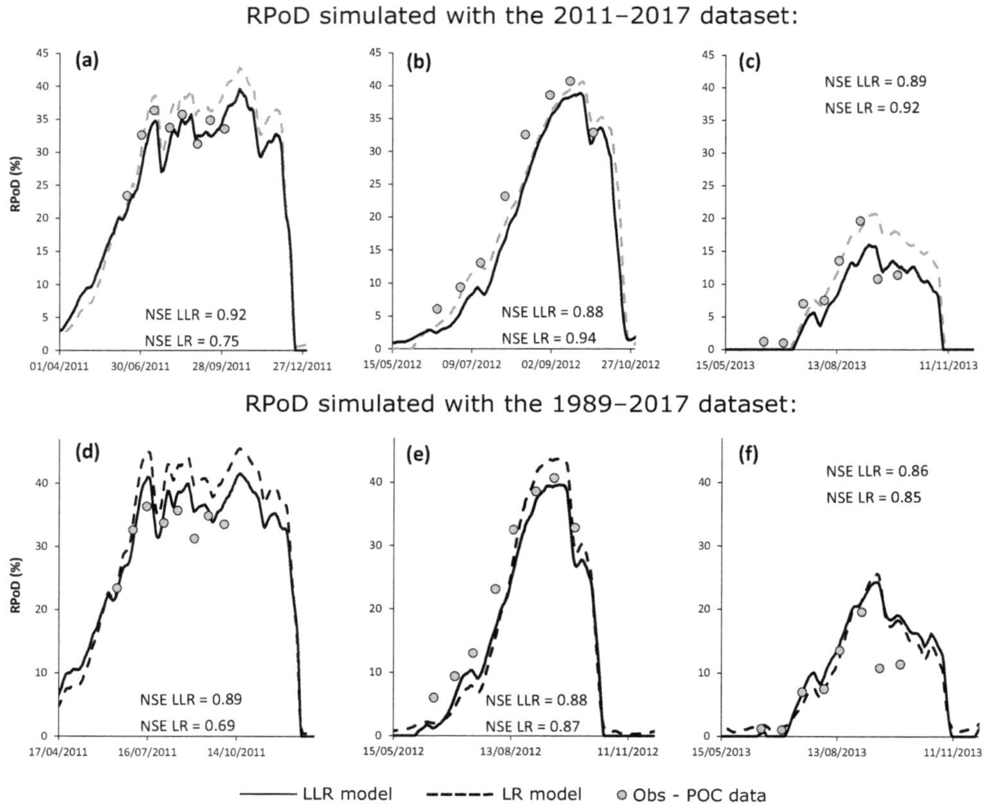

Figure 9. Comparison between observed proportion of drying $RPoD_{POC}$ and RPoD predicted by the LLR and LR models with the 2011–2017 dataset in **(a)** 2011, **(b)** 2012 and **(c)** 2013 and with the 1989–2017 dataset in **(d)** 2011, **(e)** 2012 and **(f)** 2013.

Table 2. NSE criteria obtained between 2012 and 2017 with the LLR and LR models calibrated over the period 2012–2016.

		2011–2017 dataset						1989–2017 dataset					
		Calibration					Valid.	Calibration					Valid.
		2012	2013	2014	2015	2016	2017	2012	2013	2014	2015	2016	2017
LLR model	May	0.2	0.0	0.5	0.5	0.6	0.4	0.2	0.0	0.3	0.0	0.7	0.2
	Jun	0.6	0.3	0.8	0.5	0.8	0.5	0.6	0.3	0.5	0.3	0.8	0.5
	Jul	0.7	0.5	0.6	0.6	0.8	0.7	0.7	0.5	0.5	0.4	0.8	0.6
	Aug	0.8	0.6	0.7	0.7	0.8	0.6	0.7	0.5	0.5	0.5	0.8	0.6
	Sep	0.7	0.8	0.6	0.6	0.7	0.6	0.6	0.7	0.5	0.5	0.6	0.6
	May–Sep	0.8	0.8	0.7	0.7	0.8	0.7	0.8	0.7	0.5	0.6	0.8	0.7
LR model	May	0.2	0.0	0.5	0.1	0.6	0.3	0.3	0.0	0.3	0.0	0.7	0.2
	Jun	0.6	0.5	0.8	0.5	0.8	0.4	0.6	0.4	0.5	0.3	0.7	0.4
	Jul	0.7	0.6	0.5	0.6	0.8	0.6	0.7	0.4	0.5	0.4	0.8	0.6
	Aug	0.7	0.6	0.7	0.6	0.7	0.6	0.6	0.4	0.5	0.4	0.7	0.5
	Sep	0.6	0.8	0.6	0.7	0.7	0.6	0.5	0.6	0.4	0.5	0.6	0.6
	May–Sep	0.8	0.8	0.7	0.7	0.8	0.7	0.8	0.7	0.5	0.6	0.8	0.7

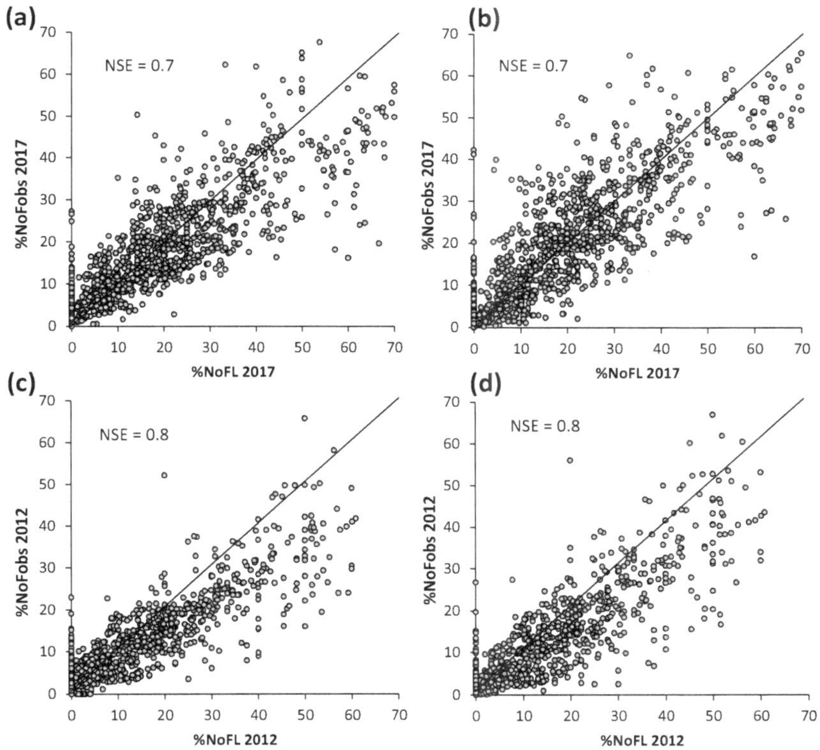

Figure 10. Scatter plot of the predicted RPoD (x axis) and drying observed at ONDE sites (y axis) in 2017 and 2012 simulated with the 2011–2017 dataset by (**a** and **c**) the LLR model and (**b** and **d**) the LR model.

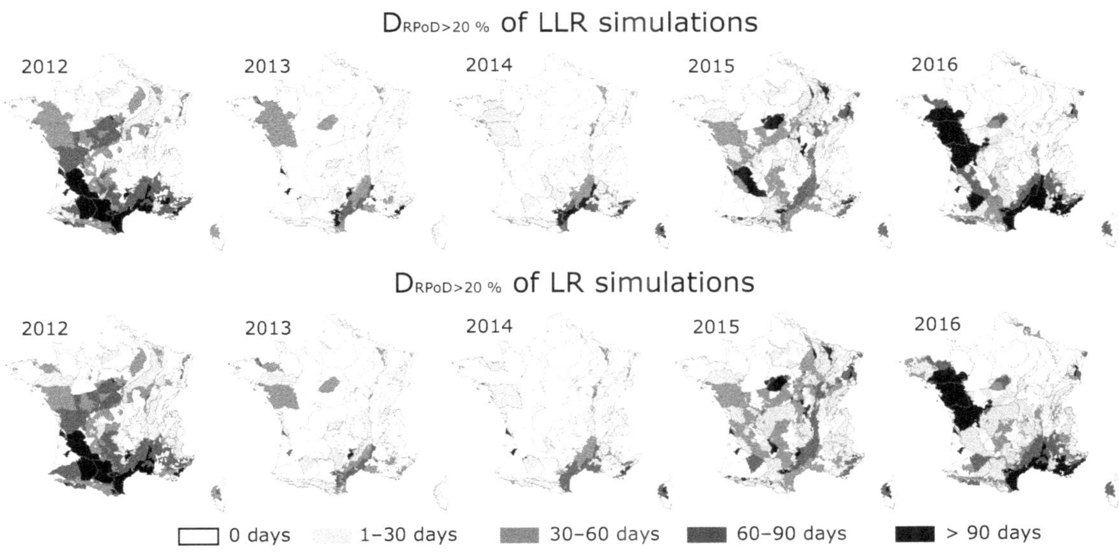

Figure 11. Maximum duration of consecutive days with RPoD higher than 20 % simulated with the LLR and LR models.

datasets. The accuracy of the predictions is dependent on the number of gauging stations, ONDE sites and piezometers available to calibrate the regressions. The highest NSEs are obtained in western sedimentary plains and south-eastern France, where a significant number of streams have dryings regardless of years (Fig. 5). The dominant river flow regime in these regions is mainly influenced by precipitation and the lowest water levels are reached in August and September,

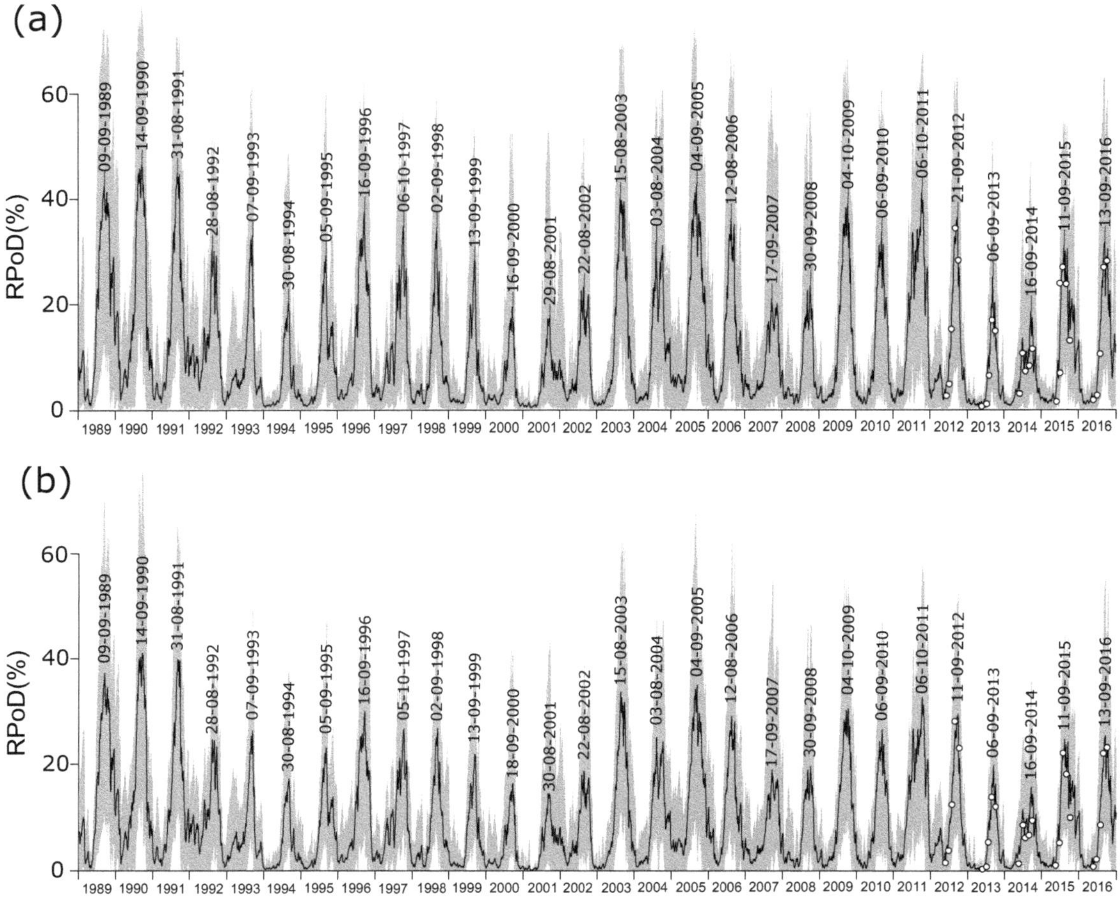

Figure 12. RPoD simulated between 1989 and 2016 with the 1989–2017 dataset with **(a)** the LR model and **(b)** the LLR model. The grey area represents the RPoD between the 90th percentile and the 10th percentile simulated on the HER2–HR combination, the black curve represents the average RPoD simulated by the HER2–HR combination and white dots represent the mean $RPoD_{ONDE}$ for each observation date. Dates mentioned correspond to the day of the maximum average RPoD simulated by the HER2–HR combination (black curve) of each year.

which corresponds to the monitoring period of the ONDE database. They benefit from a dense monitoring network (gauging stations, ONDE sites, piezometers), which allows a better representation of the hydrological functioning of streams located within the same HER2. Conversely, performance was poor in mountainous areas such as in the Alps or the Massif Central ($NSE < 0.4$) where river flow regimes are diversified combining rainfall and snowmelt influences. By construction, the area of HER2–HR combination in mountains is reduced, which leads to a limited number of monitoring stations, certainly not sufficient to fit the models. Moreover, the observation period for ONDE sites was limited between May and September and dryings can be missed, particularly for streams influenced by snow or ice melting with potential drying periods in winter. In regions potentially concerned with drying events from the May–September period, the actual ONDE monitoring strategy needs to be adapted to provide reliable temporal observations and extrapolations of drying frequencies.

We have chosen to average the non-exceedance frequencies of flows and groundwater levels in order to increase the monitoring network. If models had been calibrated using only gauging stations, performance will have been globally similar, or slightly better, in some HER2–HR combinations (Fig. 13). Therefore, we could not validate the real gain of using groundwater-level data in addition to discharge data. This is certainly due to the dominant proportion of the gauging stations compared to the piezometers. Indeed, in the 2011–2017 dataset, the proportion of gauging stations is greater than 75 % for more than 70 % of HER2–HR combinations, whereas the proportion of piezometers exceeds 70 % in only 5 % of HER2–HR combinations. Groundwater-level data thus have a small weight in regressions for this dataset. However, in the 1989–2017 dataset, the proportion

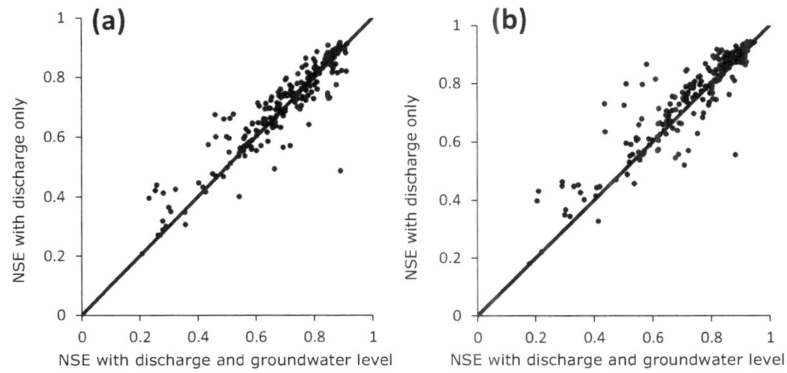

Figure 13. Comparison of NSE obtained with regression including only the discharge variable as a function of NSE obtained with discharge and groundwater-level variables in the 2011–2017 dataset with **(a)** the LLR model and **(b)** the LR model.

of piezometers is greater than 70 % in more than 30 % of HER2–HR combinations. The presence of piezometers increases the density of the monitoring network in HER2–HR combinations with few available gauging stations. Thanks to groundwater-level data, RPoD can be predicted on more HER2–HR combinations.

4.3 Interest in reconstructing the dynamic regional probability of drying

Spatio-temporal simulation of the probability of drying is crucial for advancing our understanding of IRES ecology and management. Some aquatic species can persist in a dry reach for a few days, weeks or months, while some are highly sensitive to desiccation (Datry, 2012; Storey and Quinn, 2013; Stubbington and Datry, 2013). Estimation of the total duration of days with drying at the reach scale is therefore needed to understand biological patterns in river networks (Kelso and Entrekin, 2018). To our knowledge, no study has proposed to reconstruct daily flow-state time series of headwater streams at the country scale such as France (> 500 000 km^2) using discrete observations in time and space. In the literature, studies at national scale remain focused on the detection and the mapping of IRES because these rivers are historically poorly investigated and their proportion in existing hydrographic networks remains inaccurate or misunderstood (Nadeau and Rains, 2007; Snelder et al., 2013). Recently, several studies proposed alternative methodologies in order to estimate metrics in ungauged IRES (Gallart et al., 2016) or to predict daily streamflow in river basins experiencing flow intermittence (De Girolamo et al., 2017b), but remain applicable at the local scale.

This study provides a first regional approach to use discrete data obtained from regular observations. The average non-exceedance frequency is a global hydrological statistic that only captures the hydrological conditions at the regional scale in modelling the RPoD. For rainfall-driven river flow regimes, the effect of rainfall events on flow intermittence at the HER2–HR scale is probably indirectly reflected by the daily discharge and groundwater levels used to calculate the average non-exceedance frequency. However, when more observation data are available, it is likely that including more detailed descriptors of rainfall events and local geology could improve our approach. In France, based on the 2011–2017 dataset, both models suggest the highest values of RPoD along the Mediterranean coast ($D_{RPoD > 20\,\%} > 100$ days each year). Rivers in this region are subject to a predominantly pluvial regime (Class 7; Sauquet et al., 2008), i.e. hot and dry summers followed by intense rainfall events in autumn, leading to high flows in November (Skoulikidis et al., 2017b). The catchments in this region are small and particularly reactive to environmental changes, making them highly sensitive to flow intermittence. Rivers located in the sedimentary plain in western France are also very impacted by flow intermittence. The regime is also influenced by precipitation and for the basins subject to intense agriculture significant water abstractions during summer in this region reduce water availability in rivers and in aquifers which are no longer able to support the low water levels, leading to increased flow intermittence. Regarding alteration issues in our datasets, we do not have access to the exact location and the volumes of water withdrawal for irrigation purposes. However, due to their upstream location, water availability is expected to be low, which may limit potential withdrawals and as a consequence flow alteration at ONDE sites. The alteration of groundwater levels is unknown because no information is available. However, in sedimentary plains where agricultural crops dominate the landscape, we are not sure that no human action affects low flows. It is important to note that the responses of biological communities to artificial flow intermittence are still poorly understood compared to natural IRES (Datry et al., 2014b; Skoulikidis et al., 2017a).

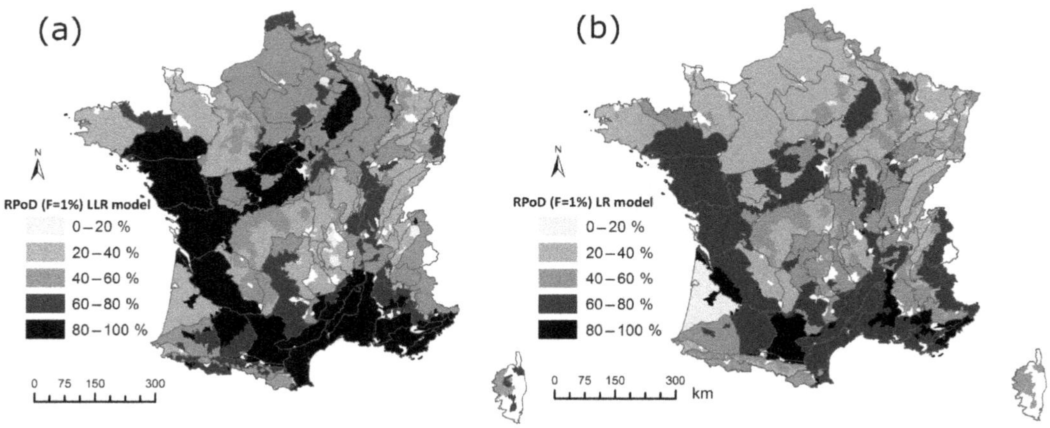

Figure 14. Regional probability of drying simulated with $F = 1\,\%$ predicted with **(a)** the LLR model and **(b)** the LR model.

4.4 Validity of historical regional probability of drying during severe low-flow period

The second application aimed at reconstructing historical RPoD over the period 1989–2016. Both models suggest the highest values of mean RPoD (> 35 %) in 1989–1991, 2003 and 2005. During these dry years, predicted values of RPoD result from extrapolation but are consistent with published studies (Mérillon and Chaperon, 1990; Moreau, 2004). For example, Mérillon (1992) estimated that for the whole of France, 11 000 km of rivers were dried at the end of the summers of 1989 and 1990. Caillouet et al. (2017) found that the low-flow event observed in 1989–1990 was particularly severe in terms of duration and affected 95 % of France. Snelder et al. (2013) showed from 628 gauging stations that the years 1989–1991, 2003 and 2005 witnessed particularly high values of duration and frequency of drying events. They found that regions with the highest probability of drying were located along the Mediterranean and Atlantic coasts, which is consistent with ONDE observations and with our results.

Both models suggest the same sequence of dry and wet years. However, the application of the LLR model leads to less contrasted RPoD than the LR model (Fig. 12).

To illustrate these differences, the RPoD has been simulated by both models with an extreme F of 1 % (Fig. 14). The $\mathrm{RPoD}_{\mathrm{LLR}}$ is significantly higher and exceeds 80 % in 30 % of the study area against only 5 % of the area with the $\mathrm{RPoD}_{\mathrm{LR}}$. On the other hand, models simulate low RPoD in HER2–HR combinations where the $\mathrm{RPoD}_{\mathrm{ONDE}}$ is very low between 2012 and 2016, even when F was 1 %, because this situation never occurred during the calibration period (Fig. 14). The logistic function of the LR model takes an S-shape which induced a decrease in the slope of the curve toward extreme values observed during the calibration period (2012–2016). The truncated logarithmic function of the LLR model is not bounded and RPoD can reach 100 % during extreme low-flow events by extrapolation. Since the ONDE network monitoring period does not include a period with drought as severe as in the 1990s, it is not currently possible to assess the relative performance of the two models. Refining extrapolated values requires additional information on headwater collected during more severe droughts than those observed during the last 5 years and then gives support to the pursuit of the ONDE network.

5 Conclusion

This paper investigates the spatial and temporal dynamics of the regional probability of drying (RPoD) of headwater streams by taking benefit from qualitative and discontinuous data provided by the ONDE network. Two models based on linear or logistic regressions have been developed and succeeded in reconstructing the temporal dynamics of RPoD. They are based on a strong relationship between the non-exceedance frequencies of discharges and groundwater levels as a function of the proportion of drying statuses observed at ONDE sites per HER2–HR combination. LLR and LR models show similar performance and perform well between 2011 and 2017. The accuracy of predictions is dependent on the number of gauging stations, ONDE sites and piezometers available to calibrate the regressions. Regions with the highest performance are located in the sedimentary plains, where the monitoring network is dense and where the RPoD is the highest. Conversely, the worst performances are obtained in the mountainous regions. Finally, both models have been used to reconstruct historical RPoD between 1989 and 2016 and suggest the highest values of mean RPoD (> 35 %) in 1989–1991, 2003 and 2005. This is consistent with other published studies, but the high density of ONDE sites makes it possible to improve the detection of drying events and lead to better capturing of the spatial distribution of IRES located at the upstream extent of the hydrographic network. Moreover, the duration of drying is of importance for ecologists for the composition and persistence of aquatic species.

From a methodological point of view, our method relating discrete drying observation obtained by citizen science networks to continuous daily gauging data seems robust across the highly diverse (climate and topography) regions of France, and provides good predictions in an independent region excluded from the calibration process (PoC). These two results suggest a potential application of our approach in other countries. Citizen science creates opportunities to overcome the lack of hydrological data, contributes to densifying the flow-state observation network (Turner and Richter, 2011; Buytaert et al., 2014) and remains less expensive than the installation of additional gauging stations to survey flow intermittence. The next step will be to use this regional approach to simulate the RPoD in future periods by taking into account effects of climate change through predicted discharge and groundwater-level data. This would allow quantification of the evolution of the probability of drying between the current period and the different climate projections provided by the latest IPCC Report (IPCC 2014) and would assist decision makers in defining protocols for restoring flows with appropriate measures to preserve aquatic ecosystems (Woelfle-Erskine, 2017).

Secondly, further work is needed to develop an approach capable of reconstructing the drying dynamics locally by differentiating each stream. Our approach remains spatially valid for estimating RPoDs at the scale of HER2–HR combinations, but does not allow characterisation of the variability of drying occurrence between nearby streams within these regions. From a methodological point of view, statistical tools such as neural networks (Breiman, 2001) have shown good ability to assess both the occurrence and extent of perennial and temporary segments (González-Ferreras and Barquín, 2017) and could be investigated as an alternative method to reconstruct locally the temporal variability of drying.

Author contributions. AB, NL and ES developed the main ideas and designed the experiments. AB implemented the algorithms and analysed the results. AB prepared the manuscript with contributions from all co-authors.

Competing interests. The authors declare that they have no conflict of interest.

Acknowledgements. The authors wish to thank Anne van Loon and Catherine Sefton for their valuable comments, suggestions and positive feedback on the manuscript. The research project was partly funded by the French Agency for Biodiversity (AFB, formerly ONEMA). This study is based upon works from COST Action CA15113 (SMIRES, Science and Management of Intermittent Rivers and Ephemeral Streams, www.smires.eu), supported by COST (European Cooperation in Science and Technology).

Edited by: Laurent Pfister

References

Acuña, V., Datry, T., Marshall, J., Barceló, D., Dahm, C. N., Ginebreda, A., McGregor, G., Sabater, S., Tockner, K., and Palmer, M. A.: Why should we care about temporary waterways?, Science, 343, 1080–1081, 2014.

Acuña, V., Hunter, M., and Ruhí, A.: Managing temporary streams and rivers as unique rather than second-class ecosystems, Biol. Conserv., 211, 12–19, https://doi.org/10.1016/j.biocon.2016.12.025, 2017.

AFB (French Agency for Biodiversity): Observatoire National des Etiages (ONDE), available at: https://onde.eaufrance.fr (last access: 23 May 2018), 2012–2018.

Benda, L., Hassan, M. A., Church, M., and May, C. L.: Geomorphology Of Steepland Headwaters: The Transition From Hillslopes To Channels1, J. Am. Water Resour. As., 41, 835–851, 2005.

Benstead, J. P. and Leigh, D. S.: An expanded role for river networks, Nat. Geosci., 5, 678–679, 2012.

Boulton, A. J.: Conservation of ephemeral streams and their ecosystem services: what are we missing?: Editorial, Aquat. Conserv., 24, 733–738, https://doi.org/10.1002/aqc.2537, 2014.

Breiman, L.: Random forests, Mach. Learn., 45, 5–32, 2001.

Brugeron, A., Allier, D., and Klinka, T.: Approche exploratoire des liens entre référentiels hydrogéologique et hydrographique: Premières identifications des piézomètres potentiellement représentatifs d'une relation nappe/rivière et contribution à leur valorisation, Rapport final BRGM/RP-61047-FR, 241 pp., 2012.

Buytaert, W., Zulkafli, Z., Grainger, S., Acosta, L., Alemie, T. C., Bastiaensen, J., De Bièvre, B., Bhusal, J., Clark, J., Dewulf, A., Foggin, M., Hannah, D. M., Hergarten, C., Isaeva, A., Karpouzoglou, T., Pandeya, B., Paudel, D., Sharma, K., Steenhuis, T., Tilahun, S., Van Hecken, G., and Zhumanova, M.: Citizen science in hydrology and water resources: opportunities for knowledge generation, ecosystem service management, and sustainable development, Front. Earth Sci., 2, 1–21, https://doi.org/10.3389/feart.2014.00026, 2014.

Caillouet, L., Vidal, J.-P., Sauquet, E., Devers, A., and Graff, B.: Ensemble reconstruction of spatio-temporal extreme low-flow events in France since 1871, Hydrol. Earth Syst. Sci., 21, 2923–2951, https://doi.org/10.5194/hess-21-2923-2017, 2017.

Clarke, A., Mac Nally, R., Bond, N., and Lake, P. S.: Macroinvertebrate diversity in headwater streams: a review, Freshwater Biol., 53, 1707–1721, https://doi.org/10.1111/j.1365-2427.2008.02041.x, 2008.

Costigan, K. H., Jaeger, K. L., Goss, C. W., Fritz, K. M., and Goebel, P. C.: Understanding controls on flow permanence in intermittent rivers to aid ecological research: integrating meteorology, geology and land cover: Integrating Science to Understand Flow Intermittence, Ecohydrology, 9, 1141–1153, https://doi.org/10.1002/eco.1712, 2016.

Datry, T.: Benthic and hyporheic invertebrate assemblages along a flow intermittence gradient: effects of duration of dry events: River drying and temporary river invertebrates, Freshwater Biol., 57, 563–574, https://doi.org/10.1111/j.1365-2427.2011.02725.x, 2012.

Datry, T., Larned, S. T., Fritz, K. M., Bogan, M. T., Wood, P. J., Meyer, E. I., and Santos, A. N.: Broad-scale patterns of invertebrate richness and community composition in temporary rivers: effects of flow intermittence, Ecography, 37, 94–104, https://doi.org/10.1111/j.1600-0587.2013.00287.x, 2014a.

Datry, T., Larned, S. T., and Tockner, K.: Intermittent Rivers: A Challenge for Freshwater Ecology, BioScience, 64, 229–235, https://doi.org/10.1093/biosci/bit027, 2014b.

Datry, T., Pella, H., Leigh, C., Bonada, N., and Hugueny, B.: A landscape approach to advance intermittent river ecology, Freshwater Biol., 61, 1200–1213, https://doi.org/10.1111/fwb.12645, 2016a.

Datry, T., Fritz, K., and Leigh, C.: Challenges, developments and perspectives in intermittent river ecology, Freshwater Biol., 61, 1171–1180, https://doi.org/10.1111/fwb.12789, 2016b.

De Girolamo, A. M., Lo Porto, A., Pappagallo, G., Tzoraki, O., and Gallart, F.: The Hydrological Status Concept: Application at a Temporary River (Candelaro, Italy): Evaluating Hydrological Status In Temporary Rivers, River Res. Appl., 31, 892–903, https://doi.org/10.1002/rra.2786, 2015.

De Girolamo, A. M., Bouraoui, F., Buffagni, A., Pappagallo, G., and Lo Porto, A.: Hydrology under climate change in a temporary river system: Potential impact on water balance and flow regime, River Res. Appl., 33, 1219–1232, https://doi.org/10.1002/rra.3165, 2017a.

De Girolamo, A. M., Barca, E., Pappagallo, G., and Lo Porto, A.: Simulating ecologically relevant hydrological indicators in a temporary river system, Agr. Water Manage., 180, 194–204, https://doi.org/10.1016/j.agwat.2016.05.034, 2017b.

Döll, P., and Schmied, H. M.: How is the impact of climate change on river flow regimes related to the impact on mean annual runoff? A global-scale analysis, Environ. Res. Lett., 7, 014037, https://doi.org/10.1088/1748-9326/7/1/014037, 2012.

Eng, K., Wolock, D. M., and Dettinger, M. D.: Sensitivity of Intermittent Streams to Climate Variations in the USA: Sensitivity of Intermittent Streams, River Res. Appl., 32, 885–895, https://doi.org/10.1002/rra.2939, 2016.

Finn, D. S., Bonada, N., Múrria, C., and Hughes, J. M.: Small but mighty: headwaters are vital to stream network biodiversity at two levels of organization, J. N. Am. Benthol. Soc., 30, 963–980, https://doi.org/10.1899/11-012.1, 2011.

Fritz, K. M., Hagenbuch, E., D'Amico, E., Reif, M., Wigington, P. J., Leibowitz, S. G., Comeleo, R. L., Ebersole, J. L., and Nadeau, T.-L.: Comparing the Extent and Permanence of Headwater Streams From Two Field Surveys to Values From Hydrographic Databases and Maps, J. Am. Water Resour. As., 49, 867–882, https://doi.org/10.1111/jawr.12040, 2013.

Gallart, F., Prat, N., García-Roger, E. M., Latron, J., Rieradevall, M., Llorens, P., Barberá, G. G., Brito, D., De Girolamo, A. M., Lo Porto, A., Buffagni, A., Erba, S., Neves, R., Nikolaidis, N. P., Perrin, J. L., Querner, E. P., Quiñonero, J. M., Tournoud, M. G., Tzoraki, O., Skoulikidis, N., Gómez, R., Sánchez-Montoya, M. M., and Froebrich, J.: A novel approach to analysing the regimes of temporary streams in relation to their controls on the composition and structure of aquatic biota, Hydrol. Earth Syst. Sci., 16, 3165–3182, https://doi.org/10.5194/hess-16-3165-2012, 2012.

Gallart, F., Llorens, P., Latron, J., Cid, N., Rieradevall, M., and Prat, N.: Validating alternative methodologies to estimate the regime of temporary rivers when flow data are unavailable, Sci. Total Environ., 565, 1001–1010, https://doi.org/10.1016/j.scitotenv.2016.05.116, 2016.

Garcia, C., Gibbins, C. N., Pardo, I., and Batalla, R. J.: Long term flow change threatens invertebrate diversity in temporary streams: Evidence from an island, Sci. Total Environ., 580, 1453–1459, https://doi.org/10.1016/j.scitotenv.2016.12.119, 2017a.

Garcia, C., Amengual, A., Homar, V., and Zamora, A.: Losing water in temporary streams on a Mediterranean island: Effects of climate and land-cover changes, Global Planet. Change, 148, 139–152, https://doi.org/10.1016/j.gloplacha.2016.11.010, 2017b.

Gómez, R., Hurtado, I., Suárez, M. L., and Vidal-Abarca, M. R.: Ramblas in south-east Spain: threatened and valuable ecosystems, Aquat. Conserv., 15, 387–402, https://doi.org/10.1002/aqc.680, 2005.

González-Ferreras, A. M. and Barquín, J.: Mapping the temporary and perennial character of whole river networks: Mapping Flow Permanence In River Network, Water Resour. Res., 53, 6709–6724, https://doi.org/10.1002/2017WR020390, 2017.

Huxter, E. H. H. and van Meerveld, H. J.: Intermittent and Perennial Streamflow Regime Characteristics in the Okanagan, Can. Water Resour. J., 37, 391–414, https://doi.org/10.4296/cwrj2012-910, 2012.

IPCC.: Climate Change 2014: Synthesis Report. Contribution of Working Groups I, II and III to the Fifth Assessment Report of the Intergovernmental Panel on Climate Change, edited by: Core Writing Team, Pachauri, R. K., and Meyer, L. A., IPCC, Geneva, Switzerland, 151 pp., 2014.

Jaeger, K. L., Olden, J. D., and Pelland, N. A.: Climate change poised to threaten hydrologic connectivity and endemic fishes in dryland streams, P. Natl. Acad. Sci. USA, 111, 13894–13899, https://doi.org/10.1073/pnas.1320890111, 2014.

Kelso, J. E. and Entrekin, S. A.: Intermittent and perennial macroinvertebrate communities had similar richness but differed in species trait composition depending on flow duration, Hydrobiologia, 807, 189–206, https://doi.org/10.1007/s10750-017-3393-y, 2018.

Larned, S. T., Datry, T., Arscott, D. B., and Tockner, K.: Emerging concepts in temporary-river ecology, Freshwater Biol., 55, 717–738, https://doi.org/10.1111/j.1365-2427.2009.02322.x, 2010.

Larue, J. P. and Giret A.: The drying up of streams in the Maine basin from 1989 to 1992, Norois, 192, 117–133, https://doi.org/10.4000/norois.944, 2004.

Lee, S.: Application of logistic regression model and its validation for landslide susceptibility mapping using GIS and remote sensing data, Int. J. Remote Sens., 26, 1477–1491, https://doi.org/10.1080/01431160412331331012, 2005.

Leigh, C. and Datry, T.: Drying as a primary hydrological determinant of biodiversity in river systems: a broad-scale analysis, Ecography, 40, 487–499, https://doi.org/10.1111/ecog.02230, 2017.

Leigh, C., Boulton, A. J., Courtwright, J. L., Fritz, K., May, C. L., Walker, R. H., and Datry, T.: Ecological research and management of intermittent rivers: an historical review and future directions, Freshwater Biol., 61, 1181–1199, https://doi.org/10.1111/fwb.12646, 2016.

Leopold, L. B.: A View of the River, Harvard University Press, Cambridge, Massachusetts, USA, 1994.

Leopold, L. B., Wolman, M. G., and Miller, J. P.: Fluvial Processes in Geomorphology, Dover Publications, New York, USA, 1964.

Mérillon, Y.: Sécheresse, gestion de l'eau et pratiques agricoles, L'Information agricole, 649, 25–26, 1992.

Mérillon, Y. and Chaperon, P.: Drought in France in 1989, Houille Blanche, 325–340, https://doi.org/10.1051/lhb/1990025, 1990.

Meyer, J. L., Strayer, D. L., Wallace, J. B., Eggert, S. L., Helfman, G. S., and Leonard, N. E.: The Contribution of Headwater

Streams to Biodiversity in River Networks1: The Contribution of Headwater Streams to Biodiversity in River Networks, J. Am. Water Resour. As., 43, 86–103, https://doi.org/10.1111/j.1752-1688.2007.00008.x, 2007.

Moreau, F.: Drought crisis management: Loire basin example, Houille Blanche, 4, 70–76, https://doi.org/10.1051/lhb:200404010, 2004.

Nadeau, T.-L. and Rains, M. C.: Hydrological Connectivity Between Headwater Streams and Downstream Waters: How Science Can Inform Policy1: Hydrological Connectivity Between Headwater Streams and Downstream Waters: How Science Can Inform Policy, J. Am. Water Resour. As., 43, 118–133, https://doi.org/10.1111/j.1752-1688.2007.00010.x, 2007.

Nash, J. E. and Sutcliffe, J. V.: River flow forecasting through conceptual models part I – A discussion of principles, J. Hydrol., 10, 282–290, https://doi.org/10.1016/0022-1694(70)90255-6, 1970.

Nowak, C. and Durozoi, B.: Observatoire National Des Etiages, Note technique, ONEMA, 2012.

Pella, H., Lejot, J., Lamouroux, N., and Snelder, T.: Le réseau hydrographique théorique (RHT) français et ses attributs environnementaux, Géomorphologie: relief, processus, environnement, 18, 317–336, 2012.

Pumo, D., Caracciolo, D., Viola, F., and Noto, L. V.: Climate change effects on the hydrological regime of small non-perennial river basins, Sci. Total Environ., 542, 76–92, https://doi.org/10.1016/j.scitotenv.2015.10.109, 2016.

Reynolds, L. V., Shafroth, P. B., and LeRoy Poff, N.: Modeled intermittency risk for small streams in the Upper Colorado River Basin under climate change, J. Hydrol., 523, 768–780, https://doi.org/10.1016/j.jhydrol.2015.02.025, 2015.

Sarremejane, R., Cañedo-Argüelles, M., Prat, N., Mykrä, H., Muotka, T., and Bonada, N.: Do metacommunities vary through time? Intermittent rivers as model systems, J. Biogeogr., 44, 2752–2763, https://doi.org/10.1111/jbi.13077, 2017.

Sauquet, E., Gottschalk, L., and Krasovskaia, I.: Estimating mean monthly runoff at ungauged locations: an application to France, Hydrol. Res., 39, 403, https://doi.org/10.2166/nh.2008.331, 2008.

Skoulikidis, N. T.: The environmental state of rivers in the Balkans—A review within the DPSIR framework, Sci. Total Environ., 407, 2501–2516, https://doi.org/10.1016/j.scitotenv.2009.01.026, 2009.

Skoulikidis, N. T., Vardakas, L., Karaouzas, I., Economou, A. N., Dimitriou, E., and Zogaris, S.: Assessing water stress in Mediterranean lotic systems: insights from an artificially intermittent river in Greece, Aquat. Sci., 73, 581–597, https://doi.org/10.1007/s00027-011-0228-1, 2011.

Skoulikidis, N. T., Vardakas, L., Amaxidis, Y., and Michalopoulos, P.: Biogeochemical processes controlling aquatic quality during drying and rewetting events in a Mediterranean non-perennial river reach, Sci. Total Environ., 575, 378–389, https://doi.org/10.1016/j.scitotenv.2016.10.015, 2017a.

Skoulikidis, N. T., Sabater, S., Datry, T., Morais, M. M., Buffagni, A., Dörflinger, G., Zogaris, S., del Mar Sánchez-Montoya, M., Bonada, N., Kalogianni, E., Rosado, J., Vardakas, L., De Girolamo, A. M., and Tockner, K.: Non-perennial Mediterranean rivers in Europe: Status, pressures, and challenges for research and management, Sci. Total Environ., 577, 1–18, https://doi.org/10.1016/j.scitotenv.2016.10.147, 2017b.

Snelder, T. H., Datry, T., Lamouroux, N., Larned, S. T., Sauquet, E., Pella, H., and Catalogne, C.: Regionalization of patterns of flow intermittence from gauging station records, Hydrol. Earth Syst. Sci., 17, 2685–2699, https://doi.org/10.5194/hess-17-2685-2013, 2013.

Storey, R. G. and Quinn, J. M.: Survival of aquatic invertebrates in dry bed sediments of intermittent streams: temperature tolerances and implications for riparian management, Freshw. Sci., 32, 250–266, https://doi.org/10.1899/12-008.1, 2013.

Stubbington, R. and Datry, T.: The macroinvertebrate seedbank promotes community persistence in temporary rivers across climate zones, Freshwater Biol., 58, 1202–1220, https://doi.org/10.1111/fwb.12121, 2013.

Turner, D. S. and Richter, H. E.: Wet/Dry Mapping: Using Citizen Scientists to Monitor the Extent of Perennial Surface Flow in Dryland Regions, Environ. Manage., 47, 497–505, https://doi.org/10.1007/s00267-010-9607-y, 2011.

Vadher, A. N., Millett, J., Stubbington, R., and Wood, P. J.: Drying duration and stream characteristics influence macroinvertebrate survivorship within the sediments of a temporary channel and exposed gravel bars of a connected perennial stream, Hydrobiologia, 814, 121–132, https://doi.org/10.1007/s10750-018-3544-9, 2018.

van Meerveld, H. J. I., Vis, M. J. P., and Seibert, J.: Information content of stream level class data for hydrological model calibration, Hydrol. Earth Syst. Sci., 21, 4895–4905, https://doi.org/10.5194/hess-21-4895-2017, 2017.

Vardakas, L., Kalogianni, E., Economou, A. N., Koutsikos, N., and Skoulikidis, N. T.: Mass mortalities and population recovery of an endemic fish assemblage in an intermittent river reach during drying and rewetting, Fund. Appl. Limnol., 190, 331–347, https://doi.org/10.1127/fal/2017/1056, 2017.

Wasson, J.-G., Chandesris, A., Pella, H., and Blanc, L.: Typology and reference conditions for surface water bodies in France: the hydro-ecoregion approach, TemaNord, 566, 37–41, 2002.

Woelfle-Erskine, C.: Collaborative Approaches to Flow Restoration in Intermittent Salmon-Bearing Streams: Salmon Creek, CA, USA, Water, 9, 217, https://doi.org/10.3390/w9030217, 2017.

8

Characterization and evaluation of controls on post-fire streamflow response across western US watersheds

Samuel Saxe[1,2], **Terri S. Hogue**[1], **and Lauren Hay**[2]

[1]Civil and Environmental Engineering and Hydrologic Science and Engineering, Colorado School of Mines, Golden, Colorado, USA

[2]National Research Program, United States Geological Survey, Lakewood, Colorado, USA

Correspondence: Samuel Saxe (ssaxe@usgs.gov)

Abstract. This research investigates the impact of wildfires on watershed flow regimes, specifically focusing on evaluation of fire events within specified hydroclimatic regions in the western United States, and evaluating the impact of climate and geophysical variables on response. Eighty-two watersheds were identified with at least 10 years of continuous pre-fire daily streamflow records and 5 years of continuous post-fire daily flow records. Percent change in annual runoff ratio, low flows, high flows, peak flows, number of zero flow days, baseflow index, and Richards–Baker flashiness index were calculated for each watershed using pre- and post-fire periods. Independent variables were identified for each watershed and fire event, including topographic, vegetation, climate, burn severity, percent area burned, and soils data.

Results show that low flows, high flows, and peak flows increase in the first 2 years following a wildfire and decrease over time. Relative response was used to scale response variables with the respective percent area of watershed burned in order to compare regional differences in watershed response. To account for variability in precipitation events, runoff ratio was used to compare runoff directly to PRISM precipitation estimates. To account for regional differences in climate patterns, watersheds were divided into nine regions, or clusters, through k-means clustering using climate data, and regression models were produced for watersheds grouped by total area burned. Watersheds in Cluster 9 (eastern California, western Nevada, Oregon) demonstrate a small negative response to observed flow regimes after fire. Cluster 8 watersheds (coastal California) display the greatest flow responses, typically within the first year following wildfire. Most other watersheds show a positive mean relative response. In addition, simple regression models show low correlation between percent watershed burned and streamflow response, implying that other watershed factors strongly influence response.

Spearman correlation identified NDVI, aridity index, percent of a watershed's precipitation that falls as rain, and slope as being positively correlated with post-fire streamflow response. This metric also suggested a negative correlation between response and the soil erodibility factor, watershed area, and percent low burn severity. Regression models identified only moderate burn severity and watershed area as being consistently positively/negatively correlated, respectively, with response. The random forest model identified only slope and percent area burned as significant watershed parameters controlling response.

Results will help inform post-fire runoff management decisions by helping to identify expected changes to flow regimes, as well as facilitate parameterization for model application in burned watersheds.

1 Introduction

The number of wildfires in the western United States (US) is increasing annually, on average costing federal agencies billions of dollars a year in suppression efforts (Whitlock, 2004) and causing an increase in flood events destructive to both life and infrastructure (Neary et al., 2005; Pausas et al., 2008). Westerling et al. (2006) and Westerling (2016) showed that the western fire regime exhibited a significant transition from infrequent and short-duration events to higher-frequency, longer-duration regimes during the mid-

1980s. The greatest increases in fire frequency were found to occur in mid-elevation forests, most commonly in the Northern Rockies, Sierra Nevada, southern Cascades, and western coast ranges in northern California and southern Oregon (Littell et al., 2009). This marked change is strongly correlated with climate change impacts, such as warmer springs and longer dry seasons, commonly in occurrence with reduced winter precipitation rates and earlier spring snowmelt. Overall, Westerling et al. (2016) determine that, though land-use history may be a significant factor in the spatial distribution of wildfires within specific forest types, changes in fire regimes in the western US are most likely attributable to recent changes in climate. Other notable research has also provided significant correlative evidence between climate change and wildfire occurrences (Littell et al., 2009; Moritz et al., 2010).

Though wildfires are a part of the natural process of vegetation dynamics, they cause wide-ranging changes to ecosystems (Neary, 2003; Santos et al., 2015) depending on numerous factors, most importantly burn severity. Studies examining the effects of wildfires on a small scale, such as in plot-sized and laboratory experiments, show high fire temperatures can result in the combustion of organic matter within soils and cause permanent alteration to the chemical structure of local clays, decreasing soil stability (Shakesby and Doerr, 2006). Water-repellent soil layers can be created in a discrete layer on or below the soil surface through chemical bonding of the combusted organic matter to mineral particles, potentially increasing overall topsoil erosion rates in burned regions (Wilkinson et al., 2009), though this hydrophobicity is highly variable, depending on fire behavior, burn severity, and soil properties (DeBano, 2000).

At larger scales, such as entire watersheds or multiple watershed systems, studies of post-fire erosion rates have shown incompatible conclusions (Moody and Martin, 2001; Owens et al., 2013; Smith et al., 2011), though this is most likely due to the variability of precipitation events and general climate patterns (Moody et al., 2013). In terms of water quality, contaminant levels can be dramatically increased for many years after a wildfire in both soil (Burke et al., 2010) and stream systems (Emelko et al., 2011; Stein et al., 2012; Burke et al., 2013), increasing the workload on source water protection organizations in communities reliant upon burned watersheds for drinking and agricultural water. Furthermore, wildfires are readily attributed as the cause of substantial increases in debris flows (Benavides-Solorio and MacDonald, 2001; Cannon et al., 2001; Meyer et al., 2001).

Studies evaluating post-fire water yield change are highly disparate owing to the transient nature of climate patterns, variations in basin geomorphology, and vegetation recovery patterns, and the resulting complex interactions (Moody et al., 2013). For example, studies in rangeland regions of the US found moderate increases in flow, infiltration, and erosion rates after major wildfires, with trends continuing for as long as 15 years (Emmerich and Cox, 1994; Pierson et al., 2009; Hester et al., 1997; Kinoshita and Hogue, 2015). Fires in chaparral environments, such as in southern California, exhibited increased flows of up to as much as 2 orders of magnitude, with much of this occurring in the dry season (Coombs and Melack, 2013; Kinoshita and Hogue, 2011, 2015; Loáiciga et al., 2001). Fires in other chaparral environments were found to also yield flow increases, such as in South Africa (Lindley et al., 1988; Scott, 1993), Cyprus (Hessling, 1999), and France (Lavabre et al., 1993). Additional increases to post-fire flow regimes were found in temperate, forested catchments as well (Neary et al., 2005; Watson et al., 2001). A concise summary of historic changes in US post-fire stream systems is found in Neary et al. (2005), documenting changes in first-year runoff and peak flows, encompassing a range of ecological regions. Conversely, several studies found limited or no significant changes to hydrologic systems post fire, or attributed fluctuations to natural annual variability (Aronica et al., 2002; Bart and Hope, 2010; Britton, 1991; Townsend and Douglas, 2000).

These discrepancies in post-fire flow response lead to the question of which watershed characteristics have the greatest influence over the observed response. Moody et al. (2013) provide a succinct summary of soil-related theories, such as reduced infiltration due to increases in soil-water repellency, increased overland flow velocities due to increased bare ground, and reduced infiltration caused by soil-sealing. Theories commonly found in the literature attribute flow changes to a wider range of factors, including reduction in interception and evapotranspiration (Lavabre et al., 1993; Scott, 1993) and increased hydrophobicity of soils (Neary et al., 2005). In regards to altered peak flows, conflicting evidence is found regarding the importance of burned watershed areas, with some studies finding an inverse correlation between peak flows and watershed size (Biggio and Cannon, 2001; Neary et al., 2005) and others finding no relation at all (Bart and Hope, 2010).

The current research builds upon prior studies but develops a more comprehensive and systematic assessment of post-fire streamflow dynamics by examining burned watersheds that encompass a wide spectrum of climatological and geophysical parameters across the western US. A variety of statistical parameters are also examined which describe changes to flow regimes at several levels. Furthermore, the variability in response by distinct regions is investigated, anticipating distinct differences influenced by regional climate. With downstream communities at risk of flooding, and also relying on catchment runoff for water supply, investigating alterations in post-fire discharge over large scales will provide critical information for regional managers on post-fire runoff mitigation. In addition, understanding factors controlling discharge response will help inform development and calibration of surface water models used for post-fire streamflow predictions.

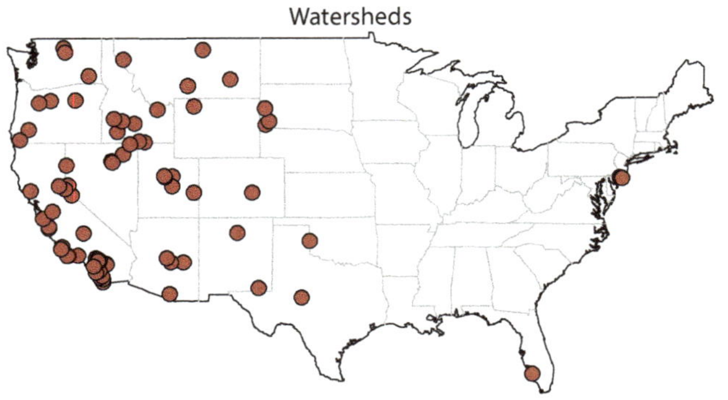

Figure 1. Continental United States-scale map of the locations of the 82 watersheds utilized in this study.

2 Study area

A total of 82 burned watersheds were utilized in the current study (Fig. 1), encompassing a range of spatial, temporal, climatological, and topographic factors (Figs. 2 and 3). Selected watersheds were limited to those with burn areas of $>5\,\%$ and adequate discharge records (continuous 15-year daily flow, including 10 years pre-fire and 5 years post-fire) (U.S. Geological Survey, 2014) identified through the GAGES-II database (Falcone, 2011). The majority of available watersheds are overwhelmingly found in the western US, predominantly in California, Oregon, and Idaho, with several located in the northeast, Florida, and Kentucky. Due to discharge and burn severity data limitations, the fires cover a temporal range from water years 1984 through 2010. A water year is the 12-month period 1 October through 30 September designated by the calendar year in which it ends. Percent area burned ranges from 5 to 97 %, with a mean of 25 %, over a range of watershed areas from 4.6 to 7000 km^2. The wide spatial distribution of the studied watersheds results in mean elevations and burn-area slopes varying from 13 to 2760 m and from 0.11 to 16 %, respectively (Fig. 2).

The most important difference between many of these watersheds is the variability in climate, the values of which were collected from the GAGES-II dataset (Falcone, 2011). GAGES-II watersheds have at least one 20-year period with continuous daily flow records. Average basin precipitation ranges from 29 to 220 $mm\,yr^{-1}$, with a mean of 72 $mm\,yr^{-1}$, and average temperature ranges from 1.4 to 23 °C, with a mean of 10 °C. Important for identifying snow dominated regions is the percent of precipitation (PPT) that falls as snow (%Snow/PPT), which ranges from 0 to 72 % for the study watersheds. Relative humidity ranges from 39 to 73 %, with a mean of 55, and potential evapotranspiration ranges from 400 to 1200 $mm\,yr^{-1}$, with a mean of 633 $mm\,yr^{-1}$.

Vegetation types vary across the watersheds as well. Evergreen forest and shrub vegetation are the dominant land cover type over most watersheds used in this study (Fig. 2). The high proportion of evergreen is due to the dominance

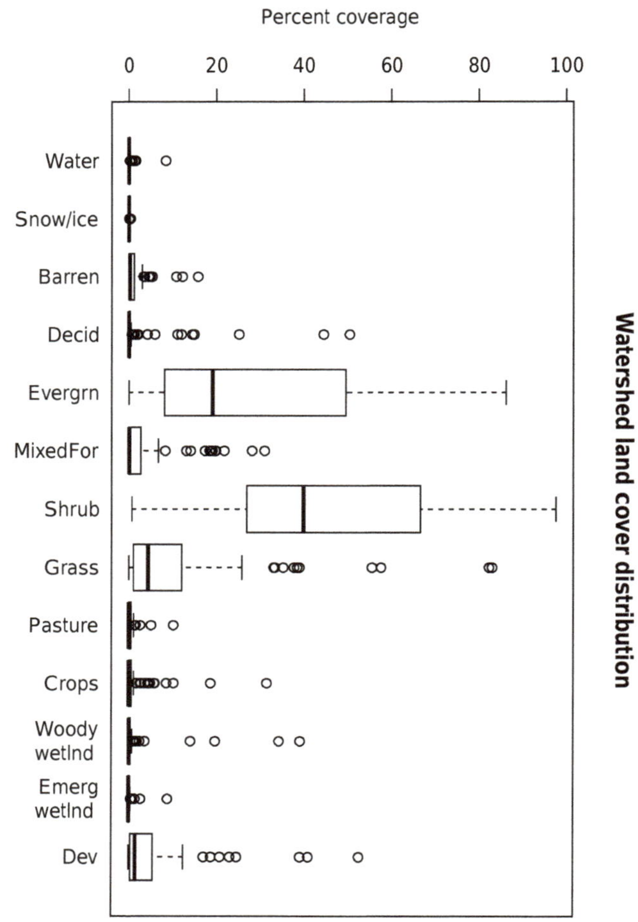

Figure 2. Boxplot of National Land Cover Database (Homer et al., 2004) distribution across study watersheds.

of this study's high elevation fires in mountainous regions and shrub prevalence is due to the abundance of study watersheds found in the chaparral regions of southern California. Grassland, mixed forest, and developed land account for a smaller proportion of land cover types. Barren land and wetland account for only a small percentage of land cover types throughout the watersheds in this study.

3 Methods

3.1 Data collection

Watershed parameters were chosen to encompass the variability in geophysical, burn area, and climatic conditions found throughout the watersheds used in this study (Appendix A). To identify spatial trends in post-fire response, watersheds were first grouped through k-means clustering based on geographic and climatological data (described in Sect. 3.3 below).

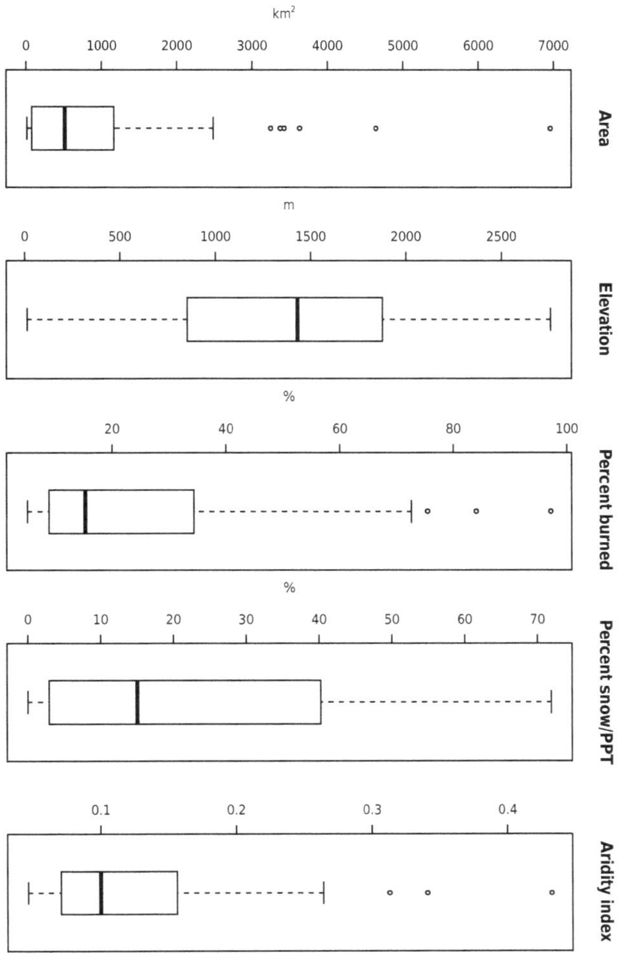

Figure 3. Boxplots summarizing the distribution of watershed area, elevation, percent area burned, percent precipitation that falls as snow, and aridity index across study watersheds.

3.1.1 Watershed selection

Selected watersheds were required to have continuous daily flow records (> 95 % of daily flow records accounted for in each year) for a minimum of 10 years pre-fire and 5 years post-fire. Using the approximately 9000 watersheds in the GAGES-II dataset (Falcone, 2011), delineated watersheds were spatially cross-referenced with the Monitoring Trends in Burn Severity (MTBS) database of historic wildfires from 1984 to 2010 (Eidenshink et al., 2007). The results were again cross-referenced with USGS daily flow records (U.S. Geological Survey, 2014) to identify watersheds with the required flow records, resulting in 263 unique watersheds in the US with greater than 5 % total burn area in a single water year. Of these, 23 contained 2–3 wildfires within the same year, burning over 5 % of the total area. The remainder contained only a single significant fire in the year of interest. Further exclusion of watersheds was based on the presence of major dams within the watershed flow regimes extracted from the GAGES-II database (Falcone, 2011), resulting in a final collection of 82 watersheds.

3.1.2 Streamflow and precipitation data

Daily flow and peak flow data were obtained from the USGS National Water Information System (U.S. Geological Survey, 2014) subset for the range of 10 years pre-fire and 5 years post-fire for each selected watershed. Monthly precipitation data were collected from the PRISM database, a nation-wide 4 km resolution gridded monthly dataset that extrapolates station climate measurements over unmonitored areas using a topographic- and climate-based algorithm (PRISM Climate Group, 2004). Monthly national precipitation rasters were averaged for each watershed for all months within the flow record period.

3.1.3 Climatological data

Watershed climatological parameters included percent of precipitation that falls as snow (%Snow/PPT) and the aridity index (AI). The %Snow/PPT for each watershed was available through the GAGES-II dataset (Falcone, 2011) and the aridity index was calculated for each watershed as

$$AI = \frac{P_{avg}}{PET_{avg}}, \quad (1)$$

where P_{avg} is average annual precipitation and PET_{avg} is average annual potential evapotranspiration, both of which were available in the GAGES-II dataset.

3.1.4 Burn severity data

Burn severity is the classification of burn areas relating visible changes in living and non-living biomass, fire byproducts, and soil exposure within one growing season, including low, moderate, and high severity categories (Eidenshink et al., 2007). Though categorization varies by region, some generalizations can be made. Typical high severity burns result in complete kills of canopy trees and almost complete consumption of surface litter and organic soil layers (Neary et al., 2005). Characteristics of moderate burn severity include partial canopy cover kill, completely charred or consumed understory vegetation, and widespread destruction of the soil organic layer. Low severity burns lightly scorch trees, char or consume surface litter, and produce little to no charring of the soil organic layer. Wildfires are almost always a patchwork of varying degrees of burn severity. More specifically, burn severity is the qualitative assessment of the heat pulse directed toward the ground during a fire, relating soil heating, fuel consumption, and mortality of buried plant parts (National Wildfire Coordinating Group, 2016). Burn severity data were obtained for each unique fire through the MTBS database (Eidenshink et al., 2007) and quantified as the percent area burned of the total watershed area. Values were categorized as unburned low severity, moderate severity, and high severity (Eidenshink et al., 2007).

3.1.5 Vegetation data

Vegetation data were identified for each burn area prior to the fire event. Due to the temporal gaps in the National Land Cover Database (Homer et al., 2004), an averaged normalized difference vegetation index (NDVI) was collected for pre-fire burn areas to quantitatively summarize vegetation conditions, similar to previous studies (Barbosa et al., 1999; Kinoshita and Hogue, 2011; Lee and Chow, 2015). NDVI is defined as

$$\mathrm{NDVI} = \frac{a_{\mathrm{nir}} - a_{\mathrm{vis}}}{a_{\mathrm{nir}} + a_{\mathrm{vis}}}, \tag{2}$$

where a_{nir} and a_{vis} are surface reflectance averaged over the ranges of wavelengths in the near infrared and visible spectrums, respectively. Despite NDVI having been shown to have accuracy issues related to atmospheric interference and variations in soil brightness (Carlson and Ripley, 1997), the extended timespan over which values were being averaged may have muted any such error responses.

Average values were collected for each watershed through national 32-day NDVI rasters hosted on Google Earth Engine (GEE) (Google, Inc., 2015), in turn calculated from Landsat5 composite satellite data freely available through the U.S. Landsat archive at the USGS Earth Resources Observation and Science (EROS) Center (Woodcock et al., 2008). Mean NDVI values were produced for the burn areas of all watersheds for 4 years pre-fire.

3.1.6 Soils data

Soils data were collected through an adapted version of the State Soil Geographic (STATSGO) database, a national collection of over 78 000 polygons containing a host of soil characteristics (Schwartz and Alexander, 1995). The soil erodibility factor (KFact) was utilized to numerically represent average soil types, as it provides a quantitative description of a soil's erodibility. The calculation included values of particle size, percent organic matter, percent clay, soil structure index, and profile-permeability class factor (Goldman et al., 1986). KFact increases as the potential erodibility of a soil increases.

3.1.7 Topographic data

Utilized topographic information included watershed area and the elevation, slope, and aspect of burn areas, calculated through a 30 m resolution continental US (CONUS) digital elevation model (U.S. Geological Survey, 2015). Percent burn areas of watersheds were calculated by intersection of watersheds with MTBS fire polygons.

3.2 Response variables

A range of response values were selected to quantify post-fire flow changes across a variety of flow conditions, including flows relating to dry season (low flows, baseflows) and wet season (high flows, peak flows).

3.2.1 Low, high, and peak flows

Low flow and high flow metrics were calculated for each of the 10 water years prior to the fire water year and averaged to produce a single value. A low flow value for water was computed as the average of mean daily flows with a 90 % exceedance. To reduce calculation bias due to zero flow days commonly found in ephemeral stream systems, zero flow days were eliminated from exceedance value calculations. High flows were defined similarly, with a 10 % exceedance threshold used to isolate larger volume flows (Kinoshita and Hogue, 2015). Changes in low flows and high flows were calculated as the post-fire percent change from the average of 10 water years pre-fire. Post-fire values were calculated for the first year, the second year, and the 5-year mean. Peak flows were defined as the largest daily mean flow measurement each water year and post-fire changes were calculated as the percent change of the first year, second year, and 5-year mean peak flow measurements from the pre-fire 10-year mean. Percent changes in the number of zero flow days were calculated similarly.

3.2.2 Runoff ratios

The runoff ratio is defined as the fraction of total annual runoff depth over total annual precipitation:

$$\mathrm{RO} = \frac{Q_{\mathrm{tot}}/A_{\mathrm{ws}}}{P_{\mathrm{tot}}}, \tag{3}$$

where P_{tot} is total annual precipitation, Q_{tot} is total annual runoff depth, and A_{ws} is watershed area. Runoff ratio was calculated for the 10 years pre-fire and 5 years post-fire using PRISM precipitation and USGS mean daily flow data. Post-fire runoff ratio response was calculated as the percent change between the first year, second year, and average 5-year values post-fire and the pre-fire 10-year mean.

3.2.3 Baseflow and Richards–Baker indices

Baseflow index, defined as the fraction of total streamflow that is baseflow (Baker et al., 2004), was calculated for each water year through the "hydrostats" R package (Bond, 2015), which applies the Lyne–Hollick filter (Ladson et al., 2013). Baseflow index response was calculated as the percent change in the first-, second-, and 5-year post-fire from the mean of the 10-year pre-fire.

The Richards–Baker index quantifies the frequency and rapidity of short-term changes in streamflow (flashiness) based on daily flow data through the equation

$$\text{R–B index} = \frac{\sum_{i=1}^{n} |q_i - q_{i-1}|}{\sum_{i=1}^{n} q_i}, \tag{4}$$

where q is mean daily flow and i is time (Baker et al., 2004). Richards–Baker response was calculated as the percent change from the average value over the 10 years pre-fire to the average 1, 2, and 5 years post-fire.

3.3 *k*-means clustering

To create region-specific regression models, watersheds were classified into unique areas through *k*-means clustering (MacQueen, 1967) based on all GAGES II watersheds. This approach partitions an *N*-dimensional population of observations into clusters with minimal variation, allowing for relatively simple similarity grouping. For the current study, the ideal ensemble of clusters was one that produced easily recognizable regions with unique climatological characteristics. Large-scale clustering methods have been applied in prior watershed classification studies, but utilized more complex streamflow and ecological indices as parameters (McManamay et al., 2014; Poff, 1996).

Wildfires in this study are primarily in western evergreen and shrub environments, so clustering using only these watersheds would likely produce regions biased by fire occurrence. To limit this impact, we applied the mclust package in R (Fraley et al., 2012) to cluster using the entire 9000 GAGES-II watershed to produce national regions. The mclust package was chosen over the standard kmeans function in R due to its inclusion of numerous model-based approaches and application of the Bayes information criterion (BIC) to determine the most accurate model and cluster count (Schwarz, 1978). Various groupings of simple parameters were used for clustering, including watershed latitude and longitude, elevation, AI, %Snow/PPT, and mean monthly and seasonal flow statistics.

3.4 Control evaluation methods

Characterization of the influence of watershed geophysical parameters on flow response was performed using three approaches. First, calculation of the Spearman correlation coefficient between independent and response variables. Second, calculation of two types of regression equations: (1) a standard multiple linear regression model, and (2) a logistic regression model where responses greater than 20 % are assigned a value of 1 and responses less than 20 % relative to pre-fire conditions are assigned a value of 0. Third, application of a conditional inference tree algorithm through the "party" package in R (Hothorn et al., 2006). This algorithm, also called the random forest method, first applies a significance test to identify the independent variables that have the strongest association with the response variable. It then searches for the best split points in those independent variables to partition the data into a tree.

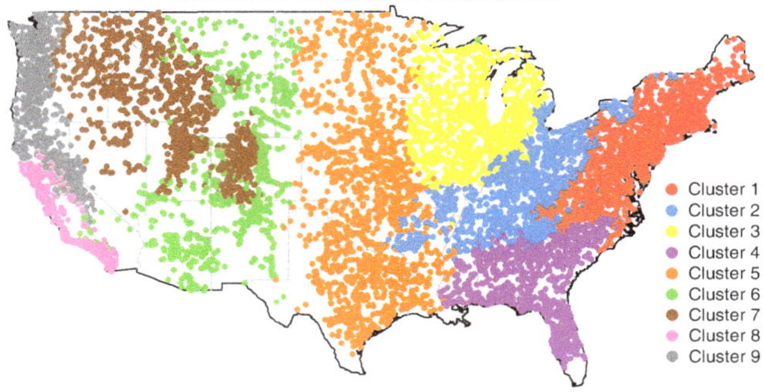

Figure 4. CONUS-scale map of the results of *k*-means clustering of the GAGES-II watershed set by climate variables.

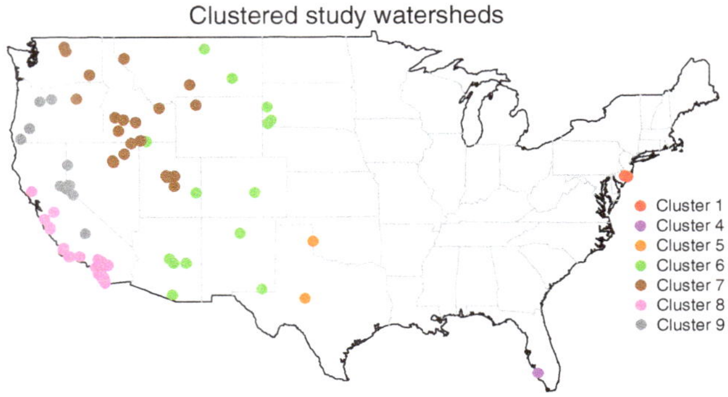

Figure 5. CONUS-scale map of the *k*-means clustered study watersheds.

4 Results and discussion

4.1 Clustering

The *k*-means clustering performed on the GAGES-II watershed set yielded nine clusters or regions (Fig. 4). The most useful clusters for the current study are 6 through 9, which assemble 77 of the 82 watersheds into unique regions (Fig. 5). These four clusters have unique characteristics (Fig. 6). Watersheds in Cluster 6, based on the median, have the highest KFact, though most are burned less than 20 %. They have relatively moderate %Snow/PPT values and the lowest AI values. Watersheds in Cluster 7 have the highest median watershed area and elevation, and accordingly the highest %Snow/PPT and the lowest NDVI. Cluster 8 contains watersheds with the widest range of relative fire sizes, including watersheds burned from as little as 10 % to as great as 97 %. Watersheds in Cluster 8 also have the lowest median elevations and watershed areas, as well as the smallest %Snow/PPT and low AI. The percent of the burned area rated as high burn severity is also the greatest, based on the me-

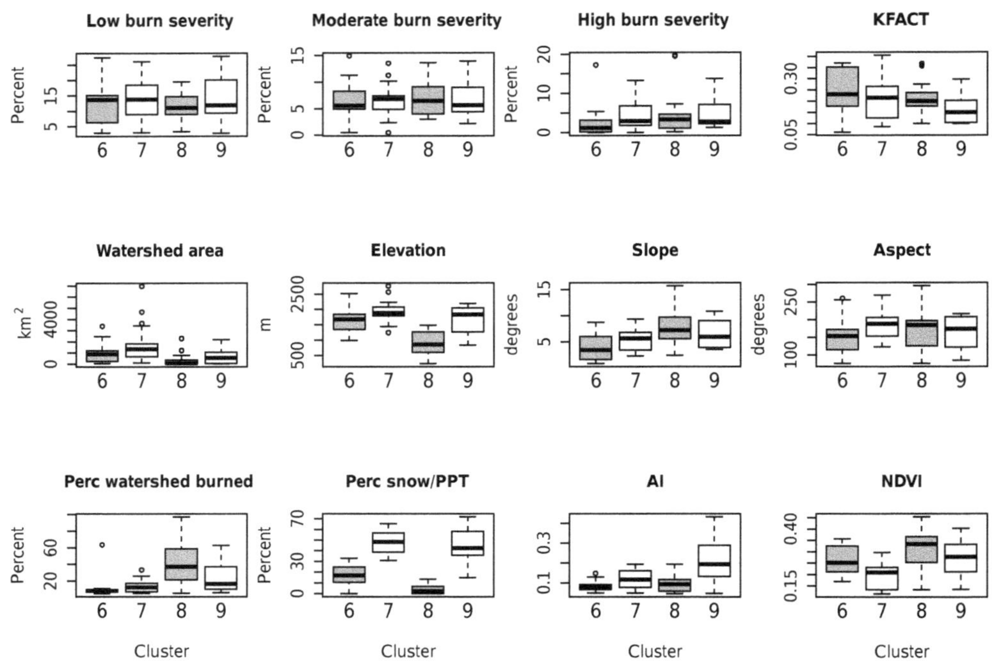

Figure 6. Boxplots of the variation in explanatory variables by cluster.

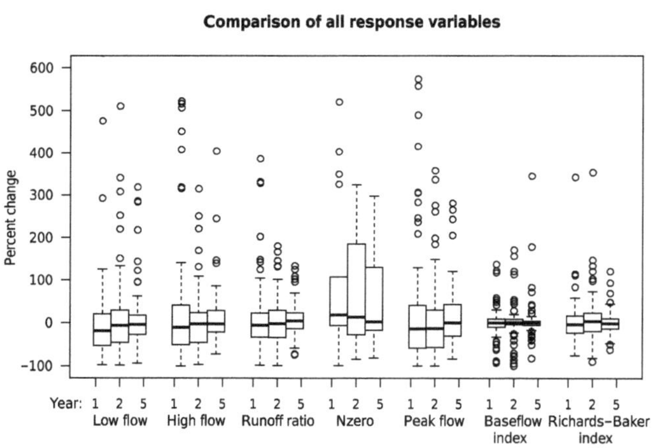

Figure 7. Boxplot of the distribution of response variables utilized in this study. Nzero refers to the number of zero flow days.

dian, in Cluster 8 watersheds. Cluster 9 watersheds have the lowest median KFact and second highest median elevations. These watersheds also have high %Snow/PPT and AI.

4.2 Response variable distribution and analyses

Calculated flow response variables indicate an extremely wide range of post-fire system responses (Fig. 7). The greatest ranges occur within variables representing changes in low flows, such as first-year change to the number of zero flow days (SD = 243 %), first-year baseflow index (236 %), and 5-year change to the number of zero flow days (201 %). The tightest ranges are typically found within mean 5-year variables where extreme changes are muted, such as 5-year Richards–Baker (28 %), runoff ratio (39 %), and baseflow index (48 %). Response variable means range from as low as 0.92 % (5-year Richards–Baker) to as great as 115 % (first-year change to the number of zero flow days). Due to the nature of response calculations, the number of zero flow day values were limited. First- and second-year zero flow day responses were found for only 22 watersheds, and for 27 watersheds for the 5-year values.

4.2.1 Trend analysis

Comparison of first-year response variables to 5-year mean variables typically results in slope values greater than one, with a mean slope of 1.9 (Fig. 8) and the greatest slopes being found in first-year low-flow, high-flow, and baseflow index response variables. Best fit lines of second-year response variables versus 5-year mean variables produce significantly lower slopes, with a mean of 0.78. We can infer from this that the greatest increases in these response variables are typically generated in the first year following a fire. Similarly, in comparing response variables to percent area burned (Fig. 9), best fit lines show the greatest slopes in first-year values for low flows, high flows, and peak flows. Slopes in second-year values are almost always less than those of the first-year values and almost always greater than those of the 5-year mean values. The exception to this are second-year numbers of zero flow day values that decrease with increased

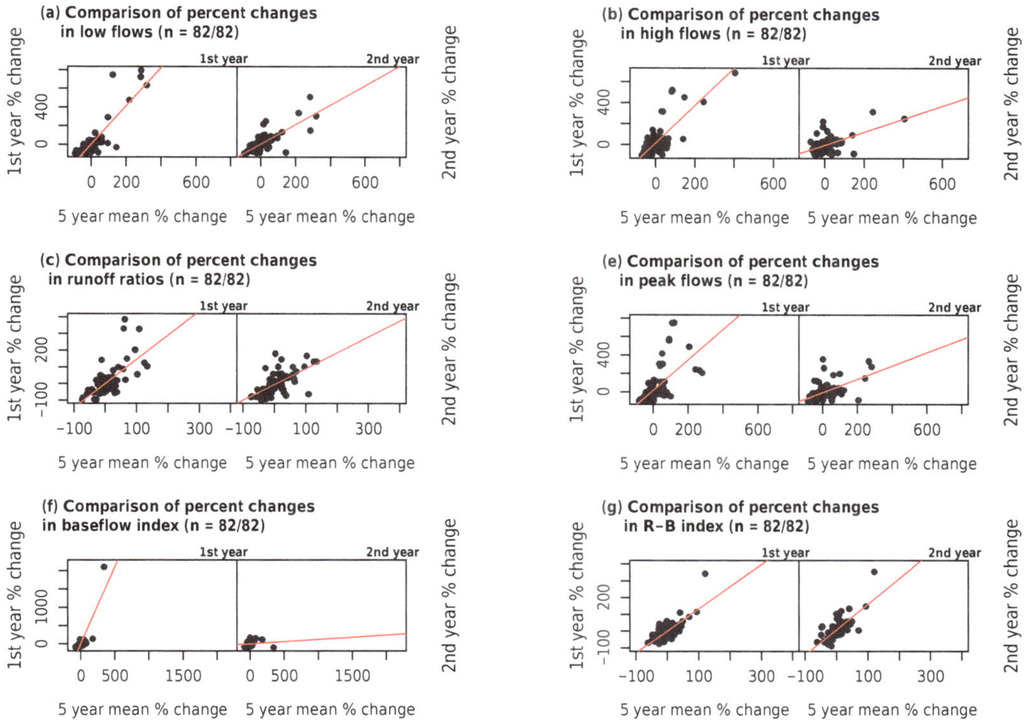

Figure 8. Scatterplots of first- and second-year response variables versus 5-year means with lines-of-best-fit.

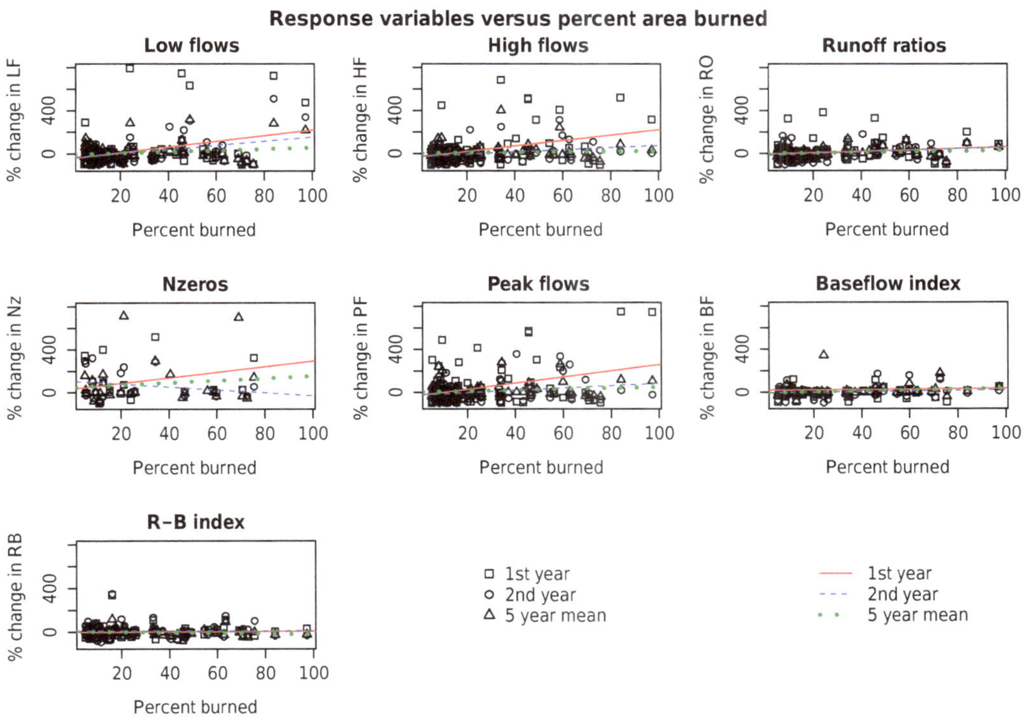

Figure 9. Scatterplots of response variables versus percent watershed burned. Best-fit lines provided to demonstrate correlation. R–B is an abbreviation for Richards–Baker and Nzero refers to the number of zero-flow days.

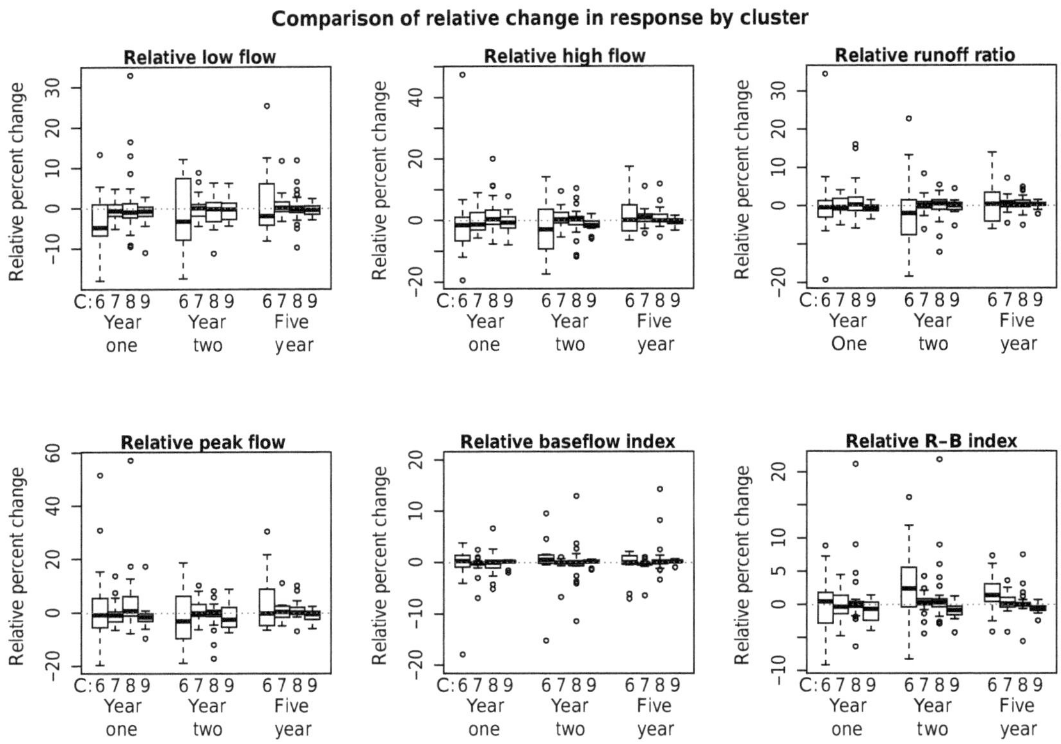

Figure 10. Boxplots of relative response variables by cluster. Relative response is actual response divided by percent area burned, in order to show response relative to fire size.

fire size. Overall, the number of zero flow days is observed to increase post-fire, though due to both a small sample size and a short time period, results are most likely unreliable. Only Richards–Baker indicates little linear correlation with burn area, yielding marginal first- and second-year slopes, and 5-year mean values decrease with percent area burned.

These findings largely confirm those of previous studies which posit that streamflow response tends to be greatest immediately following a wildfire. While there is substantial noise in the data, we find similar results. While response tends to increase overall within 5 years of a wildfire, the greatest increases are frequently found within the first year.

4.2.2 Response variability by cluster

Boxplots for response variables show differences across the four useful clusters noted above (Fig. 10). In this instance and that of the CONUS plots, all variables are scaled by dividing the variable by the percent area burned of the watershed in order to show relative response. Generalizing variability and magnitudes, dramatic differences can be established between clusters. Most noteworthy is Cluster 9, which dominantly produces negative streamflow responses post-fire. The largest magnitude relative responses are found in Cluster 6, which results in the greatest variability between variables (SD = 8.5). The most positive responses are found in Cluster 8, which frequently exhibits large positive responses and infrequent low negative responses. Cluster 7 produces low magnitude responses. Though they are, on average, positive, there are also several negative responses.

Evaluating cluster responses when they are not scaled relative to fire size percent area burned shows many similarities to the transformed results (not shown). For instance, average response in Cluster 9 is still dominantly negative with an overall low magnitude. Cluster 8 demonstrates the greatest overall variability and magnitude. On average, response in this cluster is strongly positive. Cluster 7 responses are small relative to other clusters in this study with a negative trend. Cluster 6 still exhibits high variability and low magnitude. What we surmise from these results is that watersheds in the Cluster 8 region will most likely be impacted the most from wildfire in terms of flow response. Watersheds in Clusters 7 and 9 see substantially lower responses, if any at all. Cluster 6 may prove to be the most difficult to predict due to its high variability in response.

Evidence of temporal patterns is sometimes noticeable when examining response by cluster. In Clusters 6 and 9, there are no distinct trends between first-, second-, and 5-year mean relative values (Fig. 10). In some response variables, there are increases in the values of second-year variables within Cluster 9. Cluster 7 typically shows negative relative responses in first-year values but increasing positive values in the second- and 5-year means. Cluster 8 produces

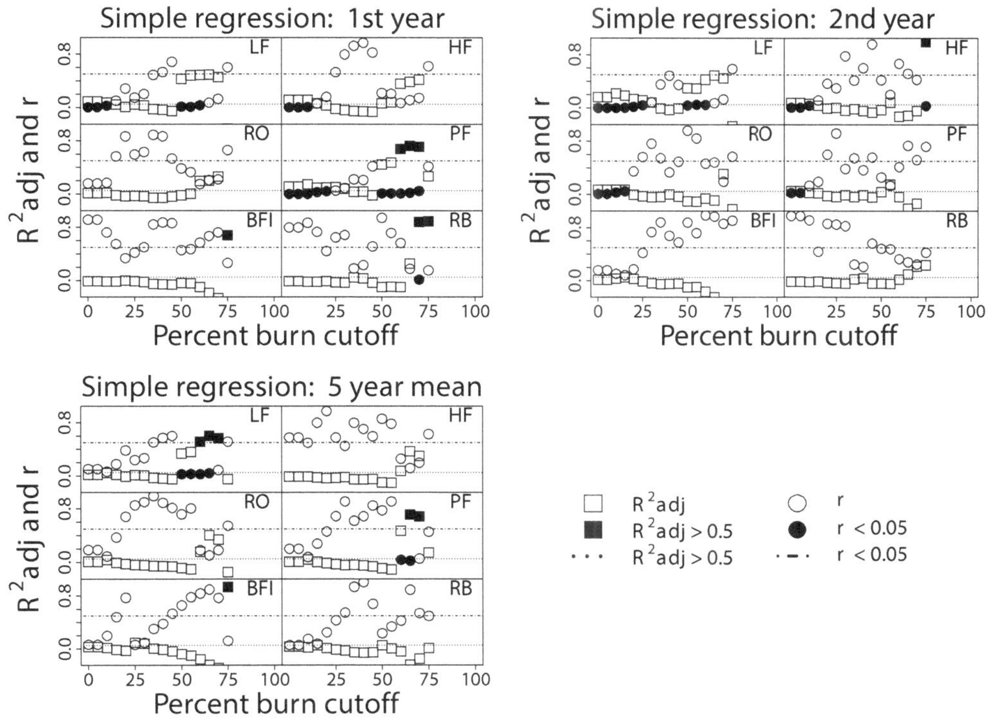

Figure 11. Scatterplots of the adjusted R^2 and p-value results (r) of simple linear regression modeling of response variables versus an increasing percent burn limitation. Response variables are abbreviated as LF (low flow), HF (high flow), RO (runoff ratio), PF (peak flow), BFI (baseflow index), and RB (Richards–Baker).

the largest response values in the first year after a fire. Relative values tend to be negative in the second year and positive over the 5-year mean. Actual values are still positive in the second year.

By examining response values by cluster, we are able to identify more intricate and robust trends than by simply examining the dataset as a whole. Spatially, we find that watersheds in Cluster 8 produce much greater and more predictable post-fire flow responses than watersheds in Clusters 6, 7, and 9. Responses in Clusters 7 and 9, overall, tend to be low magnitude and negative. Cluster 6 watersheds yield highly variable responses. Watersheds in Cluster 8 follow the temporal trend found when examining all study watersheds as a whole, with the greatest responses occurring in the first post-fire year and decreasing over time. However, the magnitude of the response of these fires skewed the results of the generalized examination at the beginning of this section. Cluster 7 watersheds, in fact, produce decreased responses in the first-year post-fire with increased flows occurring at the second-year and 5-year mean time periods. Clusters 6 and 9 exhibit little to no temporal trends at all.

Variability in response variables also generally decreases with increasing percent watershed burned. Linear regression modeling of response variables by percent watershed burned yields the error statistics found in Fig. 11 (i.e., decreasing sample size with increasing percent area burned). Figures include the adjusted R^2 and p-value significance tests ($\alpha = 0.05$). Generally, linear modeling of first-year response variables increases in accuracy as included values are limited by increasing percent burn area. First-year low-flow, high-flow, peak-flow, and Richards–Baker variables show substantial increases in adjusted R^2 once included watersheds are limited to those exceeding a 50 % burn area ($n = 12$). Included p values indicate that several of the first-year low-flow and peak-flow models are statistically significant.

Applying the same methods to second-year values shows dissimilar results, with adjusted R^2 exceeding 0.5 in only a single instance (second-year high flows), and few significant p values. However, in the cases of second-year low flow and Richards–Baker, R^2 values increase with increasing percent burn threshold. Simple regression of 5-year mean values versus percent area burned produces mixed results, with only 5-year mean low flow and peak flow allowing for adjusted R^2 values greater than 0.5, few of which are statistically significant. The 5-year mean baseflow index shows a single instance of a high adjusted R^2 value, but is statistically insignificant. Simple regression modeling was also performed on response variables by limiting included watersheds by decreasing percent area burned (i.e., decreasing sample size with decreasing percent area burned) and results demonstrated zero significant adjusted R^2 values.

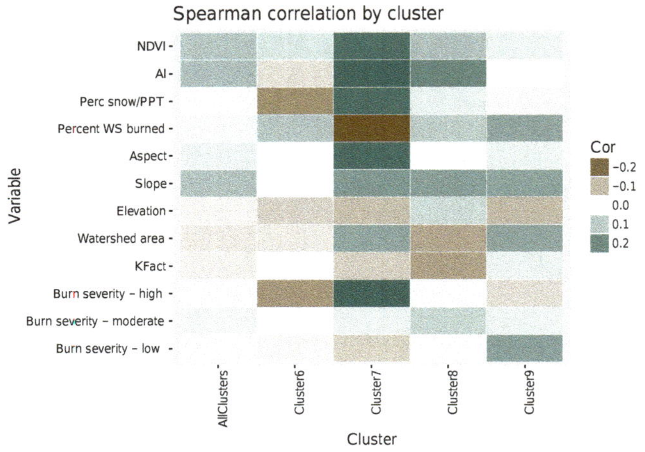

Figure 12. Heatmap of the averaged Spearman correlation coefficients of independent variables versus response variables, by cluster. WS refers to watershed.

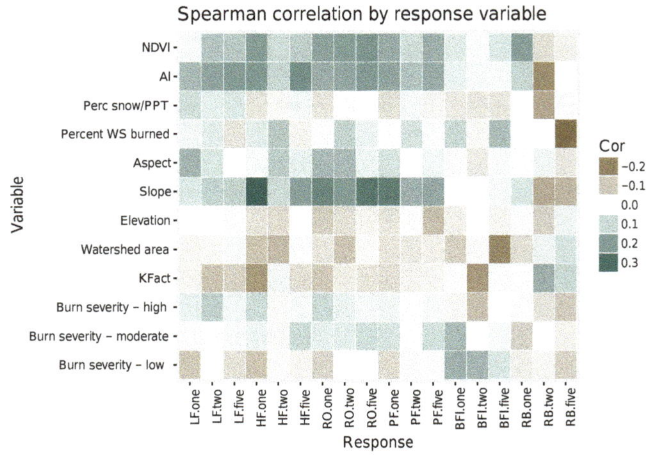

Figure 13. Heatmap of Spearman correlation coefficients of independent variables versus each response variable, for all watersheds. Response variables are abbreviated as LF (low flow), HF (high flow), RO (runoff ratio), PF (peak flow), BFI (baseflow index), and RB (Richards–Baker). "One", "Two", and "Five" refer to the year of the response variable. WS refers to watershed.

The results from Cluster 8, a dominantly chaparral environment, agree with other studies within the same region (Coombs and Melack, 2013; Kinoshita and Hogue, 2011, 2015; Loáiciga et al., 2001) and in chaparral environments outside the United States (Hessling, 1999; Lavabre et al., 1993; Lindley et al., 1988; Scott, 1993). As mentioned in the introduction, some studies find little to no change in streamflow post-fire (Aronica et al., 2002; Bart and Hope, 2010; Britton, 1991; Townsend and Douglas, 2000).

4.3 Influence of geophysical parameters on response

4.3.1 Spearman correlation

Spearman correlation between the independent and dependent variables, averaged across response variables, shows inconclusive results (Fig. 12). Generally, NDVI, AI, Snow/PPT, and slope are positively correlated with flow response. Low burn severity area, KFact, and watershed area are typically negatively correlated with flow response. Thus, as NDVI, AI, %Snow/PPT, and slope increase, response increases. As low burn severity, KFact, and watershed area increase, response decreases. More difficult to interpret are the results for other independent variables, including moderate burn severity, high burn severity, and elevation. The correlation of these variables appears to switch from positive to negative by cluster. Ascertaining a significance for their relationships may be impossible via correlation for such a small sample size.

Correlation coefficients tend to be most different within Cluster 6, where NDVI, AI, percent area burned, KFact, high burn severity, and moderate burn severity show values contradicting those produced for Clusters 7–8, as well as for the fires as a whole. This is most likely indicative of a significant difference in response patterns for the region. Unfortunately, identifying significant trends is difficult given the relatively small dataset currently available.

Figure 13 demonstrates the correlation coefficients found across all burned watersheds. The largest absolute correlation value is 0.32, only 20 % of coefficients are greater than 0.10, and only 10 correlations are statistically significant according to asymptotic p values. As relative fire size is progressively limited to larger and larger values, correlation coefficients increase. NDVI, AI, percent watershed burned, slope, high burn severity, and moderate burn severity are dominantly positively correlated with flow response. Low burn severity, KFact, and watershed area are dominantly negatively correlated with flow response. Elevation and %Snow/PPT display more varied correlations, but are generally negatively correlated with flow response. From these results, we can surmise that as the NDVI, AI, percent area burned, slope, and moderate/high burn severity area increase, flow response values will increase as well. Furthermore, as low burn severity, KFact, and watershed area increase, flow response will decrease.

4.3.2 Regression

Figure 14 shows the beta (coefficient) values produced by averaging the results of the linear and logistic regression models. The resulting betas are highly variable and thus trends are either difficult to interpret or nonexistent. To clarify results, Fig. 15 is provided with logical values, where positive and negative betas are shown as values 1 and −1, respectively. Some trends are in agreement with those of the correlation results, such as moderate burn severity being positively correlated with response. Watershed area is also in

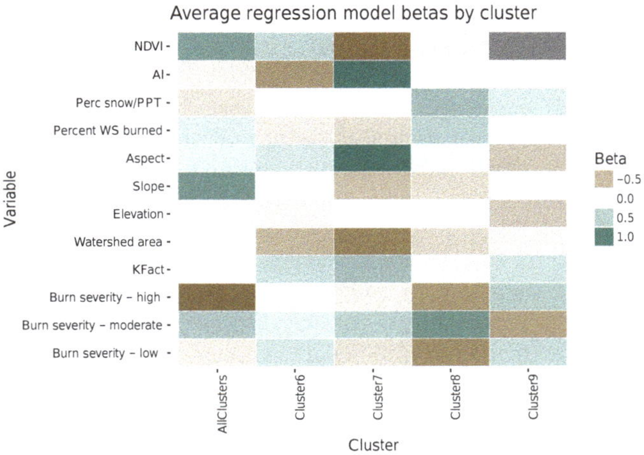

Figure 14. Heatmap of the averaged regression model beta values for each cluster. WS refers to watershed.

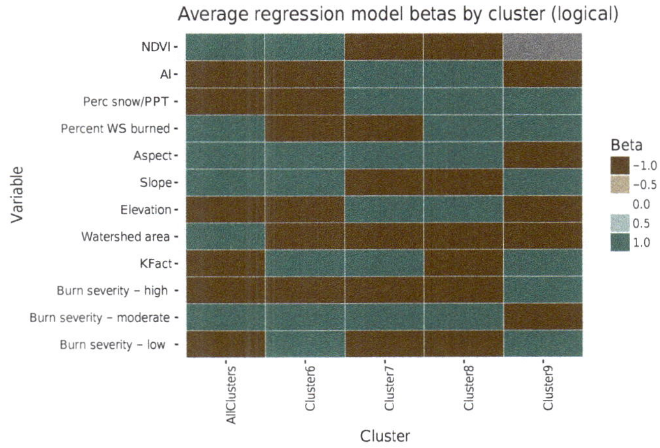

Figure 15. Heatmap of logical values of the averaged regression model beta values for each cluster shown in Fig. 14. WS refers to watershed.

agreement, showing negative correlation with response. All other independent variables that were dominant under Spearman correlation are either too variable between clusters to accurately characterize, or are opposed to the correlation results. For instance, NDVI and AI can be strongly positively or negatively correlated with response depending on cluster. Clusters 7 and 8 demonstrate an expected response of inverse correlation between NDVI and runoff; i.e., as vegetation becomes denser and ET processes recover, less water exits the watershed through streamflow.

Similarly, as AI increases, streamflow increases; i.e., in more arid watersheds post-fire vegetation is likely less dense, using less water and releasing more water through streamflow. However, in Cluster 6, NDVI appears to be positively correlated with streamflow and AI negatively correlated. Thus, as vegetation density increases and aridity decreases, streamflow increases. These unexpected deviations in trends may be due in part to climate fluctuations that can have a strong impact on results due to the small sample size. High burn severity is shown to be negatively correlated with response for Clusters 6, 7, and 8 and all clusters combined. Because the sample size of these clusters is so small (ranging from $n = 12$ to 29), regression results may not be appropriate.

4.3.3 Random forest

Application of the random forest method provided little further insight into controlling watershed parameters. Applying the algorithm to the response values resulted in significant trees for only first-year low flow, high flow, runoff ratio, and peak flow, as well as for second-year low flow. For the low-flow response variables, area burned and aspect were the dominant controlling independent variables. For first-year low flow, watersheds burned greater than 23.3 % show the greatest response. Of those burned less than that threshold, some significant responses are found when burn areas have an aspect greater than 215°. For second-year low flow, the largest responses are found in watersheds burned greater than 37 %.

First-year high flow, runoff ratio, and peak flow are identified as being significantly affected by slope. Slopes of 7.0, 7.0, and 9.8° divide response, respectively. Watersheds greater than these slopes demonstrate much greater high flow, runoff ratio, and peak flow in the first year than those with gentler slopes. This may be because steeper watersheds produce greater overland flow velocities or support shallower soil layers and thus have lower water absorption rates than in more gently sloped watersheds.

4.3.4 Discussion of geophysical parameters

Through Spearman correlation, the independent variables NDVI, AI, percent area burned, slope, moderate burn severity, and high burn severity are positively correlated with post-fire flow response. Low burn severity, KFact, and watershed area are negatively correlated with response. The results of several regression models are less clear, with inconstant relationships across all fires and clusters. Generally, they show that only moderate burn severity is dominantly positively correlated with response and that watershed area and, somehow, high burn severity are negatively correlated with response (Fig. 15). Random forest analysis shows that significant relationships can be found for the first-year low-flow, high-flow, runoff-ratio, and peak-flow response variables, as well as second-year low flow. These relationships are dominated by dependence on either percent burn area or by slope.

Overall, results of geophysical parameter characterization are somewhat inconsistent, likely due to the sample size of this study. Though it is one of the largest to date, there are still too few fire events relative to the number of geophysical parameters to produce consistent results. What can be gleaned from the various methods used in this section is that slope is frequently a strong predictor of response. Possible reasons for this are that water absorption into soil may decrease along steeper slopes due to greater overland flow velocity (i.e., less time available for absorption), or that soils along steeper slopes may be shallower, thus decreasing potential absorption volume. Both of these possibilities may account for increased water contribution to streamflow in regions with greater slope.

Support for this argument is found in the Spearman correlation coefficient analysis in Figs. 12 and 13, where slope is shown to be strongly correlated with low-flow, high-flow, peak-flow, and runoff responses, as well as in the random forest analysis where slope is one of the few independent variables to be identified as significant.

Determining the influence of independent variables on response at the cluster level seems, with this sample size, unreasonable. Furthermore, several explanatory variables are significantly correlated. Although averaged linear and logistic regression and random forest methods identify correlated values, reducing the overall number of variables in future studies may yield more understandable models. There is too much variance in flow regimes to complete a trend analysis with groupings of the size produced in this study. However, over the next decade the sample size of this study should be able to be increased significantly and provide a more robust dataset for analysis.

5 Conclusions

Post-fire changes in streamflow are found to be highly variable across regions of the western US and some trends can be difficult to discern or may be non-existent. In general, flow response for the study watersheds was found to be greatest in both magnitude and variability within the first year following a fire and shown to decrease over a 5-year period. Variations in flow response can be found by examining groups of watersheds clustered by climatic variables. We find that these general trends are dominated by the high-magnitude responses of watersheds from Cluster 8. While watersheds from Clusters 6 and 9 do not show identifiable trends, Cluster 7 shows decreased responses in the first year following fires and positive, increasing responses in just the second year and in the 5-year means, indicating that Cluster 7 watersheds have a more delayed flow response. Trends within Cluster 8 agree with other studies within the same region and in chaparral environments in other parts of the world. This study similarly identifies several regions (Clusters 6 and 9) that also do not exhibit significant distinguishable trends.

Identification of controlling watershed parameters on response yielded weak correlation metrics. Various methods showed sometimes contrary results, or none at all. Spearman correlation indicated that watershed slope, normalized difference vegetation index (NDVI), aridity index (AI), and percent of precipitation that falls as snow (%Snow/PPT) are positively correlated with flow response. Percent area low burn severity, soil erodibility factor (KFact), and watershed area were shown to be negatively correlated with response. These correlations largely agree with the positions asserted by Moody et al. (2013), Biggio and Cannon (2001), and Neary et al. (2005). Regression models showed that only moderate burn severity and watershed area are consistently positively and negatively correlated, respectively, with flow response. Random forest models indicated that percent area burned and slope are the only significant factors, and only for first-year response metrics.

The observed changes in streamflow following wildfire identified have wide-ranging implications for better understanding post-fire water budgets, downstream flood response, long-term water yield, and watershed modeling. Improved streamflow predictions will ultimately allow water resource managers in water-scarce regions to anticipate surplus volume in their budgeting forecast calculations and also help flood forecasters to identify areas at greater risk of damage and infrastructure overload. Identification of the influence of watershed geophysical parameters on post-fire streamflow should also enable improved calibration of regional models for burned watersheds.

Competing interests. The authors declare that they have no conflict of interest.

Special issue statement. This article is part of the special issue "Vegetation changes under a changing environment and the impacts on water and carbon cycling". It is not associated with a conference.

Acknowledgements. This work was partially supported by a USGS internship through the Colorado Water Science Center. We thank the reviewers for their constructive comments and insight during the review process which improved our manuscript. Any use of trade, firm, or product names is for descriptive purposes only and does not imply endorsement by the U.S. Government.

Edited by: Nadia Ursino

References

Aronica, G., Candela, A., and Santoro, M.: Changes in the hydrological response of two Sicilian basins affected by fires, in: FRIEND 2002 – Regional Hydrology: Bridging the Gap between Research and Practice, edited by: van Lanen, H. A. J. and Demuth, S., Proc. Cape Town Conf., March 2002, IAHS Press, Wallingford, UK, IAHS Publ. 274, 163–172, 2002.

Baker, D. B., Richards, R. P., Loftus, T. T., and Kramer, J. W.: A new flashiness index: Characteristics and applications to Midwestern rivers and streams, J. Am. Water Resour. As. (JAWRA), 40, 503–522, 2004.

Barbosa, P. M., Stroppiana, D., Gregoire, J. M., and Pereira, J. M. C.: An assessment of vegetation fire in Africa (1981–1991): Burned areas, burned biomass, and atmospheric emissions, Global Biogeochem. Cy., 13, 933–950, 1999.

Bart, R. and Hope, A.: Streamflow response to fire in large catchments of a Mediterranean-climate region using paired-catchment experiments, J. Hydrol., 388, 370–378, 2010.

Benavides-Solorio, J. and MacDonald, L. H.: Post-fire runoff and erosion from simulated rainfall on small plots, Colorado Front Range, Hydrol. Process., 15, 2931–2952, 2001.

Biggio, E. R. and Cannon, S. H.: Compilation of post-wildfire runoff data from the Western United States, U.S. Geological Survey, 2004-1085, 2001.

Bond, N.: hydrostats: Hydrologic indices for daily time series data, https://rdrr.io/cran/hydrostats/man/hydrostats-package.html (last access: January 2016), 2015.

Britton, D. L.: Fire and the chemistry of a South African mountain stream, Hydrobiologia, 218, 177–192, 1991.

Burke, M. P., Hogue, T. S., Ferreira, M., Mendez, C. B., Navarro, B., Lopez, S., and Jay, J. A.: The effect of wildfire on soil mercury concentrations in southern California watersheds, Water Air Soil Poll., 212, 369–385, 2010.

Burke, M. P., Hogue, T. S., Kinoshita, A., Barco, J., Wessel, C., and Stein, E.: Pre- and post-fire pollutant loads in an urban watershed in southern California, Environ. Monit. Assess., 185, 10131–10145, 2013.

Cannon, S. H., Kirkham, R. M., and Parise, M.: Wildfire-related debris-flow initiation processes, Storm King Mountain, Colorado, Geomorphology, 39, 171–188, 2001.

Carlson, T. N. and Ripley, D. A.: On the relation between NDVI, fractional vegetation cover, and leaf area index, Remote Sens. Environ., 62, 241–252, 1997.

Coombs, J. S. and Melack, J. M.: Initial impacts of a wildfire on hydrology and suspended sediment and nutrient export in California chaparral watersheds, Hydrol. Process., 27, 3842–3851, 2013.

DeBano, L. F.: The role of fire and soil heating on water repellency in wildland environments: a review, J. Hydrol., 231, 195–206, 2000.

Eidenshink, J., Schwind, B., Brewerm K., Zhu, Z.-L., Quayle, B., and Howard, S.: A project for monitoring trends in burn severity, Fire Ecology, 3, 3–21, 2007.

Emelko, M. B., Silins, U., Bladon, K. D., and Stone, M.: Implications of land disturbance on drinking water treatability in a changing climate: Demonstrating the need for "source water supply and protection" strategies, Water Res., 45, 461–472, 2011.

Emmerich, W. E. and Cox, J. R.: Changes in surface runoff and sediment production after repeated rangeland burns, Soil Sci. Soc. Am. J., 58, 199–203, 1994.

Falcone, J.: GAGES-II: Geospatial Attributes of Gages for Evaluating Streamflow, U.S. Geological Survey, Reston, Virginia, available at: https://water.usgs.gov/GIS/metadata/usgswrd/ (last access: September 2015), 2011.

Fraley, C., Raftery, A. E., Murphy, T. B., and Scrucca, L.: mclust Version 4 for R: Normal Mixture Modeling for Model-Based Clustering, Classification, and Density Estimation, Department of Statistics, University of Washington, Technical Report No. 597, 57 pp., 2012.

Goldman, S. J., Jackson, K., and Bursztynsky, T. A.: Erosion and sediment control handbook, McGraw-Hill, New York City, New York, United States, 1986.

Google, Inc.: Google Earth Engine, available at: https://earthengine.google.com/, last access: September 2015.

Hessling, M.: Hydrological modelling and a pair basin study of Mediterranean catchments, Phys. Chem. Earth Pt. B 24, 59–63, 1999.

Hester, J. W., Thurow, T. L., and Taylor Jr., C. A.: Hydrologic characteristics of vegetation types as affected by prescribed burning, J. Range Manage., 50, 199–204, 1997.

Homer, C., Huang, C., Wylie, B., and Coan, M.: Development of a 2001 National Landcover Database for the United States, Photogramm. Eng. Rem. S., 70, 829–840, 2004.

Hothorn, T, Hornik, K., and Zeileis, A.: Unbiased recursive partitioning: A conditional inference framework, J. Comput. Graph. Stat., 15, 651–674, 2006.

Kinoshita, A. M. and Hogue, T. S.: Spatial and temporal controls on post-fire hydrologic recover in southern California watersheds, Catena, 87, 240–252, 2011.

Kinoshita, A. M. and Hogue, T. S.: Increased dry season water yield in burned watersheds in Southern California, Environ. Res. Lett., 10, 014003, https://doi.org/10.1088/1748-9326/10/1/014003, 2015.

Ladson, A. R., Brown, R., Neal, B., and Nathan, R.: A standard approach to baseflow separation using the Lyne and Hollick filter, Australasian Journal of Water Resources, 17, 24–34, 2013.

Lavabre, J., Torres, D. S., and Cernesson, F.: Changes in the hydrological response of a small Mediterranean basin a year after a wildfire, J. Hydrol., 142, 273–299, 1993.

Lee, R. J. and Chow, T. E.: Post-wildfire assessment of vegetation regeneration in Bastrop, Texas, using Landsat imagery, GISci. Remote Sens., 52, 609–626, 2015.

Lindley, A. J., Bosch, J. M., and van Wyk, D. B.: Changes in water yield after fire in fynbos catchments, Water SA, 14, 7–12, 1988.

Littell, J. S., McKenzie, D., Peterson, D. L., and Westerling, A. L.: Climate and wildfire area burned in western U.S. ecoprovinces, 1916–2003, Ecol. Appl., 19, 1003–1021, 2009.

Loáiciga, H. A., Pedreros, D., and Roberts, D.: Wildfire-streamflow interactions in a chaparral watershed, Adv. Environ. Res., 5, 295–305, 2001.

MacQueen, J.: Some methods for classification and analysis of multivariate observations, Proc. Fifth Berkeley Symp. On Math. Statist. And Prob., University of California Press, 1, 281–297, 1967.

McManamay, R. A., Bevelhimer, M. S., and Kao, S.-C.: Updating the US hydrologic classification: an approach to clustering and stratifying ecohydrologic data, Ecohydrology, 7, 903–926, 2014.

Meyer, G. A., Pierce, J. L., Wood, S. H., and Jull, A. J. T.: Fire, storms, and erosional events in the Idaho batholith, Hydrol. Process., 15, 3025–3038, 2001.

Moody, J. A. and Martin, D. A.: Initial hydrologic and geomorphic response following a wildfire in the Colorado Front Range, Earth Surf. Proc. Land., 26, 1049–1070, 2001.

Moody, J. A., Shakesby, R. A., Robichaud, P. R., Cannon, S. H., and Martin, D. A.: Current research issues related to post-wildfire runoff and erosion processes, Earth-Sci. Rev., 122, 10–37, 2013.

Moritz, M. A., Moody, T. J., Krawchuk, M. A., Hughes, M., and Hall, A.: Spatial variation in extreme winds predicts large wildfire locations in chaparral ecosystems, Geophys. Res. Lett., 37, L04801, https://doi.org/10.1029/2009GL041735, 2010.

National Wildfire Coordinating Group: Glossary of Wildland Fire Terms, available at: https://www.nwcg.gov/glossary-of-wildland-fire-terminology, last access: July 2016.

Neary, D. G.: Post-wildfire watershed flood responses, in: Second International Wildland Fire Ecology and Fire Management Congress and Fifth Symposium on Fire Forest Meteorology, 16–20 November 2003, Orlando, Florida, American Meteorology Society, 2003.

Neary, D. G., Ryan, K. C., and DeBano, L. F.: Wildland fire in ecosystems: effects of fire on soils and water, U.S. Department of Agriculture, Forest Service, Rocky Mountain Research Station, General Technical mReport RMRS-GTR-42-volume 4, 2005.

Owens, P. N., Giles, T. R., Petticrew, E. L., Leggat, M. S., Moore, R. D., and Eaton, B. C.: Muted responses of streamflow and suspended sediment flux in a wildfire-affected watershed, Geomorphology, 202, 128–139, 2013.

Pausas, J. G., Llovet, J., Rodrigo, A., and Vallejo, R.: Are wildfires a disaster in the Mediterranean basin? – A review, Int. J. Wildland Fire., 17, 713–723, 2008.

Pierson, F. B., Moffet, C. A., Williams, C. J., Hardegree, S. P., and Clark, P. E.: Prescribed-fire effects on rill and interrill runoff and erosion in a mountainous sagebrush landscape, Earth Surf. Proc. Land., 34, 193–203, 2009.

Poff, N.: A hydrogeography of unregulated streams in the United States and an examination of scale-dependence in some hydrological descriptors, Freshwater Biol., 36, 71–79, 1996.

PRISM Climate Group, Oregon State University, http://prism.oregonstate.edu, last access: December 2004.

Santos, R. M. B., Sanches Fernandes, L. F., Pereira, M. G., Cortes, R. M. V., and Pacheco, F. A. L.: Water resources planning for a river basin with recurrent wildfires, Sci. Total Environ., 526, 1–13, 2015.

Saxe, S. W., Hogue, T. S., and Hay, L. E.: Post-Fire Streamflow Spatial data for Western US Watersheds: U.S. Geological Survey data release, available at: https://doi.org/10.5066/F77S7MRP, 2018.

Schwartz, G. E. and Alexander, R. B.: Soils data for the Conterminous United States Derived from the NRCS State Soil Geographic (STATSGO) Data Base, United States Geological Survey, available at: https://www.sciencebase.gov/catalog/item/535ea386e4b08e65d60fb441 (last access: June 2015), 1995.

Schwarz, G.: Estimating the dimension of a model, Ann. Stat., 6, 461–464, 1978.

Scott, D. F.: The hydrological effects of fire in South African mountain catchments, J. Hydrol., 150, 409–432, 1993.

Shakesby, R. A. and Doerr, S. H.: Wildfire as a hydrological and geomorphological agent, Earth-Sci. Rev., 74, 269–307, 2006.

Smith, H. G., Sheridan, G. J., Lane, P. N. J., Nyman, P., and Haydon, S.: Wildfire effects on water quality in forest catchments: A review with implications for water supply, J. Hydrol., 396, 170–192, 2011.

Stein, E. D., Brown, J. S., Hogue, T. S., Burke, M. P., and Kinoshita, A.: Stormwater contaminant loading following southern California wildfires, Environ. Toxicol. Chem., 31, 2625–2638, 2012.

Townsend, S. A. and Douglas, M. M.: The effect of three fire regimes on stream water quality, water yield and export coefficients in a tropical savanna (northern Australia), J. Hydrol., 229, 118–137, 2000.

U.S. Geological Survey: National Water Information System – NWISWeb, USGS, available at: http://waterdata.usgs.gov, accessed on various dates in 2014.

U.S. Geological Survey: The national map, USGS, available at: https://nationalmap.gov/elevation.html, accessed on various dates in 2015.

Watson, F., Vertessy, R., McMahon, T., Rhodes, B., and Watson, I.: Improved methods to assess water yield changes from paired-catchment studies: application to the Maroondah catchments, Forest Ecol. Manag., 143, 189–204, 2001.

Westerling, A. L.: Increasing western US forest wildfire activity: sensitivity to changes in the timing of spring, Philos. T. Roy. Soc. B, 371, 20150178, https://doi.org/10.1098/rstb.2015.0178, 2016.

Westerling, A. L., Hidalgo, H. G., Cayan, D. R., and Swetnam, T. W.: Warming and earlier spring increase western US forest wildfire activity, Science, 313, 940–943, 2006.

Wilkinson, S. N., Wallbrink, P. J., Hancock, G. J., Blake, W. H., Shakesby, R. A., and Doerr, S. H.: Fallout radionuclide tracers identify a switch in sediment sources and transport-limited sediment yield following wildfire in a eucalypt forest, Geomorphology, 110, 140–151, 2009.

Whitlock, C.: Land management: Forests, fires and climate, Nature, 432, 28–29, 2004.

Woodcock, C. E., Allen, R., Anderson, M., Belward, A., Bindschadler, R., Cohen, W., Gao, F., Goward, S. N., Helder, D., Helmer, E., Nemani, R., Oreopoulos, L., Schott, J., Thenkabail, P. S., Vermote, E. F., Vogelmann, J., Wulder, M. A., and Wynne, R.: Free access to Landsat imagery, Science, 320, 1011, https://doi.org/10.1126/science.320.5879.1011a, 2008.

9

How good are hydrological models for gap-filling streamflow data?

Yongqiang Zhang and David Post

CSIRO Land and Water, GPO Box 1700, Acton 2601, Canberra, Australia

Correspondence: Yongqiang Zhang (yongqiang.zhang@csiro.au, yongqiang.zhang2014@gmail.com)

Abstract. Gap-filling streamflow data is a critical step for most hydrological studies, such as streamflow trend, flood, and drought analysis and hydrological response variable estimates and predictions. However, there is a lack of quantitative evaluation of the gap-filled data accuracy in most hydrological studies. Here we show that when the missing data rate is less than 10 %, the gap-filled streamflow data obtained using calibrated hydrological models perform almost the same as the benchmark data (less than 1 % missing) when estimating annual trends for 217 unregulated catchments widely spread across Australia. Furthermore, the relative streamflow trend bias caused by the gap filling is not very large in very dry catchments where the hydrological model calibration is normally poor. Our results clearly demonstrate that the gap filling using hydrological modelling has little impact on the estimation of annual streamflow and its trends.

1 Introduction

Streamflow is channel runoff, i.e. the flow of water in streams and rivers and water accumulated from surface runoff from the land surface and groundwater recharge. It is one of the major water balance components in a catchment, where precipitation is partially stored in surface water, soil, and groundwater stores, and the rest is partitioned into two fluxes: evapotranspiration and streamflow. It is almost impossible to measure evapotranspiration dynamics at a catchment scale. In contrast, streamflow time series can be easily measured at a catchment outlet. Therefore, streamflow data become a fundamental dataset underpinning hydrological studies. Without such a dataset, it is hard to understand catchment hydrological processes under climate change and non-stationarity (Dai et al., 2009; Gedney et al., 2006a; Ukkola et al., 2015; Zhang et al., 2016b).

Unfortunately, streamflow data are not always continuously available and most gauges suffer from missing streamflow data issues (Dai et al., 2009). Often, the missing data rate is important when selecting streamflow gauges, especially when the data are used for annual trend analysis. To choose qualified catchments, researchers often set up a threshold for the missing data ratio, for instance 1 % (Petrone et al., 2010), 5 % (Ukkola et al., 2015), 10 % (Déry et al., 2009), 15 % (Liu and Zhang, 2017), and 20 % (Lopes et al., 2016). Only those gauges with a missing data rate less than a particular threshold are selected, and the rest are excluded for further analysis because of high missing data rates.

There are many methods used for gap-filling the missing data, including interpolation from nearby gauges (Hannaford and Buy, 2012; Lavers et al., 2010; Lopes et al., 2016), statistical methods (Gedney et al., 2006b), hydrological modelling (Dai et al., 2009; Sanderson et al., 2012), and multiple infilling methods (Harvey et al., 2012). Among them, the hydrological modelling method is widely used since it fully considers the spatial heterogeneity and temporal variability of climate forcing data, and can achieve sufficient simulations when it is calibrated against a small number of observations (Peña-Arancibia et al., 2015; Rojas-Serna et al., 2016; Seibert and Beven, 2009; Liu and Zhang, 2017). This is particularly important in Australia, where hydrological modelling is a major tool for simulating continuous streamflow at a catchment scale. More recently, the Australian Bureau of Meteorology used a hydrological model – GR4J – to infill missing daily streamflow data for 222 Hydrologic Reference Stations (http://www.bom.gov.au/water/hrs/about.shtml, last access on 8 August 2018). The gap-filled streamflow data are then used for trend analysis, which provides hydrological information to all users.

One major concern for the hydrology community is to understand how reliable the gap-filled data are. Unfortunately there are no studies in the literature to comprehensively evaluate the reliability and accuracy of the gap-filled data that are influenced by different thresholds and by patterns of missing data. Our study aims to provide a framework to evaluate the annual trends and annual variables obtained from gap-filled streamflow data using two hydrological models (GR4J and SIMHYD), together with a large streamflow dataset available across the Australian continent (Zhang et al., 2013). This can guide researchers to more sensibly define a threshold for catchment selection and hydrological analysis.

2 Data and methods

2.1 Data

We obtained a daily streamflow dataset from 780 unregulated catchments widely spread across Australia (Zhang et al., 2013). The dataset has undergone strict quality assurance and quality control, including quality code check and spike (i.e. outlier points) control, and covered the period from 1975 to 2012. This dataset has been used by modellers for various hydrological modelling and studies of extreme events (Li and Zhang, 2017; Liu and Zhang, 2017; Ukkola et al., 2016; Yang et al., 2017). The missing data rates for the pre-1980 and post-2010 periods were high. To meet our study requirements, we selected 217 catchments with a missing data rate of less than 1 % for the period 1981–2010, and the streamflow data for the 217 catchments are regarded as benchmark data (Fig. 1). Out of the 780 catchments there are 146, 91, and 61 with a missing data rate of 1 %–5 %, 5 %–10 %, and 10 %–20 % during 1981–2010, respectively (Fig. 1), and these catchments account for 38 % of total available catchments. Table 1 summarises major catchment attributes for the 217 selected catchments. The data gaps for Australian streamflow gauges mainly include the following issues: (i) non-sensible records; (ii) broken sensors; (iii) no recorded data (instrumentation removed); (iv) no existing data, and (v) no records or records lost.

Out of the 217 catchments, about half of the catchments showed a significant decreasing trend, 37 % showed a non-significant decreasing trend, and 13 % showed a non-significant increasing trend (Fig. 2), detected using Mann–Kendall trend analysis (see Sect. 2.3). This is because Australia experienced the millennium drought over the period 2001–2009, which caused a dramatic streamflow reduction in this period (van Dijk et al., 2013). Trend analysis for the 217 catchments is explained in Sect. 2.3 and trend results are summarised in Sect. 3.

Out of the 217 catchments, about 46 % of catchments have no missing data in 1981–2010, 12 % with a missing data rate of < 0.1 %, 22 % with a missing data rate of 0.1 %–0.5 %, and 20 % with a missing data rate of 0.5 %–1 % (Fig. 2).

To drive the two hydrological models, we obtained daily meteorological time series (including minimum tempera-

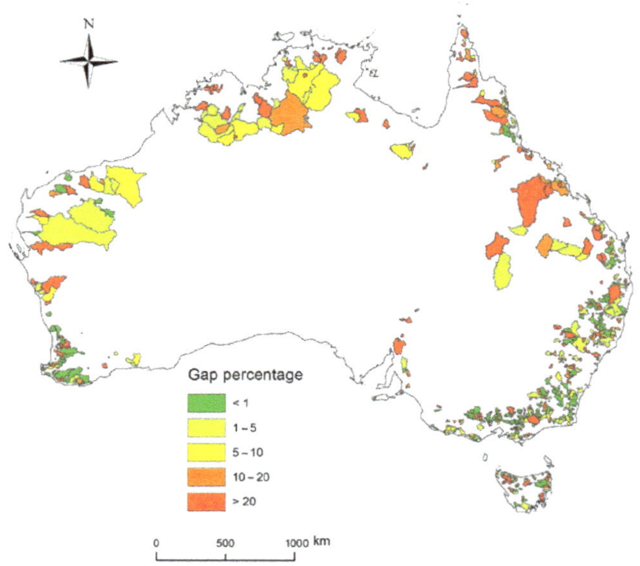

Figure 1. The 780 unregulated catchments grouped by different streamflow data gaps for the period of 1981–2010.

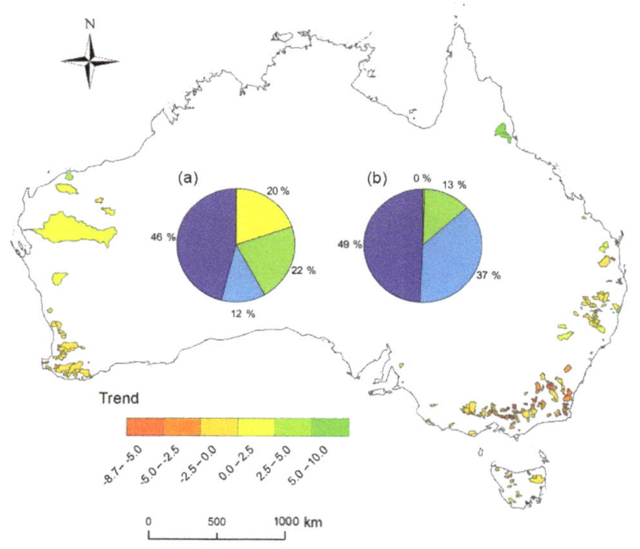

Figure 2. Trends and streamflow data summary for the 217 catchments used in this study. The trend in annual streamflow is shown in units of mm $year^{-1}$ $year^{-1}$. The chart in **(a)** indicates the catchment percentage with different missing data rates (dark blue with a missing data rate of 0 %, navy blue with a missing data rate of 0 %–0.1 %, green with a missing data rate of 0.1 %–0.5 %, and yellow with a missing data rate of 0.5 %–1.0 %). The chart in **(b)** indicates the catchment percentage with different trends (dark blue with a significant ($p \leq 0.05$) decreasing trend, navy blue with a non-significant ($p > 0.05$) decreasing trend, green with a non-significant ($p > 0.05$) increasing trend, and yellow with a significant ($p \leq 0.05$) increasing trend).

Table 1. Major catchment attributes for the 217 catchments.

Attribute	Definition	Unit	Min	2.5th	25th	Median	75th	97.5th	Max
Area	Catchment area	km^2	53	70	180	392	844	4562	72 902
Elevation	Catchment average elevation above sea level	m	46	100	278	449	753	1194	1351
Slope	Catchment mean slope	°	0.3	0.6	2.0	3.9	7.7	12.0	13.6
P	Mean annual precipitation	$mm\ year^{-1}$	256	371	703	853	1107	1966	2473
ET_P	Mean annual potential evapotranspiration	$mm\ year^{-1}$	906	968	1149	1235	1408	1791	1892
AI	Aridity index	–	0.38	0.55	1.11	1.44	1.89	4.75	6.47
Forest ratio	Ratio of forest to all land cover types	–	0.02	0.06	0.39	0.55	0.67	0.83	0.90

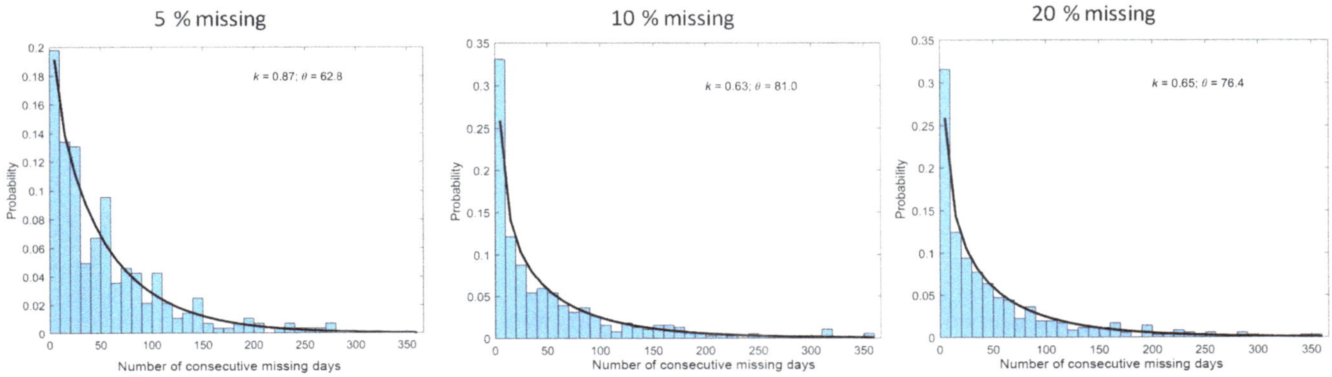

Figure 3. Missing data patterns for three groups of catchments with missing data rates of 4 %–6 %, 8 %–12 %, and 18 %–22 %, which are represented by 5 %, 10 %, and 20 % missing data rates, respectively.

ture, maximum temperature, incoming solar radiation, actual vapour pressure, and precipitation) from 1975 to 2012 at 0.05° (~ 5 km) grid resolution from the SILO Data Drill of the Queensland Department of Natural Resources and Water (https://silo.longpaddock.qld.gov.au/, last access on 8 August 2018). The data quality is reasonably good, indicated by the mean absolute error for maximum daily air temperature, minimum daily air temperature, vapour pressure, and precipitation at 1.0 °C, 1.4 °C, 0.15 kPa, and 0.40 mm day^{-1} (Jeffrey et al., 2001).

2.2 Gap-filling experiments

For thoroughly investigating the potential impacts of infilled streamflow data on annual trend accuracy, we conducted three groups of experiments to test how the missing data rates at 5 %, 10 %, and 20 % impact streamflow trends. We followed three steps for each missing data rate of the experiments.

1. *Missing data patterns were obtained using actual streamflow data.*

We selected patterns of consecutive missing days from actual data from the 780 catchments. For the 5 % group of missing data rate experiments, we selected 44 catchments with missing rates of 4 %–6 %; for the 10 % group of missing data rate experiments, we selected 39 catchments with missing data rates of 8 %–12 %; and for the 20 % group of missing data rate experiments, we selected 22 catchments with missing data rates of 18 %–22 %. Figure 3 shows the probability distribution of consecutive missing days from each group of catchments, which is skewed toward the low end. We therefore used the two-parameter Gamma distribution to simulate probability distribution of consecutive missing days (Fig. 3). The Γ distribution is expressed as

$$X \sim \Gamma(k,\theta) = \Gamma(k,\theta), \tag{1}$$

where X is the number of consecutive missing days, k is the shape parameter, and θ is the scale parameter. The corresponding probability density function in the shape-scale parameterisation is

$$f(x;k,\theta) = \frac{1}{\Gamma(k)\theta^k} x^{k-1} e^{-\frac{x}{\theta}}, \tag{2}$$

where $\Gamma(k)$ is the gamma function.

As seen from Fig. 3, the two parameters are stable under the three groups of catchments. The k parameter varies from 0.63 to 0.87 and the θ parameter changes from 62 to 81. It is noted that we removed all times when the number of consecutive missing days was > 365. We did that for a number of reasons. Firstly, gap-filling an entire year of missing data would likely impact annual trends. Secondly, the focus of this paper is on gap-filling short periods of missing data to be able

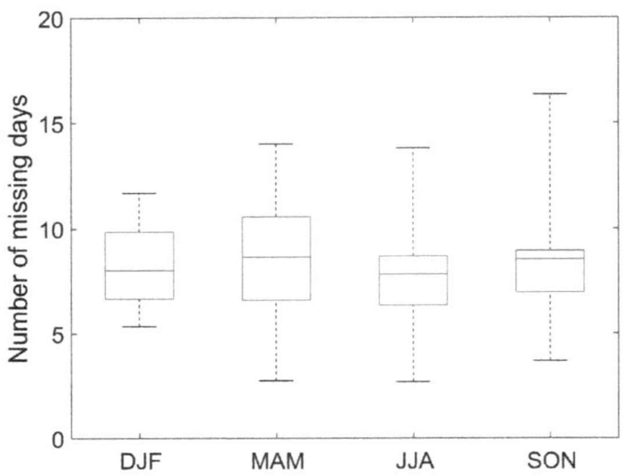

Figure 4. Distribution of number of missing days across different seasons, summarised from 39 catchments with a missing data rate ranging from 8 % to 12 % (i.e. 10 % missing data group).

to include more catchments in streamflow analyses. Thirdly, removing all periods of greater than 365 days allowed us to better fit a gamma distribution to the number of missing days.

We also checked the seasonality of missing data to see if one season were more likely to have missing data than another. As seen from Fig. 4, the missing data are more or less evenly distributed through different seasons across all the 39 catchments (with a missing data rate of 8 % to 12 %) within the 10 % missing data group. This indicates that the data gaps were not skewed toward a particular season and they occurred randomly through the year.

2. *Numbers of random consecutive missing days were generated using a random number generator (sampling without replacement) based on the gamma distribution.*

The random number generation was repeated 100 times to ensure the selected samples covered a wide range of streamflow time series.

3. *Streamflow data were gap-filled.*

The selected days were treated as missing data and the unselected data were used for hydrological model calibration. The missing data were then gap-filled using the simulated streamflow from the calibrated GR4J and SIMHYD models.

For consistent interpretation hereafter, the benchmark streamflow data are regarded as "observed" and the experiment data as "filled". For each of the three experiments, there are 100×217 (21 700) missing data time series, with 100 representing sample times using the random number generator and 217 representing the number of catchments.

2.3 Trend analysis

We used the Mann–Kendall Tau-b non-parametric test including Sen's slope method (Burn and Elnur, 2002) for annual streamflow trend analysis and significance testing for all three groups of experiments and benchmark data.

We used the following equation to quantify the trend bias:

$$B_t = T_{\text{filled}} - T_{\text{obs}}, \tag{3}$$

where B_t is the bias in the annual streamflow trend (mm year^{-1} year^{-1}), T_{filled} is the annual trend for gap-filled streamflow (mm year^{-1} year^{-1}), and T_{obs} is the annual trend in observed streamflow (mm year^{-1} year^{-1}). It measures the trend error between the infilled and observed runoff trends with $B_t \approx 0$, which indicates that the trend in observed annual runoff is almost the same as that in the infilled annual runoff.

We also defined relative trend bias (P_{Bt}) as

$$P_{B_t} = \frac{T_{\text{filled}} - T_{\text{obs}}}{T_{\text{obs}}} \times 100. \tag{4}$$

2.4 Hydrological models

Two widely used hydrological models, SIMHYD and GR4J (Chiew et al., 2002, 2010; Li et al., 2014; Oudin et al., 2008; Perrin et al., 2003; Zhang and Chiew, 2009; Zhang et al., 2016a), were used to infill daily missing streamflow data. Both models require daily precipitation and daily potential evaporation (Priestley and Taylor, 1972) as model inputs, and model outputs are daily streamflow at each gauge. The daily inputs of the maximum and minimum temperatures, incoming solar radiation, and vapour pressure data were used to calculate the Priestley–Taylor daily potential evaporation.

The two models were calibrated using a global optimiser, consisting of a genetic algorithm (The MathWorks, 2006), at each catchment, with the first 6 years (i.e. 1975–1980) for spin-up and the remainder (1981–2010) for modelling experiments. Since this study mainly evaluates the trends obtained using the gap-filled streamflow from hydrological modelling, it is crucial to predict high flow and mean flow as accurately as possible. To this end, the model calibration was to minimise the following objective function (F) (Viney et al., 2009; Zhang et al., 2016b):

$$F = (1 - \text{NSE}) + 5|\ln(1 + B)|^{2.5}, \tag{5}$$

$$B = \frac{\sum_{i=1}^{N} Q_{\text{sim},i} - \sum_{i=1}^{N} Q_{\text{obs},i}}{\sum_{i=1}^{N} Q_{\text{obs},i}}, \tag{6}$$

where NSE is the Nash–Sutcliffe efficiency of daily streamflow, B is the model bias, Q_{sim} and Q_{obs} are the simulated and observed daily runoff, i is the ith day, and N is the total number of days sampled. The NSE gives higher streamflow more weight, and varies between $-\infty$ and 1, with NSE > 0.6 indicating a good agreement (Zhang and Chiew, 2009). The B measures water balance error between the observed and

Table 2. Summary of model calibration number carried out for benchmark data and missing data experiments.

Model	Benchmark	5 % missing	10 % missing	20 % missing	Sum
GR4J	217	21 700	21 700	21 700	65 317
SIMHYD	217	21 700	21 700	21 700	65 317
Sum	434	43 400	43 400	43 400	130 634

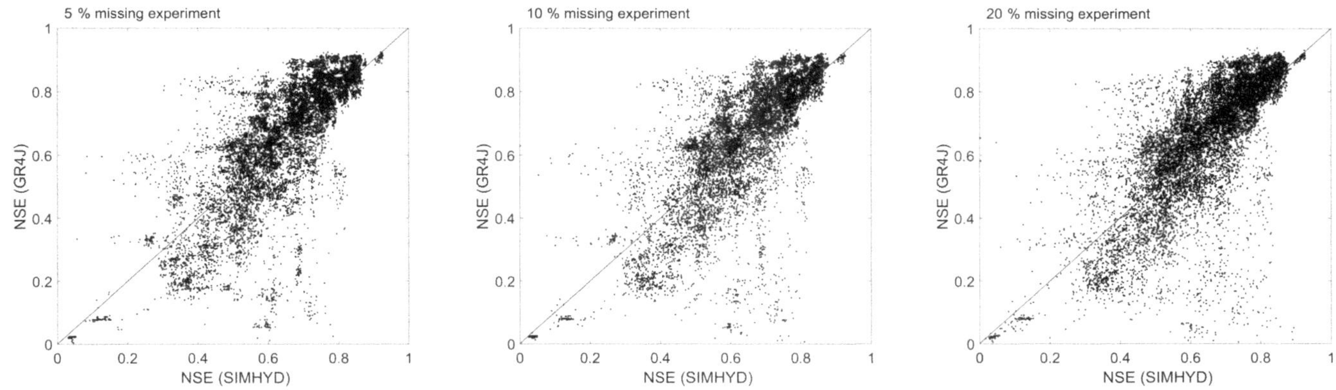

Figure 5. Comparisons between calibrated GR4J and calibrated SIMHYD for 44 catchments of the 5 % missing data experiment, 39 catchments of the 10 % missing data experiment, and 22 catchments of the 20 % missing data experiment. In each catchment, there were 100 replicates carried out.

modelled daily streamflow, with $B = 0$ indicating that the average of modelled daily streamflow is the same as the average of observed daily streamflow.

For each catchment, GR4J and SIMHYD were calibrated using benchmark data and 100 time series of streamflow data with missing data (see Sect. 2.2), respectively. For benchmark data without any missing data (46 % catchments), no gap filling is required; for the benchmark data with missing data rates less than 1 %, the calibrated continuous streamflow data were used to fill the gaps. For the missing data experiments, the calibrated continuous streamflow data for each missing replicate were used to infill the artificially made missing data. Table 2 summarises the model calibrations carried out for the benchmark and for each experiment. Finally, 130 634 model calibrations and 130 200 times of gap filling were carried out. Finally, the trends estimated from the benchmark were used to evaluate those obtained from the missing data experiments.

3 Results

The gap-filled data from the two hydrological models were evaluated against the benchmark data. Overall, the two models perform well and neither significantly outperforms the other (Fig. 5). For the three groups of gap-filling experiments, these two models perform similarly (i.e. the difference of NSE of daily runoff between the two is less than 0.02) in 18 %–19 % catchments; the SIMHYD model outperforms the GR4J model (NSE difference between the two is larger than 0.02) in 30 %–31 % catchments; and the GR4J model outperforms the SIMHYD model in 50 %–51 % catchments.

Figures 6 and 7 summarise the performance of the gap-filled data for estimating annual trends, annual streamflow, monthly streamflow, and daily streamflow, respectively. The three missing data rate experiments (5 %, 10 %, and 20 %) perform almost the same as the benchmark (Figs. 6 and 7). The coefficient of determination (r^2) between the gap-filled trends and observed trends is more than 0.98 for the three experiments and two hydrological models.

Since errors in gap-filled trends are likely to be different and time steps are different when daily infilled streamflow data are used, we further investigate how gap-filled errors are propagated from daily to monthly and to annual scales in the three gap-filling cases (5 %, 10 %, and 20 %) (Figs. 6 and 7). It is expected that daily gap-filled streamflow has a larger standard deviation from the benchmark than monthly and annual streamflow since the streamflow was gap-filled at a daily scale. This indicates that the temporal aggregation smooths the gap-filled error strongly, and it generates very reasonable monthly and annual streamflow estimates with less standard deviation. It is interesting to note that both models tend to underestimate very high flows, though they are calibrated against the NSE of daily streamflow, which gives larger weight to the correct representation of higher flows.

Figure 8 further summarises the catchments with trend direction mismatch between the benchmark and gap-filled data (i.e. change from negative to positive or change from pos-

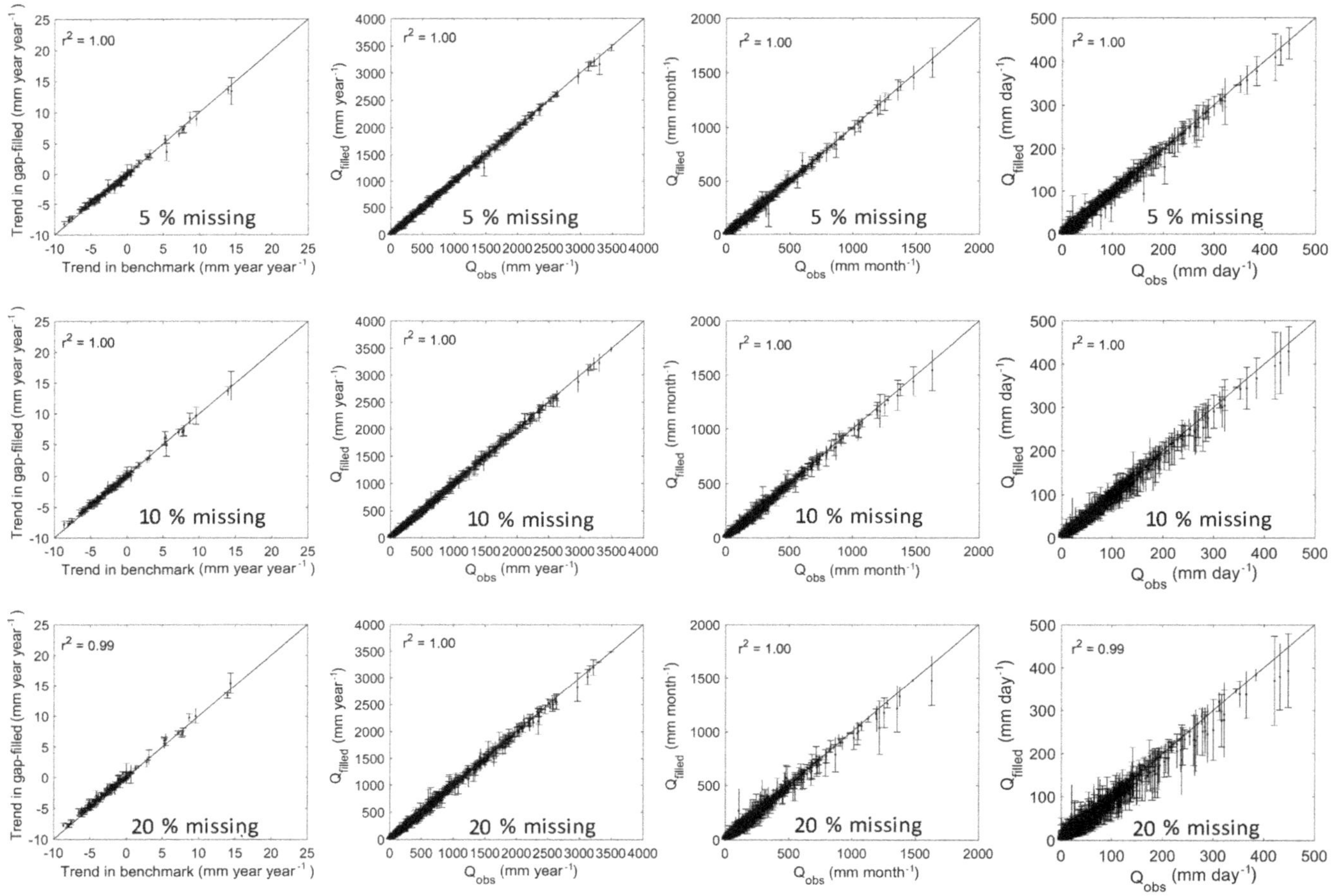

Figure 6. Comparisons between the observed streamflow (x axis) and gap-filled ones (y axis) for streamflow trend (mm year^{-1} year^{-1}, left panels), annual streamflow (mm year^{-1}, second left panels), monthly streamflow (mm month^{-1}, second right panels), and daily streamflow (mm day^{-1}, right panels). The gaps were filled using GR4J. The error bar represents standard deviation of the 100 replicates for each group of missing data experiments.

itive to negative). For the experiments with 5 % and 10 % missing data rates and for GR4J, fewer than 8 out of the 217 catchments show a trend mismatch, and almost all of them show non-significant trends ($p > 0.05$). For the experiments with a 20 % missing data rate for GR4J, fewer than 10 out of the 217 catchments show a trend mismatch, and all of them show non-significant trends. SIMHYD results are almost the same as GR4J results. All these indicate that there is very marginal influence on annual streamflow trend directions when the missing data rate is less than 20 %.

Though the three groups of experiments show small trend direction changes (Fig. 8), it is not clear how the trend bias (Eq. 3) looks. To this end, Fig. 9 further compares the trend bias between the experiments. It is clear that the trend biases between 5 % and 10 % missing data experiments are similar. For GR4J, both have a trend bias varying from -1 to 1 mm year^{-1} year^{-1}. For SIMHYD, the trend bias between the two is similar when it varies from -0.5 to 1 mm year^{-1} year^{-1}, and the trend bias of the 5 % missing data experiment is even larger than that of the 10 % missing data experiment. The trend bias of the 20 % missing data experiment is noticeably larger than that of the 10 % and 5 % missing data experiments for both models, and the underperformance is more noticeable in the SIMHYD gap-filled data than in the GR4J gap-filled data. This result suggests that the trend bias is reasonable when the missing data rate is less than 10 %, and can be large for a small number of catchments when the missing data rate is up to 20 %.

4 Discussion and conclusions

Researchers are keen to have a comprehensive understanding of rules for excluding catchments with gaps in the streamflow record. Our results indicate that when the streamflow data gaps are up to 10 %, the gap-filled data obtained using hydrological modelling are very reasonable for annual trend analysis and annual streamflow estimates. Choosing the threshold of the 10 % missing data rate will allow the use of

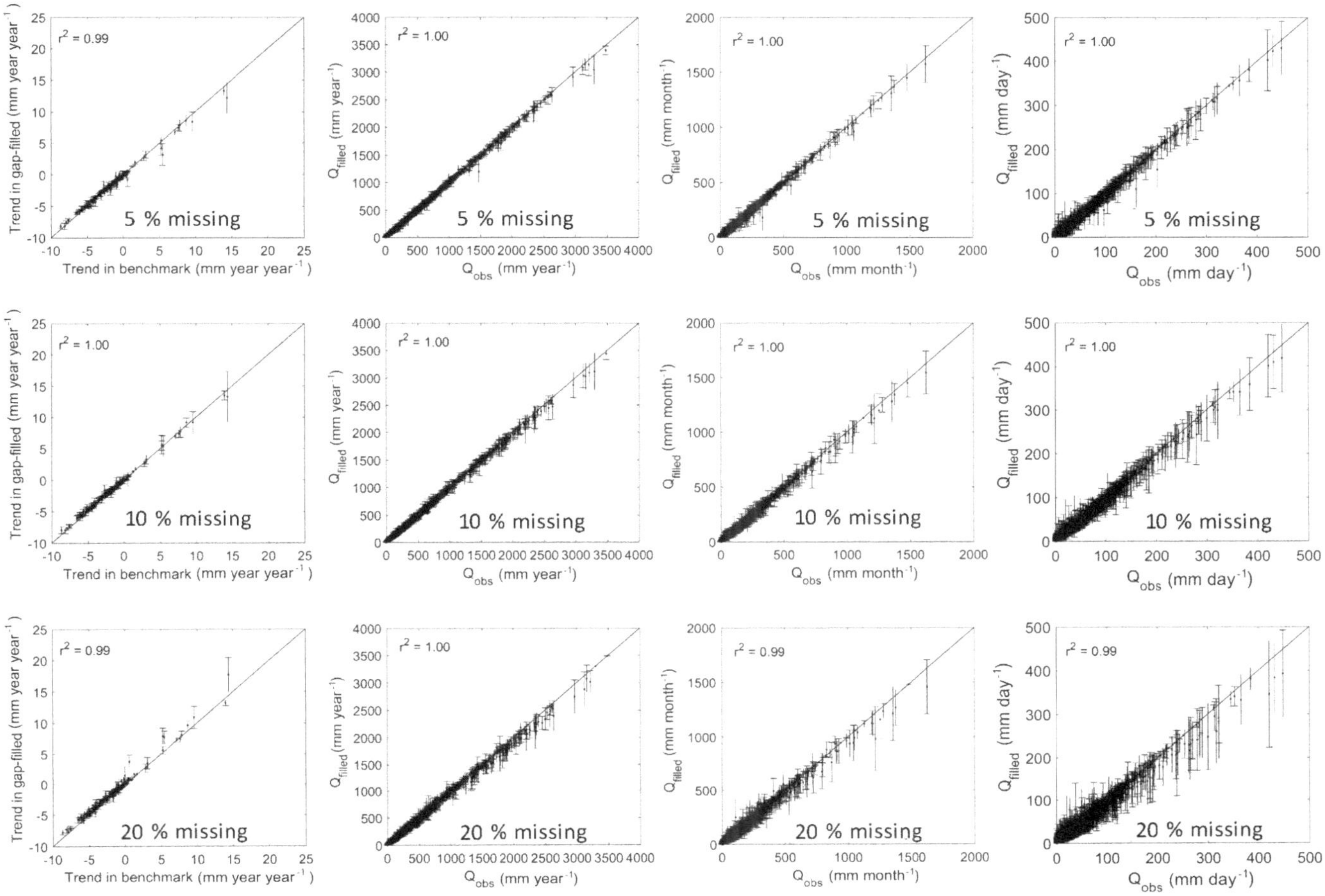

Figure 7. Same as Fig. 6 but using SIMHYD.

many more catchments in modelling and data analysis studies. For example, of the 780 unregulated Australian catchments available for modelling studies (Zhang et al., 2013), there are 237 catchments with a missing data rate of 1 %–10 % during 1981–2010, accounting for 38 % of total available catchments (Fig. 1). Of these 237, 67 (∼ 28 %) also have gaps lasting more than 1 year (which we did not consider in this analysis), and therefore these may not be suitable for use. With an increased number of catchments, more reliable large-scale hydrological modelling studies can be carried out (Beck et al., 2016; Parajka et al., 2013; Zhang et al., 2016a).

The missing rate experiments designed in this study are based on the actual missing data patterns obtained from the 780 catchments. In most cases, the number of consecutive missing days is less than 10, as indicated by Fig. 3, indicating brief periods of gauge malfunctions. It is, however, interesting to note that there are streamflow gaps lasting much longer than this in many catchments, with gaps of many months in some cases, noting that we excluded gaps lasting 1 year or more. It is highly likely that filling a gap of 1 year or more will result in biases larger than those presented here.

Furthermore, we also tested the quality of random gap-filled daily streamflow. In that case, the missing data patterns were randomly selected using a random number generator. The results obtained from the random gap filling (not shown) are similar to the results presented here. Thus, it is likely that the length of the gaps (as long as it is less than 1 year) is unlikely to impact the results of the gap-filling experiment. We conclude from this that the use of hydrologic modelling for filling the substantially gapped data (up to a 10 % missing data rate) described here for Australia will not impact annual trends of streamflow. Impacts on other streamflow characteristics also need to be examined, as well as seeing if the results obtained in Australia are comparable with those in other parts of the world, where the length of observational gaps may be quite different to those shown in Fig. 3.

It is possible that data gaps may only exist during high flow or low flow conditions, although that is not what we observed here, with the majority of missing data being more or less evenly distributed throughout the year (Fig. 4). We did, however, test the impact of filling streamflow data in high flow or low flow conditions (results not shown here). In these cases, the missing data patterns were selected using only high

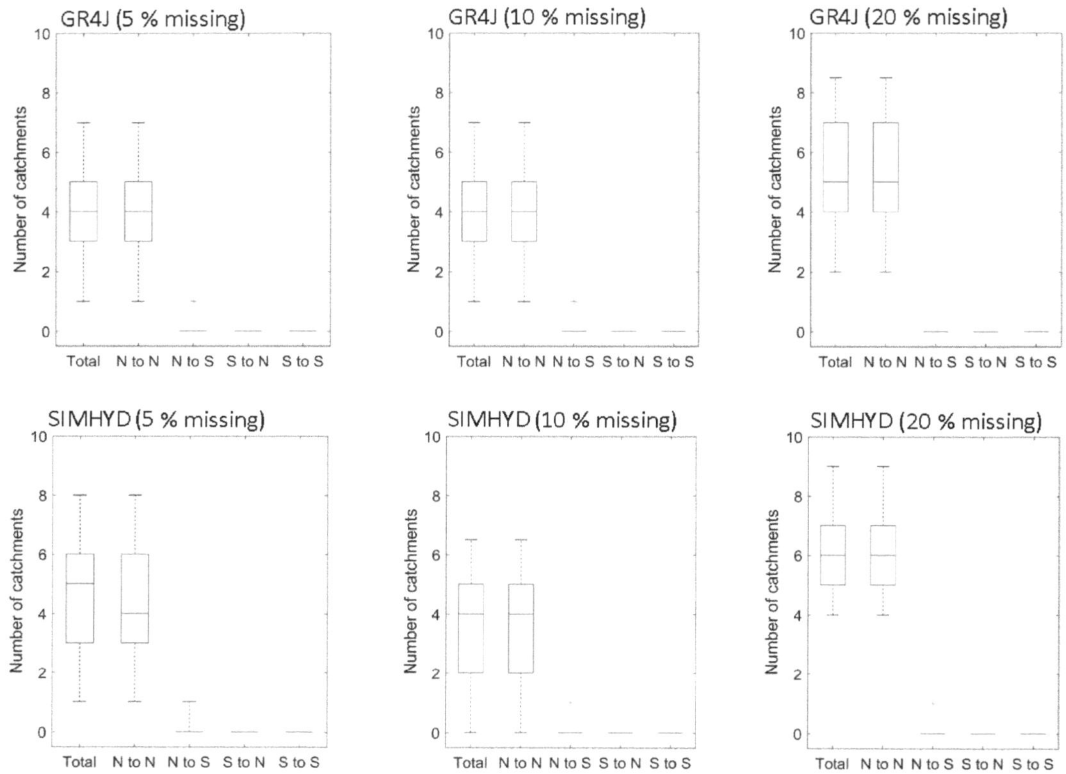

Figure 8. Trend mismatch analysis between the gap-filled and benchmark data. "Total" refers to all the mismatch catchments, "N" denotes non-significant trends ($p > 0.05$), and "S" denotes significant trends ($p \leq 0.05$). The bottom, middle, and top of each box show the 25th, 50th, and 75th percentiles, and the bottom and top whiskers are the 5th and 95th percentiles.

flow (> 95th percentile) or low flow (< 50th percentile) data. The results obtained from the low flow gap filling indicates that there is only a negligible influence on annual streamflow trend estimates when the missing data rate is less than 50 %. In contrast, the high flow gap-filled data show a noticeable change in annual streamflow trend when the missing data rate is 5 %. This is understandable since high flow is usually several orders of magnitude higher than low flow, and errors in filling high flow could have large impacts on annual flow and its trends (Slater and Villarini, 2017).

To understand if the quality of gap-filled streamflow is related to catchment attributes and calibration accuracy, we conducted further analysis among the trend bias, model calibration efficiency (i.e. NSE), and catchment aridity index (mean annual potential evaporation divided by mean annual precipitation) (Fig. 10). The model calibration results in dry catchments are normally poorer than those in wet catchments. However, the trend bias (mm year^{-1} year^{-1}) obtained from dry catchments is usually smaller. The large biases are observed from the catchments with an aridity index less than 2 and with the calibrated NSE being larger than 0.60. In part, this is to be expected since the streamflow is also lower in more arid catchments, meaning that the trend bias is also likely to be lower.

Figure 11 shows the relationship between the relative trend bias (%, Eq. 4) and aridity index. It shows that not only is the actual trend bias lower in drier catchments, but so too is the relative (%) trend. This result suggests that the large bias in annual trends as a result of gap filling is observed in relatively wet catchments, where model calibrations are reasonably good. This result seems counter-intuitive and requires further exploration, which is beyond the scope of the current paper.

This study focuses on evaluating annual streamflow and its trends. Therefore, we used the Nash–Sutcliffe efficiency plus model bias (Eqs. 5 and 6) to calibrate the two hydrological models. If other hydrological response variables such as low flow metrics are required, other model calibration schemes should be used since the NSE model calibration scheme gives more weight to the reproduction of high flows at the expense of low flows (Zhang et al., 2014). Low flow metrics have important ecological implications (Mackay et al., 2014; Smakhtin, 2001). In general, however, it is challenging to use hydrological modelling for low flow simulations and predictions (Pushpalatha et al., 2012; Staudinger et al., 2011). To obtain credible low flow gap filling, model calibrations should use an objective function that gives more weight to low flows, such as the NSE of daily inverse streamflow and

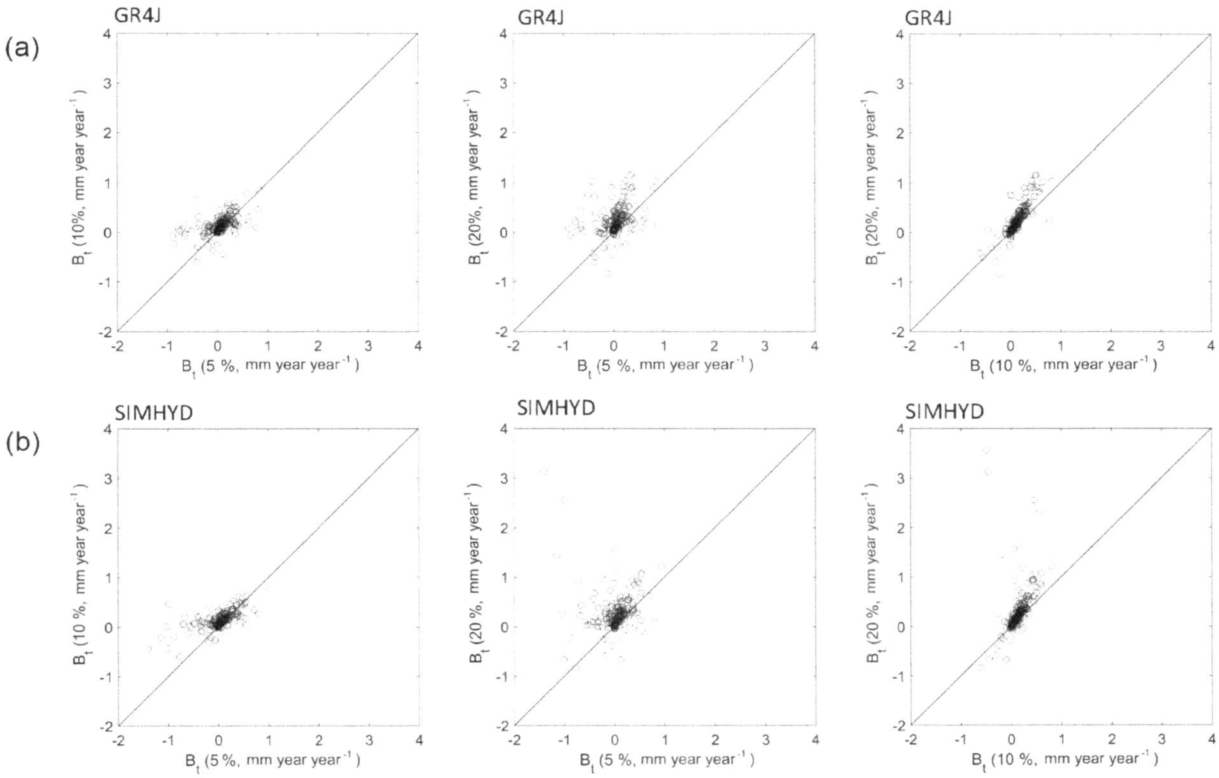

Figure 9. Trend biases comparison between the three groups of gap-filling experiments (5 %, 10 % and 20 %). **(a)** is for GR4J and **(b)** is for SIMHYD.

the direct low flow metrics. Another possible method is to combine hydrological modelling with other methods of gap filling, such as using nearby gauges (Lopes et al., 2016) and statistical methods (Gedney et al., 2006b).

It is noted that the infilled data purely refer to the missing data. All streamflow gauges only measure to a certain flow. Once the flow exceeds that level during flooding, the results are interpolated using stage–discharge relationships (Peña-Arancibia et al., 2015). These interpolations could be a major source of observation error. However, investigating high flow interpolation and data quality is beyond the scope of this study.

The modelling experiments and findings from this study could have important implications for other parts of the world as well as Australia. First, to develop appropriate gap-filling modelling experiments, it is necessary to evaluate the distribution of consecutive missing data. The probability distribution of consecutive missing data is skewed toward the low end, which can be nicely simulated using the gamma distribution (Eq. 1). This distribution should be very useful for similar missing data patterns in other regions. Second, hydrological modelling is a very good tool for filling gaps since it can fully take advantage of climate forcing and non-gap streamflow data and obtain the best possible daily simulations. Third, the threshold of 10 % identified in this study should be applicable to regions/catchments with similar missing data patterns. However, if the data gaps continue for seasons or years, the threshold may not hold.

It would also be interesting to compare hydrological modelling to other approaches for filling streamflow data gaps. Hydrological modelling is a most useful method used in Australia for predicting daily streamflow in ungauged catchments (Chiew et al., 2009; Li and Zhang, 2017; Zhang and Chiew, 2009; Viney et al., 2009). It has been used operationally by the Australian Bureau of Meteorology for filling daily streamflow data gaps for many years. In the future, this operational method could further be comprehensively evaluated against other approaches, such as interpolation or correlation with nearby gauging sites.

In summary, our results clearly demonstrate that the gap-filled data are most accurate when examining trends at the annual scale, followed by the monthly scale, and with least satisfaction at the daily scale. This gives researchers confidence in annual trend analysis, a hot topic in hydrological and climate sciences. Our results also clearly indicate that the gap filling of Australian streamflow data using hydrological models is very reasonable when the missing data rate is less than 10 %, with only a small number of catchments showing a large trend bias when the missing data rate is up to 20 %. The results also indicate that gap-filling drier catch-

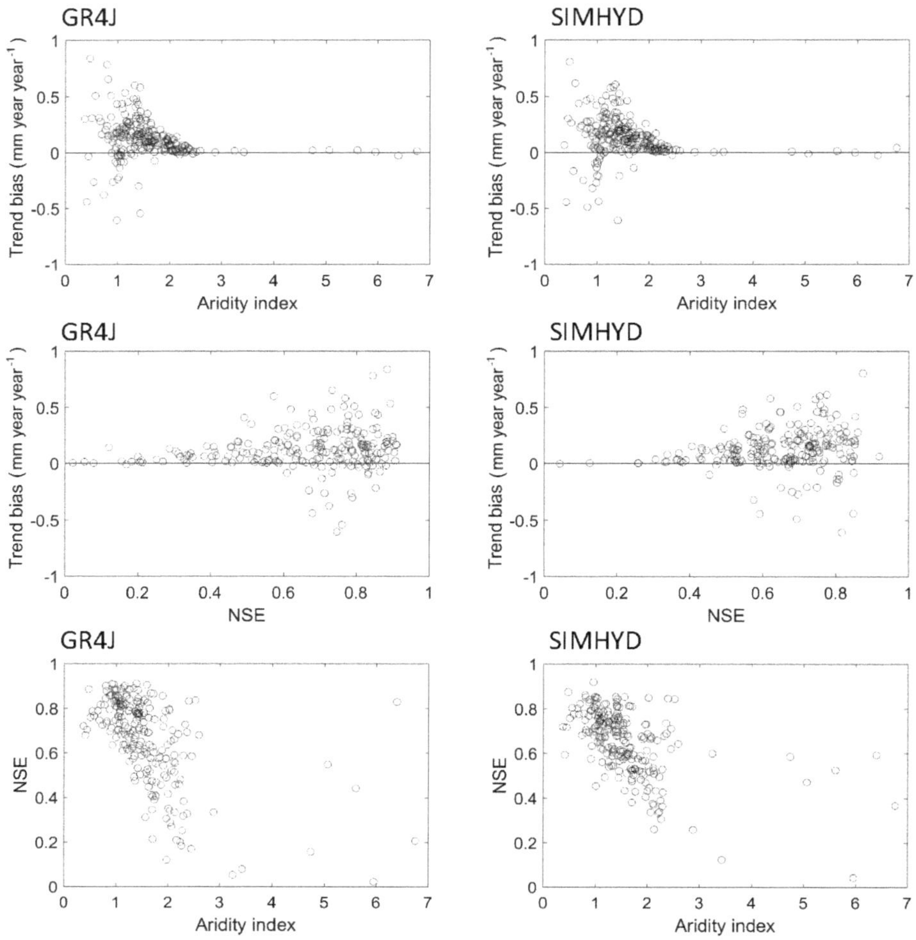

Figure 10. Relationships among trend bias (mm year^{-1} year^{-1}), model calibration Nash–Sutcliffe efficiency and aridity index for each catchment and for the experiment of the 10 % missing data rate.

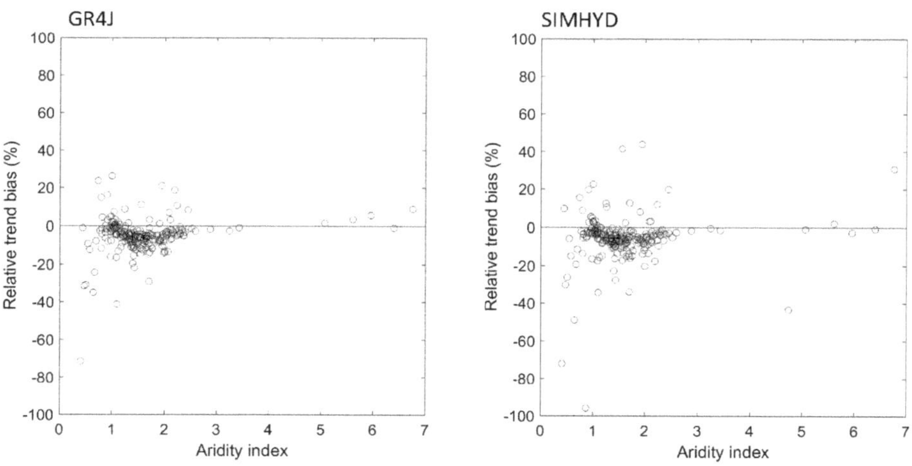

Figure 11. Relationships between relative trend bias (mm year^{-1} year^{-1}) and aridity index for each catchment and for the experiment of the 10 % missing data rate.

ments appears to be more successful than gap-filling wetter catchments.

Author contributions. YQZ conceived this study and conducted modelling experiments and data analysis. YQZ finished the first draft and DP modified it. YQZ and DP addressed comments from reviewers.

Competing interests. The authors declare that they have no conflict of interest.

Acknowledgements. This study was supported by the CSIRO strategic project "Next generation methods and capability for multi-scale cumulative impact assessment". The authors thank constructive suggestions from the editor, Louse Slater, and two reviewers, Maxine Zaidman and Juraj Parajka.

Edited by: Louise Slater

References

Beck, H. E., van Dijk, A. I. J. M., de Roo, A., Miralles, D. G., McVicar, T. R., Schellekens, J., and Bruijnzeel, L. A.: Global-scale regionalization of hydrologic model parameters, Water Resour. Res., 52, 3599–3622, https://doi.org/10.1002/2015wr018247, 2016.

Burn, D. H. and Elnur, M. A. H.: Detection of hydrologic trends and variability, J. Hydrol., 255, 107–122, 2002.

Chiew, F. H. S., Peel, M. C., and Western, A. W.: Application and testing of the simple rainfall-runoff model SIMHYD, in: Mathematical Models of Small Watershed Hydrology and Applications, edited by: Singh, V. P. and Frevert, D. K., Water resources Publication, Littleton, Colorado, USA, 335–367, 2002.

Chiew, F. H. S., Teng, J., Vaze, J., Post, D. A., Perraud, J. M., Kirono, D. G. C., and Viney N. R.: Estimating climate change impact on runoff across southeast Australia: Method, results, and implications of the modeling method, Water Resour. Res.. 45, W10414, https://doi.org/10.1029/2008WR007338, 2009.

Chiew, F. H. S., Kirono, D. G. C., Kent, D. M., Frost, A. J., Charles, S. P., Timbal, B., Nguyen, K. C., and Fu, G.: Comparison of runoff modelled using rainfall from different downscaling methods for historical and future climates, J. Hydrol., 387, 10–23, https://doi.org/10.1016/j.jhydrol.2010.03.025, 2010.

Dai, A., Qian, T., Trenberth, K. E., and Milliman, J. D.: Changes in Continental Freshwater Discharge from 1948 to 2004, J. Climate, 22, 2773–2792, https://doi.org/10.1175/2008jcli2592.1, 2009.

Déry, S. J., Hernández-Henríquez, M. A., Burford, J. E., and Wood, E. F.: Observational evidence of an intensifying hydrological cycle in northern Canada, Geophys. Res. Lett., 36, 1–5, https://doi.org/10.1029/2009gl038852, 2009.

Gedney, N., Cox, P. M., Betts, R. A., Boucher, O., Huntingford, C., and Stott, P. A.: Detection of a direct carbon dioxide effect in continental river runoff records, Nature, 439, 835–838, https://doi.org/10.1038/nature04504, 2006a.

Gedney, N., Cox, P. M., Betts, R. A., Boucher, O., Huntingford, C., and Stott, P. A.: Continental runoff – A quality-controlled global runoff data set – Reply, Nature, 444, E14–E15, https://doi.org/10.1038/nature05481, 2006b.

Hannaford, J. and Buys, G.: Trends in seasonal river flow regimes in the UK, J. Hydrol., 475, 158–174, https://doi.org/10.1016/j.jhydrol.2012.09.044, 2012.

Harvey, C. L., Dixon, H., and Hannaford, J.: An appraisal of the performance of data-infilling methods for application to daily mean river flow records in the UK, Hydrol. Res., 43, 618–636, https://doi.org/10.2166/nh.2012.110, 2012.

Jeffrey, S. J., Carter, J. O., Moodie, K. B., and Beswick, A. R.: Using spatial interpolation to construct a comprehensive archive of Australian climate data, Environ. Model. Softw., 16, 309–330, 2001.

Lavers, D., Prudhomme, C., and Hannah, D. M.: Large-scale climate, precipitation and British river flows Identifying hydroclimatological connections and dynamics, J. Hydrol., 395, 242–255, https://doi.org/10.1016/j.jhydrol.2010.10.036, 2010.

Li, F., Zhang, Y., Xu, Z., Liu, C., Zhou, Y., and Liu, W.: Runoff predictions in ungauged catchments in southeast Tibetan Plateau, J. Hydrol., 511, 28–38, https://doi.org/10.1016/j.jhydro1.2014.01.014, 2014.

Li, H. and Zhang, Y.: Regionalising rainfall-runoff modelling for predicting daily runoff: Comparing gridded spatial proximity and gridded integrated similarity approaches against their lumped counterparts, J. Hydrol., 550, 279–293, https://doi.org/10.1016/j.jhydrol.2017.05.015, 2017.

Liu, J. and Zhang, Y.: Multi-temporal clustering of continental floods and associated atmospheric circulations, J. Hydrol., 555, 744–759, https://doi.org/10.1016/j.jhydrol.2017.10.072, 2017.

Lopes, A. V., Chiang, J. C. H., Thompson, S. A., and Dracup, J. A.: Trend and uncertainty in spatial-temporal patterns of hydrological droughts in the Amazon basin, Geophys. Res. Lett., 43, 3307–3316, https://doi.org/10.1002/2016gl067738, 2016.

Mackay, S. J., Arthington, A. H., and James, C. S.: Classification and comparison of natural and altered flow regimes to support an Australian trial of the Ecological Limits of Hydrologic Alteration framework, Ecohydrology, 7, 1485–1507, https://doi.org/10.1002/eco.1473, 2014.

Oudin, L., Andreassian, V., Perrin, C., Michel, C., and Le Moine, N.: Spatial proximity, physical similarity, regression and ungaged catchments: A comparison of regionalization approaches based on 913 French catchments, Water Resour. Res., 44, W03413, https://doi.org/10.1029/2007WR006240, 2008.

Parajka, J., Viglione, A., Rogger, M., Salinas, J. L., Sivapalan, M., and Blöschl, G.: Comparative assessment of predictions in ungauged basins – Part 1: Runoff-hydrograph studies, Hydrol. Earth Syst. Sci., 17, 1783–1795, https://doi.org/10.5194/hess-17-1783-2013, 2013.

Peña-Arancibia, J. L., Zhang, Y., Pagendam, D. E., Viney, N. R., Lerat, J., van Dijk, A. I. J. M., Vaze, J., and Frost, A. J.: Streamflow rating uncertainty: Characterisation and impacts on model calibration and performance, Environ. Model. Softw., 63, 32–44, https://doi.org/10.1016/j.envsoft.2014.09.011, 2015.

Perrin, C., Michel, C., and Andreassian, V.: Improvement of a parsimonious model for streamflow simulation, J. Hydrol., 279, 275–289, https://doi.org/10.1016/s0022-1694(03)00225-7, 2003.

Petrone, K. C., Hughes, J. D., Van Niel, T. G., and Silberstein, R. P.: Streamflow decline in southwestern Australia, 1950–2008, Geophys. Res. Lett., 37, 1–7, https://doi.org/10.1029/2010gl043102, 2010.

Priestley, C. H. B. and Taylor, R. J.: On the assessment of surface heat flux and evaporation using large-scale parameters, Mon. Weather Rev., 100, 81–92, 1972.

Pushpalatha, R., Perrin, C., Le Moine, N., and Andreassian, V.: A review of efficiency criteria suitable for evaluating low-flow simulations, J. Hydrol., 420, 171–182, https://doi.org/10.1016/j.jhydrol.2011.11.055, 2012.

Rojas-Serna, C., Lebecherel, L., Perrin, C., Andreassian, V., and Oudin, L: How should a rainfall-runoff model beparameterized in an almost ungaugedcatchment? A methodology tested on 609 catchments, Water Resour. Res., 52, 4765–4784, https://doi.org/10.1002/2015WR018549, 2016.

Sanderson, M. G., Wiltshire, A. J., and Betts, R. A.: Projected changes in water availability in the United Kingdom, Water Resour. Res., 48, W08512, https://doi.org/10.1029/2012wr011881, 2012.

Seibert, J. and Beven, K. J.: Gauging the ungauged basin: how many discharge measurements are needed?, Hydrol. Earth Syst. Sci., 13, 883–892, https://doi.org/10.5194/hess-13-883-2009, 2009.

Slater, L. and Villarini, G.: On the impact of gaps on trend detection in extreme streamflow time series, Int. J. Climatol., 37, 3976–3983, https://doi.org/10.1002/joc.4954, 2017.

Smakhtin, V. U.: Low flow hydrology: a review, J. Hydrol., 240, 147–186, https://doi.org/10.1016/s0022-1694(00)00340-1, 2001.

Staudinger, M., Stahl, K., Seibert, J., Clark, M. P., and Tallaksen, L. M.: Comparison of hydrological model structures based on recession and low flow simulations, Hydrol. Earth Syst. Sci., 15, 3447–3459, https://doi.org/10.5194/hess-15-3447-2011, 2011.

The MathWorks, Inc.: Genetic algorithm and direct search toolbox 2 user's guide, Natick, Mass, 5.2–5.13, 2006.

Ukkola, A. M., Prentice, I. C., Keenan, T. F., van Dijk, A. I. J. M., Viney, N. R., Myneni, R. B., and Bi, J.: Reduced streamflow in water-stressed climates consistent with CO_2 effects on vegetation, Nat. Clim. Change, 6, 75–78, https://doi.org/10.1038/nclimate2831, 2015.

Ukkola, A. M., Keenan, T. F., Kelley, D. I., and Prentice, I. C.: Vegetation plays an important role in mediating future water resources, Environ. Res. Lett., 11, 094022, https://doi.org/10.1088/1748-9326/11/9/094022, 2016.

van Dijk, A. I. J. M., Beck, H. E., Crosbie, R. S., de Jeu, R. A. M., Liu, Y. Y., Podger, G. M., Timbal, B., and Viney, N. R.: The Millennium Drought in southeast Australia (2001–2009): Natural and human causes and implications for water resources, ecosystems, economy, and society, Water Resour. Res., 49, 1040–1057, https://doi.org/10.1002/wrcr.20123, 2013.

Viney, N. R., Perraud, J., Vaze, J., Chiew, F. H. S., Post, D. A., and Yang, A.: The usefulness of bias constraints in model calibration for regionalisation to ungauged catchments, 18th World IMACS Congress and Modsim09 International Congress on Modelling and Simulation: Interfacing Modelling and Simulation with Mathematical and Computational Sciences, Cairns, Australia, 13–17 July 2009, 3421–3427, 2009.

Yang, Y., McVicar, T. R., Donohue, R. J., Zhang, Y., Roderick, M. L., Chiew, F. H. S., Zhang, L., and Zhang, J.: Lags in hydrologic recovery following an extreme drought: Assessing the roles of climate and catchment characteristics, Water Resour. Res., 53, 4821–4837, https://doi.org/10.1002/2017wr020683, 2017.

Zhang, Y. and Chiew, F. H. S.: Relative merits of different methods for runoff predictions in ungauged catchments, Water Resour. Res., 45, W07412, https://doi.org/10.1029/2008WR007504, 2009.

Zhang, Y., Vaze J., Chiew, F. H. S., Teng, J., and Li, M.: Predicting hydrological signatures in ungauged catchments using spatial interpolation, index model, and rainfall–runoff modelling, J. Hydrol., 517, 936–948, 2014.

Zhang, Y., Pena-Arancibia, J. L., McVicar, T. R., Chiew, F. H., Vaze, J., Liu, C., Lu, X., Zheng, H., Wang, Y., Liu, Y. Y., Miralles, D. G., and Pan, M.: Multi-decadal trends in global terrestrial evapotranspiration and its components, Sci. Rep., 6, 19124, https://doi.org/10.1038/srep19124, 2016a.

Zhang, Y., Zheng, H., Chiew, F. H. S., Pena-Arancibia, J., and Zhou, X.: Evaluating Regional and Global Hydrological Models against Streamflow and Evapotranspiration Measurements, J. Hydrometeorol., 17, 995–1010, https://doi.org/10.1175/jhm-d-15-0107.1, 2016b.

Zhang, Y. Q., Viney, N. R., Frost, A., Oke, A., Brooks, M., Chen, Y., and Campbell, N.: Collation of Australian modeller's streamflow dataset for 780 unregulated Australian catchments, CSIRO: Water for a Healthy Country National Research Flagship, 117 pp., 2013.

10

State updating and calibration period selection to improve dynamic monthly streamflow forecasts for an environmental flow management application

Matthew S. Gibbs[1,2], David McInerney[1], Greer Humphrey[1], Mark A. Thyer[1], Holger R. Maier[1], Graeme C. Dandy[1], and Dmitri Kavetski[1]

[1]School of Civil, Environmental and Mining Engineering, The University of Adelaide, North Terrace, Adelaide, South Australia, 5005, Australia
[2]Department of Environment, Water and Natural Resources, Government of South Australia, P.O. Box 1047, Adelaide, 5000, Australia

Correspondence: Matthew S. Gibbs (matthew.gibbs@adelaide.edu.au)

Abstract. Monthly to seasonal streamflow forecasts provide useful information for a range of water resource management and planning applications. This work focuses on improving such forecasts by considering the following two aspects: (1) state updating to force the models to match observations from the start of the forecast period, and (2) selection of a shorter calibration period that is more representative of the forecast period, compared to a longer calibration period traditionally used. The analysis is undertaken in the context of using streamflow forecasts for environmental flow water management of an open channel drainage network in southern Australia. Forecasts of monthly streamflow are obtained using a conceptual rainfall–runoff model combined with a post-processor error model for uncertainty analysis. This model set-up is applied to two catchments, one with stronger evidence of non-stationarity than the other. A range of metrics are used to assess different aspects of predictive performance, including reliability, sharpness, bias and accuracy. The results indicate that, for most scenarios and metrics, state updating improves predictive performance for both observed rainfall and forecast rainfall sources. Using the shorter calibration period also improves predictive performance, particularly for the catchment with stronger evidence of non-stationarity. The results highlight that a traditional approach of using a long calibration period can degrade predictive performance when there is evidence of non-stationarity. The techniques presented can form the basis for operational monthly streamflow forecasting systems and provide support for environmental decision-making.

1 Introduction

Predictions of streamflow a month or a season ahead are essential information required by water resource managers for subsequent planning (Wang et al., 2011). This is particularly true in unregulated catchments with no capacity for storage and a highly variable flow regime that can be difficult to predict from historical data. A number of approaches have been developed to provide streamflow predictions with lead times from a month to a season ahead. These include "dynamic" hydrological modelling approaches (Demargne et al., 2014; Wood and Schaake, 2008), statistical approaches (Bennett et al., 2014; Robertson and Wang, 2013), or a combination of the two (Robertson et al., 2013).

In this work, a dynamic hydrological modelling based approach is adopted to provide streamflow forecasts for an environmental management application. The dynamic approach can often better capture catchment dynamics than statistical models based on simple climatic indices (Robertson et al., 2013). In forecast mode, a hydrological model calibrated using historical data is run forward in time, with input data provided by forecast climate forcings. The following three major factors control forecasting performance (Luo et al., 2012):

(1) the ability of the hydrological model to predict streamflow with actual forcings; (2) the accuracy of the assumed initial conditions (e.g. soil moisture stores); and (3) the accuracy of the forecasts of the climate inputs. The focus of this paper is on the first two factors, in the context of a user need for monthly streamflow forecasts to support environmental management and decision-making.

Conceptual rainfall–runoff (CRR) models are widely used to simulate streamflow, due to their simplicity and accuracy (Li et al., 2015a; Tuteja et al., 2011). The parameters of these models have a limited relationship with measurable catchment attributes (e.g. soil horizon depth; Fenicia et al., 2014), and typically require calibration to observed streamflow data (noting that physical models also require some calibration; Mount et al., 2016; Pappenberger and Beven, 2006). The use of long calibration periods assumes time-invariant catchment characteristics and processes, and that the parameter values derived from the calibration period are representative of the prediction period (Vaze et al., 2010). It is generally considered that longer calibration periods produce more robust parameter estimates, as a longer period exposes the model to a more diverse range of catchment conditions and flow events (Wu et al., 2013); however, this is not always the case (for example, Brigode et al., 2013).

The assumption that parameters are constant in time can result in decreased model performance if the conditions encountered in the forecast period are different from those in the calibration period (Bowden et al., 2012; Coron et al., 2012). In this work, the term "non-stationary" is used to refer to situations where physical changes are expected to have occurred in a catchment, and where there is evidence to reject the hypothesis of stationarity. In practice, catchments may have different "degrees" of non-stationarity, depending on the evidence available to reject the hypothesis of stationarity, the degree of change in a catchment, and the timescales over which the changes take place. Examples of catchment non-stationarity that can be expected to change the rainfall–runoff relationship include changes in land use or land cover (e.g. deforestation, urbanisation), land drainage, interception (e.g. dams, diversions), groundwater abstractions or responses to changes in climate (Milly et al., 2015). This definition of catchment non-stationarity can be contrasted to a broader definition of "hydrological model non-stationarity", which refers to temporal changes in hydrological model parameters for any reason (e.g. systematic data errors, poor calibration procedures, model structural deficiencies); see, for example, Westra et al. (2014).

The degradation in model predictive performance due to catchment non-stationarity can impact on the decisions informed by these forecasts. To address this concern, a number of studies have calibrated model parameters to subsets of the available data, by attempting to find periods in the historical record that are analogous to conditions expected in the prediction time period, and by tailoring the time period selection to compensate for deficiencies in the model structure or input data (Brigode et al., 2013; de Vos et al., 2010; Luo et al., 2012; Vaze et al., 2010; Wu et al., 2013; Zhang et al., 2011). Often there is a trade-off between the benefits of a longer calibration period, which exposes the model to a more diverse range of conditions and tends to improve parameter identifiability, versus the benefits of a shorter calibration period, which exposes the model to the most recent – and hence often the most relevant – dynamics in the catchment. Demonstrating and understanding the impact of this trade-off on model predictive performance is a key research gap pursued in this study.

Predictive uncertainty quantification is another major aspect of practical streamflow prediction. Many approaches are available to quantify predictive uncertainty, from approaches that identify a range of model parameters that represent the behaviour of the catchment using approaches such as generalised likelihood uncertainty estimation (GLUE; Beven and Binley, 1992), to post-processor approaches (e.g. Krzysztofowicz and Maranzano, 2004) and disaggregation approaches that attempt to characterise each individual source of error explicitly (e.g. Kavetski et al., 2003; Vrugt et al., 2005). In this work, predictive uncertainty is estimated using an aggregated post-processor residual error model. The residual error model represents the differences between the hydrological model predictions and observed data, without trying to identify the contributing sources (Evin et al., 2014). The post-processor approach is chosen because it can lead to more robust estimates of predictive uncertainty compared to joint calibration of all parameters (i.e. estimating CRR model and error model parameters concurrently; Evin et al., 2014).

Much of the skill in seasonal streamflow forecasts over periods following rainy seasons is commonly attributed to accurately representing initial catchment conditions (Koster et al., 2010; Pagano et al., 2004; Wang et al., 2009). In contrast, forecast skill over periods following dry seasons is generally attributed to both initial catchment conditions and meteorological inputs (Maurer and Lettenmaier, 2003; Wood and Lettenmaier, 2008). The impact of the initial catchment conditions is particularly pronounced when forecasting over short lead times, typically up to 1 month (Li et al., 2009; Wang et al., 2011), although this time frame is generally catchment-dependent.

In CRR models, catchment conditions are represented by (usually multiple) model storages, referred to as "state variables". The values of these storages at the start of a forecast period are typically determined using a warm-up period, which allows the internal model states to reach reasonable values. Given the expected influence of the initial conditions on the simulated streamflow, observed data can be assimilated into the model to update the state of the model storages. The most commonly used approaches in hydrological data assimilation include direct updating of storages (for example, Demirel et al., 2013), Kalman filtering, particle filtering, and variational data assimilation (see Liu and Gupta, 2007). Berthet (2010) considered a number of tests for differ-

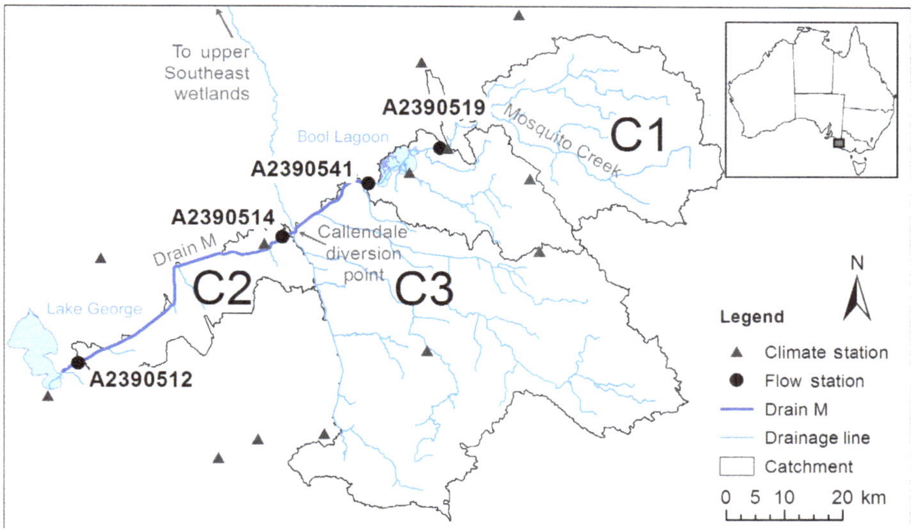

Figure 1. Map of the case study region, in southern Australia.

ent updating approaches for the GRP model, a CRR model commonly used in short-term streamflow forecasting applications in France.

Updating the states of conceptual rainfall–runoff models is not straightforward, as any environmental model is at best an approximate representation of the real catchment (Berthet et al., 2009). A number of observed data sources can be used to update model storages, including observed streamflow and in situ or remotely sensed soil moisture. From these options, Li et al. (2015b) suggest that gauged discharge data assimilation is a more effective way to improve short-term forecasts and is still preferred for operational streamflow forecasting purposes.

Studies on observed data assimilation and CRR model state updating have focused primarily on flood forecasting with short lead times. The benefits at longer lead times (e.g. monthly to seasonal) to forecast water availability have received less attention in the published literature.

1.1 Study aims

This work focuses on determining the degree to which state updating and the selection of calibration period length can enhance monthly streamflow predictions in the context of an environmental flow management application. More specifically, the aims of this study are to

1. evaluate the ability of state updating in a daily CRR model to improve predictive performance when forecasting streamflow volume for the upcoming month; and

2. assess the degree to which using a shorter calibration period that is more representative of the forecast period can improve predictive performance, in particular when there is evidence of catchment non-stationarity.

The paper is organised as follows. Section 2 outlines the user need for monthly forecasts to manage a drainage network for environmental and social outcomes in southern Australia, and describes the case study catchments and data available. Section 3 describes the model set-up and forecasting framework, as well as the methodology designed to achieve the aims above. Sections 4 and 5 present and discuss the case study results, and Sect. 6 summarises the key conclusions.

2 Environmental flow management case study

2.1 Catchment location and characteristics

The location considered in this study is a component of an extensive drainage network (exceeding 2500 km of open channels) in southern Australia (Fig. 1). Historically, runoff flowed in a northerly direction, along the watercourses adjacent to ranges, parallel to the coastline. Over the past 150 years, these flow paths have been diverted through a series of cross-country drains, constructed to provide flood relief and improve the agricultural productivity of the region by draining water in a south-westerly direction, creating outlets to the ocean. The largest of these cross-country drains is Drain M (Fig. 1), which conveys water from Bool Lagoon to Lake George. Monthly runoff volumes from Drain M are highly variable, ranging from close to zero to more than is required to support Lake George, with the historical volumes varying over 3–4 orders of magnitude for a given month (Fig. 2). This variability makes it difficult to maximise the use of water, as the seasonal pattern described by the historical record alone provides little guidance.

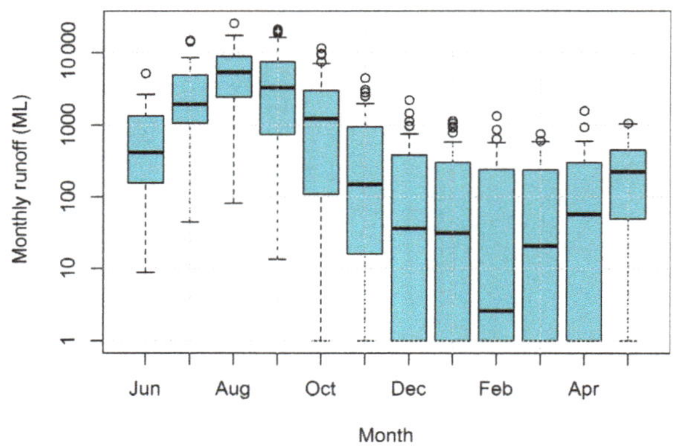

Figure 2. Variability in monthly runoff in Drain M at the location at flow station A2390512.

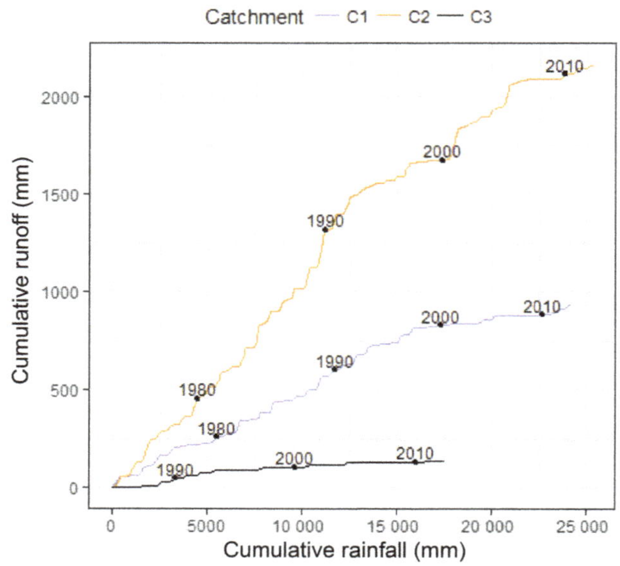

Figure 3. Double mass plot of the rainfall–runoff data in the three main catchments contributing to Drain M. It can be seen that (1) the volume of runoff for the same volume of rainfall has reduced in the latter decade, and (2) very little runoff is generated from catchment C3.

The streamflow in the case study region is seasonal to ephemeral, with very low flow over the summer and autumn months (Fig. 2). Runoff coefficients are low, with annual runoff in the range of 0.01–0.1 of annual rainfall (Gibbs et al., 2012). The predominant land use in the region is dry land pasture with some flood irrigation as well as plantation forestry; there is no major urbanisation in the catchments. The topography of the region is very flat, with mainstream slopes of the order of 0.005. The hydrogeology of the catchment includes shallow aquifers with major karstification of limestone, which may be suggestive of non-conservative catchments with appreciable groundwater exchanges across their boundaries.

Mosquito Creek flows into Bool Lagoon (catchment C1 in Fig. 1, area 1002 km^2). Drain M commences at the outlet of Bool Lagoon, and a large catchment flows into Drain M between Bool Lagoon and a diversion point at Callendale (catchment C3 with an area of 2200 km^2). Finally, the Drain M local catchment contributes flow downstream of the Callendale diversion point, flowing into Lake George (catchment C2, area 383 km^2).

In the region where the case study catchments are located, plantation forestry expanded substantially in the late 1990s. Changes in the relationship between rainfall and runoff also occurred during this period, evidenced by the reduced slope in the plot of cumulative runoff against cumulative rainfall (double-mass analysis) in Fig. 3 (Searcy et al., 1960; Yihdego and Webb, 2013). The runoff ratio in catchment C1 is approximately 0.045 before year 2000, but reduces by 70 % to 0.013 after 2000. The runoff ratio in catchment C2 is around 0.088 before year 2000, but reduces by 30 % to 0.061 after 2000. This comparison provides stronger evidence of non-stationarity in catchment C1 than in catchment C2. Other studies have also investigated the link between changes in the hydrology and changes in land use in the region (Avey and Harvey, 2014; Brookes et al., 2017). These changes have implications for the choice of calibration data period, as data from the 1970s may not be representative of hydrological conditions in the 2000s.

It is evident from Fig. 3 that catchment C3, despite having the largest catchment area (2200 km^2) of the three catchments, generates very little runoff. This behaviour is due to a number of factors, including the very flat terrain and depression storage, substantial vegetation cover (both plantation and natural) and irrigation extractions from the shallow underlying aquifer. Given its limited streamflow volume, catchment C3 is excluded from further analysis in this study. From a practical perspective, it is assumed that in the years where there is substantial yield from this catchment there will already be surplus flow from the upstream catchments.

2.2 Management issues

Drain M serves multiple competing demands on the water resources available in this catchment system. These demands influence the decision to use the regulators along the system.

a. Bool Lagoon has water requirements that influence releases from the lagoon into Drain M.

b. Lake George has water requirements to maintain the estuarine ecology of the lake, and to support its significance as a biological resource and as a resource for recreational fishing.

c. The ocean outlet requires some flow to prevent sediment from entering Lake George and to maintain connectivity to the sea (which allows fish movement and aids fish recruitment). However, high flows may impact on sea grasses, due to their low salinity and high nutrient load.

d. The wetlands of the upper south-east to the north typically benefit from as much water as possible from the Drain M system.

Decisions to undertake diversions from Drain M must be made throughout the year (mainly in the high-flow season from late winter and throughout spring). It is expected that forecasts of future flows at key locations will assist in maximising the environmental and social outcomes achieved from the available water. Forecasts of monthly volume with a lead time of 1 month ahead are considered most appropriate for this application, because (1) the main quantities of interest in this application are volume and the overall water balance, rather than the size or timing of daily peak flows, and (2) a 1-month lead time provides sufficient time to undertake any changes in diversions to satisfy the competing demands on the system.

2.3 Climate data

The mean annual rainfall for the region varies from 600 mm in the north to 675 mm in the south. The mean annual FAO56 potential evapotranspiration (PET; Allen et al., 1998) is approximately 1000 mm. The highest rainfalls are experienced in the winter months, with rainfall exceeding evapotranspiration in May–September. The SILO Patched Point Dataset (Jeffrey et al., 2001) was used for the observed rainfall and the FAO56 evapotranspiration data were adopted, with the climate stations used shown in Fig. 1. Time series of rainfall and evapotranspiration in each catchment were obtained using a Thiessen polygon approach. This weighting approach is considered appropriate for the region, due to the flat terrain being unlikely to lead to significant topographic effects on the spatial distribution of rainfall.

Rainfall forecasts from the Australian Bureau of Meteorology's seasonal forecast system, POAMA-2 (Hudson et al., 2011), were used. POAMA-2 is a dynamical climate forecasting system designed to produce multi-week to seasonal forecasts of climate for Australia based on a coupled ocean–atmosphere model and ocean–atmosphere–land observation assimilation systems. In this paper, we use a 30-member ensemble of monthly/multi-week forecasts from version 2.4 of POAMA-2. POAMA-2 predictions have a coarse spatial resolution (~ 250 km), which does not capture the spatial variability in catchment-scale rainfall. For the purposes of this application, the POAMA-2 rainfall hindcasts (i.e. forecasts developed by applying the modelling system to the historical period) at the relevant pixel were downscaled to each climate station in the study region (Fig. 1) using the statistical downscaling method detailed in Shao and Li (2013). Further details of the downscaling approach are provided in Humphrey et al. (2016).

2.4 Streamflow data

Daily streamflow data are available from the South Australian Department of Environment, Water and Natural Resources Surface Water Archive (https://www.waterconnect.sa.gov.au/Systems/swd), with the flow stations used shown in Fig. 1. Three of the flow stations have data available from the early 1970s, with the exception being the station at the outlet of Bool Lagoon (site A2390541), where data were available from 1985. Travel times along Drain M between flow stations are typically less than 1 day. To determine the flow generated within catchment C2, the daily flows recorded at upstream flow station A2390514 were subtracted from downstream flow station A2390512.

The identification of high-quality data is important because biases and systematic changes in the measurement of hydrological data can significantly affect model calibration and lead to non-stationarity in the estimated model parameters (Westra et al., 2014). Analysis of the data and monitoring stations suggested that streamflow data uncertainty is expected to be low, given the regular cross sections of the weirs used for monitoring stage and upstream drains, and the high number of gaugings (between 78 and 166 flow gaugings at each flow station) available to develop stage–discharge relationships.

3 Methodology

3.1 CRR model

The GR4J model (Perrin et al., 2003) is a parsimonious daily CRR model, selected for this study because it explicitly accounts for non-conservative (or "leaky") catchments (relevant for the study area; see Sect. 2.1) and has demonstrated good performance for Australian conditions (Coron et al., 2012; Guo et al., 2017; Westra et al., 2014). The standard form of the GR4J model has four calibration parameters: the maximum capacity of a production (soil) store, $X1$, a catchment water exchange coefficient, $X2$, the maximum capacity of a routing store, $X3$, and a time base for a unit hydrograph, $X4$. Further details of the model structure and parameters can be found in Perrin et al. (2003).

Note that the catchments considered have a relatively slow streamflow response. Consequently, the pre-specified split to the routing store of 0.9 in the original specification of the GR4J model may be too low for these catchments. To mitigate this potential deficiency, we have modified the GR4J model so that the split between the routing store and the direct runoff is included as an explicit calibration parameter termed *split*.

3.2 Parameter estimation

The GR4J parameters are inferred using Bayes' equation. The posterior probability density of the parameters given

Table 1. Bounds adopted for the uniform prior distribution on the GR4J parameters.

Parameter	Name	Lower bound	Upper bound
$X1$	production store maximal capacity (mm)	100	600
$X2$	catchment water exchange coefficient (mm)	−15	5
$X3$	1-day maximal capacity of the routing reservoir (mm)	1	300
$X4$	unit hydrograph time base (days)	0.5	6
split	proportion of flow directed to the routing store	0.6	0.99

daily observed streamflow data $\widetilde{q}$ and climate data $\mathbf{X}$, $p(\theta|\widetilde{q}\mathbf{X})$, is given by

$$p(\theta|\widetilde{q},\mathbf{X}) \propto p(\widetilde{q},|\theta X)\, p(\theta), \tag{1}$$

where $p(\theta)$ is the prior distribution and $p\,(\widetilde{q}|\theta,\mathbf{X})$ is the likelihood function.

A standard least squares likelihood function is adopted (see, for example, Thyer et al., 2009), which is derived from a residual error model that assumes independent, homoscedastic residuals. This likelihood function is adopted for the calibration of the daily hydrological model because it provides a better fit to the high daily flows (Wright et al., 2015), which make an important contribution to monthly volumes of interest in our study. Uniform prior distributions are used for all parameters, with bounds given in Table 1.

The posterior distribution in Eq. (1) is sampled using the DiffeRential Evolution Adaptive Metropolis (DREAM) algorithm (Vrugt et al., 2009). The sampled parameter sets are then used to approximate the posterior parameter distribution for a given calibration period. Computations were carried out using the Hydromad R package implementation of the DREAM algorithm and the GR4J model (Andrews et al., 2011). A total of 25 000 iterations of the DREAM algorithm were carried out, including a "burn-in" period of 6250 iterations to allow the Markov chain to stabilise. The number of parallel chains was set equal to the number of parameters (Vrugt et al., 2009), which, for the modified GR4J model used in this work (Sect. 3.1), led to five parallel chains being used.

The posterior distributions obtained for different calibration time periods are investigated for evidence of trends and changes over time. For the purposes of developing streamflow predictions using the post-processing approach (Sect. 3.5), only the single parameter set resulting in the maximum posterior probability is used.

3.3 Calibration approach

A rolling calibration approach is used to account for the impact of non-stationarity on the inferred CRR model parameters. This rolling calibration approach is similar to the approach used by Luo et al. (2012) and Wagener et al. (2003). It consists of choosing a calibration length and then moving it forward year by year, while recalibrating the model parameters to each such calibration "window". The calibrated parameter values are used to simulate the following 1 year of data, before recalibrating the model and repeating the process. This methodology allows the identification of changes in parameter distributions over time, without the need to identify specific periods when changes in the rainfall–runoff response may have occurred.

Calibration period lengths of CPL = 10 and CPL = 20 years are considered, to assess the trade-off between using a longer calibration period to expose the model to more diverse catchment conditions and improve parameter identifiability, versus using a shorter calibration period length to expose the model to more recent hydrological dynamics.

As an example, consider a 10-year calibration period from 1 May 1995 to 30 April 2005, after a 1-year warm-up period. Predictions are computed for the following 1-year "prediction period", i.e. 1 May 2005 to 30 April 2006. The process is then repeated each year, i.e. the next calibration period is 1 May 1996 to 30 April 2006, and the calibrated model is used to predict the period 1 May 2006 to 30 April 2007. The starting month of May corresponds to the start of the flow season (Fig. 2).

3.4 State updating in GR4J

The approach used for the state updating of GR4J is similar to the approach of Crochemore et al. (2016) and Demirel et al. (2013). State updating is set to take place at the start of each month within the 1-year prediction period, using the observed streamflow at the start of each month. GR4J has two stores, namely the production store and the routing store. Following the procedure of Demirel et al. (2013), the routing store level is updated such that the GR4J simulation of streamflow matches the observed flow. This procedure is undertaken after accounting for the modelled direct flow from the production store (Demirel et al., 2013).

More specifically, the following procedure is used. In GR4J, the total simulated streamflow on a given day q_t^θ is defined by the sum of the direct flow from the production store (after applying a unit hydrograph), $q_{t,d}^\theta$, and the flow from the routing store, $q^{\theta}_{t,r}$,

$$q^{\theta}_{t} = q^{\theta}_{t,d} + q^{\theta}_{t,r}. \tag{2}$$

Let $q^{\mathrm{SU}}_{t,r}$ denote the flow from the routing store that yields q^{θ}_{t} equal to the observed flow $\widetilde{q}_t$. This quantity is calculated as

$$q^{\mathrm{SU}}_{t,r} = \max(\widetilde{q}_t - q^{\theta}_{t,d}, 0). \tag{3}$$

The routing store level, R, can then be obtained by setting $q^{\theta}_{t,r} = q^{\mathrm{SU}}_{t,r}$, and solving (using the bisection method) the equation used by the GR4J model to calculate the outflow from this storage:

$$q^{\theta}_{t,r} = R\left(1 - \left(1 + \left(\frac{R}{X3}\right)^4\right)^{-1/4}\right). \tag{4}$$

Equations (2)–(4) can be used to update R given the observed streamflow $\widetilde{q}_t$.

3.5 Estimation of predictive uncertainty

The monthly streamflow forecasts are obtained by aggregating the daily GR4J simulations. In order to quantify predictive uncertainty using a residual error model, the monthly aggregated GR4J simulations, $\boldsymbol{Q}^{\theta}$, are compared to observed monthly streamflow volumes, $\widetilde{\boldsymbol{Q}}$. The quantification of error is based on residual errors, defined by the differences between observed and simulated monthly streamflow. Separate error models are estimated for the GR4J predictions for each catchment and for each type of forcing data (observed or forecast rainfall), as follows.

When observed rainfall is used as input to GR4J, the daily streamflow time series simulated using GR4J are aggregated to produce monthly time series of hydrological model predictions, $\boldsymbol{Q}^{\theta}$.

When forecast rainfall is used as input to GR4J, an ensemble of daily streamflow forecasts is produced (with a single GR4J streamflow time series per rainfall forecast time series). Each such "individual" daily GR4J time series is then aggregated to a monthly time step. The time series $\boldsymbol{Q}^{\theta}$ is constructed from the time series of medians of the individual monthly streamflow time series. Although the use of aggregation approaches for single-valued streamflow forecast from ensemble predictions has been seen in operational applications (see, for example, Lerat et al., 2015), we note that this approach may result in some information loss.

The heteroscedasticity (i.e. larger residuals for larger flows) and skewness of forecast errors is accounted for using the Box–Cox transformation, by defining normalised residuals as

$$\eta_t = Z(\widetilde{Q}_t) - Z(Q^{\theta}_t) \tag{5}$$

where

$$Z(Q; \lambda, A) = \begin{cases} \dfrac{(Q+A)^{\lambda} - 1}{\lambda} & \text{if } \lambda \neq 0 \\ \log(Q+A) & \text{otherwise} \end{cases} \tag{6}$$

with a transformation parameter λ and an offset parameter A (often important when transforming low flows).

$\lambda = 0.5$ was used, as this setting was shown to produce good predictive performance (especially in terms of sharpness and bias) in ephemeral catchments by McInerney et al. (2017). The offset is set as $A = 1 \times 10^{-5}$ mm month^{-1}.

The normalised residuals η_t in Eq. (7) are assumed to be Gaussian with mean μ_{η} and variance σ^2_{η}, i.e.

$$\eta_t \sim N(\mu_{\eta}, \sigma^2_{\eta}). \tag{7}$$

The parameters μ_{η} and σ_{η} are estimated using the method of moments, i.e. as the sample mean and sample standard deviation of the time series of η. The same rolling calibration approach outlined in Sect. 3.3 for the GR4J model is also applied for the calibration of the post-processor error models.

Once the residual error model is calibrated, replicates from the predictive distribution, $\boldsymbol{Q}^{(r)}$ for $r = 1 \ldots N_r$, can be generated for any time period of interest, as follows.

1. Sample the normalised residual at time step

$$t, \eta^{(r)}_t \leftarrow N(\mu_{\eta}, \sigma^2_{\eta}). \tag{8}$$

2. Rearrange Eq. (6) to yield

$$Q^{(r)}_t = Z^{-1}\left(Z\left(Q^{\theta}_t\right) + \eta^{(r)}_t\right). \tag{9}$$

3. Truncate negative values to zero.

Equations (5)–(8) are used to generate replicates from the predictive distribution (PD) of the forecasts for each month ($\boldsymbol{Q}_t$).

The assumptions of the post-processor residual error model used to estimate predictive uncertainty for monthly volumes are different to the assumptions of the residual error model used in the likelihood function for calibrating the daily GR4J model. As outlined in Sect. 3.2, the GR4J model is calibrated at the daily scale to observed streamflow using the standard least squares likelihood function, because it better captures the high daily flows, important for estimating the monthly volumes. The post-processing error model for the monthly volumes is designed to capture the predictive uncertainty in these monthly volumes, in particular the heteroscedasticity and skew of the residuals (McInerney et al., 2017; Refsgaard, 1997). These choices of residual error models at the daily and monthly timescales contribute to the study objectives of reliable forecasts at the monthly timescale (see another example in Lerat et al., 2015).

3.6 Model configurations and implementation

Two options for state updating (with versus without) and two options for calibration period length (CPL = 10 years versus CPL = 20 years) are considered. The combination of these options leads to four model configurations. Four different cases are considered for each model configuration, given by the combinations of two catchments (C1 and C2) and two sources of climate data (observed and forecast). This results in a total of 16 scenarios considered.

Twelve sets of 1-month ahead predictions are generated during the 1-year prediction period. For all scenarios, observed rainfall is used as input to the hydrological model prior to the start of each set of 1-month ahead predictions. When state updating is used, the GR4J state is updated at the start of this month using the procedure outlined in Sect. 3.4. During the 1-month ahead predictions, either observed or forecast rainfall is used, depending on the scenario considered.

3.7 Performance metrics

Five metrics are used to evaluate distinct aspects of predictive performance. All metrics are calculated on the accumulated 1-year prediction period following each rolling calibration period. These include metrics for reliability, sharpness, volumetric bias, the cumulative ranked probability score (CRPS) and the Nash–Sutcliffe efficiency (NSE).

Reliability refers to the degree to which the observations (of streamflow) over a series of time steps can be considered to be statistically consistent with the predictive distribution. In this work, reliability is assessed using predictive quantile–quantile (PQQ) plots, and quantified using the reliability metric of Renard et al. (2010) based on the area between the PQQ plot and the 1 : 1 line. A value of 0 represents perfect reliability, while a value of 1 represents the worst reliability, i.e. all observations lying outside (above or below) the PD.

Sharpness refers to the width of the predictive distribution, and can otherwise be known as “resolution” or “precision”. Typically, sharpness is determined using the predicted values only. In this work a measure of sharpness (as the sum of the standard deviation of the predictions each time step) is normalised by the sum of the observed values, to enable a comparison of this metric across catchments with different magnitudes of flow. As such, sharpness is quantified using the following metric from McInerney et al. (2017):

$$\text{Sharpness}_{\text{Norm}} = \sum_{t=1}^{N} \text{sdev}\left(\boldsymbol{Q}_t\right) / \sum_{t=1}^{N} \widetilde{Q}_t, \tag{10}$$

where N is the number of months and sdev is the sample standard deviation, $\boldsymbol{Q}_t$ is the predictive distribution of streamflow for month t, and $\widetilde{Q}_t$ is the observed streamflow for this month (as described in Sect. 3.5).

Volumetric bias measures the overall water balance error of the predictions relative to the observations. It is calculated as

$$\text{VolBias} = \left| \frac{\sum_{t=1}^{N} \text{mean}(\boldsymbol{Q}_t) - \sum_{t=1}^{N} \widetilde{Q}_t}{\sum_{t=1}^{N} \widetilde{Q}_t} \right|, \tag{11}$$

where mean is the sample mean.

CRPS is a widely used probabilistic performance metric that combines in a single measure multiple aspects of predictive performance, including reliability, sharpness and bias (Hersbach, 2000). The CRPS is calculated by comparing the cumulative distribution of the predictions with the cumulative distribution of the observation at each time step. At a single time step, the CRPS is defined as

$$\text{CRPS} = \int_{-\infty}^{\infty} \left[F_{\text{p},t}(Q) - F_{\text{o},t}(Q)\right]^2 \text{d}Q, \tag{12}$$

where $F_{\text{p},t}$ and $F_{\text{o},t}$ are the cumulative distributions of the streamflow predictions ($\boldsymbol{Q}_t$) and observation ($\widetilde{Q}_t$), at time step t. The average value of the CRPS is then calculated over all time steps t. Note that the cumulative distribution of the observations is a step function. A CRPS of 0 corresponds to the perfect prediction, while larger CRPS values correspond to worse performance.

To normalise CRPS metric values across catchments, the CRPS metric for the predictions (CRPS_P) is expressed as a skill score with respect to the CRPS metric of a “reference” distribution for that catchment (CRPS_R):

$$\text{CRPS}_{\text{SS}} = \frac{\text{CRPS}_\text{R} - \text{CRPS}_\text{P}}{\text{CRPS}_\text{R}}. \tag{13}$$

CRPS_{SS} values below 0 indicate forecasts with worse performance than the reference distribution, a CRPS_{SS} of 0 corresponds to the predictions with the same performance as the reference distribution, and a CRPS_{SS} of 1 corresponds to a perfect prediction.

The reference distribution for each month is calculated as the empirical distribution of all observed data in that month, using the entire set of observed data (including data from the prediction period). This approach provides a stringent baseline for the CRPS normalisation in Eq. (13).

NSE is a commonly used metric for the assessment of the accuracy of deterministic hydrological model predictions, and is calculated as

$$\text{NSE} = 1 - \frac{\sum_{t=1}^{N} \left(Q_t^\theta - \widetilde{Q}_t\right)^2}{\sum_{t=1}^{N} \left(\widetilde{Q}_t - \text{mean}(\widetilde{Q})\right)^2}, \tag{14}$$

Table 2. Best and worst values for each predictive performance metric across all model configurations. For $CRPS_{SS}$ and NSE, higher values denote better performance; for the other metrics lower values denote better performance. The values in this table should be interpreted alongside Fig. 4, where the worst and best values reported here correspond to metric values of 0 and 1, respectively.

	Reliability	Sharpness	Bias	$CRPS_{SS}$	NSE
Worst	0.41	2.21	1.49	−0.65	−1.01
Best	0.07	0.45	0.11	0.57	0.88

where Q_t^θ is the monthly aggregated GR4J prediction for month t (as described in Sect. 3.5). The NSE can range from $-\infty$ to 1, with NSE = 1 corresponding to perfectly accurate predictions of the observed data, and NSE < 0 indicating the observed mean is a better predictor than the model.

To ensure a consistent comparison of multiple model scenarios, the metrics are computed as follows:

- the same period is used to calculate the metrics in all cases. This period was determined by the availability of the forecast rainfall, from May 2001 to April 2011.
- the performance metrics are normalised by linearly scaling the worst value to a value of 0 and the best value to 1:

$$M_r = \frac{M - M_\mathrm{w}}{M_\mathrm{b} - M_\mathrm{w}}, \tag{15}$$

where the worst and best values for each metric, M_w and M_b, respectively, are listed in Table 2. The remainder of the presentation, in particular Fig. 4, reports the normalised metrics computed using Eq. (15).

4 Results

The performance metrics for all model configurations are summarised in Fig. 4. First the predictive performance of model configurations with and without state updating is compared (Aim 1), and then the influence of calibration period length in the context of catchment non-stationarity is investigated (Aim 2), considering changes in both the predictive performance and changes in CRR parameter values over time.

4.1 Impact of state updating

The impact of state updating on predictive performance can be seen in Fig. 4, by comparing the red and blue bars (darker shading indicating results for the 10-year calibration period length, and lighter shading indicating results for the 20-year calibration period length). It is clear that state updating improves the sharpness, bias, $CRPS_{SS}$ and NSE metrics.

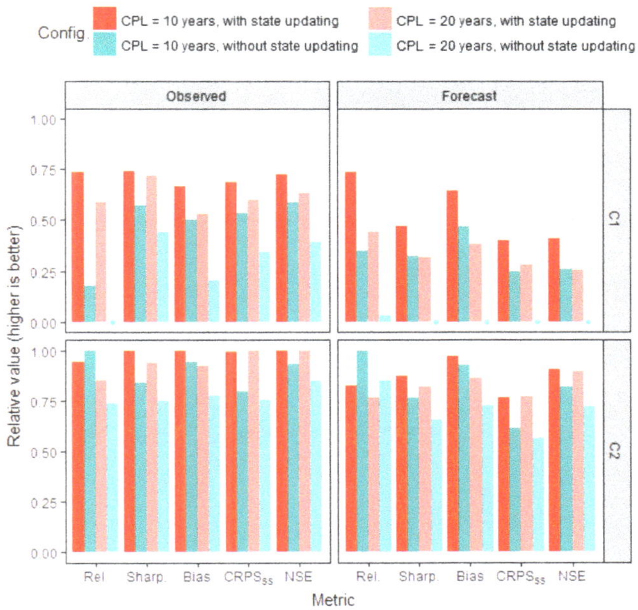

Figure 4. Predictive performance metrics for the two case study catchments (C1 and C2) and the two sources of rainfall forcing data (observed and forecast). Relative metric values are presented (Sect. 3.7 and Table 2); higher values represent better performance. The impact of state updating can be seen by comparing the red versus blue bars. The change in performance due to different calibration period lengths (CPLs) can be seen by comparing the bars with darker versus lighter shading.

The improvement in predictive performance achieved by state updating to the observed flow data is tentatively attributed to being able to correct the model for any systematic overestimation of simulated streamflow. Consider Figs. 5 and 6, which show the 90th percentile predictive limits for each model configuration, for catchments C1 and C2, respectively. The longer 20-year calibration period length without state updating is considered the "typical approach", and is shown in grey in each panel. A representative time period is shown, with the full time series for each case provided in the Supplement. Figures 5 and 6 show that state updating sharpens the predictive limits, especially during low-flow months. For example, this behaviour can be seen for the 20-year CPL by comparing the predictions in panels (a) to (b) for the case of forecast rainfall and the predictions in panels (e) to (f) for the case of observed rainfall.

In terms of reliability, Fig. 4 shows that state updating provides improved predictions for catchment C1. However, for catchment C2, Fig. 4 shows that the reliability of all model configurations is relatively high compared to the reliability achieved in catchment C1, and state updating can lead to a slight loss of reliability.

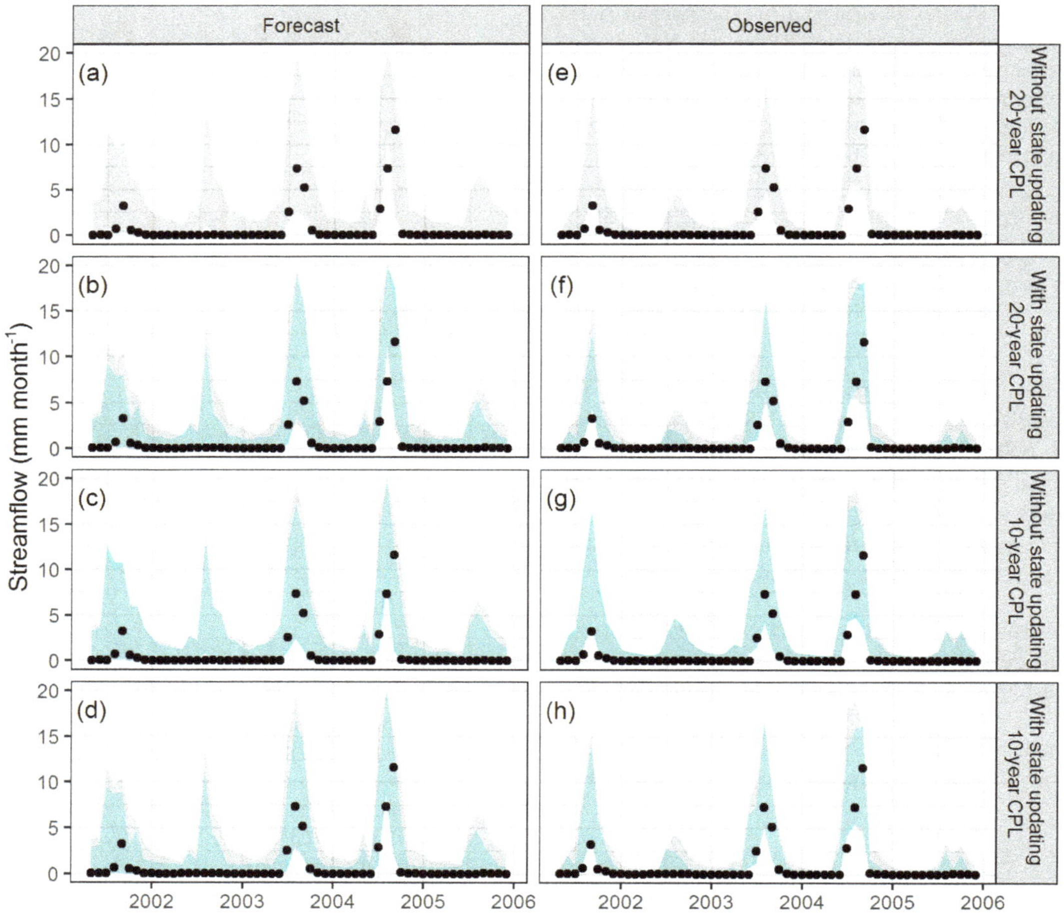

Figure 5. Representative streamflow time series in catchment C1 obtained using forecast rainfall **(a, b, c, d)** and observed rainfall **(e, f, g, h)**. The shaded area represents the 90th percentile prediction limits and the black dots the observed values. The "traditional approach" of the 20-year calibration period length (CPL) without state updating is shown in grey in each panel.

4.2 Impact of calibration period length

4.2.1 Differences in predictive distribution

The changes in the predictive distribution due to changes in the calibration period length can be seen in Fig. 4, by comparing the darker and lighter shades of each colour (darker colour for 10-year calibration period length, lighter colour for 20-year calibration period length). The following findings can be seen.

- When state updating is not used (comparing dark blue versus light blue in Fig. 4), all metrics improved when the shorter 10-year calibration period length was used.
- When state updating is used (comparing the dark red versus light red in Fig. 4), the impact of the shorter 10-year calibration period length depends on the catchment. In catchment C1, which provided stronger evidence of non-stationarity than catchment C2 (Sect. 2.1), the use of the 10-year calibration period length improves all metrics compared to the use of the 20-year calibration period length. In contrast, in catchment C2, the length of the calibration period had little impact on the NSE and $CRPS_{SS}$ values, and only small improvements in the reliability, sharpness and bias metrics are obtained when the 10-year period is used.

The differences between the streamflow predictions obtained in the two catchments C1 and C2 (for the case of GR4J forced with observed rainfall) are illustrated in Fig. 7 for the most recent period: 2009–2011. In catchment C1, using a longer calibration period length tends to yield wider prediction limits and an overestimation of the observed flow in 2009 and 2010, whereas using the shorter calibration length provides a better capture of the catchment response in these 2 years. In contrast, in catchment C2, which has less evidence of non-stationarity (Sect. 2.1), the calibration period length makes very little difference to the resulting streamflow predictions.

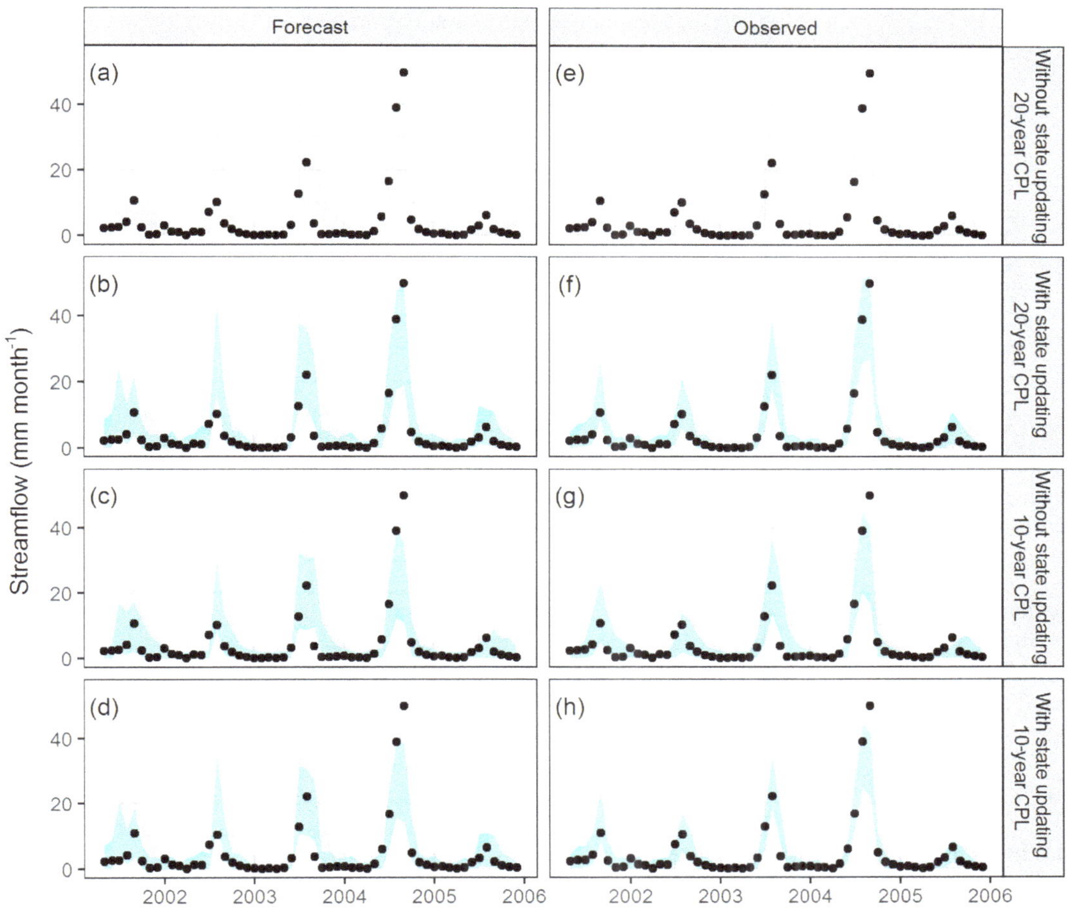

Figure 6. Representative streamflow time series in catchment C2 obtained using forecast rainfall **(a, b, c, d)** and observed rainfall **(e, f, g, h)**. The shaded area represents the 90th percentile prediction limits and the black dots the observed values. The "traditional approach" of the 20-year calibration period length (CPL) without state updating is shown in grey in each panel.

4.3 Differences in trends in parameter values

The rolling calibration approach (see Sect. 3.3) enables temporal trends in the parameter distributions to be investigated. Figure 8 presents the median and 90th percentile prediction limits of these distributions for each parameter for each catchment, with the 10-year and 20-year calibration period lengths shown in different colours.

In catchment C1, up until year 2005 (representing models calibrated from 1995 to 2004 for the 10-year calibration period length), the calibration period length has little impact on the median value for each parameter. Slightly wider parameter bounds are obtained when the shorter calibration period length is used, likely due to the reduced data available to infer representative parameter values. Post-2005, the parameter values obtained using the shorter calibration period length respond to the distinct non-stationarity of the catchment discussed in Sect. 2.1. The more pronounced negative values of the groundwater exchange coefficient $X2$ estimated in the 1994–2005 calibration period are consistent with the reduced runoff ratio in the period post-2000. In contrast, parameter values estimated from the longer calibration period length, which includes data from the 1980s even when predicting the 2000s, do not exhibit this distinct change.

In catchment C2, the median values of parameters estimated from each calibration period length were similar over the record. This result agrees with the lack of strong evidence of non-stationarity in this catchment. However, there is some evidence of a reduction in streamflow in this catchment, with the post-2000 period being characterised by a reduction in the runoff ratio from 0.088 to 0.061 (Sect. 2.1). This reduction is weaker in catchment C2 than in catchment C1, yet appears to be supported by the trends in the median parameter values. Analysis of results from the 20-year calibration period length suggests statistically significant trends ($p < 0.05$) in the median values of the model parameters, namely $\Delta X1 = 3.96$ and $\Delta X3 = -5.17\,\text{mm}\,\text{yr}^{-1}$. An exception to the pattern of the median parameter values being insensitive to calibration period lengths can be seen in 1999, where the use of the 10-year calibration period length pro-

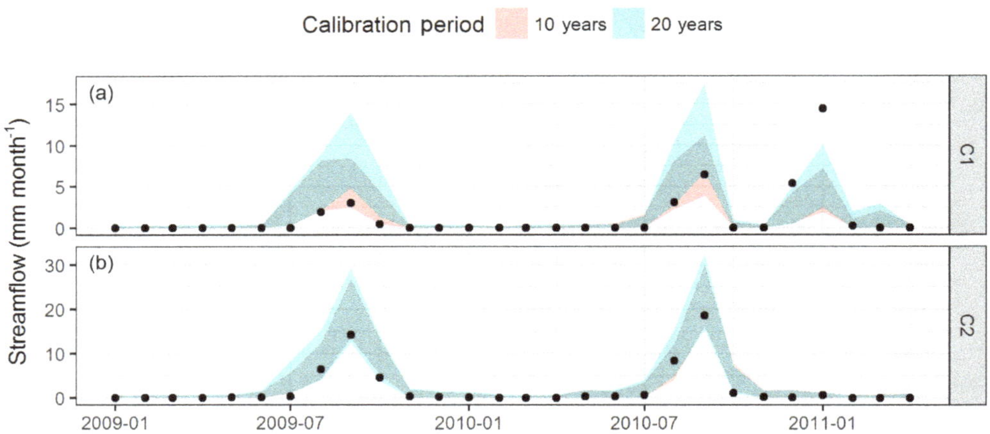

Figure 7. Streamflow predictions for catchments C1 **(a)** and C2 **(b)** for the period 2009–2011 using observed rainfall. The shaded area represents the 90th percentile prediction limits and the black dots the observed values. For catchment C1, using shorter calibration periods (red) can be seen to produce lower streamflow predictions than using longer calibration periods (blue).

duces higher values of $X4$ and lower values of $X2$ and the *split* introduced in this study (Sect. 3.1). This exception could represent a model fitting anomaly resulting from a shorter calibration period length.

5 Discussion

5.1 Beneficial impact of state updating on forecast performance

Most previous studies have used state updating in a short-term flood forecasting context, and found a limited effect of the initial conditions after a number of days (e.g. Berthet et al., 2009; Randrianasolo et al., 2014; Sun et al., 2017). However, forecasting of flood peak and timing is a different application to the forecasting of streamflow volumes. A number of data-driven modelling studies have demonstrated that monthly streamflow lagged by 1 month (or more) provided some useful information for forecasting at a 1-month lead time (e.g. Bennett et al., 2014; Humphrey et al., 2016; Yang et al., 2017). This study demonstrates that these benefits also hold when CRR models, rather than data-driven approaches, are used as the forecasting model.

State updating is found to improve predictive performance in both catchments considered, for the majority of the multiple performance metrics considered. State updating is expected to reduce predictive bias, as errors in the simulated streamflow during the warm-up period are corrected at the start of the forecast period. State updating is also expected to increase the sharpness of the predictive distribution, as the range of model predictions is generally tightened by forcing the model to simulate the observed streamflow at the start of the forecast period.

The only metric where state updating did not show an improvement is for the reliability of predictions for catchment C2. However, the reliability of all model configurations in this catchment is already relatively high without state updating. All other metrics (sharpness, bias, CRPS and NSE) show improvements from state updating in catchment C2, suggesting potential trade-offs in performance, similar to that found by Crochemore et al. (2016) and McInerney et al. (2017). This slight reduction in reliability is not considered to have a significant detrimental impact on the PD produced for this practical application.

5.2 Importance of choosing a calibration period that is representative of current catchment conditions

Traditionally, long calibration periods are used to maximise the use of available data and increase parameter identifiability. The empirical results in this study suggest that the shorter calibration period can provide better (or at least not worse) predictive performance. The reduction in performance seen when the longer calibration period is used is likely due to the calibration data representing catchment conditions that are substantially different to those in the prediction period. For example, when the prediction period is 2009 (as shown for catchment C1 in Fig. 7), a 20-year calibration period length corresponds to the period of 1989–2008, which includes a large portion of the pre-2000 period when catchment C1 displayed a much higher runoff coefficient (Sect. 2.1). In contrast, a 10-year calibration period length corresponds to a calibration period of 1999–2008, which is likely to be more representative of the lower runoff hydrological regime seen in the post-2000 period.

The reported improvement in model performance with the 10-year calibration period length does not imply that shorter calibration periods would result in further improvements. Shorter calibration period lengths will eventually reduce parameter identifiability (e.g. as manifested by greater parameter uncertainty in Fig. 8), and may produce poor parameter

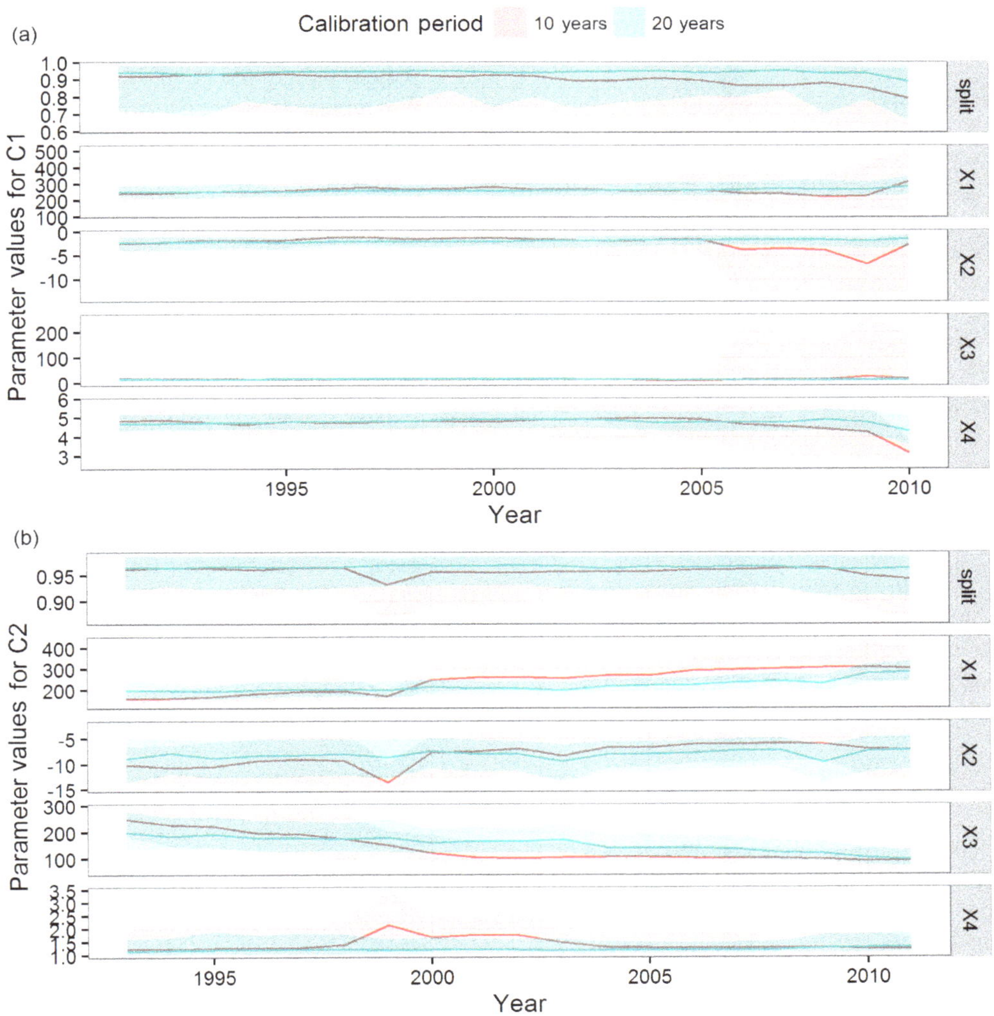

Figure 8. Temporal trends in posterior parameter distributions, for catchments C1 **(a)** and C2 **(b)**. The median values are shown as the solid lines and the shaded areas represent the 90th percentile prediction limits.

estimates due to fitting only a small number of events and hence being unable to represent the full range of flow conditions.

The empirical findings highlight the benefits of identifying a calibration period of data that is representative of conditions of interest for a given model application, which is a task often overlooked in practical applications. Suitable representative periods can be identified through techniques such as trend analysis, using knowledge of changes in a catchment (e.g. land use data, abstraction volumes), and testing predictive performance for different calibration period lengths (as done in this work). The empirical results indicate that, if the selection of calibration data is poorly implemented, and/or if the modeller naively assumes that longer calibration periods are inherently better for model development, predictive performance can degrade.

5.3 Value of forecasts for improving water management

The forecasting approaches developed in this work can support improved water management in the drainage system considered. The approach currently used by the management authority is very conservative: streamflow forecasts are not attempted, and changes in water management are made only once downstream requirements have been met. With the forecasting models and methods developed in this work, it becomes possible to produce streamflow forecasts with a high reliability, improved sharpness and reduced bias. Thus it becomes possible to provide useful probabilistic estimates of how likely it is that the downstream flow requirements will be met in the next month. With this information, managers can more confidently consider increasing the frequency and duration of inundation for many of the wetlands in the region, and can make decisions on management changes much earlier in the season.

5.4 Future research work

The enhancements to predictive performance of streamflow forecasts from state updating and a shorter calibration period have been demonstrated on two catchments. These catchments were selected based on an established user need for monthly forecasts to improve the water management of a channel drainage system with multiple competing demands. Importantly, the case study catchments in this work are ephemeral and dry, with low runoff ratios. These types of catchments are known to be challenging to model (McInerney et al., 2017; Ye et al., 1997). For example, the models predict a streamflow response in 2002 and 2005 in Fig. 5 that did not occur in the observations, even when observed rainfall and state updating were used. Some of this difference may be due to errors in the input rainfall data, but this result highlights the difficulty in representing streamflow generation in low-yielding, ephemeral catchments, such as those considered. Future work will evaluate the proposed monthly streamflow forecasting techniques over a wider range of catchments and environmental conditions.

6 Conclusions

This work has focused on improving monthly streamflow forecasts by considering two aspects: (1) state updating to force the GR4J hydrological model to match observations from the start of the forecast period, and (2) investigating the trade-offs between using shorter versus longer calibration periods. The analysis was applied to two ephemeral catchments in southern Australia, which are part of a drainage network with competing environmental management demands.

The major findings from the empirical analysis are as follows.

1. State updating improves predictive performance in the case study catchments, for the majority of the multiple performance metrics considered. Previous studies focusing on the forecasting of flood peak and timing have typically found a limited effect of initial conditions on predictive performance after a number of days. This study demonstrates that, when forecasting streamflow volumes, using state updating to more accurately represent initial conditions can have a benefit even at a 1-month lead time.

2. The length of the calibration period has a major impact on the predictive performance of a hydrological model. In the case study catchments, the shorter calibration period typically improves predictive performance, especially in the case study catchment with stronger evidence of non-stationarity. The benefits of a shorter calibration length appear contrary to the standard approach of using as much data as possible for model calibration. The reduction in performance for the longer calibration period is likely due to the model being calibrated to data that represent higher-yielding conditions from the past which no longer hold true in the forecast period. This finding highlights that identifying a data set that is representative of the forecast period, through trend analysis and other knowledge of a catchment, is an important step in model development. If this step is ignored, and it is naively assumed that longer calibration data are inherently better for model development, all aspects of predictive performance may suffer.

The conclusions of this empirical study are limited by the small number of catchments and single hydrological model used. Further work will consider a larger sample of catchments and a wider range of hydrological model structures. In general, we expect the techniques of state updating, post-processing uncertainty estimation, and usage of shorter calibration period length representative of future forecast conditions to be of value to hydrologists and environmental modellers seeking to improve the predictive performance of their modelling systems.

Author contributions. MG performed the analysis and produced the manuscript, with contributions from all co-authors. HM and GD assisted with the design of the project. DM undertook the post-processor error modelling and analysis, with help from MT and DK. GH implemented the climate model forecast downscaling to generate the inputs for the hydrological models.

Competing interests. The authors declare that they have no conflict of interest.

Special issue statement. This article is part of the special issue "Sub-seasonal to seasonal hydrological forecasting". This article is part of the special issue "Sub-seasonal to seasonal hydrological forecasting". It is not associated with a conference.

Acknowledgements. Matthew S. Gibbs and Greer Humphrey were supported by the Goyder Institute for Water Research, project E.2.4. David McInerney was supported by Australian Research Council grant LP140100978 with the Australian Bureau of Meteorology and South East Queensland Water. Input from South East Water Conservation and Drainage Board staff, in particular Senior Environmental Officer, Mark DeJong, is gratefully acknowledged. The authors would like to thank the three anonymous reviewers for their comments and suggestions, which improved the clarity and contribution of the manuscript.

Edited by: Maria-Helena Ramos

References

Allen, R. G., Pereira, L. S., Raes, D., and Smith, M.: Crop evapotranspiration-Guidelines for computing crop water requirements, FAO, RomeFAO Irrigation and drainage paper 56, 300 pp., 1998.

Andrews, F. T., Croke, B. F. W., and Jakeman, A. J.: An open software environment for hydrological model assessment and development, Environ. Modell. Softw., 26, 1171–1185, 2011.

Avey, S. and Harvey, D.: How water scientists and lawyers can work together: A 'down under' solution to a water resource management problem, Journal of Water Law, 24, 25-61, 2014.

Bennett, J. C., Wang, Q. J., Pokhrel, P., and Robertson, D. E.: The challenge of forecasting high streamflows 1–3 months in advance with lagged climate indices in southeast Australia, Nat. Hazards Earth Syst. Sci., 14, 219–233, https://doi.org/10.5194/nhess-14-219-2014, 2014.

Berthet, L.: Prévision des crues au pas de temps horaire : pour une meilleure assimilation de l'information de débit dans un modèle hydrologique, AgroParisTech, 2010.

Berthet, L., Andréassian, V., Perrin, C., and Javelle, P.: How crucial is it to account for the antecedent moisture conditions in flood forecasting? Comparison of event-based and continuous approaches on 178 catchments, Hydrol. Earth Syst. Sci., 13, 819–831, https://doi.org/10.5194/hess-13-819-2009, 2009.

Beven, K. and Binley, A.: The future of distributed models: Model calibration and uncertainty prediction, Hydrol. Process., 6, 279–298, 1992.

Bowden, G. J., Maier, H. R., and Dandy, G. C.: Real-time deployment of artificial neural network forecasting models: Understanding the range of applicability, Water Resour. Res., 48, W10549, https://doi.org/10.1029/2012WR011984, 2012.

Brigode, P., Oudin, L., and Perrin, C.: Hydrological model parameter instability: A source of additional uncertainty in estimating the hydrological impacts of climate change?, J. Hydrol., 476, 410–425, 2013.

Brookes, J. D., Aldridge, K., Dalby, P., Oemcke, D., Cooling, M., Daniel, T., Deane, D., Johnson, A., Harding, C., Gibbs, M., Ganf, G., Simonic, M., and Wood, C.: Integrated science informs forest and water allocation policies in the South East of Australia, Inland Waters, 7, 358–371, 2017.

Coron, L., Andreassian, V., Perrin, C., Lerat, J., Vaze, J., Bourqui, M., and Hendrickx, F.: Crash testing hydrological models in contrasted climate conditions: An experiment on 216 Australian catchments, Water Resour. Res., 48, W05552, https://doi.org/10.1029/2011WR011721, 2012.

Crochemore, L., Ramos, M.-H., and Pappenberger, F.: Bias correcting precipitation forecasts to improve the skill of seasonal streamflow forecasts, Hydrol. Earth Syst. Sci., 20, 3601–3618, https://doi.org/10.5194/hess-20-3601-2016, 2016.

de Vos, N. J., Rientjes, T. H. M., and Gupta, H. V.: Diagnostic evaluation of conceptual rainfall–runoff models using temporal clustering, Hydrol. Process., 24, 2840–2850, 2010.

Demargne, J., Wu, L. M., Regonda, S. K., Brown, J. D., Lee, H., He, M. X., Seo, D. J., Hartman, R., Herr, H. D., Fresch, M., Schaake, J., and Zhu, Y. J.: The Science of NOAA's Operational Hydrologic Ensemble Forecast Service, B. Am. Meteorol. Soc., 95, 79–98, 2014.

Demirel, M. C., Booij, M. J., and Hoekstra, A. Y.: Effect of different uncertainty sources on the skill of 10 day ensemble low flow forecasts for two hydrological models, Water Resour. Res., 49, 4035–4053, 2013.

Evin, G., Thyer, M., Kavetski, D., McInerney, D., and Kuczera, G.: Comparison of joint versus postprocessor approaches for hydrological uncertainty estimation accounting for error autocorrelation and heteroscedasticity, Water Resour. Res., 50, 2350–2375, 2014.

Fenicia, F., Kavetski, D., Savenije, H. H. G., Clark, M. P., Schoups, G., Pfister, L., and Freer, J.: Catchment properties, function, and conceptual model representation: is there a correspondence?, Hydrol. Process., 28, 2451–2467, 2014.

Gibbs, M. S., Maier, H. R., and Dandy, G. C.: A generic framework for regression regionalization in ungauged catchments, Environ. Modell. Softw., 27–28, 1–14, 2012.

Guo, D., Westra, S., and Maier, H. R.: Impact of evapotranspiration process representation on runoff projections from conceptual rainfall-runoff models, Water Resour. Res., 53, 435–454, https://doi.org/10.1002/2016WR019627, 2017.

Hersbach, H.: Decomposition of the Continuous Ranked Probability Score for Ensemble Prediction Systems, Weather Forecast., 15, 559–570, https://doi.org/10.1175/1520-0434(2000)015<0559:DOTCRP>2.0.CO;2, 2010.

Hudson, D., Alves, O., Hendon, H. H., and Marshall, A. G.: Bridging the gap between weather and seasonal forecasting: intraseasonal forecasting for Australia, Q. J. Roy. Meteor. Soc., 137, 673–689, 2011.

Humphrey, G. B., Gibbs, M. S., Dandy, G. C., and Maier, H. R.: A hybrid approach to monthly streamflow forecasting: Integrating hydrological model outputs into a Bayesian artificial neural network, J. Hydrol., 540, 623–640, 2016.

Jeffrey, S. J., Carter, J. O., Moodie, K. B., and Beswick, A. R.: Using spatial interpolation to construct a comprehensive archive of Australian climate data, Environ. Modell. Softw., 16, 309–330, 2001.

Kavetski, D., Franks, S. W., and Kuczera, G.: Confronting Input Uncertainty in Environmental Modelling, in: Calibration of Watershed Models, American Geophysical Union, 49–68, 2003.

Koster, R. D., Mahanama, S. P. P., Livneh, B., Lettenmaier, D. P., and Reichle, R. H.: Skill in streamflow forecasts derived from large-scale estimates of soil moisture and snow, Nat. Geosci., 3, 613–616, 2010.

Krzysztofowicz, R. and Maranzano, C. J.: Hydrologic uncertainty processor for probabilistic stage transition forecasting, J. Hydrol., 293, 57–73, 2004.

Lerat, J., Pickett-Heaps, C., Shin, D., Zhou, S., Feikema, P., Khan, U., Laugesen, R., Tuteja, N., Kuczera, G. T. M., and Kavetski, D.: Dynamic streamflow forecasts within an uncertainty framework for 100 catchments in Australia, 36th Hydrology and Water Resources Symposium: The art and science of water, Barton, ACT, 1396–1403, 2015.

Li, H. B., Luo, L. F., Wood, E. F., and Schaake, J.: The role of initial conditions and forcing uncertainties in seasonal hydrologic forecasting, J. Geophys. Res.-Atmos., 114, D04114, https://doi.org/10.1029/2008JD010969, 2009.

Li, L., Lambert, M. F., Maier, H. R., Partington, D., and Simmons, C. T.: Assessment of the internal dynamics of the Australian Water Balance Model under different calibration regimes, Environ. Modell. Softw., 66, 57–68, 2015a.

Li, Y., Ryu, D., Western, A. W., and Wang, Q. J.: Assimilation of stream discharge for flood forecasting: Updating a semidistributed model with an integrated data assimilation scheme, Water Resour. Res., 51, 3238–3258, 2015b.

Liu, Y. and Gupta, H. V.: Uncertainty in hydrologic modeling: Toward an integrated data assimilation framework, Water Resour. Res., 43, W07401, https://doi.org/10.1029/2006WR005756, 2007.

Luo, J., Wang, E., Shen, S., Zheng, H., and Zhang, Y.: Effects of conditional parameterization on performance of rainfall-runoff model regarding hydrologic non-stationarity, Hydrol. Process., 26, 3953–3961, 2012.

Maurer, E. P. and Lettenmaier, D. P.: Predictability of seasonal runoff in the Mississippi River basin, J. Geophys. Res.-Atmos., 108, 8607, https://doi.org/10.1029/2002JD002555, 2003.

McInerney, D., Thyer, M., Kavetski, D., Lerat, J., and Kuczera, G.: Improving probabilistic prediction of daily streamflow by identifying Pareto optimal approaches for modeling heteroscedastic residual errors, Water Resour. Res., 53, 2199–2239, 2017.

Milly, P. C. D., Betancourt, J., Falkenmark, M., Hirsch, R. M., Kundzewicz, Z. W., Lettenmaier, D. P., Stouffer, R. J., Dettinger, M. D., and Krysanova, V.: On Critiques of “Stationarity is Dead: Whither Water Management?”, Water Resour. Res., 51, 7785–7789, 2015.

Mount, N. J., Maier, H. R., Toth, E., Elshorbagy, A., Solomatine, D., Chang, F. J., and Abrahart, R. J.: Data-driven modelling approaches for socio-hydrology: opportunities and challenges within the Panta Rhei Science Plan, Hydrolog. Sci. J., 61, 1192–1208, https://doi.org/10.1080/02626667.2016.1159683, 2016.

Pagano, T., Garen, D., and Sorooshian, S.: Evaluation of official western US seasonal water supply outlooks, 1922–2002, J. Hydrometeorol., 5, 896–909, 2004.

Pappenberger, F. and Beven, K. J.: Ignorance is bliss: Or seven reasons not to use uncertainty analysis, Water Resour. Res., 42, W05302, https://doi.org/10.1029/2005WR004820, 2006.

Perrin, C., Michel, C., and Andreassian, V.: Improvement of a parsimonious model for streamflow simulation, J. Hydrol., 279, 275–289, https://doi.org/10.1016/S0022-1694(03)00225-7, 2003.

Randrianasolo, A., Thirel, G., Ramos, M. H., and Martin, E.: Impact of streamflow data assimilation and length of the verification period on the quality of short-term ensemble hydrologic forecasts, J. Hydrol., 519, 2676–2691, 2014.

Refsgaard, J. C.: Validation and Intercomparison of Different Updating Procedures for Real-Time Forecasting, Hydrol. Res., 28, 65–84, 1997.

Renard, B., Kavetski, D., Kuczera, G., Thyer, M., and Franks, S. W.: Understanding predictive uncertainty in hydrologic modeling: The challenge of identifying input and structural errors, Water Resour. Res., 46, W05521, https://doi.org/10.1029/2009WR008328, 2010.

Robertson, D. E. and Wang, Q. J.: Seasonal Forecasts of Unregulated Inflows into the Murray River, Australia, Water Resour. Manag., 27, 2747–2769, 2013.

Robertson, D. E., Pokhrel, P., and Wang, Q. J.: Improving statistical forecasts of seasonal streamflows using hydrological model output, Hydrol. Earth Syst. Sci., 17, 579–593, https://doi.org/10.5194/hess-17-579-2013, 2013.

Searcy, J. K., Hardison, C. H., and Langein, W. B.: Double-mass curves; with a section fitting curves to cyclic data, Report 1541B, 1960.

Shao, Q. and Li, M.: An improved statistical analogue downscaling procedure for seasonal precipitation forecast, Stoch. Env. Res. Risk A, 27, 819–830, 2013.

Sun, L., Seidou, O., and Nistor, I.: Data Assimilation for Streamflow Forecasting: State-Parameter Assimilation versus Output Assimilation, J. Hydrol. Eng., 22, 04016060, https://doi.org/10.1061/(ASCE)HE.1943-5584.0001475, 2017.

Thyer, M., Renard, B., Kavetski, D., Kuczera, G., Franks, S. W., and Srikanthan, S.: Critical evaluation of parameter consistency and predictive uncertainty in hydrological modeling: A case study using Bayesian total error analysis, Water Resour. Res., 45, W00B14, https://doi.org/10.1029/2008WR006825, 2009.

Tuteja, N. K., Shin, D., Laugesen, R., Khan, U., Shao, Q., Li, M., Zheng, H., Kuczera, G., Kavetski, D., Evin, G., Thyer, M. A., MacDonald, A., Chia, T., and Le, B.: Experimental evaluation of the dynamic seasonal streamflow forecasting approach, Bureau of Meteorology, Australia, 99 pp., 2011.

Vaze, J., Post, D. A., Chiew, F. H. S., Perraud, J. M., Viney, N. R., and Teng, J.: Climate non-stationarity – Validity of calibrated rainfall–runoff models for use in climate change studies, J. Hydrol., 394, 447–457, 2010.

Vrugt, J. A., Diks, C. G. H., Gupta, H. V., Bouten, W., and Verstraten, J. M.: Improved treatment of uncertainty in hydrologic modeling: Combining the strengths of global optimization and data assimilation, Water Resour. Res., 41, W01017, https://doi.org/10.1029/2004WR003059, 2005.

Vrugt, J. A., ter Braak, C. J. F., Diks, C. G. H., Robinson, B. A., Hyman, J. M., and Higdon, D.: Accelerating Markov Chain Monte Carlo Simulation by Differential Evolution with Self-Adaptive Randomized Subspace Sampling, Int. J. Nonlin. Sci. Num., 10, 273–290, 2009.

Wagener, T., McIntyre, N., Lees, M. J., Wheater, H. S., and Gupta, H. V.: Towards reduced uncertainty in conceptual rainfall-runoff modelling: dynamic identifiability analysis, Hydrol. Process., 17, 455–476, 2003.

Wang, E., Zheng, H., Chiew, F., Shao, Q., Luo, J., and Wang, Q. J.: Monthly and seasonal streamflow forecasts using rainfall-runoff modeling and POAMA predictions, 19th International Congress on Modelling and Simulation (Modsim2011), 3441–3447, 2011.

Wang, Q. J., Robertson, D. E., and Chiew, F. H. S.: A Bayesian joint probability modeling approach for seasonal forecasting of streamflows at multiple sites, Water Resour. Res., 45, W05407, https://doi.org/10.1029/2008WR007355, 2009.

Westra, S., Thyer, M., Leonard, M., Kavetski, D., and Lambert, M.: A strategy for diagnosing and interpreting hydrological model nonstationarity, Water Resour. Res., 50, 5090–5113, https://doi.org/10.1002/2013WR014719, 2014. 2014.

Wood, A. W. and Lettenmaier, D. P.: An ensemble approach for attribution of hydrologic prediction uncertainty, Geophys. Res. Lett., 35, L14401, https://doi.org/10.1029/2008GL034648, 2008.

Wood, A. W. and Schaake, J. C.: Correcting errors in streamflow forecast ensemble mean and spread, J. Hydrometeorol., 9, 132–148, 2008.

Wright, D. P., Thyer, M., and Westra, S.: Influential point detection diagnostics in the context of hydrological model calibration, J. Hydrol., 527, 1161–1172, 2015.

Wu, W., May, R. J., Maier, H. R., and Dandy, G. C.: A benchmarking approach for comparing data splitting methods for modeling water resources parameters using artificial neural networks, Water Resour. Res., 49, 7598–7614, 2013.

Yang, T., Asanjan, A. A., Welles, E., Gao, X., Sorooshian, S., and Liu, X.: Developing reservoir monthly inflow forecasts using artificial intelligence and climate phenomenon information, Water Resour. Res., 53, 2786–2812, 2017.

Ye, W., Bates, B. C., Viney, N. R., Sivapalan, M., and Jakeman, A. J.: Performance of conceptual rainfall-runoff models in low-yielding ephemeral catchments, Water Resour. Res., 33, 153–166, 1997.

Yihdego, Y. and Webb, J.: An Empirical Water Budget Model As a Tool to Identify the Impact of Land-use Change in Stream Flow in Southeastern Australia, Water Resour. Manag., 27, 4941–4958, 2013.

Zhang, H., Huang, G. H., Wang, D., and Zhang, X.: Multi-period calibration of a semi-distributed hydrological model based on hydroclimatic clustering, Adv. Water Resour., 34, 1292–1303, 2011.

11

A retrospective streamflow ensemble forecast for an extreme hydrologic event: a case study of Hurricane Irene and on the Hudson River basin

Firas Saleh, Venkatsundar Ramaswamy, Nickitas Georgas, Alan F. Blumberg, and Julie Pullen

Stevens Institute of Technology, Davidson Laboratory, Department of Civil, Environmental and Ocean Engineering, Hoboken NJ, 07030, USA

Correspondence to: Firas Saleh (fsaleh@stevens.edu)

Abstract. This paper investigates the uncertainties in hourly streamflow ensemble forecasts for an extreme hydrological event using a hydrological model forced with short-range ensemble weather prediction models. A state-of-the art, automated, short-term hydrologic prediction framework was implemented using GIS and a regional scale hydrological model (HEC-HMS). The hydrologic framework was applied to the Hudson River basin ($\sim 36\,000\,\mathrm{km}^2$) in the United States using gridded precipitation data from the National Centers for Environmental Prediction (NCEP) North American Regional Reanalysis (NARR) and was validated against streamflow observations from the United States Geologic Survey (USGS). Finally, 21 precipitation ensemble members of the latest Global Ensemble Forecast System (GEFS/R) were forced into HEC-HMS to generate a retrospective streamflow ensemble forecast for an extreme hydrological event, Hurricane Irene. The work shows that ensemble stream discharge forecasts provide improved predictions and useful information about associated uncertainties, thus improving the assessment of risks when compared with deterministic forecasts. The uncertainties in weather inputs may result in false warnings and missed river flooding events, reducing the potential to effectively mitigate flood damage. The findings demonstrate how errors in the ensemble median streamflow forecast and time of peak, as well as the ensemble spread (uncertainty) are reduced 48 h pre-event by utilizing the ensemble framework. The methodology and implications of this work benefit efforts of short-term streamflow forecasts at regional scales, notably regarding the peak timing of an extreme hydrologic event when combined with a flood threshold exceedance diagram. Although the modeling framework was implemented on the Hudson River basin, it is flexible and applicable in other parts of the world where atmospheric reanalysis products and streamflow data are available.

1 Introduction

Riverine floods are known to adversely impact affected communities by causing casualties, inflicting damage to physical property, temporarily disrupting social and economic activities, and forcing a community to take emergency measures (IFRC, 2013). In the United States, for example, floods are recognized as the main natural disaster with USD 7.96 billion in flood-related damages per year and 82 fatalities per year, averaged over the past 30 years (NWS, 2014). It is reported that 78 % of emergencies are weather related (Weaver et al., 2014; Hoss and Fischbeck, 2016).

The increase in global averaged temperatures enhanced the potential for severe to extreme weather events (Becker and Grunewald, 2003; WMO, 2003). As the world warms, northern regions and mountainous areas are experiencing more precipitation falling as rain rather than snow, with a pronounced increase in precipitation being observed in the area of eastern North America (Karl, 2009). The special report on "Managing the Risks of Extreme Events and Disasters to Advance Climate Change Adaptation" of the Intergovernmental Panel on Climate Change (IPCC), critically assessed recent scientific literature on climate change and the impacts from extreme events. They reported that increased frequency and intensity of rainfall, based on climate models, substantially contributed in local flooding (Kundzewicz et al., 2014). Studies addressing flood damage in the United States show that the impact of floods to a community has increased over time as a result of both climate factors, such as precipitation, and societal factors, such as increased population and urban development (Pielke Jr., 2000; Changnon et al., 2001). Furthermore, studies report that associated monetary damages from flooding are also likely to go up in the 21st century and beyond (Milly et al., 2002; Allamano et al., 2009; Pall et al., 2011). The rise in the number of extreme weather events in recent years has spurred the need to better predict floods and mitigate flood damage. Such advanced warnings are not only important for economic losses but can mean the difference between life and death (NWS, 2012). Studies suggest that as little as 1 h of lead time can result in a 10 % reduction in flood damages if forecast information is communicated in a timely manner (McEnery et al., 2005). Flood modeling and prediction has greatly advanced in recent years with the advent of geographic information systems (GIS), high-resolution digital elevation models (DEMs), distributed hydrologic and weather models and better delivery systems on the internet (McEnery et al., 2005). However, despite the advancement in hydrological prediction systems, such systems remain plagued by uncertainty from numerical weather prediction models (Clark and Hay, 2004) and, hydrologic model and structure parameters (Krzysztofowicz, 2001a; Gupta, 2005).

In terms of atmospheric forcing, the main source of uncertainty in streamflow forecasts arises from precipitation forecast errors which include errors arising from parameterizations of physical processes in atmospheric models, resolution and initial conditions (Krzysztofowicz, 2001a; Bartholmes and Todini, 2005; Cuo et al., 2011). In this context, ensemble hydrological forecasts using every member in the ensemble are appealing to account for the uncertainty in numerical weather prediction (NWP) model forecasts (Buizza et al., 1999; Krzysztofowicz, 2001b; Bowler et al., 2008; Hamill et al., 2008; Cloke and Pappenberger, 2009).

Recent studies show the promise of adopting streamflow ensemble forecast techniques due to advantages over deterministic forecasts (Habets et al., 2004; Younis et al., 2008; Boucher et al., 2011; Schellekens et al., 2011; Verkade and Werner, 2011; Alfieri et al., 2013) as well as a way of accounting for uncertainties in hydrological forecasting (Chen and Yu, 2007; Demeritt et al., 2007; Davolio et al., 2008; Pappenberger et al., 2008; Reggiani and Weerts, 2008; Cloke and Pappenberger, 2009; Bao et al., 2011; Bogner and Pappenberger, 2011; Cuo et al., 2011; Schellekens et al., 2011; Alfieri et al., 2012; Amengual et al., 2015). Other advantages include the ability to distinguish between an extreme event forecast that is more or less likely to occur within the model's forecast horizon (Buizza, 2008; Golding, 2009) and better decision making with respect to operational hydrological concerns (Ramos et al., 2007; McCollor and Stull, 2008; Boucher et al., 2012). Furthermore, ensemble-based streamflow forecasts tend to be more consistent between successive forecasts (Pappenberger et al., 2011). Fan et al. (2014) showed benefits in the use of ensembles, particularly for reservoir inflows on flooding events, and in comparison to the deterministic values given by the control member of the ensemble and by the ensemble mean.

Komma et al. (2007) reported that for longer lead forecast times, the variability of the precipitation ensemble is amplified as it propagates through the catchment system as a result of non-linear catchment response. They also showed that the ensemble spread is a useful indicator to assess potential forecast errors for lead time greater than 12 h.

A numbers of existing techniques are currently used by the hydrologic community to account for uncertainty in hydrological forecast systems. For instance, Krzysztofowicz (2001a) implemented a method to combine uncertainties from both hydrological models and precipitation forecasts using a Bayesian forecasting system (BFS). This approach was based on decomposing the total uncertainty in the river stage into precipitation uncertainty and hydrologic uncertainty, which are quantified independently and then integrated into a predictive distribution of the river stage. Montanari and Grossi (2008) indirectly related the forecast error to the sources of uncertainty in the forecasting procedure through a probabilistic link with the current forecast issued by the hydrologic model, the past forecast error and the past rainfall. Olsson and Lindström (2008) performed analysis on separating the contributions of the precipitation forecast errors and the hydrological simulation errors. Weerts et al. (2011) used quantile regressions to assess the relationship between the hydrological forecast and the associated forecast error. Renard et al. (2010) addressed the total predictive uncertainty and separated it into input and structural components under different inference scenarios. They highlighted the inherent limitations of inferring inaccurate hydrologic models using rainfall–runoff data with large uncertainties. Brown (2015) quantified the total uncertainty in future streamflow as a combination of the meteorological forcing uncertainties and the hydrologic modeling uncertainties. He implemented a Meteorological Ensemble Forecast Processor (MEFP) to quantify the meteorological uncertainties and correct biases in the forcing inputs to the streamflow forecasts modeling.

In spite of the advancements and advantages in streamflow ensemble forecasts reported in the literature there are a number of key scientific questions that need better understanding. These include how the meteorological forecast uncertainty reflects in the ensemble streamflow forecast; how the degree of spread and hydrological response correlates with the lead time of the forecast and the scale of application; and how effective streamflow ensemble forecasting is during an extreme hydrological event. The present work investigates these questions by retrospectively forecasting streamflow of an extreme event (Hurricane Irene) in the Hudson River basin, USA. Hurricane Irene had strong hydrological

Figure 1. Geostationary Operational Environmental Satellite (GOES) east image of Hurricane Irene making landfall on 28 August 2011 (image source: National Oceanic and Atmospheric Administration (NOAA), 2011).

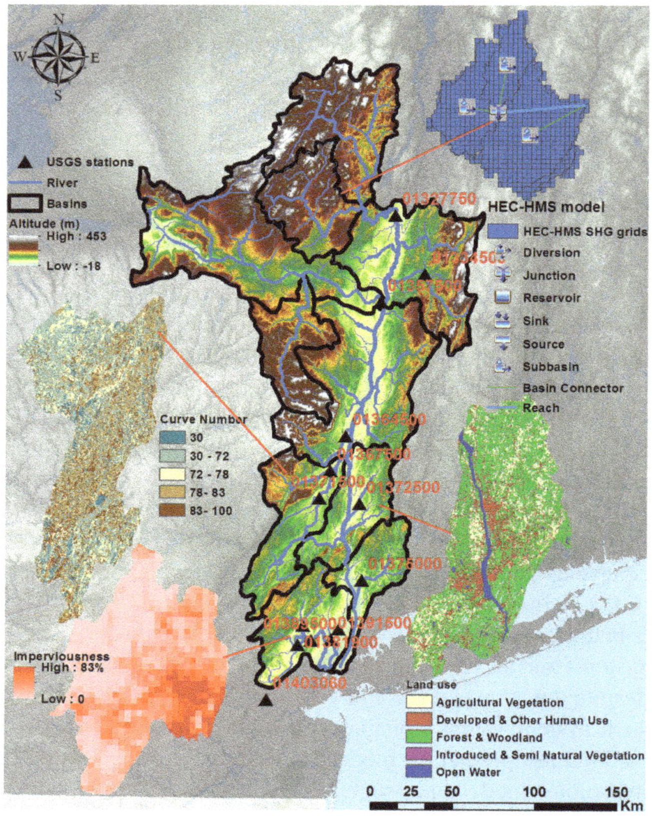

Figure 2. Map showing the Hudson River basin topography including basin divisions (thin black lines) and hydrographic network. Examples of land use, curve number and imperviousness data sets (zoomed) that were used in HEC-GeoHMS to construct the hydrological model are also shown. The upper right side of the figure exhibits an example of the HEC-HMS model structure and its subbasins using standard hydrologic grids (SHG (2×2 km)).

effects from high moisture content that brought very heavy rainfall rates to the US east coast including the Hudson River basin (Coch, 2012) (Fig. 1). The total estimated damage from Hurricane Irene was around USD 15.8 billion. This includes about USD 7.2 billion from inland flooding and storm surges (Avila and Cangialosi, 2011).

In this paper, we first describe the case study area and context. We then summarize the main data sets that were used to implement the hydrologic framework. Subsequently, we provide a detailed quantitative analysis and discussion of the uncertainties in the streamflow forecasts associated with the forcing from weather models and demonstrate how uncertainties in streamflow forecast median, time of peak and spread are reduced approaching a given event.

2 Materials and methods

2.1 Study area and context

The study encompasses the Hudson River basin (USA) which originates from the Adirondack Mountains of upstate New York and drains into the Atlantic Ocean (Fig. 2). The drainage area of the basin is approximately 36 000 km^2, covering 25 % of New York State and other portions of the states of New Jersey, Connecticut, Massachusetts and Vermont. The basin is considered one of the largest drainage areas in the eastern seaboard of the United States. According to a national water quality assessment study conducted by the United States Geological Survey (USGS), nearly 60 % of the water supplied in the basin is for commercial or industrial use. Several reservoirs within the Hudson River basin contribute to the New York City water-supply system, which supplies water to about 8 million people.

In 2011, Hurricane Irene caused severe damage and widespread destruction that affected the east coast of the United States. The storm made landfall as a strong tropical storm at Little Egg Inlet in New Jersey on 28 August 2011 (Fig. 1). The total precipitation accumulation from Hurricane Irene during 27–30 August 2011 was more than 300 mm in certain areas of the Hudson River basin (Fig. 4). It inundated streams throughout New Jersey resulting in peak stream flows exceeding the 100-year recurrence interval at many stream gages and causing heavy property and road damage. For instance, the Passaic and Hackensack River basins in northern New Jersey (south of the Hudson River basin) witnessed new record peaks at a number of streamflow-gauging stations.

President Obama issued a Major Disaster Declaration for counties in New York and New Jersey impacted by Hurricane Irene. In total, 38 counties across New York State were

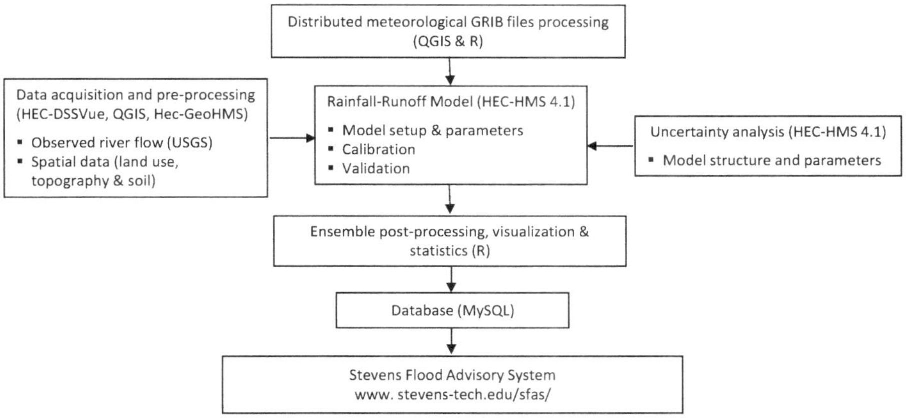

Figure 3. Framework design steps including spatial data and USGS discharge time series. Processing of general regularly distributed information in binary GRIB precipitation files data was carried out in Qgis (Qgis, 2011), R and Python, and exported to the HEC-DSSvue storage system. Basin parameters were derived for the study area and the hydrological model was run using these inputs.

impacted with an estimated USD 1.5 billion in FEMA public assistance costs and 10 deaths (FEMA, 2011). In New Jersey, the property damage was estimated to be USD 1 billion. In addition to high monetary damages, millions of people across the state were evacuated and seven deaths were reported (Watson et al., 2013; NJOEM, 2014).

2.2 Modeling framework description

The operational framework diagram is shown in Fig. 3 and the data sets used in constructing the Hudson River basin regional scale hydrological model are depicted in Fig. 2. The framework was validated using the National Centers for Environmental Prediction (NCEP) North American Regional Reanalysis (NARR) precipitation data (Mesinger et al., 2006). A retrospective forecast of Hurricane Irene utilizing the 21 ensemble members from NOAA's Global Ensemble Forecast System Reforecast (GEFS/R) was then used (Hamill et al., 2015).

Apart from this work, the framework is currently operational and fully automated on the Pharos (Greek for "lighthouse") Linux supercomputer at Stevens Institute of Technology. It produces four forecast cycles of ensemble river discharge per day, simulated at hourly time steps, which feed into the New York Harbor Observing and Prediction System (NYHOPS) (Bruno et al., 2006; Georgas et al., 2007, 2014). NYHOPS was developed at Stevens Institute of Technology's Davidson Laboratory to generate forecasts of the Atlantic coast, New York Harbor and Hudson River region through in situ monitoring equipment and hydrodynamic modeling (Blumberg et al., 2015).

2.2.1 HEC-HMS model description

The Hudson River basin was modeled using the latest Hydrologic Engineering Center's Hydrologic Modeling System (HEC-HMS), version 4.1 (USACE, 2015). HEC-HMS, developed by the US Army Corps of Engineers, is a conceptual semi-distributed hydrological model that has been used extensively in rainfall–runoff modeling and other related hydrological studies (Anderson et al., 2002; Neary et al., 2004; Knebl et al., 2005; Amengual et al., 2009; Chu and Steinman, 2009; Halwatura and Najim, 2013; Meenu et al., 2013; Seyoum et al., 2013; Zhang et al., 2013; Yang and Yang, 2014). The model uses a number of adjustable empirically derived parameters that describe the overall structure of the basin including parameters for runoff, baseflow and river routing (Feldman, 2000). In this work, the modified Clark (ModClark) distributed method (Kull and Feldman, 1998) was used to account for the spatial variability and characteristics of the basin. The gridded precipitation inputs were used to enable spatially distributed infiltration calculations at all regions of the basin. Infiltration capacity in the model was quantified using the gridded curve number (CN) methodology derived by the Soil Conservation Service (SCS) (USDA, 1986; Mishra and Singh, 2013). The SCS CN method estimates precipitation excess as a function of cumulative precipitation, soil type, antecedent soil moisture content, land use, total length of the river and the elevation of the catchment area (Scharffenberg, 2015). The baseflow component of the model includes the initial flow and the recession constant to account for groundwater contributions to stream flow (Chow, 1959; Maidment, 1992; Feldman, 2000). The river flow routing was based on the Muskingum equations of the HEC-HMS model (USACE, 2015).

2.2.2 Hudson River basin hydrological model data sets

The ArcGIS HEC-GeoHMS 10.2 extension (Fleming and Doan, 2013) was used to prepare and import the geographical information system (GIS) data into HEC-HMS (Johnson et al., 2001). The regional model data sets shown in Fig. 2

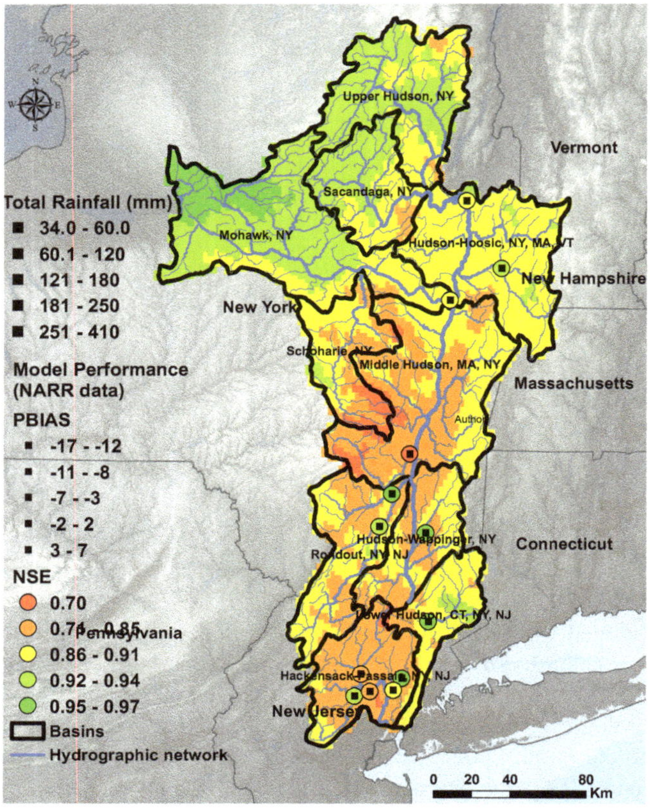

Figure 4. Summary of HEC-HMS performances at the USGS streamflow stations. The circles represent the criteria of Nash and Sutcliffe (NSE) while the squares represent the BIAS (%). The statistical criteria are computed at hourly time steps. The model performances are illustrated for the NARR precipitation forcing. The map also shows the total observed accumulated precipitation received in 48 h during 28–29 August 2011 (data from National Oceanic and Atmospheric Administration, 2011). The time series for selected flow stations are presented in Fig. 5.

include topography obtained from the USGS National Elevation Dataset (NED) (Gesch et al., 2002), land surface cover obtained from the US Department of Agriculture National Resource Conservation Service (NRCS) and soil data for New York State and New Jersey gathered from the State Soil Geographic Database (STATSGO) (Miller and White, 1998). Land use data sets were obtained from the USGS National Land Cover Dataset (NLCD) (Homer et al., 2012).

The Hudson River basin was first delineated into sub-basins based on flow direction and accumulation derived from a digital elevation model (DEM) using HEC-GeoHMS (Fleming and Doan, 2013), and then each sub-basin was subdivided into hydrologic response units, each of which has a gridded curve number representing its runoff response rate based on its unique combination of land use, soil type and slope (Gassman et al., 2007). The gridded SCS curve number was obtained by intersecting land use and land cover with the soil data using the CN-grid tool in HEC-GeoHMS. An example of the curve number grid GIS layer over a sub-basin of the Hudson River is shown in Fig. 2. It is important to point out that model runoff estimation can be very sensitive to soil moisture, and using static SCS curve number parameters may introduce limitations when the model is run operationally. To overcome such limitations, the framework utilizes a look-up table for the initial abstraction parameters based on the hindcast and the continuous run of the model with the NARR data. Other future techniques that involve integrating machine learning techniques to automatically select the optimal initial abstraction parameters on the fly can be advantageous as well. The sub-basins' storage coefficients and imperviousness percentage were derived from the GIS data sets described earlier this section using HEC-GeoHMS 10.2 (Fig. 2). The observed streamflow data for 15 USGS gauging stations, located in the Hudson River basin (Fig. 2), were automatically retrieved using the R dataRetrieval package (Hirsch and De Cicco, 2015).The observed discharge data were recorded at 15 min time intervals and imported to the USACE Data Storage System Visual Utility Engine (HEC-DSSVue) (HEC, 2009).

The model's baseflow recession constants were derived using an automated base flow separation technique based on the R low flow statistics package "lfstat" (Gustard and Demuth, 2009; Koffler and Laaha, 2012). In total, 25 years of observed historical flow data were used to derive the baseflow recession constants for each sub-basin in order to better simulate the falling limb of the simulated discharge hydrograph. The optimal recession constant values for the sub-basins ranged from 0.67 to 0.90 depending on the sub-basin that was considered. The calculated recession constant values were consistent and in agreement with the ones reported in the literature (Pilgrim and Cordery, 1993; Feldman, 2000).

For the initial baseflow, we used observed conditions in the gauging stations to reduce uncertainties in the model as the model was intended to forecast short-term extreme events, within a 96 h forecast horizon, and not long-term simulations. The model was forced with gridded precipitation, discussed in detail in the next sections of this paper, to work with the ModClark transform method (Kull and Feldman, 1998).

2.3 Model meteorological data sets

2.3.1 North American Regional Reanalysis

NARR is a long-term, dynamically consistent, high-resolution, high-frequency, atmospheric and land surface hydrology data set for the North American domain (Mesinger et al., 2006). NARR was developed as a major improvement upon the earlier National Centers for Environmental Prediction – National Center for Atmospheric Research Global Reanalysis 1 (NCEP-NCAR GR1). NARR data have successfully assimilated high-quality and detailed precipitation observations into the atmospheric analysis to create a long-term, consistent, high-resolution climate data set for the North American domain. The temporal resolution of the NARR data is 3 h and the spatial resolution is 32 km (Mesinger et al., 2006). The NARR precipitation data have been used in a number of hydrological studies. For instance, Choi et al. (2009) used the NARR data sets to successfully calibrate the semi-distributed land-use-based runoff processes (SLURP) model. Solaiman and Simonovic (2010) used the NARR data in a regional hydrological basin and reported satisfactory performance of such data in scarce regions.

In this work, NARR precipitation data, 26–31 August 2011, corresponding to Hurricane Irene was used in the hydrological model applied to the Hudson River basin in a 2 km common hydrologic gridded format using bicubic interpolation. Table 1 displays the rainfall accumulation totals extracted from the NARR data for selected sub-basins of the Hudson River. The hydrological simulation using this data set of precipitation was considered as the simulation of reference and was compared with the ensemble forecast that is reported in the next section.

2.3.2 Global Ensemble Forecast System Reforecast (GEFS/R)

The Global Ensemble Forecast System (GEFS) is a weather forecast model made up of 21 ensemble members (Hamill et al., 2013, 2015). The GEFS accounts for the amount of uncertainty in a forecast by generating an ensemble of multiple forecasts, each minutely different, or perturbed, from the control forecast. The GEFS 1° horizontal resolution data reforecasts used initial conditions obtained from high-quality reanalyses data and the same assimilation system that is used operationally. These reforecasts have been shown to be particularly useful for the calibration of relatively uncommon phenomena such as heavy precipitation (Hagedorn, 2008; Hamill et al., 2008). In relation to hydrology, reforecasts help produce quantitative probabilistic estimates of river streamflow that are as sharp and reliable as possible (Schaake et al., 2007).

2.4 Statistical criteria used to assess models' performance

The performance of the models was statistically evaluated using the criteria of Nash–Sutcliffe efficiency, referred to hereafter as NSE (Eq. 1), and bias (in %) between simulations and observations, referred to hereafter as BIAS (Eq. 2). The NSE measures the fraction of the variance of the observed flows explained by the model in terms of the relative magnitude of the residual variance (noise) to the variance of the flows (information); the optimal value is 1.0 and values should be larger than 0.0 to indicate minimally acceptable performance (Nash and Sutcliffe, 1970; O'Connell et al., 1970).

$$\mathrm{NSE} = 1 - \frac{\sum_{i=1}^{N}(P_i - O_i)^2}{\sum_{i=1}^{N}(O_i - \overline{O}_i)^2}, \tag{1}$$

where N is the number of compared values, P_i is the simulated (forecast) value, O_i is the observed value and $\overline{O}_i$ is the average of O_i time series.

The BIAS measures the average tendency of the simulated values to be larger or smaller than the observed ones. The optimal value of BIAS is 0.0, with low-magnitude values indicating accurate model simulation. Positive values indicate overestimation, whereas negative values indicate underestimation (Yapo et al., 1996).

$$\mathrm{BIAS}(\%) = 100\frac{\sum_{i=1}^{N}(P_i - O_i)}{\sum_{i=1}^{N} O_i}. \tag{2}$$

Model BIAS is reported relative to the mean observation magnitude (Eq. 2), in percentage (%).

3 Results

3.1 HEC-HMS model calibration using NARR precipitation data

Upon implementing the model setup, we calibrated the hydrological model to assess its ability to reproduce the Hurricane Irene event using the NARR gridded precipitation data (Mesinger et al., 2006). The HEC-HMS model was run on a 2×2 km standard hydrologic grid resolution (SHG) (Maidment and Djokic, 2000) at hourly time steps. The simulated flow hydrographs were calibrated against hourly river flow observations to obtain optimal performance in terms of both runoff volume and peak flow. The hydrological parameters were modified to produce a best-fit model using a root mean square error (RMSE) objective function within the HEC-HMS model's Nelder–Mead optimization method (Barati, 2011; Seyoum et al., 2013), aiming at maximizing the fit between simulated streamflow and observations at 15 US Geological Survey (USGS) gauging stations (Fig. 2). The calibration was also carried out by visual and statistical comparison to produce an accurate simulation of discharge at the Hudson River gauging stations (Fig. 2). We also used the HEC-HMS uncertainty function which, through a variant of Latin hypercube sampling, varies sensitive model parameters within a defined range, thereby producing an estimate of best-fit parameters after multiple iterations (Mousavi et al., 2012). For example, the hydrologic Muskingum routing

Table 1. Percentiles of total accumulated precipitation from the 21 GEFS ensemble members along with the peak discharge for forecasts issued at lead times of 72–24 h. The drainage area of each basin is also reported.

Station USGS ID/name	Basin area (km^2)	Forecast date	NARR precipitation (mm) over the whole event	GEFS precipitation (mm) over the whole event		Simulated peak flow ($m^3\,s^{-1}$)	
				2nd	98th	2nd	98th
01391500 Saddle River at Lodi, NJ	141	26 Aug 11	143	11	206	3	242
		27 Aug 11		110	201	104	229
		28 Aug 11		108	178	105	200
01378500 Hackensack River at New Milford, NJ	293	26 Aug 11	143	11	206	20	525
		27 Aug 11		108	200	196	514
		28 Aug 11		111	183	225	442
01371500 Wallkill River at Gardiner, NY	1800	26 Aug 11	106	3	190	22	1990
		27 Aug 11		80	187	475	1857
		28 Aug 11		82	154	558	1585
01388500 Pompton River at Pompton Plains, NJ	329	26 Aug 11	130	7	202	156	1208
		27 Aug 11		101	200	474	1150
		28 Aug 11		98	168	490	1024
01375000 Croton River on Hudson, NY	979	26 Aug 11	126	13	192	2	1255
		27 Aug 11		123	190	604	1205
		28 Aug 11		112	182	503	1205

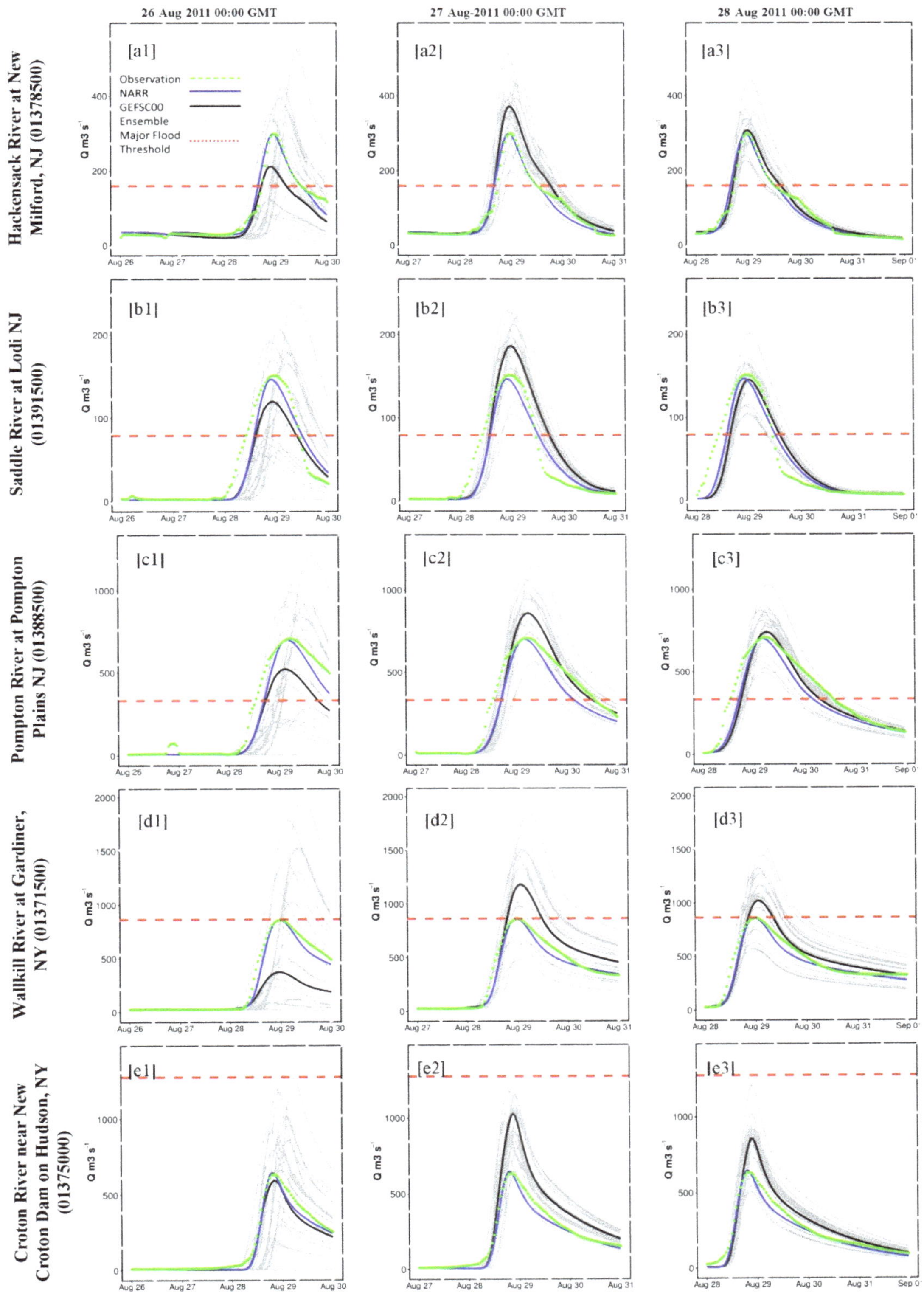

Figure 5. Ensemble streamflow forecasts compared to observed streamflow in selected stations at lead times of 72, 48 and 24 h from the observed peak flow; reported time is in GMT.

parameters were modified to include a greater ratio of attenuation to translation of runoff in the sub-basins which significantly improved the model results (Fig. 5). To assess and evaluate the parameter uncertainties we performed a Monte Carlo-based uncertainty analysis available in HEC-HMS 4.1 (Scharffenberg et al., 2015).

Table 1 lists the NARR accumulated precipitation for each sub-basin of the Hudson River while the summary of the model fit represented by the NSE and BIAS criteria is shown in Fig. 4. The NARR hydrological model results (also referred to as the simulation of reference) showed reasonable fit between model and observations in all the selected sub-basins. The lowest NSE was 0.75 and certain sub-basins had NSE values higher than 0.90 (Fig. 4). Amongst the 15 flow stations examined in the study, 13 stations had a BIAS below 10 % while 2 stations had a BIAS higher than 10 %. It was observed that the stations with a higher BIAS were located in the upstream parts of the basin, notably the Hoosic River sub-basin (USGS ID 01334500) (Fig. 4). Importantly, the hourly hydrograph shape and timing of peaks accurately replicated the observations as illustrated in Fig. 5. Overall, the reference simulation exhibited a representative fit to observations. Thus, the developed calibrated framework showed promising results for generating streamflow forecasts with a 96 h lead time using ensemble member weather forecast forcing.

3.2 Ensemble river discharge retrospective forecast

We forced the HEC-HMS model with precipitation fields from the GEFS-retrospective forecast ensemble members to examine the variations in simulated discharge among the ensemble members. More specifically, we fed the Hudson River basin hydrological model with every single GEFS member of the 21 available members. The resulting sets of streamflow forecasts were then analyzed to better understand the uncertainty of the streamflow forecasts that arises from the hydrologic framework's response to uncertainties in the meteorological forcing. The spread of ensemble members is considered as a useful measure of forecast uncertainty (Pappenberger et al., 2005).

To assess the skill of the forecasts, we compared, at lead times of 72, 48 and 24 h, observations with results of the individual ensemble members, the median of all members, the ensemble control member (GEFSC00) and the NARR simulation of reference (Fig. 5). We chose to include the control member in these comparisons as a proxy for the single deterministic models used when ensemble forecasts are not considered. The hydrological parameters and baseflow conditions calibrated in the NARR simulation of reference were retained in all simulations. The 98th and 2nd percentiles of accumulated precipitation from the 21 ensemble members at each forecast are reported in Table 1.

In the reforecast issued 72 h prior to the event, the high uncertainty in the NWP ensemble inputs translated to a high uncertainty in the simulated streamflow when simulated by the hydrological model. For instance, in station ID 01391500 (Saddle River at Lodi, NJ), there was a 20-fold spread in the accumulated precipitation amongst the 21 ensemble members. For the same station the peak flow ranged from base flow ($3\,m^3\,s^{-1}$) to $242\,m^3\,s^{-1}$ (exceeding the major flood threshold) (Fig. 5). At the Hackensack River at New Milford the peak flow was ranging anywhere between 20 and $525\,m^3\,s^{-1}$. The magnitude of spread in other stations was similar (Fig. 5). At that point of the retrospective forecast, the streamflow simulated using the control member (GEFSC00) underestimated the observed flow in all stations except for station ID 01375000 (Croton River on Hudson, NY) where the flow was simulated correctly (Fig. 5). However, the spread between the individual ensemble members remained very high as reported above. In terms of estimated time of peak, one notes that the GEFSC00 control member correctly projected the peak time of the event in all stations with an error of ± 3 h when compared with observations. However, other individual members had an offset of up to 24 h between simulated and observed peaks in certain stations (Fig. 5-a1, -b1, -c1, -d1 and -e1). For instance, at station ID 01375000 (Croton River on Hudson, NY), one individual member was projecting an estimated peak on the evening of 28 August while another individual member was projecting it at noon of the following day. In this particular station, the observed peak was around the midnight of 28 August. This finding offers an interesting perspective in terms of precisely extracting hydrograph features as one may use the control member at this stage of the forecasts to precisely project the time of the peak with a temporal error of ± 3 h. In terms of statistical evaluation, Figs. 6 and 7 report the NSE and BIAS (%) for selected stations. For station ID 01375000, the control member (GEFSC00) predicted the flow accurately with a NSE of approximately 0.95, however the flow is underestimated by about 20 % in the forecast issued 72 h prior to the event. Overall, only seven ensemble members at this station had a NSE above 0.70 while more than 60 % of the members underestimated or overestimated the flow hydrograph by more than 30 % (Fig. 7). Although uncertainties in baseflow initial conditions and model hydrological parameters are not addressed in this work, it was noted that uncertainties from precipitation inputs have a substantial impact on the prediction compared to remaining uncertainties in the initial conditions and parameters of the calibrated hydrological model.

In the next reforecast, issued 48 h before the event, the spread (or the uncertainty envelope) was substantially reduced, notably in the projected time of the peak (Fig. 5-a2, -b2, -c2, -d2 and -e2). This is primarily due to the decrease in accumulated precipitation ensemble spread (Table 1). The magnitude of the peak ensemble spread was between 104 and $229\,m^3\,s^{-1}$ for the Saddle River. The same river had a spread on the order of 20-fold in the prior forecast. This improvement was observed at other stations as well. At 48 h lead time, the control member systematically overestimated the flow at all stations (Fig. 5-a2, -b2, -c2, -d2 and -e2), but

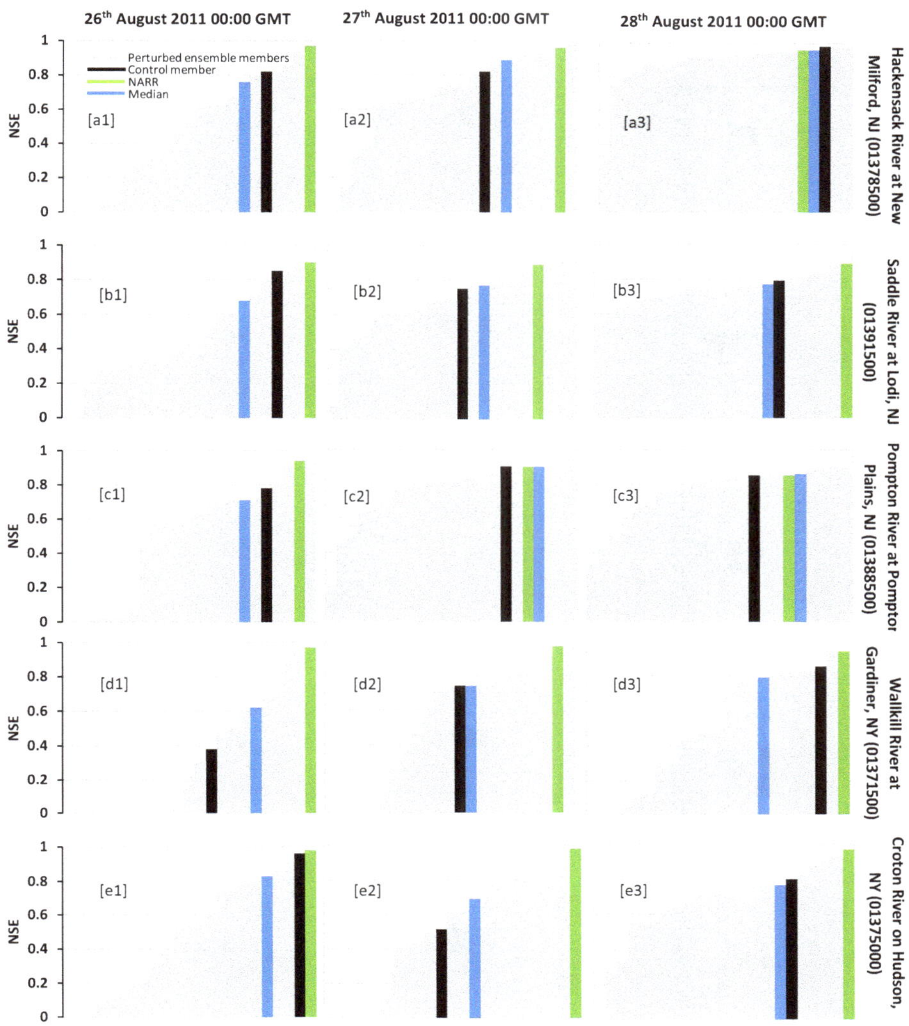

Figure 6. Models' performance represented by Nash–Sutcliffe efficiency (NSE) at lead times of 72, 48 and 24 h from the observed peak flow. The metrics are showing the NARR model outputs, the median of ensemble members and the GEFS control member. The GEFS perturbed members are shown in grey.

it consistently predicted the time of peak, as was the case in the forecast issued 72 h before the event. The uncertainty in the time of the peak also decreased, with most of the ensemble members predicting the peaks within ± 3 h from the observed ones. This suggests that the peak ensemble spread should not be the only metric considered to quantify potential uncertainty, it suggests that other features, such as the peak timing and magnitude, should also be examined. The control member (GEFSC00), which predicted the flow correctly for Croton River on Hudson (ID 01375000) 72 h before the event, however, at 48 h lead time showed a marked overestimation of the flow 65 %. This suggests that the accuracy of the predictions also varies temporally and no single member can be relied upon consistently as a perfect forecast. Overall, there was a significant improvement in NSE for most of the ensemble members in the forecasts issued 48 h before the event.

In the final reforecast considered, issued 24 h before the event, the spread was further reduced with all the members predicting the rising and the falling limbs of the hydrograph more accurately. The control member presented an almost perfect forecast for Hackensack River at New Milford (Fig. 6). The peak ensemble spread was reduced by 57 %, 60 % and 48 % compared to the 72 h lead time predicted peak ensemble spread at the Hackensack River, Saddle River and Wallkill River stations, respectively. Also, the peaks were predicted to occur within ± 3 h of the observed event peak in all cases. The control member seemed to be consistent with the observations in terms of the peak time at this stage of the forecast and the uncertainties are less than that projected 48 h before the event. At this stage of the forecast 75 % of the en-

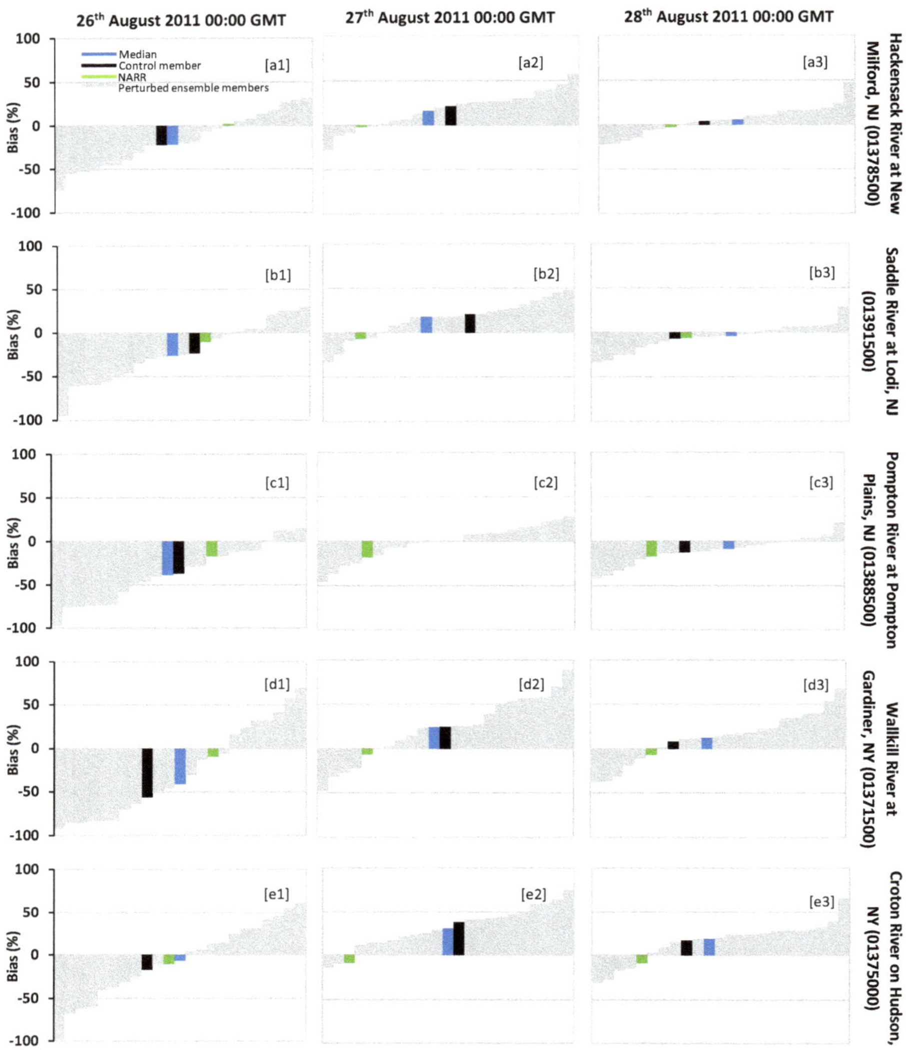

Figure 7. Models' performance represented by BIAS (%) at lead times of 72, 48 and 24 h from the observed peak flow. The metrics are showing the NARR model outputs, the median of ensemble members and the GEFS control member. The GEFS perturbed members are shown in grey.

semble members had a NSE higher than 0.75 (Fig. 6). This is consistent with findings in recent ensemble streamflow studies that used different modeling frameworks (Thielen et al., 2009; Fan et al., 2014; Yang and Yang, 2014). The results also suggest that the magnitude of spread between the ensemble members depends significantly on the sub-basin drainage area. For example, at station ID 01391500, Saddle River at Lodi (140 km^2) a peak discharge of 242 $m^3\,s^{-1}$ corresponds to a maximum accumulated precipitation of 206 mm. Thus, there is peak flow of about 1 $m^3\,s^{-1}$ for 1 mm of accumulated precipitation. This ratio, however, increases for a basin with a larger area. For example, at Wallkill River at Gardiner, NY (ID 01371500, 1800 km^2) there was a peak flow of approximately 10 $m^3\,s^{-1}$ for 1 mm of accumulated precipitation as it propagates (non-linearly) through the drainage area, thus there is a correlation between the spread in the precipitation data and the area of the sub-basin.

To have an overall assessment of the forecast skill, we calculated the range and average NSE and BIAS (%) across all stations (Fig. 8) in the Hudson River basin for the different reforecast ensemble members 24 h before the event. The figure depicts how a member that has a good NSE and BIAS at one station can have a very poor performance in other parts of the basin that may partly be due to the statistical downscaling of precipitation from 1° resolution to 2 km, and the associated uncertainty in the spatial distribution of precipitation. The median of all the members and the control member showed good performance with an average NSE of 0.75 compared to each of the members. The results show that there is no "one size fits all" solution for selecting an ensemble

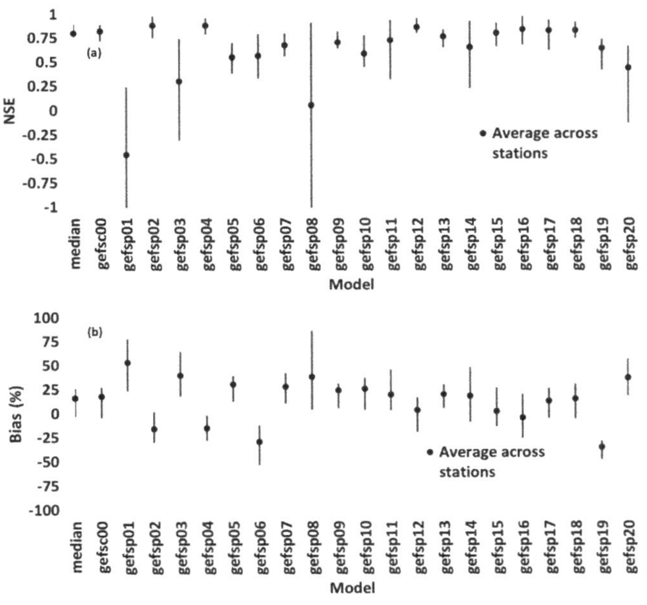

Figure 8. Overall NSE and BIAS (%) across the stations for forecast issued 24 h before Hurricane Irene.

member, noting that each sub-basin has its own distinct set of characteristics manifested in local conditions, such as the size of the basin and land use. This finding calls for further work involving higher-resolution precipitation models to assess the effect of basin size and meteorological forcing resolutions.

3.2.1 Threshold exceedance persistence diagram

In addition to the statistical metrics and visual comparison of stream-flow, an assessment of the forecasts skill was carried out for Hurricane Irene using a threshold exceedance persistence diagram at 6 h time intervals of the forecasts streamflow time series (Fig. 9). The exceedance diagram used in this study is an adaptation of the operational European Flood Alert System (Bartholmes et al., 2009; Thielen et al., 2009). Such quantitative diagrams give an idea about the forecast persistence and support operational flood management decisions and human judgment. The major flood threshold for each station was obtained from the NOAA National Weather Service (2011). It is defined as the category of flood where extensive inundation of structures and roads is expected, which calls for significant evacuation of people and transfer of property to higher elevations (NWS, 2012).

The major thresholds for observed and simulated discharge were transformed into dichotomous time series of 1 (the major threshold is exceeded) and 0 (the major threshold is not exceeded). The information of each cell in the matrix diagram is the probability of exceeding the major flooding value, calculated using all the ensemble members at a given forecast. For instance, a probability of 100 % suggests that all 21 ensemble members used in the work are projecting a major flood. Figure 9 exhibits the exceedance diagram results at selected stations of the Hudson River basin in which observations exceeded the major flood threshold during this event. The results suggest that one may establish a highly reliable streamflow forecast 48 h prior to the event. For instance, the 72 h lead time reforecasts (issued on 26 August 2011, 00:00 GMT) for station ID 01381900 were projecting a 52 % probability (11 members out of 21) of having a major flood event 78 h out. However, the probability increased to 100 % in the 27 August 2011, 00:00 GMT (48 h before the event) reforecast for the same station. The average across all stations for the day 3 (72 h before the event) reforecast was showing a 60 % probability of having a major flood on 29 August 2011 (between 00:00 and 06:00 GMT), while the day 2 (48 h before the event) reforecasts had a 99 % average chance of exceeding the major flood threshold. This highlights the increase in reliability as the event approaches. The diagram also suggests that the persistence of the event occurrence among subsequent forecasts is a good indicator to trigger a major flood warning, especially when the proportion of members above the threshold exceeds 71 % (15 or more out of the 21 ensemble members). By contrast, in station ID 0137500, both observations and flow ensemble members did not exceed the major flood threshold (Fig. 9). This is particularly important for the reliability and validation of the operational forecast system. However, caution should be practiced when interpreting the persistence exceeding diagrams as time series will have to be examined in parallel to confirm any potential discrepancy between the models.

4 Summary and discussion

The first part of this work consisted of implementing a regional scale hydrological modeling framework on the Hudson River basin using the HEC-HMS model (USACE, 2015) forced with NARR gridded precipitation inputs (Mesinger et al., 2006). The second part investigated the use of GEFS (Hamill et al., 2013, 2015) ensemble inputs to retrospectively forecast an extreme hydrological event, Hurricane Irene, with a 96 h time horizon at hourly time steps. In total, 21 GEFS ensemble members were tested on the Hudson River basin with reforecasts issued 72, 48 and 24 h prior to the event.

In terms of assessment, we visually and statistically quantified the results of the GEFS individual members, the deterministic forecasts represented by the control member (GEFSC00) and the median of all members. We also used the persistence exceedance diagram of established thresholds to project the possibility of major floods based on all the individual members. The comparison showed that ensemble forecasts are advantageous in quantifying uncertainty of forecasts lead time and raising reliability from an operational perspective. This work did not address uncertainty in hydrological initial conditions and model parameters, which were

Station & major flood threshold	Time / Forecast h	8/26/2011				8/27/2011				8/28/2011				8/29/2011				8/30/2011				8/31/2011			
		0:00	6:00	12:00	18:00	0:00	6:00	12:00	18:00	0:00	6:00	12:00	18:00	0:00	6:00	12:00	18:00	0:00	6:00	12:00	18:00	0:00	6:00	12:00	18:00
Hoosic River at Eagle Bridge, NY (01334500) $886\ m^3\ s^{-1}$	08/26/2011 t00	0	0	0	0	0	0	0	0	0	0	0	14	52	71	62	29								
	08/27/2011 t00					0	0	0	0	0	0	0	62	100	100	86	29	0	0	0	0				
	08/28/2011 t00									0	0	0	71	100	100	90	0	0	0	0	0	0	0	0	0
	Observed																								
Wappinger Creek at Wappingers Falls, NY (01372500) $286\ m^3\ s^{-1}$	08/26/2011 t00	0	0	0	0	0	0	0	0	0	0	0	5	29	52	52	38								
	08/27/2011 t00					0	0	0	0	0	0	0	10	86	90	86	48	5	0	0	0				
	08/28/2011 t00									0	0	0	5	86	90	86	62	0	0	0	0	0	0	0	0
	Observed																								
Hackensack River at New Milford, NJ (01378500) $158\ m^3\ s^{-1}$	08/26/2011 t00	0	0	0	0	0	0	0	0	0	0	14	52	67	62	48	33								
	08/27/2011 t00					0	0	0	0	0	0	62	100	100	95	86	48	5	0	0	0				
	08/28/2011 t00									0	0	43	100	100	100	81	14	0	0	0	0	0	0	0	0
	Observed																								
Passaic River at Little Falls, NJ (01389500) $283\ m^3\ s^{-1}$	08/26/2011 t00	0	0	0	0	0	0	0	0	0	0	5	24	67	67	67	67								
	08/27/2011 t00					0	0	0	0	0	0	19	90	100	100	100	100	100	100	100	100				
	08/28/2011 t00									0	0	5	86	100	100	100	100	100	100	100	100	100	100	100	100
	Observed																								
Rondout Creek at Rosendale, NJ (01367500) $708\ m^3\ s^{-1}$	08/26/2011 t00	0	0	0	0	0	0	0	0	0	0	19	48	57	48	38	19								
	08/27/2011 t00					0	0	0	0	0	0	76	95	100	81	29	5	0	0	0	0				
	08/28/2011 t00									0	0	71	100	95	86	29	0	0	0	0	0	0	0	0	0
	Observed																								
Saddle River at Lodi, NJ (01391500) $80\ m^3\ s^{-1}$	08/26/2011 t00	0	0	0	0	0	0	0	0	0	5	29	67	71	71	52	38								
	08/27/2011 t00					0	0	0	0	0	10	90	100	100	100	90	48	0	0	0	0				
	08/28/2011 t00									0	0	81	100	100	100	81	5	0	0	0	0	0	0	0	0
	Observed																								
Passaic River at Pine Brook, NJ (01381900) $141\ m^3\ s^{-1}$	08/26/2011 t00	0	0	0	0	0	0	0	0	0	0	0	19	52	52	57	57								
	08/27/2011 t00					0	0	0	0	0	0	19	81	100	100	100	100	100	100	100	100				
	08/28/2011 t00									0	0	5	48	95	100	100	100	100	100	100	100	100	100	100	100
	Observed																								
Pompton River at Pompton Plains, NJ (01388500) $329\ m^3\ s^{-1}$	08/26/2011 t00	0	0	0	0	0	0	0	0	0	0	14	43	71	67	67	62								
	08/27/2011 t00					0	0	0	0	0	0	57	95	100	100	100	100	90	76	57	19				
	08/28/2011 t00									0	0	29	100	100	100	100	100	81	76	19	5	0	0	0	0
	Observed																								
Raritan River at Bound Brook, NJ (01403060) $923\ m^3\ s^{-1}$	08/26/2011 t00	0	0	0	0	0	0	0	0	0	33	67	90	90	81	71	52								
	08/27/2011 t00					0	0	0	0	0	62	100	100	100	100	100	100	62	5	0	0				
	08/28/2011 t00									0	0	100	100	100	100	100	81	5	0	0	0	0	0	0	0
	Observed																								
Esopus Creek at Mount Marion, NY (01364500) $547\ m^3\ s^{-1}$	08/26/2011 t00	0	0	0	0	0	0	0	0	0	14	29	52	52	48	48	48								
	08/27/2011 t00					0	0	0	0	0	38	95	95	95	95	95	90	81	67	57	48				
	08/28/2011 t00									0	33	100	100	100	90	86	86	71	62	57	48	29	14	5	0
	Observed																								
Mohawk River at Cohoes, NY (01375000) $1274\ m^3\ s^{-1}$	08/26/2011 t00	0	0	0	0	0	0	0	0	0	0	0	0	0	0	0	0								
	08/27/2011 t00					0	0	0	0	0	0	0	0	0	0	0	0	0	0	0	0				
	08/28/2011 t00									0	0	0	0	0	0	0	0	0	0	0	0	0	0	0	0
	Observed																								

Figure 9. Color-coded threshold exceedance diagram for Hurricane Irene forecasts at 6 h intervals using the major flood threshold for each USGS station. The flow values below the station ID are the major flood threshold for a given station. The x axis represents the 96 h forecast horizon from the simulation date shown on the left column. The time period, in which the major event hydrological record exceeds the equivalent alert threshold, is indicated using a dark red cell, while the cell values refer to the percentage of ensemble members that were projecting a major flood. For instance, if the value is 100 then all 21 flow ensemble members are projecting a major event within a given time interval, while 0 means none of the ensemble members are exceeding the major event threshold. The observed occurrence of the threshold is exhibited at the last row for each station, the red color code indicates an observed flow higher than the major flood threshold, while the green cell indicates flow below the major threshold. Reported time is in GMT; date format is mm/dd/yyyy.

optimized in the NARR-based calibration, but rather was focused on how dominant the precipitation inputs are on the results of the hydrological streamflow forecasts. The findings of this work confirm that using ensemble streamflow predictions has advantages over deterministic forecasts in terms of better representing the uncertainty, particularly in an extreme hydrologic event such as the one presented in this work. Furthermore, while the general perception is that hydrological uncertainty is reduced with lead time, there are no studies that quantitatively characterize this aspect using the GEFS retrospective data and in the event of an extreme flood event such as Hurricane Irene.

The work shows that streamflow forecasts are highly dependent on the meteorological inputs and reflect uncertainties associated with these inputs. Clearly, the relatively small area of the sub-basin and resolution of the weather model was of importance for the spread overestimation due to the precipitation inputs. In this context, higher-resolution weather models such as the European Centre for Medium-Range Weather Forecasts (ECMWF) (Molteni et al., 1996; Thiemig et al., 2015) or the Coupled Ocean-Atmosphere Mesoscale Prediction System (COAMPS)® (Pullen et al., 2015) may be of particular interest in smaller scale subcatchments in order to provide higher accuracy. The operational GEFS ensemble forecast data sets were upgraded to 0.5° resolution on 2 December 2015. More detailed weather model forcing may lead to a reduction in the streamflow spread, and reduce the uncertainty of the forecast.

The modeling of the Hudson River basin demonstrated that a given streamflow ensemble member may produce streamflow that is highly in agreement with observations, accounting for uncertainty in precipitation and in the hydrological model parameters. However, there is no "one size fits all" solution for selecting an ensemble member, noting that each sub-basin has its own distinct set of characteristics manifested in local conditions such as the area of the drainage basin and land use. Furthermore, caution should be exercised in forecast models that only use the control member from the weather models or the average ensemble member as it may lead to considerable deviation in river discharge forecasts from observations resulting in false warnings and missed flooding events (Fig. 5).

The findings also suggest that higher confidence in the river discharge forecasts may be attained as we approach a major event by approximately 48 h. The outcomes of this work provide interesting perspectives for future ensemble postprocessing techniques and features extraction, notably regarding the peak timing of an extreme hydrologic event when combined with the major flood persistence diagram.

The operational framework presented in this work offers an improvement over the available NOAA's Advanced Hydrological Prediction System (AHPS) in this particular region (McEnery et al., 2005) and is operationally running 125 ensemble members including (in addition to GEFS) the ECMWF, ECMWF-HRES, the Short-Range Ensemble Forecast (SREF), the Canadian Meteorological Centre (CMC) and the North American Mesoscale Forecast System (NAM). The AHPS streamflow forecasts are at 6 h time intervals using one weather deterministic forecast as input and with a lead time that is less than 60 h in this region (Adams, 2015).

The framework is highly flexible and directly handles GRIB1, GRIB2 and NetCDF meteorological input formats. This operational flexibility in the framework allows it to use other sources of meteorological data instead of NARR, which is not available in areas outside North America. Examples of atmospheric reanalysis products that may be used include the European Centre for Medium-Range Weather Forecasting (ECMWF), National Centers for Environmental Prediction (NCEP) and National Center for Atmospheric Research (NCAR). As for framework validation, if applied to watersheds without gauging stations (or with only a few), it is possible to use remote-sensing river discharge data to calibrate and validate the modeling outputs. Despite that fact that such data have uncertainties, there have been many advancements in this field and there is potential for future applications (e.g., the Surface Water and Ocean Topography (SWOT) satellite mission; Saleh et al., 2012; Biancamaria et al., 2015). Thus, this framework shows promise for operational streamflow forecasting in other parts of the world.

The computational time required to run the Hudson River basin for the 21 GEFS ensemble members was approximately 30 min. This includes processing the GRIB input files and postprocessing of the ensemble outputs. The total time required to run the entire 125 ensemble members is approximately 5.5 h; this includes preprocessing inputs and updating the database and the Stevens Flood Advisory System (SFAS) website. The forecasts are updated every 6 h, with a lead time of 87 h, which is sufficient for issuing a flood warning.

The current work maybe potentially expanded to integrate a hydrodynamic model, such as HEC-RAS (Brunner, 2002), that has advantages in terms of simulating the water levels and flood extents in addition to streamflow. Such hydrodynamic modeling requires flow boundary conditions that can be obtained from the hydrologic forecasting model. This modeling will also require detailed representation of rivers' cross sections that may be obtained using lidar data at site-specific locations; however, it is not always available at regional scale (Saleh et al., 2011).

In terms of applications, the Hudson River basin regional scale model may be used to continuously forecast the overall variability of the water resources (Saleh et al., 2011; Pryet et al., 2015; Saraiva Okello et al., 2015), predicting fate and transport of water quality such as nitrate (Schoonover and Lockaby, 2006; Wang et al., 2012; Bastola and Misra, 2015; Schuetz et al., 2016) and climate change scenarios (Ducharne et al., 2007, 2010; Graham et al., 2007; Quintana Seguí et al., 2010; Habets et al., 2013). Moreover, socio-economic analysis may be used to weigh how such improved forecasts can prevent loss of life and minimize the damage to property, with the aid of effective communication and social media.

Acknowledgements. This work was funded by a research task agreement entered between the Trustees of the Stevens Institute of Technology and the Port Authority of New York and New Jersey, effective 19 August 2014. We would like to first thank the reviewers for their valuable comments and suggestions to improve the quality of the paper. We are grateful to William Scharffenberg and his team at the USACE Hydrologic Engineering Center for their help in operationally running HEC-HMS and HEC-DSSvue on Linux. We are grateful to Sarah R. Hatala from New Jersey Department of Environmental Protection Office of Engineering and Construction Bureau of Dam Safety and Flood Control for providing reservoir data. We are thankful to Alon Dominitz from New York State Department of Environmental Conservation for providing us with sources of dams elevation-storage data. We are also thankful to Jacob Isleib from the USDA-Natural Resources Conservation Service for his help in extracting the soil data for New York and New Jersey from the State Soil Geographic Database (STATSGO).

We are thankful to Wesley Ebisuzaki from the NOAA Climate Prediction Center for his help with the NARR data. The GEFS Retrospective Forecasts were graciously provided by Yuejian Zhu, Justin Cooke, Xiaqiong Zhou and Rebecca Cosgrove of NOAA EMC.

Edited by: F. Pappenberger

References

Adams, T.: Verification of the NOAA/NWS MMEFS Operational Hydrologic Ensemble Forecasting System in the Ohio River Valley, World Environmental and Water Resources Congress 2015 at Floods, Droughts, and Ecosystems, 2688–2700, 2015.

Alfieri, L., Salamon, P., Pappenberger, F., Wetterhall, F., and Thielen, J.: Operational early warning systems for water-related hazards in Europe, Environ. Sci. Pol., 21, 35–49, 2012.

Alfieri, L., Burek, P., Dutra, E., Krzeminski, B., Muraro, D., Thielen, J., and Pappenberger, F.: GloFAS – global ensemble streamflow forecasting and flood early warning, Hydrol. Earth Syst. Sci., 17, 1161–1175, doi:10.5194/hess-17-1161-2013, 2013.

Allamano, P., P. Claps, and F. Laio, Global warming increases flood risk in mountainous areas, Geophys. Res. Lett., 36, L24404, doi:10.1029/2009GL041395, 2009.

Amengual, A., Romero, R., Vich, M., and Alonso, S.: Inclusion of potential vorticity uncertainties into a hydrometeorological forecasting chain: Application to a medium size basin of Mediterranean Spain, Hydrol. Earth Syst. Sci., 13, 793–811, doi:10.5194/hess-13-793-2009, 2009.

Amengual, A., Homar, V., and Jaume, O.: Potential of a probabilistic hydrometeorological forecasting approach for the 28 September 2012 extreme flash flood in Murcia, Spain, Atmos. Res., 166, 10–23, 2015.

Anderson, M. L., Chen, Z. Q., Kavvas, M. L., and Feldman, A.: Coupling HEC-HMS with atmospheric models for prediction of watershed runoff, J. Hydrol. Eng., 7, 312–318, 2002.

Avila, A. L. and Cangialosi, J.: Tropical Cyclone Report-Hurricane Irene (Technical Report AL092011), 21–28 August 2011, National Hurricane Center, http://www.nhc.noaa.gov/data/tcr/AL092011_Irene.pdf (last access: 4 February 2016), 2011.

Bao, H.-J., Zhao, L.-N., He, Y., Li, Z.-J., Wetterhall, F., Cloke, H. L., Pappenberger, F., and Manful, D.: Coupling ensemble weather predictions based on TIGGE database with Grid-Xinanjiang model for flood forecast, Adv. Geosci., 29, 61–67, doi:10.5194/adgeo-29-61-2011, 2011.

Barati, R.: Parameter estimation of nonlinear Muskingum models using Nelder-Mead simplex algorithm, J. Hydrol. Eng., 16, 946–954, 2011.

Bartholmes, J. and Todini, E.: Coupling meteorological and hydrological models for flood forecasting, Hydrol. Earth Syst. Sci., 9, 333–346, doi:10.5194/hess-9-333-2005, 2005.

Bartholmes, J. C., Thielen, J., Ramos, M. H., and Gentilini, S.: The european flood alert system EFAS – Part 2: Statistical skill assessment of probabilistic and deterministic operational forecasts, Hydrol. Earth Syst. Sci., 13, 141–153, doi:10.5194/adgeo-13-141-2009, 2009.

Bastola, S. and Misra, V.: Seasonal hydrological and nutrient loading forecasts for watersheds over the Southeastern United States, Environ. Model. Softw., 73, 90–102, 2015.

Becker, A. and Grunewald, U.: Flood risk in central Europe, Science, 300, 1099–1099, 2003.

Biancamaria, S., Lettenmaier, D. P., and Pavelsky, T. M.: The SWOT Mission and Its Capabilities for Land Hydrology, Surv. Geophys., 1–31, 2015.

Blumberg, A. F., Georgas, N., Yin, L., Herrington, T. O., and Orton, P. M.: Street-scale modeling of storm surge inundation along the New Jersey Hudson river waterfront, J. Atmos. Ocean. Technol., 32, 1486–1497, 2015.

Bogner, K. and Pappenberger, F.: Multiscale error analysis, correction, and predictive uncertainty estimation in a flood forecasting system, Water Resour. Res., 47, W07524, doi:10.1029/2010WR009137, 2011.

Boucher, M.-A., Anctil, F., Perreault, L., and Tremblay, D.: A comparison between ensemble and deterministic hydrological forecasts in an operational context, Adv. Geosci., 29, 85–94, 2011.

Boucher, M.-A., Tremblay, D., Delorme, L., Perreault, L., and Anctil, F.: Hydro-economic assessment of hydrological forecasting systems, J. Hydrol., 416, 133–144, 2012.

Bowler, N. E., Arribas, A., Mylne, K. R., Robertson, K. B., and Beare, S. E.: The MOGREPS short-range ensemble prediction system, Q. J. Roy. Meteor. Soc., 134, 703–722, 2008.

Brown, J.: An evaluation of the minimum requirements for meteorological reforecasts from the Global Ensemble Forecast System (GEFS) of the U.S. National Weather Service (NWS) in support of the calibration and validation of the NWS Hydrologic Ensemble Forecast Service (HEFS), National Weather Service (NWS), technical report, 2015.

Brunner, G. W.: HEC-RAS River Analysis System: User's Manual, US Army Corps of Engineers, Institute for Water Resources, Hydrologic Engineering Center, 2002.

Bruno, M. S., Blumberg, A. F., and Herrington, T. O.: The urban ocean observatory-coastal ocean observations and forecasting in the New York Bight, Proceedings-Institute of Marine Engineering Science and Technology Part C, Journal of Marine Science and Environment, 4, 31, Institute of Marine Engineering Science and Technology, 2006.

Buizza, R., Hollingsworth, A., Lalaurette, F., and Ghelli, A.: Probabilistic predictions of precipitation using the ECMWF Ensemble Prediction System, Weather Forecast., 14, 168–189, 1999.

Buizza, R.: The value of probabilistic prediction, Atmos. Sci. Lett., 9, 36–42, 2008.

Changnon, S. A., Kunkel, K. E., and Andsager, K.: Causes for record high flood losses in the central United States, Water Int., 26, 223–230, 2001.

Chen, S.-T. and Yu, P.-S.: Real-time probabilistic forecasting of flood stages, J. Hydrol., 340, 63–77, 2007.

Choi, W., Kim, S. J., Rasmussen, P. F., and Moore, A. R.: Use of the North American Regional Reanalysis for hydrological modelling in Manitoba, Canad. Water Resour. J., 34, 17–36, 2009.

Chow, V. T.: Open channel hydraulics, McGraw-Hill Book Company Inc, New York, p. 680, 1959.

Chu, X. and Steinman, A.: Event and continuous hydrologic modeling with HEC-HMS, J. Irr. Drain. Eng., 135, 119–124, 2009.

Clark, M. P. and Hay, L. E.: Use of Medium-Range Numerical Weather Prediction Model Output to Produce Forecasts of Streamflow, J. Hydrometeorol., 5, 15–32, 2004.

Cloke, H. and Pappenberger, F.: Ensemble flood forecasting: a review, J. Hydrol., 375, 613–626, 2009.

Coch, N. K.: Hurricane Irene (2011) – A Hydrological Catastrophe In The Northeast US, American Institute of Professional Geologist, 47–52, 2012.

Cuo, L., Pagano, T. C., and Wang, Q. J.: A Review of Quantitative Precipitation Forecasts and Their Use in Short- to Medium-Range Streamflow Forecasting, J. Hydrometeorol., 12, 713–728, 2011.

Davolio, S., Miglietta, M. M., Diomede, T., Marsigli, C., Morgillo, A., and Moscatello, A.: A meteo-hydrological prediction system based on a multi-model approach for precipitation forecasting, Nat. Hazards Earth Syst. Sci., 8, 143–159, doi:10.5194/nhess-8-143-2008, 2008.

Demeritt, D., Cloke, H., Pappenberger, F., Thielen, J., Bartholmes, J., and Ramos, M.-H.: Ensemble predictions and perceptions of risk, uncertainty, and error in flood forecasting, Environ. Hazards, 7, 115–127, 2007.

Ducharne, A., Baubion, C., Beaudoin, N., Benoît, M., Billen, G., Brisson, N., Garnier, J., Kieken, H., Lebonvallet, S., and Ledoux, E.: Long term prospective of the Seine River system: Confronting climatic and direct anthropogenic changes, Sci. Total Environ., 375, 292–311, 2007.

Ducharne, A., Habets, F., Pagé, C., Sauquet, E., Viennot, P., Déqué, M., Gascoin, S., Hachour, A., Martin, E., and Oudin, L.: Climate change impacts on water resources and hydrological extremes in northern France, XVIII Conference on Computational Methods in Water Resources, 2010.

Fan, F. M., Collischonn, W., Meller, A., and Botelho, L. C. M.: Ensemble streamflow forecasting experiments in a tropical basin: The São Francisco river case study, J. Hydrol., 519, 2906–2919, 2014.

Feldman, A. D.: Hydrologic modeling system HEC-HMS: technical reference manual, US Army Corps of Engineers, Hydrologic Engineering Center, 2000.

FEMA: Federal Emergency Management Agency Technical Report on New York-Hurricane Irene (DR-4020), available at: http://www.fema.gov/ar/disaster/4020 (last access: 4 February 2016), 2011.

Fleming, M. and Doan, J.: HEC-GeoHMS geospatial hydrologic modelling extension: user's manual version 10.2, US Army Corps of Engineers, Institute for Water Resources, Hydrologic Engineering Centre, Davis, CA, http://www.hec.usace.army.mil/software/hec-geohms/downloads.aspx, 2013.

Gassman, P. W., Reyes, M. R., Green, C. H., and Arnold, J. G.: The soil and water assessment tool: historical development, applications, and future research directions, T. ASABE, 50, 1211–1250, 2007.

Georgas, N., Blumberg, A. F., and Herrington, T. O.: An operational coastal wave forecasting model for New Jersey and Long Island waters, Shore Beach, 75, p. 30, 2007.

Georgas, N., Orton, P., Blumberg, A., Cohen, L., Zarrilli, D., and Yin, L.: The impact of tidal phase on Hurricane Sandy's flooding around New York City and Long Island Sound, J. Extr. Events, 1, 1450006, 2014.

Gesch, D., Oimoen, M., Greenlee, S., Nelson, C., Steuck, M., and Tyler, D.: The national elevation dataset, Photogram. Eng. Remote Sens., 68, 5–32, http://www.ned.usgs.gov/, 2002.

Golding, B.: Long lead time flood warnings: reality or fantasy?, Meteorol. Appl., 16, 3–12, 2009.

Graham, L. P., Andréasson, J., and Carlsson, B.: Assessing climate change impacts on hydrology from an ensemble of regional climate models, model scales and linking methods – a case study on the Lule River basin, Climatic Change, 81, 293–307, 2007.

Gupta, H. V., Beven, K. J., and Wagener, T.: Model Calibration and uncertainty estimation, in: Encyclopedia of Hydrological Sciences, John Wiley and Sons, 2015–2032, 2005.

Gustard, A. and Demuth, S.: Manual on low-flow estimation and prediction, Operational Hydrology Report No. 50, World Meteorological Organization, WMO-No. 1029, 2009.

Habets, F., LeMoigne, P., and Noilhan, J.: On the utility of operational precipitation forecasts to served as input for streamflow forecasting, J. Hydrol., 293, 270–288, 2004.

Habets, F., Boé, J., Déqué, M., Ducharne, A., Gascoin, S., Hachour, A., Martin, E., Pagé, C., Sauquet, E., Terray, L., Thiéry, D., Oudin, L., and Viennot, P.: Impact of climate change on the hydrogeology of two basins in northern France, Clim. Change, 121, 771–785, 2013.

Halwatura, D. and Najim, M. M. M.: Application of the HEC-HMS model for runoff simulation in a tropical catchment, Environ. Model. Softw., 46, 155–162, 2013.

Hamill, T. M., Hagedorn, R., and Whitaker, J. S.: Probabilistic forecast calibration using ECMWF and GFS ensemble reforecasts. Part II: Precipitation, Mon. Weather Rev., 136, 2620–2632, 2008.

Hamill, T. M., Bates, G. T., Whitaker, J. S., Murray, D. R., Fiorino, M., Galarneau Jr, T. J., Zhu, Y., and Lapenta, W.: NOAA's second-generation global medium-range ensemble reforecast dataset, B. Am. Meteorol. Soc., 94, 1553–1565, http://para.nomads.ncep.noaa.gov/, 2013.

Hamill, T. M., Scheuerer, M., and Bates, G. T.: Analog Probabilistic Precipitation Forecasts Using GEFS Reforecasts and Climatology-Calibrated Precipitation Analyses, Mon. Weather Rev., 143, 3300–3309, 2015.

HEC: Data storage system visual utility engine user's manual version 2.0. Davis, CA: US Army Corps of Engineers, Institute for Water resource, Hydrologic Engineering Center, 2009.

Hirsch, R. M. and De Cicco, L. A.: User guide to Exploration and Graphics for RivEr Trends (EGRET) and dataRetrieval: R packages for hydrologic data, US Geological Survey 2328-7055, https://github.com/USGS-R/dataRetrieval, 2015.

Homer, C. H., Fry, J. A., and Barnes, C. A.: The national land cover database, US Geological Survey Fact Sheet, 3020, 1–4, http://www.mrlc.gov/finddata.php, 2012.

Hoss, F. and Fischbeck, P.: Increasing the Value of Uncertain Weather and River Forecasts for Emergency Managers, B. Am. Meteorol. Soc., 97, 85–97, 2016.

IFRC: World disaster report, Geneva: International Federation Of Red Cross And Red Crescent Societies (www. ifrc. Org), 2013.

Johnson, C. A., Yung, A. C., Nixon, K. R., and Legates, D. R.: The Use of HEC-GeoHMS and HEC-HMS to Perform Grid-based Hydrologic Analysis of a Watershed, Proceedings of the 25th Annual Conference of the Association of State Floodplain Managers. Charlotte, North Carolina: The Association of State Floodplain Managers, Inc., 2002.

Karl, T. R.: Global Climate Change Impacts in the United States, Cambridge University Press, p. 17, 2009.

Knebl, M. R., Yang, Z. L., Hutchison, K., and Maidment, D. R.: Regional scale flood modeling using NEXRAD rainfall, GIS, and HEC-HMS/ RAS: A case study for the San Antonio River Basin Summer 2002 storm event, J. Environ. Manage., 75, 325–336, 2005.

Koffler, D., and Laaha, G.: LFSTAT-an R-package for low-flow analysis, EGU General Assembly Conference Abstracts, 14, 8940, https://cran.r-project.org/web/packages/lfstat/index.html, 2012.

Komma, J., Reszler, C., Blöschl, G., and Haiden, T.: Ensemble prediction of floods – catchment non-linearity and forecast probabilities, Nat. Hazards Earth Syst. Sci., 7, 431–444, doi:10.5194/nhess-7-431-2007, 2007.

Krzysztofowicz, R.: Integrator of uncertainties for probabilistic river stage forecasting: precipitation-dependent model, J. Hydrol., 69–85, 2001a.

Krzysztofowicz, R.: The case for probabilistic forecasting in hydrology, J. Hydrol., 249, 2–9, 2001b.

Kull, D. W. and Feldman, A. D.: Evolution of Clark's unit graph method to spatially distributed runoff, J. Hydrol. Eng., 3, 9–19, 1998.

Kundzewicz, Z. W., Kanae, S., Seneviratne, S. I., Handmer, J., Nicholls, N., Peduzzi, P., Mechler, R., Bouwer, L. M., Arnell, N., and Mach, K.: Flood risk and climate change: global and regional perspectives, Hydrol. Sci. J., 59, 1–28, 2014.

Land surface cover from US Department of Agriculture National Resource Conservation Service (NRCS), https://gdg.sc.egov.usda.gov/, 2016.

Maidment, D. R.: Handbook of hydrology, McGraw-Hill Inc., 21.1–21.39, 1992.

Maidment, D. R., and Djokic, D.: Hydrologic and hydraulic modeling support: With geographic information systems, ESRI, Inc., p. 20, 2000.

McCollor, D., and Stull, R.: Hydrometeorological short-range ensemble forecasts in complex terrain. Part II: Economic evaluation, Weather Forecast., 23, 557–574, 2008.

McEnery, J., Ingram, J., Duan, Q., Adams, T., and Anderson, L.: NOAA'S Advanced Hydrologic Prediction Service: Building Pathways for Better Science in Water Forecasting, B. Am. Meteorol. Soc., 86, 375–385, 2005.

Meenu, R., Rehana, S., and Mujumdar, P. P.: Assessment of hydrologic impacts of climate change in Tunga-Bhadra river basin, India with HEC-HMS and SDSM, Hydrological Processes, 27, 1572–1589, 2013.

Mesinger, F., DiMego, G., Kalnay, E., Mitchell, K., Shafran, P. C., Ebisuzaki, W., Jović, D., Woollen, J., Rogers, E., Berbery, E. H., Ek, M. B., Fan, Y., Grumbine, R., Higgins, W., Li, H., Lin, Y., Manikin, G., Parrish, D., and Shi, W.: North American Regional Reanalysis, B. Am. Meteorol. Soc., 87, 343–360, http://nomads.ncdc.noaa.gov/data.php?name=access#narr_datasets, 2006.

Miller, D. A. and White, R. A.: A conterminous United States multilayer soil characteristics dataset for regional climate and hydrology modeling, Earth Interact., 2, 1–26, http://www.nrcs.usda.gov/wps/portal/nrcs/detail/soils/survey/geo/?cid=nrcs142p2_053629, 1998.

Milly, P. C. D., Wetherald, R. T., Dunne, K., and Delworth, T. L.: Increasing risk of great floods in a changing climate, Nature, 415, 514–517, 2002.

Mishra, S. K., and Singh, V.: Soil conservation service curve number (SCS-CN) methodology, Springer Science & Business Media, 278–322, 2013.

Molteni, F., Buizza, R., Palmer, T. N., and Petroliagis, T.: The ECMWF ensemble prediction system: Methodology and validation, Q. J. Roy. Meteor. Soc., 122, 73–119, 1996.

Montanari, A. and Grossi, G.: Estimating the uncertainty of hydrological forecasts: A statistical approach, Water Resour. Res., 44, W00B08, doi:10.1029/2008WR006897, 2008.

Mousavi, S. J., Abbaspour, K., Kamali, B., Amini, M., and Yang, H.: Uncertainty-based automatic calibration of HEC-HMS model using sequential uncertainty fitting approach, J. Hydroinform., 14, 286–309, 2012.

Nash, J. E. and Sutcliffe, J. V.: River flow forecasting through conceptual models part I – A discussion of principles, J. Hydrol., 10, 282–290, 1970.

Neary, V. S., Habib, E., and Fleming, M.: Hydrologic modeling with NEXRAD precipitation in Middle Tennessee, J. Hydrol. Eng., 9, 339–349, 2004.

NJOEM: New Jersey Office of Emergency Management, New Jersey State Hazard Mitigation Plan Technical Report 2014, 5.8-2. http://www.ready.nj.gov/programs/mitigation_plan2014.html (last access: 1 February 2016), 2014.

NOAA: National Oceanic and Atmospheric Administration, 2011m, National Weather Service, Advanced Hydrologic Prediction Center, daily precipitation totals for New York, 26–29 April, 28–29 August, and 6–9 September 2011, http://water.weather.gov/precip/ (last access: 29 January 2016), 2011.

NWS: National Weather Service Manual 10-950, Operations and Services Hydrologic Services Program, NWSPD 10-9, http://www.nws.noaa.gov/directives/, 4 December 2012.

NWS: United States Flood Loss Report – Water Year 2014 http://www.nws.noaa.gov/hic/summaries/WY2014.pdf (last access: 1 February 2016), 2014.

O'Connell, P. E., Nash, J. E., and Farrell, J. P.: River flow forecasting through conceptual models part II – The Brosna catchment at Ferbane, J. Hydrol., 10, 317–329, 1970.

Olsson, J. and Lindström, G.: Evaluation and calibration of operational hydrological ensemble forecasts in Sweden, J. Hydrol., 350, 14–24, 2008.

Pall, P., Aina, T., Stone, D. A., Stott, P. A., Nozawa, T., Hilberts, A. G., Lohmann, D., and Allen, M. R.: Anthropogenic greenhouse gas contribution to flood risk in England and Wales in autumn 2000, Nature, 470, 382–385, 2011.

Pappenberger, F., Beven, K. J., Hunter, N. M., Bates, P. D., Gouweleeuw, B. T., Thielen, J., and de Roo, A. P. J.: Cascading model uncertainty from medium range weather forecasts (10 days) through a rainfall-runoff model to flood inundation predictions within the European Flood Forecasting System (EFFS), Hydrol. Earth Syst. Sci., 9, 381–393, doi:10.5194/hess-9-381-2005, 2005.

Pappenberger, F., Bartholmes, J., Thielen, J., Cloke, H. L., Buizza, R. and de Roo, A., New dimensions in early flood warning across the globe using grand-ensemble weather predictions, Geophys. Res. Lett., 35, L10404, doi:10.1029/2008GL033837, 2008.

Pappenberger, F., Cloke, H., Persson, A., and Demeritt, D.: HESS Opinions "On forecast (in) consistency in a hydro-meteorological chain: curse or blessing?", Hydrol. Earth Syst. Sci., 15, 2391–2400, doi:10.5194/hess-15-2391-2011, 2011.

Pielke, R. A. and Downton, M. W.: Precipitation and Damaging Floods: Trends in the United States, J. Climate, 3625–3637, 2000. Pilgrim, D. H. and Cordery, I.: Flood runoff, chap. 9, edited by: Maidment, D. R., Handbook of hydrology, New York, McGraw-Hill, 9.1–9.42, 1993.

Pryet, A., Labarthe, B., Saleh, F., Akopian, M., and Flipo, N.: Reporting of stream-aquifer flow distribution at the regional scale with a distributed process-based model, Water Resour. Manage., 29, 139–159, 2015.

Pullen, J., Gordon, A. L., Flatau, M., Doyle, J. D., Villanoy, C., and Cabrera, O.: Multiscale influences on extreme winter rainfall in the Philippines, J. Geophys. Res.-Atmos., 120, 3292–3309, doi:10.1002/2014JD022645, 2015.

Qgis, D.: Quantum GIS geographic information system, Open Source Geospatial Foundation Project, http://qgis.osgeo.org, 2011.

Quintana Seguí, P., Ribes, A., Martin, E., Habets, F., and Boé, J.: Comparison of three downscaling methods in simulating the impact of climate change on the hydrology of Mediterranean basins, J. Hydrol., 383, 111–124, 2010.

Ramos, M. H., Bartholmes, J., and Thielen-del Pozo, J.: Development of decision support products based on ensemble forecasts in the European flood alert system, Atmos. Sci. Lett., 8, 113–119, 2007.

Reggiani, P., and Weerts, A.: Probabilistic quantitative precipitation forecast for flood prediction: An application, J. Hydrometeorol., 9, 76–95, 2008.

Renard, B., Kavetski, D., Kuczera, G., Thyer, M., and Franks, S. W.: Understanding predictive uncertainty in hydrologic modeling: The challenge of identifying input and structural errors, Water Resour. Res., 46, W05521, doi:10.1029/2009WR008328, 2010.

Saleh, F., Flipo, N., Habets, F., Ducharne, A., Oudin, L., Viennot, P., and Ledoux, E.: Modeling the impact of in-stream water level fluctuations on stream-aquifer interactions at the regional scale, J. Hydrol., 400, 490–500, 2011.

Saleh, F., Flipo, N., Biancamaria, S., Habets, F., Rodríguez, E., and Mognard, N.: The Contribution of the Future Mission SWOT to improve Simulations of River Stages and Stream-aquifer Interactions at Regional scale, 20 years of Progress in Radar Altimetry, Venice, Italy, http://www.altimetry2012.org/, 2012.

Saraiva Okello, A., Masih, I., Uhlenbrook, S., Jewitt, G., Van der Zaag, P., and Riddell, E.: Drivers of spatial and temporal variability of streamflow in the Incomati River basin, Hydrol. Earth Syst. Sci., 19, 657–673, 19, doi:10.5194/hess-19-657-2015, 2015.

Schaake, J. C., Hamill, T. M., Buizza, R., and Clark, M.: HEPEX: the hydrological ensemble prediction experiment, B. Am. Meteorol. Soc., 88, 1541–1547, 2007.

Scharffenberg, W.: Hydrologic Modeling System HEC-HMS User's Manual, 2015.

Scharffenberg, W., Duren, A., and Fleming, M.: Uncertainty Analysis Using Monte Carlo Techniques in the Hydrologic Modeling System, Joint Federal Interagency Conference, Reno, Nevada USA, http://www.hec.usace.army.mil/software/hec-hms/downloads.aspx, 2015.

Schellekens, J., Weerts, A., Moore, R., Pierce, C., and Hildon, S.: The use of MOGREPS ensemble rainfall forecasts in operational flood forecasting systems across England and Wales, Adv. Geosci., 29, 77–84, 2011.

Schoonover, J. E. and Lockaby, B. G.: Land cover impacts on stream nutrients and fecal coliform in the lower Piedmont of West Georgia, J. Hydrol., 331, 371–382, 2006.

Schuetz, T., Gascuel-Odoux, C., Durand, P., and Weiler, M.: Nitrate sinks and sources as controls of spatio-temporal water quality dynamics in an agricultural headwater catchment, Hydrol. Earth Syst. Sci., 20, 843–857, doi:10.5194/hess-20-843-2016, 2016.

Seyoum, M., van Andel, S. J., Xuan, Y., and Amare, K.: Precipitation forecasts for rainfall runoff predictions. A case study in poorly gauged Ribb and Gumara catchments, upper Blue Nile, Ethiopia, Phys. Chem. Earth, 61–62, 43–51, 2013.

Solaiman, T. A. and Simonovic, S. P.: National Centers for Environmental Prediction-National Center for Atmos. Res. (NCEP-NCAR) reanalyses data for hydrologic modelling on a basin scale, Canad. J. Civil Eng., 37, 611–623, 2010.

Thielen, J., Bartholmes, J., Ramos, M.-H., and de Roo, A.: The European Flood Alert System – Part 1: Concept and development, Hydrol. Earth Syst. Sci., 13, 125–140, doi:10.5194/hess-13-125-2009, 2009.

Thiemig, V., Bisselink, B., Pappenberger, F., and Thielen, J.: A pan-African medium-range ensemble flood forecast system, Hydrol. Earth Syst. Sci., 19, 3365–3385, doi:10.5194/hess-19-3365-2015, 2015.

USACE: Hydrologic Modeling System: User's Manual Version 4.1", US Army Corps of Engineers, Hydrologic Engineering Center, Davis, CA, USA, 2015.

USDA: Urban hydrology for small watersheds, Technical release, 55, 2–6, 1986.

Verkade, J. S. and Werner, M. G. F.: Estimating the benefits of single value and probability forecasting for flood warning, Hydrol. Earth Syst. Sci., 15, 3751–3765, doi:10.5194/hess-15-3751-2011, 2011.

Wang, L., Stuart, M. E., Bloomfield, J. P., Butcher, A. S., Gooddy, D. C., McKenzie, A. A., Lewis, M. A., and Williams, A. T.: Prediction of the arrival of peak nitrate concentrations at the water table at the regional scale in Great Britain, Hydrological Processes, 26, 226–239, 2012.

Watson, K. M., Collenburg, J. V., and Reiser, R. G.: Hurricane Irene and associated floods of August 27–30, 2011, in New Jersey, US Geological Survey 2328-0328, 2013.

Weaver, J., Harkabus, L. C., Braun, J., Miller, S., Cox, R., Griffith, J., and Mazur, R. J.: An Overview of a Demographic Study of United States Emergency Managers, B. Am. Meteorol. Soc., 95, 199–203, 2014.

Weerts, A. H., Winsemius, H. C., and Verkade, J. S.: Estimation of predictive hydrological uncertainty using quantile regression: examples from the National Flood Forecasting System (England and Wales), Hydrol. Earth Syst. Sci., 15, 255–265, doi:10.5194/hess-15-255-2011, 2011.

WMO: World Meteorological Organization, Our Future Climate, Technical Report 92-63-10952-4 ISBN 92-63-10952-4, 2003.

Yang, S.-C. and Yang, T.-H.: Uncertainty assessment: reservoir inflow forecasting with ensemble precipitation forecasts and HEC-HMS, Adv. Meteorol., 2014, 1–11, doi:10.1155/2014/581756, 2014.

Yapo, P. O., Gupta, H. V., and Sorooshian, S.: Automatic calibration of conceptual rainfall-runoff models: sensitivity to calibration data, J. Hydrol., 181, 23–48, 1996.

Younis, J., Ramos, M. H., and Thielen, J.: EFAS forecasts for the March–April 2006 flood in the Czech part of the Elbe River Basin—A case study, Atmos. Sci. Lett., 9, 88–94, 2008.

Zhang, H. L., Wang, Y. J., Wang, Y. Q., Li, D. X., and Wang, X. K.: The effect of watershed scale on HEC-HMS calibrated parameters: a case study in the Clear Creek watershed in Iowa, US, Hydrol. Earth Syst. Sci., 17, 2735–2745, doi:10.5194/hess-17-2735-2013, 2013.

12

Value of uncertain streamflow observations for hydrological modelling

Simon Etter[1], **Barbara Strobl**[1], **Jan Seibert**[1,2], **and H. J. Ilja van Meerveld**[1]

[1]Department of Geography, University of Zurich, Winterthurerstrasse 190, 8057 Zurich, Switzerland
[2]Department of Aquatic Sciences and Assessment, Swedish University of Agricultural Sciences, P.O. Box 7050, 75007 Uppsala, Sweden.

Correspondence: Simon Etter (simon.etter@geo.uzh.ch)

Abstract. Previous studies have shown that hydrological models can be parameterised using a limited number of streamflow measurements. Citizen science projects can collect such data for otherwise ungauged catchments but an important question is whether these observations are informative given that these streamflow estimates will be uncertain. We assess the value of inaccurate streamflow estimates for calibration of a simple bucket-type runoff model for six Swiss catchments. We pretended that only a few observations were available and that these were affected by different levels of inaccuracy. The level of inaccuracy was based on a log-normal error distribution that was fitted to streamflow estimates of 136 citizens for medium-sized streams. Two additional levels of inaccuracy, for which the standard deviation of the error distribution was divided by 2 and 4, were used as well. Based on these error distributions, random errors were added to the measured hourly streamflow data. New time series with different temporal resolutions were created from these synthetic streamflow time series. These included scenarios with one observation each week or month, as well as scenarios that are more realistic for crowdsourced data that generally have an irregular distribution of data points throughout the year, or focus on a particular season. The model was then calibrated for the six catchments using the synthetic time series for a dry, an average and a wet year. The performance of the calibrated models was evaluated based on the measured hourly streamflow time series. The results indicate that streamflow estimates from untrained citizens are not informative for model calibration. However, if the errors can be reduced, the estimates are informative and useful for model calibration. As expected, the model performance increased when the number of observations used for calibration increased. The model performance was also better when the observations were more evenly distributed throughout the year. This study indicates that uncertain streamflow estimates can be useful for model calibration but that the estimates by citizen scientists need to be improved by training or more advanced data filtering before they are useful for model calibration.

1 Introduction

The application of hydrological models usually requires several years of precipitation, temperature and streamflow data for calibration, but these data are only available for a limited number of catchments. Therefore, several studies have addressed the question: how many data points are needed to calibrate a model for a catchment? Yapo et al. (1996) and Vrugt et al. (2006), using stable parameters as a criterion for satisfying model performance, concluded that most of the information to calibrate a model is contained in 2–3 years of continuous streamflow data and that no more value is added when using more than 8 years of data. Perrin et al. (2007), using the Nash–Sutcliffe efficiency criterion (NSE), showed that streamflow data for 350 randomly sampled days out of a 39-year period were sufficient to obtain robust model parameter values for two bucket-type models, TOPMO, which is derived from TOPMODEL concepts (Michel et al., 2003), and GR4J (Perrin et al., 2003). Brath et al. (2004), using the volume error, relative peak error and time-to-peak error, concluded that at least 3 months of continuous data were required to obtain a reliable calibration. Other studies have shown that discontinuous streamflow data can be informative for constraining model parameters (Juston et al., 2009; Pool et al., 2017; Seibert and Beven, 2009; Seibert and McDonnell, 2015). Juston et al. (2009) used a multi-objective calibration that included groundwater data and concluded that the information content of a subset of 53 days of streamflow data was the same as for the 1065 days of data from which the subset was drawn. Seibert and Beven (2009), using the NSE criterion, found that model performance reached a plateau for 8–16 streamflow measurements collected throughout a 1-year period. They furthermore showed that the use of streamflow data for one event and the corresponding recession resulted in a similar calibration performance as the six highest measured streamflow values during a 2-month period.

These studies had different foci and used different model performance metrics, but nevertheless their results are encouraging for the calibration of hydrological models for ungauged basins based on a limited number of high-quality measurements. However, the question remains: how informative are low(er)-quality data? An alternative approach to high-quality streamflow measurements in ungauged catchments is to use citizen science. Citizen science has been proven to be a valuable tool to collect (Dickinson et al., 2010) or analyse (Koch and Stisen, 2017) various kinds of environmental data, including hydrological data (Buytaert et al., 2014). Citizen science approaches use simple methods to enable a large number of citizens to collect data and allow local communities to contribute data to support science and environmental management. Citizen science approaches can be particularly useful in light of the declining stream gauging networks (Ruhi et al., 2018; Shiklomanov et al., 2002) and to complement the existing monitoring networks. However, citizen science projects that collect streamflow or stream level data in flowing water bodies are still rare. Examples are the CrowdHydrology project (Lowry and Fienen, 2013), SmartPhones4Water in Nepal (Davids et al., 2018) and a project in Kenya (Weeser et al., 2018), which all ask citizens to read stream levels at staff gauges and to send these via an app or as a text message to a central database. Estimating streamflow is obviously more challenging than reading levels from a staff gauge but citizens can apply the stick or float method, where they measure the time it takes for a floating object (e.g. a small stick) to travel a given distance to estimate the flow velocity. Combined with estimates for the width and the average depth of the stream, this allows them to obtain a rough estimate of the streamflow. However, these streamflow estimates may be so inaccurate that they are not useful for model calibration. It is therefore necessary to not only evaluate the requirements of hydrological models in terms of the amount and temporal resolution of data, but also in terms of the achievable quality by the citizen scientists before starting a citizen science project.

The effects of rating curve uncertainty on model calibration (e.g. McMillan et al., 2010; Horner et al., 2018) and the value of sparse datasets (Davids et al., 2017) have been quantified in recent studies. However, the potential value of sparse datasets in combination with large uncertainties (such as those from crowdsourced streamflow estimates) has not been evaluated so far. Therefore, the aim of this study was to determine the effects of observation inaccuracies on the calibration of bucket-type hydrological models when only a limited number of observations are available. The specific objectives of this paper are to determine (i) whether the streamflow estimates from citizen scientists are informative for model calibration or if these errors need to be reduced (e.g. through training) to become useful and (ii) how the timing of the streamflow observations affects the calibration of a hydrological model. The latter is important for citizen science projects, as it provides guidance on whether it is useful to encourage citizens to contribute streamflow observations during a specific time of the year.

2 Methods

To assess the potential value of crowdsourced streamflow estimates for hydrological model calibration, the HBV (Hydrologiska Byråns Vattenbalansavdelning) model (Bergström, 1976) was calibrated against streamflow time series for six Swiss catchments, as well as for different subsets of the data that represent citizen science data in terms of errors and temporal resolution. Similar to the approach used in several recent studies (Ewen et al., 2008; Finger et al., 2015; Fitzner et al., 2013; Haberlandt and Sester, 2010; Seibert and Beven, 2009), we pretended that only a small subset of the data were available for model calibration. In addition, various degrees of inaccuracy were assumed. The value of these data for model calibration was then evaluated by comparing the model performance for these subsets of data to the performance of the model calibrated with the complete measured streamflow time series.

2.1 HBV model

The HBV model was originally developed at the Hydrologiska Byråns Vattenbalansavdelning unit at the Swedish Meteorological and Hydrological Institute (SMHI) by Bergström (1976). The HBV model is a bucket-type model that represents snow, soil, groundwater and stream routing processes in separate routines. In this study, we used the version HBV-light (Seibert and Vis, 2012).

2.2 Catchments

The HBV-light model was set up for six 24–186 km^2 catchments in Switzerland (Table 1 and Fig. 1). The catchments were selected based on the following criteria: (i) there is little anthropogenic influence, (ii) they are gauged at a single location, (iii) they have reliable streamflow data during high flow and low flow conditions (i.e. no complete freezing dur-

Table 1. Characteristics of the six Swiss catchments used in this study. For the location of the study catchments, see Fig. 1. Long-term averages are for the period 1974–2014, except for Verzasca for which the long-term average is for the 1990–2014 period. Regime types are classified according to Aschwanden and Weingartner (1985).

Catchment		Murg	Guerbe	Allenbach	Riale di Calneggia	Mentue	Verzasca
Gauging station (FOEN station number)		Waengi (2126)	Belp Mülimatt (2159)	Adelboden (2232)	Cavergno, Pontit (2356)	Yvonand La Mauguettaz (2369)	Lavertezzo, Campiòi (2605)
Area (km^2)		79	117	29	24	105	186
Elevation (m a.s.l.)	Min	465	522	1297	885	445	490
	Max	1035	2176	2762	2921	927	2864
Regime type		Pluvial-inférieur	Pluvial-superieur	Nival-alpin	Nival-méridional	Pluvial-jurassien	Nivo-pluvial-méridional
Min–max Pardé coefficients	Dry year	0.29–1.61	0.44–1.93	0.40–2.48	0.13–3.22	0.22–2.37	0.16–2.92
	Average year	0.58–2.16	0.61–1.65	0.39–2.44	0.09–2.84	0.23–2.66	0.23–3.17
	Wet year	0.34–1.69	0.42–2.14	0.32–2.12	0.10–3.48	0.35–2.39	0.26–2.64
	Long-term	0.68–1.34	0.77–1.39	0.35–2.70	0.14–2.70	0.46–1.57	0.23–2.22
Annual runoff : rainfall ratio	Dry year	0.72	0.37	0.86	1.30[1]	0.41	0.98
	Average year	0.55	0.48	1.73[1]	1.38[1]	0.52	0.66
	Wet year	0.56	0.54	0.78	0.98	0.50	1.32[1]
	Long-term	0.56	0.57	0.94	1.06[1]	0.38	0.9
Long-term mean annual streamflow (m^3 s^{-1})		1.84	2.75	1.23	1.43	1.64	10.76
Weather stations		Aadorf-Taenikon, Hörnli	Plaffeien, Bern-Zollikofen	Adelboden	Robiei	Mathod, Pully	Acquarossa, Cimetta, Magadino, Piotta

[1] In Verzasca, Allenbach and Riale die Calneggia there are some streamflow : rainfall ratios > 1 because the weather stations are located outside the catchment and precipitation is highly variable in alpine terrain.

ing winter and a cross section that allows accurate streamflow measurement at low flows) and (iv) there are no glaciers. The six selected catchments (Table 1) represent different streamflow regime types (Aschwanden and Weingartner, 1985). The snow-dominated highest elevation catchments (Allenbach and Riale di Calneggia) have the largest seasonality in streamflow, i.e. the biggest differences between the long-term maximum and minimum Pardé coefficients, followed by the rain- and snow-dominated Verzasca catchment. The rain-dominated catchments (Murg, Guerbe and Mentue) have the lowest seasonal variability in streamflow (Table 1). The mean elevation of the catchments varies from 652 to 2003 m a.s.l. (Table 1). The elevation range of each individual catchment was divided into 100 m elevation bands for the simulations.

2.3 Measured data

Hourly runoff time series (based on 10 min measurements) for the six study catchments were obtained from the Federal Office for the Environment (FOEN; see Table 1 for the gauging station numbers). The average hourly areal precipitation amounts were extracted for each study catchment from the gridded CombiPrecip dataset from MeteoSwiss (Sideris et al., 2014). This dataset combines gauge and radar precipitation measurements at an hourly timescale and 1 km^2 spatial resolution and is available for the time period since 2005.

We used hourly temperature data from the automatic monitoring network of MeteoSwiss (see Table 1 for the stations) and applied a gradient of −6 °C per 1000 m to adjust the temperature of each weather station to the mean elevation of the catchment. Within the HBV model, the temperature was then adjusted for the different elevation bands using a calibrated lapse rate.

As recommended by Oudin et al. (2005), potential evapotranspiration was calculated using the temperature-based potential evapotranspiration model of McGuinness and Bordne (1972) using the day of the year, the latitude and the temperature. This rather simplistic approach was considered

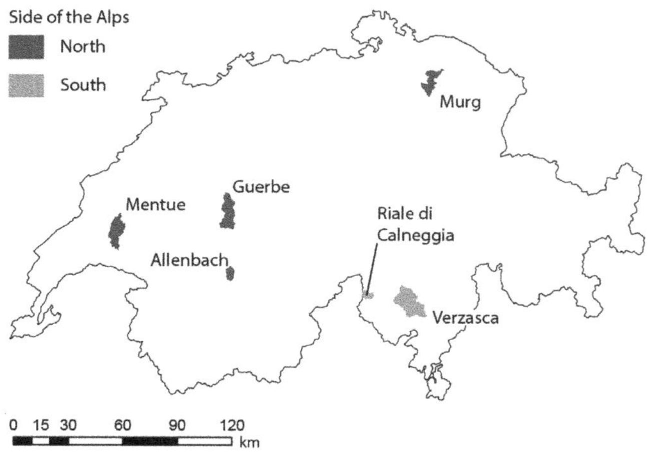

Figure 1. Location of the six study catchments in Switzerland. Shading indicates whether the catchment is located on the north or south side of the Alps. See Table 1 for the characteristics of the study catchments.

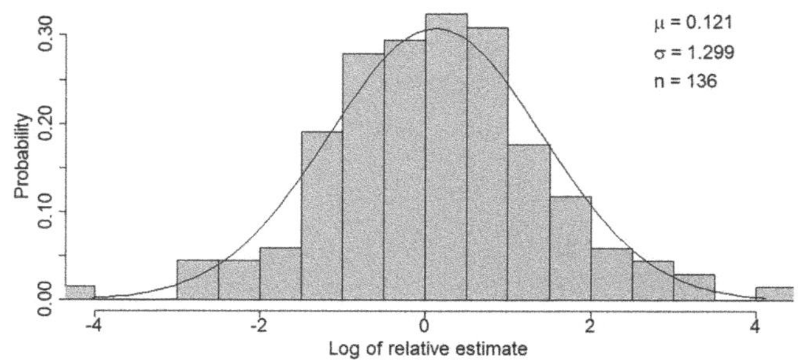

Figure 2. Fit of the normal distribution to the frequency distribution of the log-transformed relative streamflow estimates (ratio of the estimated streamflow and the measured streamflow).

sufficient because this study focused on differences in model performance relative to a benchmark calibration.

2.4 Selection of years for model calibration and validation

The model was calibrated for an average, a dry and a wet year to investigate the influence of wetness conditions and the amount of streamflow on the calibration results. The years were selected based on the total streamflow during summer (July–September). The driest and the wettest years of the period 2006–2014 were selected based on the smallest and largest sum of streamflow during the summer. The average streamflow years were selected based on the proximity to the mean summer streamflow for all the years 1974–2014 (1990–2014 for Verzasca). For each catchment the years that were the 2nd-closest to the mean summer streamflow for all years, as well as the years with the second lowest and second highest streamflow sum were chosen for model calibration (see Table 2). We did this separately for each catchment because for each catchment a different year was dry, average or wet. For the validation, we chose the year closest to the mean summer streamflow and the years with the lowest and the highest total summer streamflow (see Table 2). We used each of the parameter sets obtained from calibration for the dry, average or wet years to validate the model for each of the three validation years, resulting in nine validation combinations for each catchment (and each dataset, as described below).

2.5 Transformation of datasets to resemble citizen science data quality

2.5.1 Errors in crowdsourced streamflow observations

Strobl et al. (2018) asked 517 participants to estimate streamflow based on the stick method at 10 streams in Switzerland. Here we use the estimates for the medium-sized streams Töss, Sihl and Schanzengraben in the Canton of Zurich and the Magliasina in Ticino ($n = 136$), which had a similar streamflow range at the time of the estimations (2.6–28 $m^3 s^{-1}$) as the mean annual streamflow of the six streams used for this study (1.2–10.8 $m^3 s^{-1}$). We calculated the streamflow from the estimated width, depth and flow velocities using a factor of 0.8 to adjust the surface flow velocity to the average velocity (Harrelson et al., 1994). The resulting streamflow estimates were normalised by dividing them by the measured streamflow. We then combined the normalised estimates of all four rivers and log-transformed the relative estimates. A normal distribution with a mean of 0.12 and a standard deviation of 1.30 fits the distribution of the log-transformed relative estimates well (standard error of the mean: 0.11, standard error of the standard deviation: 0.08; Fig. 2).

To create synthetic datasets with data quality characteristics that represent the observed crowdsourced streamflow estimates, we assumed that the errors in the streamflow estimates are uncorrelated (as they are likely provided by different people). For each time step, we randomly selected a relative error value from the log-normal distribution of the relative estimates (Fig. 2) and multiplied the measured streamflow with this relative error. To simulate the effect of training and to obtain time series with different data quality, two additional streamflow time series were created using a standard deviation divided by 2 (standard deviation of 0.65) and by 4 (standard deviation of 0.33). This reduces the spread in the data (but does not change the small systematic overestimation of the streamflow), so large outliers are still possible, but are less likely. To summarise, we tested the following four cases.

Table 2. The calibration years (second most extreme and second closest to average years) and validation years (most extreme and closest to average years) for each catchment. The numbers in parentheses are the ranks over the period 1974–2014 (or 1990–2014 for Verzasca).

Year character	Murg	Guerbe	Allenbach	Riale di Calneggia	Mentue	Verzasca
Calibration						
Wet	2007 (3)	2007 (2)	2007 (4)	2009 (11)	2014 (7)	2011 (4)
Dry	2013 (8)	2011 (8)	2009 (11)	2012 (8)	2010 (4)	2013 (5)
Average	2008 (6)	2008 (17)	2013 (7)	2013 (2)	2006 (6)	2007 (7)
Validation						
Wet	2014 (1)	2014 (1)	2014 (1)	2008 (9)	2007 (1)	2008 (1)
Dry	2009 (7)	2013 (5)	2012 (9)	2006 (5)	2009 (3)	2010 (4)
Average	2011 (4)	2006 (13)	2011 (6)	2011 (1)	2013 (2)	2006 (4)

- *No error*: the data measured by the FOEN, assumed to be (almost) error-free, the benchmark in terms of quality.
- *Small error*: random errors according to the log-normal distribution of the snapshot campaigns with the standard deviation divided by 4.
- *Medium error*: random errors according to the log-normal distribution of the surveys with the standard deviation divided by 2.
- *Large error*: typical errors of citizen scientists, i.e. random errors according to the log-normal distribution of errors from the surveys.

2.5.2 Filtering of extreme outliers

Usually some form of quality control is used before citizen science data are analysed. Here, we used a very simple check to remove unrealistic outliers from the synthetic datasets. This check was based on the likely minimum and maximum streamflow for a given catchment area. We defined an upper limit of possible streamflow values as a function of the catchment area using the dataset of maximum streamflow from 1500 Swiss catchments provided by Scherrer AG, Hydrologie und Hochwasserschutz (2017). To account for the different precipitation intensities north and south of the Alps, different curves were created for the catchments on each side of the Alps. All streamflow observations, i.e. modified streamflow measurements, above the maximum observed streamflow for a particular catchment size including a 20 % buffer (Fig. S1), were replaced by the value of the maximum streamflow for a catchment of that size. This affected less than 0.5 % of all data points. A similar procedure was used for low flows based on a dataset of the FOEN with the lowest recorded mean streamflows over 7 days but this resulted in no replacements.

Table 3. Weights assigned to specific seasons, days and times of the day for the random selection of data points for Crowd52 and Crowd12. The weights for each hour were multiplied and normalised. We then used them as probabilities for the individual hours. For times without daylight the probability was set to zero.

Variable		Weight
Season		
December–February		2
March–May/September–November		6
June–August		10
Day		
Saturdays–Sundays		3
Monday–Friday		1
Time		
Times when people have breaks	06:00–08:00, 12:00–13:00, 17:00–21:00	3
Times with daylight in winter (December–February)	08:00–16:00	1
Times with daylight in spring/fall (March–May/September–November):	07:00–19:00	1
Times with daylight in summer (June–August)	06:00–21:00	1
Other times (depending on season)		0

2.5.3 Temporal resolution of the observations

Data entries from citizen scientists are not as regular as data from sensors with a fixed temporal resolution. Therefore, we decided to test eight scenarios with a different temporal resolution and distribution of the data throughout the year to simulate different patterns in citizen contributions.

- *Hourly*: one data point per hour ($8760 \leq n \leq 8784$, depending on the year).

- *Weekly*: one data point per week, every Saturday, randomly between 06:00 and 20:00 ($52 \leq n \leq 53$).
- *Monthly*: one data point per month on the 15th of the month, randomly between 06:00 and 20:00 ($n = 12$).
- *IntenseSummer*: one data point every other day from July until September, randomly between 06:00 and 20:00 ($\sim$ 15 observations per month, $n = 46$).
- *WeekendSummer*: one data point each Saturday and each Sunday between May and October, randomly between 06:00 and 20:00 ($52 \leq n \leq 54$).
- *WeekendSpring*: one data point on each Saturday and each Sunday between March and August, randomly between 06:00 and 20:00 ($52 \leq n \leq 54$).
- *Crowd52*: 52 random data points during daylight (in order to be comparable to the Weekly, IntenseSummer, WeekendSummer and WeekendSpring time series).
- *Crowd12*: 12 random data points during daylight (comparable to the Monthly data).

Except for the hourly data, these scenarios were based on our own experiences within the CrowdWater project (https://www.crowdwater.ch, last access: 3 October 2018) and information from the CrowdHydrology project (Lowry and Fienen, 2013). The hourly dataset was included to test the effect of errors when the temporal resolution of the data is optimal (i.e. by comparing simulations for the models calibrated with the hourly FOEN data and those calibrated with hourly data with errors). In the two scenarios Crowd52 and Crowd12, with random intervals between data points, we assigned higher probabilities for periods when people are more likely to be outdoors (i.e. higher probabilities for summer than winter, higher probabilities for weekends than weekdays, higher probabilities outside office hours; Table 3). Times without daylight (dependent on the season) were always excluded. We used the same selection of days, including the same times of the day for each of the four different error groups, years and catchments to allow comparison of the different model results.

2.6 Model calibration

For each of the 1728 cases (6 catchments, 3 calibration years, 4 error groups, 8 temporal resolutions), the HBV model was calibrated by optimising the overall consistency performance P_{OA} (Finger et al., 2011) using a genetic optimisation algorithm (Seibert, 2000). The overall consistency performance P_{OA} is the mean of four objective functions with an optimum value of 1: (i) NSE, (ii) the NSE for the logarithm of streamflow, (iii) the volume error and (iv) the mean absolute relative error (MARE). The parameters were calibrated within their typical ranges (see Table S1 in the Supplement.).

To consider parameter uncertainty, the calibration was performed 100 times, which resulted in 100 parameter sets for each case. For each case, the preceding year was used for the warm-up period. For the Crowd52 and Crowd12 time series, we used 100 different random selections of times, whereas for the regularly spaced time series the same times were used for each case.

2.7 Model validation and analysis of the model results

The 100 parameters from the calibration for each case were used to run the model for the validation years (Table 2). For each case (i.e. each catchment, year, error magnitude and temporal resolution), we determined the median validation P_{OA} for the 100 parameter sets for each validation year. We analysed the validation results of all years combined and for all nine combinations of dry, mean and wet years separately.

Because the focus of this study was on the value of limited inaccurate streamflow observations for model calibration, i.e. the difference in the performance of the models calibrated with the synthetic data series compared to the performance of the models calibrated with hourly FOEN data, all model validation performances are expressed relative to the average P_{OA} of the model calibrated with the hourly FOEN data (our upper benchmark, representing the fully informed case when continuous high quality streamflow data are available). A relative P_{OA} of 1 indicates that the model performance is as good as the performance of the model calibrated with the hourly FOEN data, whereas lower P_{OA} values indicate a poorer performance.

In humid climates, the input data (precipitation and temperature) often dictate that model simulations can not be too far off as long as the water balance is respected (Seibert et al., 2018). To assess the value of limited inaccurate streamflow data for model calibration compared to a situation without any streamflow data, a lower benchmark (Seibert et al., 2018) was used. Here, the lower benchmark was defined as the median performance of the model ran with 1000 random parameters sets. By running the model with 1000 randomly chosen parameter sets, we represent a situation where no streamflow data for calibration are available and the model is driven only by the temperature and precipitation data. We used 1000 different parameter sets to cover most of the model variability due to the different parameter combinations. The Mann–Whitney U test was used to evaluate whether the median P_{OA} for a specific error group and temporal resolution of the data was significantly different from the median P_{OA} for the lower benchmark (i.e. the model runs with random parameters). We furthermore checked for differences in model performance for models calibrated with the same data errors but different temporal resolutions using a Kruskal–Wallis test. By applying a Dunn–Bonferroni post hoc test (Bonferroni, 1936; Dunn, 1959, 1961), we analysed which of the validation results were significantly different from each other.

The random generation of the 100 crowdsourced-like datasets (i.e. the Crowd52 and Crowd12 scenario) for each of the catchments and year characteristics resulted in time series with a different number of high flow estimates. In order to find out whether the inclusion of more high flow values resulted in a better validation performance, we defined the threshold for high flows as the streamflow value that was exceeded 10 % of the time in the hourly FOEN streamflow dataset. The Crowd52 and Crowd12 datasets were then divided into a group that had more than the expected 10 % high flow observations and a group that had fewer high flow observations. To determine if more high flow data improve model performance, the Mann–Whitney U test was used to compare the relative median P_{OA} of the two groups.

3 Results

3.1 Upper benchmark results

The model was able to reproduce the measured streamflow reasonably well when the complete and unchanged hourly FOEN datasets were used for calibration, although there were also a few exceptions. The average validation P_{OA} was 0.61 (range: 0.19–0.83; Table 4). The validation performance was poorest for the Guerbe (validation $P_{OA} = 0.19$) because several high flow peaks were missed or underestimated by the model for the wet validation year. Similarly, the validation for the Mentue for the dry validation year resulted in a low P_{OA} (0.23) because a very distinct peak at the end of the year was missed and summer low flows were overestimated. The third lowest P_{OA} value was also for the Guerbe (dry validation year) but already had a P_{OA} of 0.35. Six out of the nine lowest P_{OA} values were for dry validation years. Validation for wet years for the models calibrated with data from wet years resulted in the best validation results (i.e. highest P_{OA} values; Table 4).

3.2 Effect of errors on the model validation results

Not surprisingly, increasing the errors in the streamflow data used for model calibration led to a decrease in the model performance (Fig. 4). For the small error category, the median validation performance was better than the lower benchmark for all temporal resolutions (Fig. 4 and Table S2). For the medium error category, the median validation performance was also better than the lower benchmark for all scenarios, except for the Crowd12 dataset. For the model calibrated with the dataset with large errors, only the Hourly dataset was significantly better than the lower benchmark (Table 5).

3.3 Effect of the data resolution on the model validation results

The Hourly measurement scenario resulted in the best validation performance for each error group, followed by the Weekly data, and then usually the Crowd52 data (Fig. 4). Although the median validation performance of the models calibrated with the Weekly datasets was better than for the Crowd52 dataset for all error cases, the difference was only statistically significant for the no error category (Fig. 5).

The validation performance of the models calibrated with the Weekly and Crowd52 datasets was better than for the scenarios focused on spring and summer observations (WeekendSpring, WeekendSummer and IntenseSummer). The median model performance for the Weekly dataset was significantly better than the datasets focusing on spring and summer for the no, small and medium error groups. The median performance of the Crowd52 dataset was only significantly better than all three measurement scenarios focusing on spring or summer for the small error case (Fig. 5). The model validation performance for the WeekendSummer and IntenseSummer scenarios decreased faster with increasing errors compared to the Weekly, Crowd52 or WeekendSpring datasets (Fig. 4). The median validation P_{OA} for the models calibrated with the WeekendSpring observations was better than for the models calibrated with the WeekendSummer and IntenseSummer datasets but the differences were only significant for the small, medium and large error groups. The differences in the model performance results for the observation strategies that focussed on summer (IntenseSummer and WeekendSummer) were not significant for any of the error groups (Fig. 5).

The median model performance for the regularly spaced Monthly datasets with 12 observations was similar to the median performance for the three datasets focusing on summer with 46–54 measurements (WeekendSpring, WeekendSummer and IntenseSummer), except for the case of large errors for which the monthly dataset performed worse. The irregularly spaced Crowd12 time series resulted in the worst model performance for each error group, but the difference from the performance for the regularly spaced Monthly data was only significant for the dataset with large errors (Fig. 5).

3.4 Effect of errors and data resolution on the parameter ranges

For most parameters the spread in the optimised parameter values was smallest for the upper benchmark. The spread in the parameter values increased with increasing errors in the data used for calibration, particularly for MAXBAS (the routing parameter) but also for some other parameters (e.g. TCALT, TT and BETA). However, for some parameters (e.g. CFMAX, FC, and SFCF) the range in the optimised parameter values was mainly affected by the temporal resolution of the data and the number of data points used for calibration. It should be noted though that the changes in the range of model parameters differed significantly for the different catchments and the trends were not very clear.

Table 4. Median and the full range of the overall consistency performance P_{OA} scores for the upper benchmark (hourly FOEN data). The P_{OA} values for the dry, average and wet calibration years were used as the upper benchmarks for the evaluation based on the year character (Figs. 6 and S2 in the Supplement); the values in the "overall median" column were used as the benchmarks in the overall median performance evaluation shown in Fig. 4.

Calibration year	Dry	Average	Wet	Overall median
Validation wet year				
Upper benchmark	0.63 (0.19–0.79)	0.65 (0.36–0.8)	0.66 (0.45–0.8)	
Lower benchmark		0.34 (−0.02–0.47)		Upper benchmark
Validation average year				0.61
Upper benchmark	0.59 (0.49–0.64)	0.61 (0.45–0.78)	0.53 (0.36–0.77)	(0.19–0.83)
Lower benchmark		0.36 (0.03–0.59)		Lower benchmark 0.34
Validation dry year				(−0.02–0.59)
Upper benchmark	0.51 (0.35–0.71)	0.59 (0.41–0.83)	0.53 (0.23–0.74)	
Lower benchmark		0.35 (0.09–0.52)		

3.5 Influence of the calibration and validation year and number of high flow data points on the model performance

The influence of the validation year on the model performance was larger than the effect of the calibration year (Figs. 6 and S2). In general model performance was poorest for the dry validation years. The model performances of all datasets with fewer observations or bigger errors than the Hourly datasets without errors were not significantly better than the lower benchmark for the dry validation years, except for Crowd52 in the no error group when calibrated with data from a wet year. However, even for the wet validation years some observation scenarios of the no error and small error group did not lead to significantly better model validation results compared to the median validation performance for the random parameters. Interestingly, the IntenseSummer dataset in the no error group resulted in a very good performance when the model was calibrated for a dry year and also validated in a dry year compared to its performance in the other calibration and validation year combinations. The median model performance was however not significantly better than the lower benchmark due to the low performance for the Guerbe and Allenbach (outliers beyond figure margins in Fig. 6). The validation results for these two catchments were the worst for all the no error–IntenseSummer datasets for all calibration and validation year combinations.

For 13 out of the 18 catchment and year combinations, the Crowd52 datasets with fewer than 10 % high streamflow data points led to a better validation performance than the Crowd52 datasets with more high streamflow data points. For six of them, the difference in model performance was significant. For none of the five cases where more high flow data points led to a better model performance was the difference significant. Also when the results were analysed by year character or catchment, there was no improvement when more high flow values were included in the calibration dataset.

4 Discussion

4.1 Usefulness of inaccurate streamflow data for hydrological model calibration

In this study, we evaluated the information content of streamflow estimates by citizen scientists for calibration of a bucket-type hydrological model for six Swiss catchments. While the hydroclimatic conditions, the model or the calibration approach might be different in other studies, these results should be applicable for a wide range of cases. However, for physically based spatially distributed models that are usually not calibrated automatically, the use of limited streamflow data would probably benefit from a different calibration approach. Furthermore, our results might not be applicable in arid catchments where rivers become dry for some periods of the year because the linear reservoirs used in the HBV model are not appropriate for such systems.

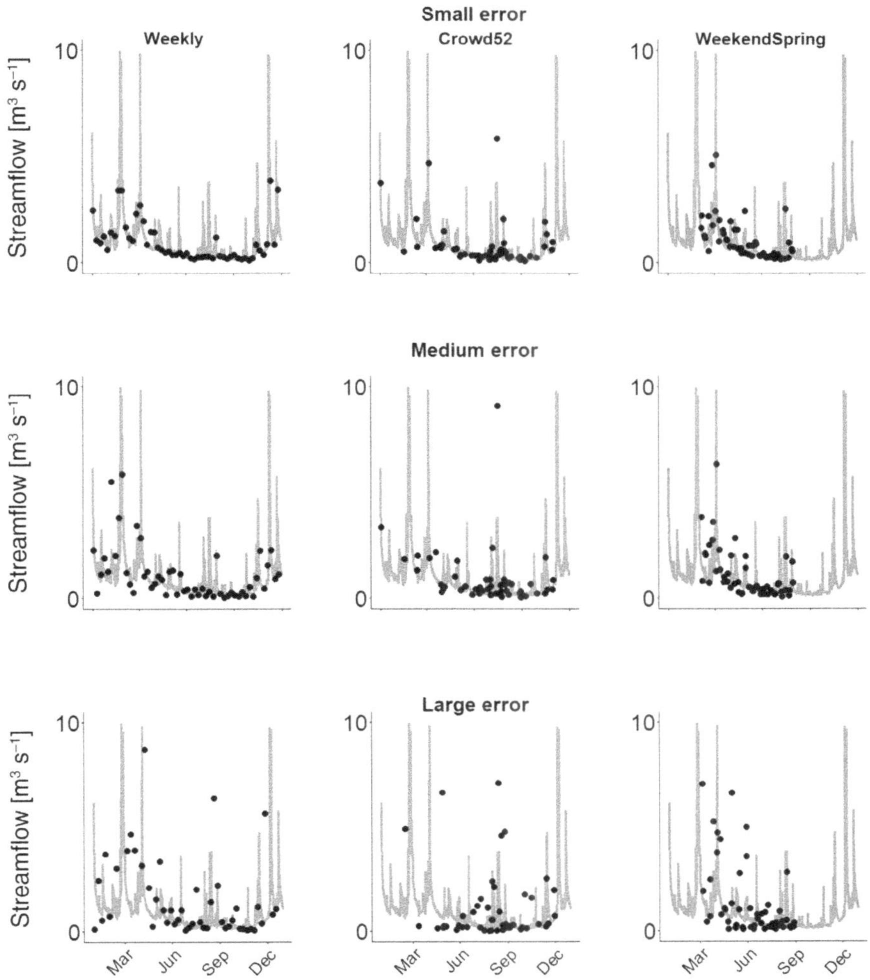

Figure 3. Examples of streamflow time series used for calibration with small, medium and large errors and different temporal resolutions (Weekly, Crowd52 and WeekendSpring) for the Mentue in 2010. Large error: adjusted FOEN data with errors resulting from the log-normal distribution fitted to the streamflow estimates from citizen scientists (see Fig. 2). Medium error: same as large error, but the standard deviation of the log-normal distribution was divided by 2. Small error: same as the large error, but the standard deviation of the log-normal distribution was divided by 4. The grey line represents the measured streamflow, and the dots the derived time series of streamflow observations. Note that especially in the large error category some dots lie outside the figure margins.

Streamflow estimates by citizens are sometimes very different from the measured values, and the individual estimates can be disinformative for model calibration (Beven, 2016; Beven and Westerberg, 2011). The results show that if the streamflow estimates by citizen scientists were available at a high temporal resolution (hourly), these data would still be informative for the calibration of a bucket-type hydrological model despite their high uncertainties. However, observations with such a high resolution are very unlikely to be obtained in practice. All scenarios with error distributions that represent the estimates from citizen scientists with fewer observations were no better than the lower benchmark (using random parameters). With medium errors, however, and one data point per week on average or regularly spaced monthly data, the data were informative for model parameterisation. Reducing the standard deviation of the error distribution by a factor of 4 led to a significantly improved model performance compared to the lower benchmark for all the observation scenarios.

A reduction in the errors of the streamflow estimates could be achieved by training of citizen scientists (e.g. videos), improved information about feasible ranges for stream depth, width and velocity, or examples of streamflow values for well-known streams. Filtering of extreme outliers can also reduce the spread of the estimates. This could be done with existing knowledge of feasible streamflow values for a catchment of a given area or the amount of rainfall right before the estimate is made to determine if streamflow is likely to be higher or lower than for the previous estimate. More de-

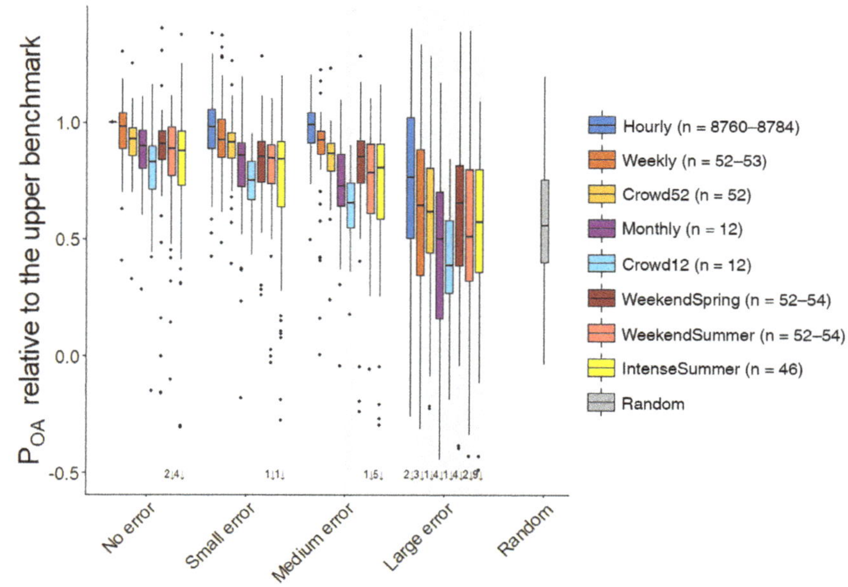

Figure 4. Box plots of the median model performance relative to the upper benchmark for all datasets. The grey rectangles around the boxes indicate non-significant differences in median model performance compared to the lower benchmark with random parameter sets. The box represents the 25th and 75th percentile, the thick horizontal line represents the median, the whiskers extend to 1.5 times the interquartile range below the 25th percentile and above the 75th percentile and the dots represent the outliers. The numbers at the bottom indicate the number of outliers beyond the figure margins; *n* is the number of streamflow observations used for model calibration. The result of the hourly benchmark FOEN dataset has some spread because the results of the 100 parameters sets were divided by their median performance. A relative P_{OA} of 1 indicates that the model performance is as good as the performance of the model calibrated with the hourly FOEN data (upper benchmark).

tailed research is necessary to test the effectiveness of such methods.

Le Coz et al. (2014) reported an uncertainty in stage–discharge streamflow measurements of around 5 %–20 %. McMillan et al. (2012) summarised streamflow uncertainties from stage–discharge relationships in a more detailed review and gave a range of ±50 %–100 % for low flows, ±10 %–20 % for medium or high (in-bank) flows and ±40 % for out-of-bank flows. The errors for the most extreme outliers in the citizen estimates are considerably higher, and could differ up to a factor of 10 000 from the measured value in the most extreme but rare cases (Fig. 2). Even with reduced standard deviations of the error distribution by a factor of 2 or 4, the observations in the most extreme cases can still differ by a factor of 100 and 10. The percentage of data points that differed from the measured value by more than 200 % was 33 % for the large error group, 19 % for the medium error group and 4 % for the small error group. Only 3 % of the data points were more than 90 % below the measured value in the large error group and 0 % for both in the medium and small error classes. If such observations are used for model calibration without filtering, they are seen as extreme floods or droughts, even if the actual conditions may be close to average flow. Beven and Westerberg (2011) suggest isolating periods of disinformative data. It is therefore beneficial to identify such extreme outliers, independent of a model, e.g. with knowledge of feasible maximum and minimum streamflow quantities, as used in this study, with the help of the maximum regionalised specific streamflow values for a given catchment area.

4.2 Number of streamflow estimates required for model calibration

In general, one would assume that the calibration of a model becomes better when there are more data (Perrin et al., 2007), although others have shown that the increase in model performance plateaus after a certain number of measurements (Juston et al., 2009; Pool et al., 2017; Seibert and Beven, 2009; Seibert and McDonnell, 2015). In this study, we limited the length of the calibration period to 1 year because in practice it may be possible to obtain a limited number of measurements during a 1-year period for ungauged catchments before the model results are needed for a certain application, as has been assumed in previous studies (Pool et al., 2017; Seibert and McDonnell, 2015). While a limited number of observations (12) was informative for model calibration when the data uncertainties were limited, the results of this study also suggest that the performance of bucket-type models decreases faster with increasing errors when fewer data points are available (i.e. there was a faster decline in model performance with in-

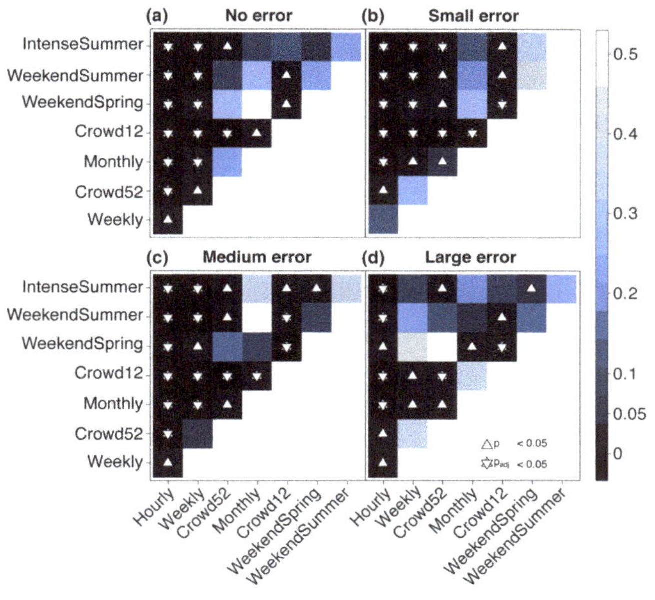

Figure 5. Results (p values) of the Kruskal–Wallis with Bonferroni post hoc test to determine the significance of the difference in the median model performance for the data with different temporal resolutions within each data quality group (no error **a**, small error **b**, medium error **c**, and large error **d**). Blue shades represent the p values. White triangles indicate p values < 0.05 and white stars indicate p values that, when adjusted for multiple comparisons, are still < 0.05.

creasing errors for models calibrated with 12 data points than for the models calibrated with 48–52 data points). This finding was most pronounced when comparing the model performance for the small and medium error groups (Fig. 4). These findings can be explained by the compensating effect of the number of observations and their accuracy because the random errors for the inaccurate data average out when a large number of observations are used, as long as the data do not have a large bias.

4.3 Best timing of streamflow estimates for model calibration

The performance of the parameter sets depended on the timing and the error distribution of the data used for model calibration. The model performance was generally better if the observations were more evenly spread throughout the year. For example, for the cases of no and small errors, the performance of the model calibrated with the Monthly dataset with 12 observations was better than for the IntenseSummer and WeekendSummer scenarios with 46–54 observations. Similarly, the less clustered observation scenarios performed better than the more clustered scenarios (i.e. Weekly vs. Crowd52, Monthly vs. Crowd12, Crowd52 vs. IntenseSummer, etc.). This suggests that more regularly distributed data over the year lead to a better model calibration. Juston et al. (2009) compared different subsamples of hydrological data for a 5.6 km^2 Swedish catchment and found that including inter-annual variability in the data used for the calibration of the HBV model reduced the model uncertainties. More evenly distributed observations throughout the year might represent more of the within-year streamflow variability and therefore result in improved model performance. This is good news for using citizen science data for model calibration as it suggests that the timing is not as important as the number of observations because it is likely much easier to get observations throughout the year than during specific periods or flow conditions.

When comparing the WeekendSpring, WeekendSummer and IntenseSummer datasets, it seems that it was in most cases more beneficial to include data from spring rather than summer. This tendency was more pronounced with increasing data errors. The reason for this might be that the WeekendSpring scenario includes more snowmelt or rain-on-snow event peaks, in addition to usually higher baseflow, and therefore contains more information on the inter-annual variability in streamflow.

By comparing different variations of 12 data points to calibrate the HBV model, Pool et al. (2017) found that a dataset that contains a combination of different maximum (monthly, yearly etc.) and other flows in model calibration led to the best model performance but also that the differences in performance for the different datasets covering the range of flows were small. In our study we did not specifically focus on the high or low flow data points, and therefore did not have datasets that contained only high flow estimates, which would be very difficult to obtain with citizen science data. However, our findings similarly show that for model calibration for catchments with seasonal variability in streamflow it is beneficial to obtain data for different magnitudes of flow. Furthermore, we found that data points during relatively dry periods are beneficial for validation or prediction in another year and might even be beneficial for years with the same characteristics, as was shown for the improved validation performance of the IntenseSummer dataset compared to the other datasets when data from dry years were used for calibration (Fig. 6).

4.4 Effects of different types of years on model calibration and validation

The calibration year, i.e. the year in which the observations were made, was not decisive for the model performance. Therefore, a model calibrated with data from a dry year can still be useful for simulations for an average or wet year. This also means that data in citizen science projects can be collected during any year and that these data are useful for simulating streamflow for most years, except the driest years. However, model performance did vary significantly for the different validation years. The results during dry validation years were almost never significantly better than the

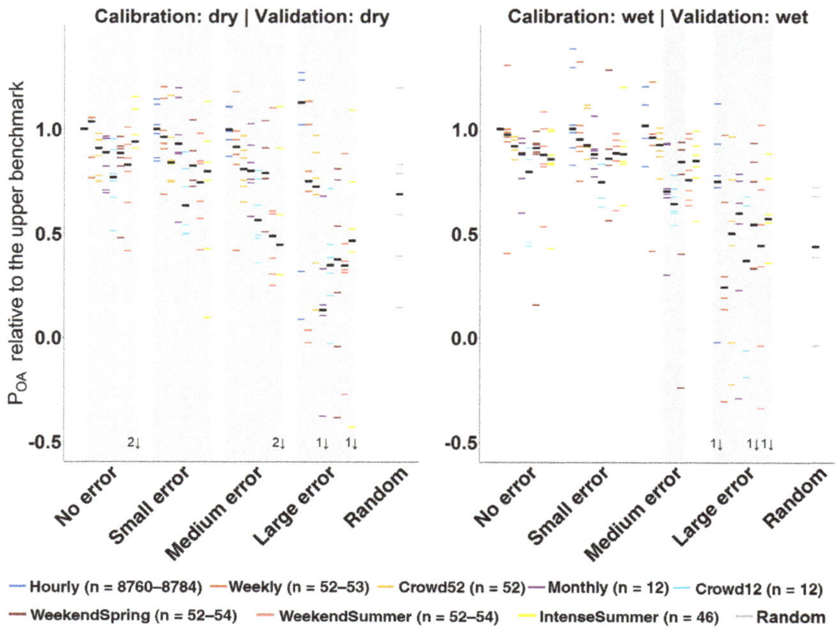

Figure 6. Median model validation performance for the datasets calibrated and validated both in a dry year and in a wet year. Each horizontal line represents the median model performance for one catchment. The black bold line represents the median for the six catchments. The grey rectangles around the boxes indicate non-significant differences in median model performance for the six catchments compared to the lower benchmark with random parameters. The numbers at the bottom indicate the number of outliers beyond the figure margins. For the individual P_{OA} values of the upper benchmark (no error–Hourly dataset) in the different calibration and validation years, see Table 4.

lower benchmark (Fig. S2). This might be due to the objective function that was used in this study. Especially the NSE was lower for dry years because the flow variance (i.e. the denominator in the equation) is smaller when there is a larger variation in streamflow. Also, these results are based on six median model performances, and therefore, outliers have a big influence on the significance of results (Fig. S2).

Lidén and Harlin (2000) used the HBV-96 model by Lindström et al. (1997) with changes suggested by Bergström et al. (1997) for four catchments in Europe, Africa and South America. They achieved better model results for wetter catchments and argued that during dry years evapotranspiration plays a bigger role and therefore the model performance is more sensitive to inaccuracies in the simulation of the evapotranspiration processes. The fact that we used a very simple method to calculate the potential evapotranspiration (McGuinness and Bordne, 1972) might also explain why the model performed less well during dry years.

The model parameterisation, obtained from calibration using the IntenseSummer dataset, resulted in a surprisingly good performance for the validation for a more extreme dry year for four out of the six catchments. For the two catchments for which the performance for the IntenseSummer dataset was poor (Guerbe and Allenbach), the weather stations are located outside the catchment boundaries. Especially during dry periods missed streamflow peaks due to misrepresentation of precipitation can affect model performance a lot. The fact that always one of these two catchments had the worst model performance for all the no error–IntenseSummer runs furthermore indicates that the July–September period might not be suitable to represent characteristic runoff events for these catchments. The bad performance for these two catchments for the IntenseSummer–no error run with calibration and validation in the dry year resulted in the insignificant improvement in model performance compared to the lower benchmark. Because the wetness of a year was based on the summer streamflow, these findings suggest that data obtained during times of low flow result in improved validation performance during dry years compared to data collected during other times (Fig. S2). This suggests that if the interest is in understanding the streamflow response during very dry years, it is important to obtain data during the dry period. To test this hypothesis, more detailed analyses are needed.

4.5 Recommendations for citizen science projects

Our results show that streamflow estimates from citizens are not informative for hydrological model calibration, unless the errors in the estimates can be reduced through training or advanced filtering of the data to reduce the errors (i.e. to reduce the number of extreme outliers). In order to make streamflow estimates useful, the standard deviation of the error distribution of the estimates needs to be reduced by a factor of 2. Gibson and Bergman (1954) suggest that errors in distance estimates can be reduced from 33 % to 14 % with very little training. These findings are encouraging, although their tests covered distances larger than 365 m (400 yards) and the widths of the medium-sized rivers for which the streamflow was estimated were less than 40 m (Strobl et al., 2018). Options for training might be tutorial videos, as well as lists with values for the width, average depth and flow velocity of well-known streams (Strobl et al., 2018). In order to determine the effect of training on streamflow estimates, further research has to be done because especially the depth estimates were inaccurate (Strobl et al., 2018).

The findings of this study suggest the following recommendations for citizen science projects that want to use streamflow estimates:

- *Collect as many data points as possible.* In this study hourly data always led to the best model performance. It is therefore beneficial to collect as many data points as possible. Because it is unlikely that hourly data are obtained, we suggest to aim for (on average) one observation per week. Provided that the standard deviation of the streamflow estimates can be reduced by a factor of 2, 52 observations (as in the Crowd52 data series) are informative for model calibration. Therefore, it is essential to invest in advertisement of a project and to find suitable locations where many people can potentially contribute, as well as to communicate to the citizen scientists that it is beneficial to submit observations regularly.

- *Encourage observations throughout the year.* To further improve the model performance, or to allow for greater errors, it is beneficial to have observations at all types of flow conditions during the year, rather than during a certain season.

Observations during high streamflow conditions were in most cases not more informative than flows during other times of the year. Efforts to ask citizens to submit observations during specific flow conditions (e.g. by sending reminders to the citizen observers) do not seem to be very effective in light of the above findings. It is rather more beneficial to remind them to submit observations regularly.

Instead of focussing on training to reduce the errors in the streamflow estimates, an alternative approach for citizen science projects is to switch to a parameter that is easier to estimate, such as stream levels (Lowry and Fienen, 2013). Recent studies successfully used daily stream-level data (Seibert and Vis, 2016) and stream-level class data (van Meerveld et al. 2017) to calibrate hydrological models, and other studies have demonstrated the potential value of crowdsourced stream level data for providing information on, e.g. baseflow (Lowry and Fienen, 2013), or for improving flood forecasts (Mazzoleni et al., 2017). However, further research is needed to determine if real crowdsourced stream-level (class) data are informative for the calibration of hydrological models.

5 Conclusions

The results of this study extend previous studies on the value of limited hydrological data for hydrological model calibration or the best timing of streamflow measurements for model calibration (Juston et al., 2009; Pool et al., 2017; Seibert and McDonnell, 2015) that did not consider observation errors. This is an important aspect, especially when considering citizen science approaches to obtain streamflow data. Our results show that inaccurate streamflow data can be useful for model calibration, as long as the errors are not too large. When the distribution of errors in the streamflow data represented the distribution of the errors in the streamflow estimates from citizen scientists, this information was not informative for model calibration (i.e. the median performance of the models calibrated with these data was not significantly better than the median performance of the models with random parameter values). However, if the standard deviation of the estimates is reduced by a factor of 2, then the (less) inaccurate data would be informative for model calibration. We furthermore demonstrated that realistic frequencies for citizen science projects (one observation on average per week or month) can be informative for model calibration. The findings of studies such as the one presented here provide important guidance on the design of citizen science projects as well as other observation approaches.

Author contributions. While JS and IvM had the initial idea, the concrete study design was based on input from all authors. SE and BS conducted the field surveys to determine the typical errors in streamflow estimates. The simulations and analyses were performed by SE. The writing of the manuscript was led by SE; all co-authors contributed to the writing.

Competing interests. The authors declare that they have no conflict of interest.

Acknowledgements. We thank all citizen scientists who participated in the field surveys, as well as the Swiss Federal Office for the Environment for providing the streamflow data, MeteoSwiss for providing the weather data, Maria Staudinger, Jan Schwanbeck and Scherrer AG for the permission to use their datasets and the reviewers for the useful comments. This project was funded by the Swiss National Science Foundation (project CrowdWater).

Edited by: Nadav Peleg

References

Aschwanden, H. and Weingartner, R.: Die Abflussregimes der Schweiz, Geographisches Institut der Universität Bern, Abteilung Physikalische Geographie, Gewässerkunde, Bern, Switzerland, 1985.

Bergström, S.: Development and application of a conceptual runoff model for Scandinavian catchments, Sveriges Meteorologiska och Hydrologiska Institut (SMHI), Norrköping, Sweden, available at: https://www.researchgate.net/publication/255274162_Development_and_Application_of_a_Conceptual_Runoff_Model_for_Scandinavian_Catchments (last access: 3 October 2018), 1976.

Bergström, S., Carlsson, B., Grahn, G., and Johansson, B.: A More Consistent Approach to Watershed Response in the HBV Model, Vannet i Nord., 4, 1997.

Beven, K.: Facets of uncertainty: epistemic uncertainty, non-stationarity, likelihood, hypothesis testing, and communication, Hydrol. Sci. J., 61, 1652–1665, https://doi.org/10.1080/02626667.2015.1031761, 2016.

Beven, K. and Westerberg, I.: On red herrings and real herrings: disinformation and information in hydrological inference, Hydrol. Process., 25, 1676–1680, https://doi.org/10.1002/hyp.7963, 2011.

Bonferroni, C. E.: Teoria statistica delle classi e calcolo delle probabilità, st. Super. di Sci. Econom. e Commerciali di Firenze, Istituto superiore di scienze economiche e commerciali, Florence, Italy, 62 pp., 1936.

Brath, A., Montanari, A., and Toth, E.: Analysis of the effects of different scenarios of historical data availability on the calibration of a spatially-distributed hydrological model, J. Hydrol., 291, 232–253, https://doi.org/10.1016/j.jhydrol.2003.12.044, 2004.

Buytaert, W., Zulkafli, Z., Grainger, S., Acosta, L., Alemie, T. C., Bastiaensen, J., De BiÃv¨re, B., Bhusal, J., Clark, J., Dewulf, A., Foggin, M., Hannah, D. M., Hergarten, C., Isaeva, A., Karpouzoglou, T., Pandeya, B., Paudel, D., Sharma, K., Steenhuis, T., Tilahun, S., Van Hecken, G., and Zhumanova, M.: Citizen science in hydrology and water resources: opportunities for knowledge generation, ecosystem service management, and sustainable development, Front. Earth Sci., 2, 21 pp., https://doi.org/10.3389/feart.2014.00026, 2014.

Davids, J. C., van de Giesen, N., and Rutten, M.: Continuity vs. the Crowd – Tradeoffs Between Continuous and Intermittent Citizen Hydrology Streamflow Observations, Environ. Manage., 60, 12–29, https://doi.org/10.1007/s00267-017-0872-x, 2017.

Davids, J. C., Rutten, M. M., Shah, R. D. T., Shah, D. N., Devkota, N., Izeboud, P., Pandey, A., and van de Giesen, N.: Quantifying the connections – linkages between land-use and water in the Kathmandu Valley, Nepal, Environ. Monit. Assess., 190, 17 pp., https://doi.org/10.1007/s10661-018-6687-2, 2018.

Dickinson, J. L., Zuckerberg, B., and Bonter, D. N.: Citizen Science as an Ecological Research Tool: Challenges and Benefits, Annu. Rev. Ecol. Evol. Syst., 41, 149–172, https://doi.org/10.1146/annurev-ecolsys-102209-144636, 2010.

Dunn, O. J.: Estimation of the Medians for Dependent Variables, Ann. Math. Stat., 30, 192–197, https://doi.org/10.1214/aoms/1177706374, 1959.

Dunn, O. J.: Multiple Comparisons among Means, J. Am. Stat. Assoc., 56, 52–64, https://doi.org/10.1080/01621459.1961.10482090, 1961.

Ewen, T., Brönnimann, S., and Annis, J.: An extended Pacific-North American index from upper-air historical data back to 1922, J. Climate, 21, 1295–1308, https://doi.org/10.1175/2007JCLI1951.1, 2008.

Finger, D., Pellicciotti, F., Konz, M., Rimkus, S., and Burlando, P.: The value of glacier mass balance, satellite snow cover images, and hourly discharge for improving the performance of a physically based distributed hydrological model, Water Resour. Res., 47, 14 pp., https://doi.org/10.1029/2010WR009824, 2011.

Finger, D., Vis, M., Huss, M., and Seibert, J.: The value of multiple data set calibration versus model complexity for improving the performance of hydrological models in mountain catchments, Water Resour. Res., 51, 1939–1958, https://doi.org/10.1002/2014WR015712, 2015.

Fitzner, D., Sester, M., Haberlandt, U., and Rabiei, E.: Rainfall Estimation with a Geosensor Network of Cars – Theoretical Considerations and First Results, Photogramm. Fernerkun., 2013, 93–103, https://doi.org/10.1127/1432-8364/2013/0161, 2013.

Gibson, E. J. and Bergman, R.: The effect of training on absolute estimation of distance over the ground, J. Exp. Psychol., 48, 473–482, https://doi.org/10.1037/h0055007, 1954.

Haberlandt, U. and Sester, M.: Areal rainfall estimation using moving cars as rain gauges – a modelling study, Hydrol. Earth Syst. Sci., 14, 1139–1151, https://doi.org/10.5194/hess-14-1139-2010, 2010.

Harrelson, C. C., Rawlins, C. L., and Potyondy, J. P.: Stream channel reference sites: an illustrated guide to field technique, Department of Agriculture, Forest Service, Rocky Mountain Forest and Range Experiment Station location, Fort Collins, CO, US, 1994.

Horner, I., Renard, B., Le Coz, J., Branger, F., McMillan, H. K., and Pierrefeu, G.: Impact of Stage Measurement Errors on Streamflow Uncertainty, Water Resour. Res., 54, 1952–1976, https://doi.org/10.1002/2017WR022039, 2018.

Juston, J., Seibert, J., and Johansson, P.: Temporal sampling strategies and uncertainty in calibrating a conceptual hydrological model for a small boreal catchment, Hydrol. Process., 23, 3093–3109, https://doi.org/10.1002/hyp.7421, 2009.

Koch, J. and Stisen, S.: Citizen science: A new perspective to advance spatial pattern evaluation in hydrology, PLoS One, 12, 1–20, https://doi.org/10.1371/journal.pone.0178165, 2017.

Le Coz, J., Renard, B., Bonnifait, L., Branger, F., and Le Boursicaud, R.: Combining hydraulic knowledge and uncertain gaugings in the estimation of hydrometric rating curves: A Bayesian approach, J. Hydrol., 509, 573–587, https://doi.org/10.1016/j.jhydrol.2013.11.016, 2014.

Lidén, R. and Harlin, J.: Analysis of conceptual rainfall–runoff modelling performance in different climates, J. Hydrol., 238, 231–247, https://doi.org/10.1016/S0022-1694(00)00330-9, 2000.

Lindström, G., Johansson, B., Persson, M., Gardelin, M., and Bergström, S.: Development and test of the distributed HBV-96 hydrological model, J. Hydrol., 201, 272–288, https://doi.org/10.1016/S0022-1694(97)00041-3, 1997.

Lowry, C. S. and Fienen, M. N.: CrowdHydrology: Crowdsourcing Hydrologic Data and Engaging Citizen Scientists, Ground Water, 51, 151–156, https://doi.org/10.1111/j.1745-6584.2012.00956.x, 2013.

Mazzoleni, M., Verlaan, M., Alfonso, L., Monego, M., Norbiato, D., Ferri, M., and Solomatine, D. P.: Can assimilation of crowdsourced data in hydrological modelling improve flood prediction?, Hydrol. Earth Syst. Sci., 21, 839–861, https://doi.org/10.5194/hess-21-839-2017, 2017.

McGuinness, J. and Bordne, E.: A comparison of lysimeter-derived potential evapotranspiration with computed values, Agricultural Research Service – United States Department of Agriculture Location, Washington D.C., 1972.

McMillan, H., Freer, J., Pappenberger, F., Krueger, T., and Clark, M.: Impacts of uncertain river flow data on rainfall-runoff model calibration and discharge predictions, Hydrol. Process., 24, 1270–1284, https://doi.org/10.1002/hyp.7587, 2010.

McMillan, H., Krueger, T., and Freer, J.: Benchmarking observational uncertainties for hydrology: rainfall, river discharge and water quality, Hydrol. Process., 26, 4078–4111, https://doi.org/10.1002/hyp.9384, 2012.

Michel, C., Perrin, C., and Andreassian, V.: The exponential store: a correct formulation for rainfall – runoff modelling, Hydrol. Sci. J., 48, 109–124, https://doi.org/10.1623/hysj.48.1.109.43484, 2003.

Oudin, L., Hervieu, F., Michel, C., Perrin, C., Andréassian, V., Anctil, F., and Loumagne, C.: Which potential evapotranspiration input for a lumped rainfall-runoff model?, J. Hydrol., 303, 290–306, https://doi.org/10.1016/j.jhydrol.2004.08.026, 2005.

Perrin, C., Michel, C., and Andréassian, V.: Improvement of a parsimonious model for streamflow simulation, J. Hydrol., 279, 275–289, https://doi.org/10.1016/S0022-1694(03)00225-7, 2003.

Perrin, C., Ouding, L., Andreassian, V., Rojas-Serna, C., Michel, C., and Mathevet, T.: Impact of limited streamflow data on the efficiency and the parameters of rainfall-runoff models, Hydrol. Sci. J., 52, 131–151, https://doi.org/10.1623/hysj.52.1.131, 2007.

Pool, S., Viviroli, D., and Seibert, J.: Prediction of hydrographs and flow-duration curves in almost ungauged catchments: Which runoff measurements are most informative for model calibration?, J. Hydrol., 554, 613–622, https://doi.org/10.1016/j.jhydrol.2017.09.037, 2017.

Ruhi, A., Messager, M. L., and Olden, J. D.: Tracking the pulse of the Earth's fresh waters, Nat. Sustain., 1, 198–203, https://doi.org/10.1038/s41893-018-0047-7, 2018.

Scherrer AG: Verzeichnis grosser Hochwasserabflüsse in schweizerischen Einzugsgebieten, Auftraggeber: Bundesamt für Umwelt (BAFU), Abteilung Hydrologie, Reinach, 2017.

Seibert, J.: Multi-criteria calibration of a conceptual runoff model using a genetic algorithm, Hydrol. Earth Syst. Sci., 4, 215–224, https://doi.org/10.5194/hess-4-215-2000, 2000.

Seibert, J. and Beven, K. J.: Gauging the ungauged basin: how many discharge measurements are needed?, Hydrol. Earth Syst. Sci., 13, 883–892, https://doi.org/10.5194/hess-13-883-2009, 2009.

Seibert, J. and McDonnell, J. J.: Gauging the Ungauged Basin?: Relative Value of Soft and Hard Data, J. Hydrol. Eng., 20, A4014004-1–6, https://doi.org/10.1061/(ASCE)HE.1943-5584.0000861, 2015.

Seibert, J. and Vis, M. J. P.: Teaching hydrological modeling with a user-friendly catchment-runoff-model software package, Hydrol. Earth Syst. Sci., 16, 3315–3325, https://doi.org/10.5194/hess-16-3315-2012, 2012.

Seibert, J. and Vis, M. J. P.: How informative are stream level observations in different geographic regions?, Hydrol. Process., 30, 2498–2508, https://doi.org/10.1002/hyp.10887, 2016.

Seibert, J., Vis, M. J. P., Lewis, E., and van Meerveld, H. J.: Upper and lower benchmarks in hydrological modelling, Hydrol. Process., 32, 1120–1125, https://doi.org/10.1002/hyp.11476, 2018.

Shiklomanov, A. I., Lammers, R. B., and Vörösmarty, C. J.: Widespread decline in hydrological monitoring threatens Pan-Arctic Research, Eos, Trans. Am. Geophys. Union, 83, 13–17, https://doi.org/10.1029/2002EO000007, 2002.

Sideris, I. V., Gabella, M., Erdin, R., and Germann, U.: Real-time radar-rain-gauge merging using spatio-temporal co-kriging with external drift in the alpine terrain of Switzerland, Q. J. Roy. Meteor. Soc., 140, 1097–1111, https://doi.org/10.1002/qj.2188, 2014.

Strobl, B., Etter, S., van Meerveld, I., and Seibert, J.: Accuracy of Crowdsourced Streamflow and Stream Level Class Estimates, Hydrol. Sci. J., (special issue on hydrological data: opportunities and barriers), in review, 2018.

van Meerveld, H. J. I., Vis, M. J. P., and Seibert, J.: Information content of stream level class data for hydrological model calibration, Hydrol. Earth Syst. Sci., 21, 4895–4905, https://doi.org/10.5194/hess-21-4895-2017, 2017.

Vrugt, J. A., Gupta, H. V., Dekker, S. C., Sorooshian, S., Wagener, T., and Bouten, W.: Application of stochastic parameter optimization to the Sacramento Soil Moisture Accounting model, J. Hydrol., 325, 288–307, https://doi.org/10.1016/j.jhydrol.2005.10.041, 2006.

Weeser, B., Stenfert Kroese, J., Jacobs, S. R., Njue, N., Kemboi, Z., Ran, A., Rufino, M. C., and Breuer, L.: Citizen science pioneers in Kenya – A crowdsourced approach for hydrological monitoring, Sci. Total Environ., 631–632, 1590–1599, https://doi.org/10.1016/j.scitotenv.2018.03.130, 2018.

Yapo, P. O., Gupta, H. V., and Sorooshian, S.: Automatic calibration of conceptual rainfall-runoff models: sensitivity to calibration data, J. Hydrol., 181, 23–48, https://doi.org/10.1016/0022-1694(95)02918-4, 1996.

13

Monthly streamflow forecasting at varying spatial scales in the Rhine basin

Simon Schick[1,2], **Ole Rössler**[1,2], **and Rolf Weingartner**[1,2]

[1]Institute of Geography, University of Bern, Bern, Switzerland
[2]Oeschger Centre for Climate Change Research, University of Bern, Bern, Switzerland

Correspondence: Simon Schick (simon.schick@giub.unibe.ch)

Abstract. Model output statistics (MOS) methods can be used to empirically relate an environmental variable of interest to predictions from earth system models (ESMs). This variable often belongs to a spatial scale not resolved by the ESM. Here, using the linear model fitted by least squares, we regress monthly mean streamflow of the Rhine River at Lobith and Basel against seasonal predictions of precipitation, surface air temperature, and runoff from the European Centre for Medium-Range Weather Forecasts. To address potential effects of a scale mismatch between the ESM's horizontal grid resolution and the hydrological application, the MOS method is further tested with an experiment conducted at the subcatchment scale. This experiment applies the MOS method to 133 additional gauging stations located within the Rhine basin and combines the forecasts from the subcatchments to predict streamflow at Lobith and Basel. In doing so, the MOS method is tested for catchments areas covering 4 orders of magnitude. Using data from the period 1981–2011, the results show that skill, with respect to climatology, is restricted on average to the first month ahead. This result holds for both the predictor combination that mimics the initial conditions and the predictor combinations that additionally include the dynamical seasonal predictions. The latter, however, reduce the mean absolute error of the former in the range of 5 to 12 %, which is consistently reproduced at the subcatchment scale. An additional experiment conducted for 5-day mean streamflow indicates that the dynamical predictions help to reduce uncertainties up to about 20 days ahead, but it also reveals some shortcomings of the present MOS method.

1 Introduction

Environmental forecasting at the subseasonal to seasonal timescale promises a basis for planning in e.g. energy production, agriculture, shipping, or water resources management. While the uncertainties of these forecasts are inherently large, they can be reduced when the quantity of interest is controlled by slowly varying and predictable phenomena. For example, the El Niño–Southern Oscillation plays an important role in predicting the atmosphere, and snow accumulation and melting often forms the backbone in predicting hydrological variables of the land surface (National Academies, 2016).

In the case of streamflow forecasting, the ESP-revESP experiment proposed by Wood and Lettenmaier (2008) provides a methodological framework to disentangle forecast uncertainty with respect to the initial conditions and the meteorological forcings. Being a retrospective simulation, the experiment consists of model runs where the initial conditions are assumed to be known and the meteorological forcing series are randomly drawn (ESP, Ensemble Streamflow Prediction) and vice versa (revESP, reverse Ensemble Streamflow Prediction). In this context the initial conditions refer to the spatial distribution, volume, and phase of water in the catchment at the date of prediction.

The framework allows for the estimation of the time range at which the initial conditions control the generation of streamflow: when the prediction error of the ESP simulation exceeds that of the revESP simulation, the meteorological forcings start to dominate the streamflow generation. Similarly, when the prediction error of the ESP simulation approaches the prediction error of the climatology (i.e. average streamflow used as naive prediction strategy), the initial conditions no longer control the streamflow generation.

In both cases this time range depends on the interplay between climatological features (e.g. transitions between wet and dry or cold and warm seasons) and catchment-specific hydrological storages (e.g. surface water bodies, soils, aquifers, and snow) and can vary from 0 up to several months (van Dijk et al., 2013; Shukla et al., 2013; Yossef et al., 2013). Indeed, this source of predictability is the rationale behind the application of the ESP approach in operational forecast settings, and it can be further exploited by conditioning on climate precursors (e.g. Beckers et al., 2016).

An emerging option for streamflow forecasting is the integration of seasonal predictions from earth system models (ESMs), i.e. coupled atmosphere–ocean–land general circulation models (Yuan et al., 2015b). Predictions from an ESM can be used threefold towards the aim of streamflow forecasting by

1. forcing a hydrological model with the predicted evolution of the atmosphere;
2. employing runoff simulated by the land surface model;
3. using the predicted states of the atmosphere, ocean, or land surface in a perfect prognosis or model output statistics context with the streamflow as the predictand.

The first approach requires a calibrated hydrological model for the region of interest. In order to correct a potential bias and to match the spatial and temporal resolution of the hydrological model, it further involves a postprocessing of the atmospheric fields. A postprocessing might also be applied to the streamflow forecasts to account for deficiencies of the hydrological model. See e.g. Yuan et al. (2015a) or Bennett et al. (2016) for recent implementations of such a model chain.

In the second approach the land surface model takes the hydrological model's place with the difference that the atmosphere and land surface are fully coupled. Since the land surface component of ESMs often represents groundwater dynamics and the river routing in a simplified way (Clark et al., 2015), the simulated runoff might be fed to a routing model as e.g. in Pappenberger et al. (2010). To the best of our knowledge, this approach has not yet been tested with a specific focus on subseasonal or seasonal streamflow forecasting.

The third approach deals with developing an empirical prediction rule for streamflow. If the model-building procedure is based on observations only, the approach is commonly referred to as perfect prognosis (PP). On the other hand, the model might be built using the hindcast archive of a particular ESM (model output statistics, MOS). In both cases the final prediction rule is applied to the actual ESM outcome to forecast the quantity of interest. Therefore, MOS methods require the presence of a hindcast archive of the ESM involved, but can take systematic errors of the ESM into account (Brunet et al., 1988).

Studies that map ESM output to streamflow with PP or MOS methods include multiple linear regression (Marcos et al., 2017), principal components regression and canonical correlation analysis (Foster and Uvo, 2010; Sahu et al., 2016), generalised linear models (Slater et al., 2017), or artificial neural networks (Humphrey et al., 2016). Whatever the selected predictors, PP and MOS methods generally conduct the mapping across spatial scales. For example, if the catchment of interest falls below the grid scale of the ESM, PP and MOS methods implicitly perform a downscaling step. If the catchment covers several grid points, the method implicitly performs an upscaling.

The present study aims to take up this scale bridging and to test a MOS-based approach for monthly mean streamflow forecasting and a range of catchment areas. To analyse the limits of predictability and to aid interpretation, we first define predictor combinations motivated by the ESP-revESP framework. Next, seasonal predictions of precipitation, surface air temperature, and runoff from the European Centre for Medium-Range Weather Forecasts (ECMWF) are entered into the regression equation and the resulting forecast skill is estimated with respect to the ESP-inspired regression model.

The variation of the catchment area is borrowed from the concept of the "working scale" (Blöschl and Sivapalan, 1995): given a particular target catchment, the regression models are applied at the catchment scale as well as at two levels of subcatchment scales. In the case of the latter, the resulting forecasts are combined in order to get a forecast at the outlet of the target catchment. By validating the combined forecasts of the subcatchments at the main outlet, any differences in the forecast quality can be attributed to the working scales.

This experiment is conducted for the Rhine River at Lobith and Basel in western Europe. Studies using subseasonal or seasonal climate predictions indicate for several parts of the Rhine basin moderate skill beyond the lead time of traditional weather forecasts. These studies apply the model chain as outlined above in approach number one: concerning catchments of the Alpine and High Rhine, Orth and Seneviratne (2013) estimate the skillful lead time for daily mean streamflow to lie between 1 and 2 weeks, which increases to about 1 month when focusing on low flows (Fundel et al., 2013; Jörg-Hess et al., 2015). Also for daily low flow Demirel et al. (2015) report a sharp decrease in skill after 30 days for the Moselle River. For a set of French catchments Crochemore et al. (2016) show that weekly streamflow forecasts are improved for lead times up to about 1 month when using postprocessed precipitation predictions. Singla et al. (2012) advance spring mean streamflow forecasts for the French part of the Rhine basin with seasonal predictions of precipitation and surface air temperature.

As a compromise between skillful lead time and temporal resolution, we decide to focus on monthly mean streamflow at lead times of 0, 1, and 2 months. In order to resolve the monthly timescale and to test the MOS method at shorter time intervals, an experiment is further conducted for 5-day mean streamflow. Here, 0 lead time refers to forecasting one time interval ahead, while e.g. a 1-month lead time denotes a temporal gap of 1 month between the release of a forecast and its time of validity.

Strictly speaking, the present study deals with hindcasts or retrospective forecasts. However, for the sake of readability we use the terms forecast, hindcast, and prediction interchangeably. Below, Sect. 2 introduces the study region, Sect. 3 describes the data set, Sect. 4 exposes the methodology in more detail, and in Sects. 5 and 6 the results are presented and discussed, respectively.

2 Study region

The Rhine River is situated in western Europe and discharges into the North Sea; in the south its basin is defined by the Alps. About 58 million people use the Rhine water for the purpose of navigation, hydropower, industry, agriculture, drinking water supply, and leisure (ICPR, 2009). The present study focuses on two gauging stations: the first is located in Lobith near the Dutch–German border, the second in Basel in the tri-border region of France, Germany, and Switzerland.

Table 1 lists some geographical attributes. The Rhine at Basel covers an area of approximately one-fifth of the Rhine at Lobith, whereas the mean elevation halves when going from Basel to Lobith. The negative minimum elevation of the Rhine at Lobith is due to a coal mine. Dominant land use classes are farmed areas and forests, but the Rhine at Basel proportionately includes more grassland, wasteland, surface water, and glacier.

Concerning the climatology of the period 1981–2011 (Fig. 1), we observe that streamflow peaks at Lobith in winter and at Basel in early summer. Streamflow at Basel is dominated by snow accumulation in winter, subsequent snow melting in spring, and high precipitation in summer. At Lobith precipitation exhibits less variability and higher surface air temperature intensifies evaporation. Based on recent climate projections, it is expected that streamflow in the Rhine basin increases in winter, decreases in summer, and slightly decreases in its annual mean in the last third of the 21th century (Bosshard et al., 2014).

3 Data

Observations of river streamflow and gridded precipitation, surface air temperature, and runoff of the period 1981–2011 in daily resolution constitute the data set. Throughout the study gridded quantities get aggregated to (sub)catchment area averages.

Table 1. Geography of the Rhine River at Basel and Lobith according to CORINE (2016), EU-DEM (2016), and GRDC (2016).

	Lobith	Basel
Area (km^2)	159 700	36 000
Gauging station (m a.s.l.)	20	250
Elevation min (m a.s.l.)	−230	250
Elevation max (m a.s.l.)	4060	4060
Elevation mean (m a.s.l.)	490	1050
Farmed area (%)	47.7	36.8
Forest (%)	35.8	31.6
Grassland (%)	3.4	11.4
Urban area (%)	9.6	7.0
Wasteland (%)	1.8	8.2
Surface water (%)	1.4	4.0
Glacier (%)	0.3	1.0

3.1 Observations

The streamflow observations consist of a set of 135 time series in m^3 s^{-1}. These series as well as the corresponding catchment boundaries are provided by several public authorities and the Global Runoff Data Centre (GRDC, 2016; see also Data Availability section) and belong to catchments with nearly natural to heavily regulated streamflow.

The ENSEMBLES gridded observational data set in Europe (E-OBS, version 16.0) provides precipitation and surface air temperature on a 0.25° regular grid (Haylock et al., 2008; E-OBS, 2017). These fields are based upon the interpolation of station data and are subject to inhomogeneities and biases. However, a comparison against meteorological fields derived from denser station networks attests to a high correlation (Hofstra et al., 2009). In the case of the Rhine basin an E-OBS tile approximately covers an area of 500 km^2.

3.2 Dynamical seasonal predictions

Precipitation, surface air temperature, and runoff from ECMWF's seasonal forecast system 4 (S4) archive are on a 0.75° regular grid, amounting in the case of the Rhine basin to a tile area of about 4500 km^2. The hindcast set consists of 15 members of which we take the ensemble mean. Runs of the coupled atmosphere–ocean–land model are initialised on the first day of each month and simulate the subsequent 7 months. Up to 2010, initial conditions are from ERA-Interim, and the year 2011 is based on the operational analysis.

The atmospheric model (IFS cycle 36r4) consists of 91 vertical levels with the top level at 0.01 hPa in the mesosphere. The horizontal resolution is truncated at TL255 and the temporal discretisation equals 45 min. The NEMO ocean model has 42 levels with a horizontal resolution of about 1°. Sea ice is considered by using its actual extent from the analysis and relaxing it towards the climatology of the past 5 years (Molteni et al., 2011).

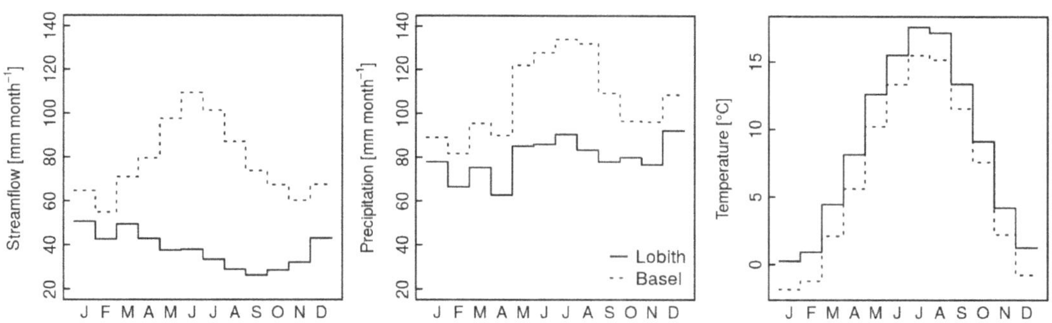

Figure 1. Monthly area averages of streamflow, precipitation, and surface air temperature for the Rhine at Lobith and Basel with respect to the period 1981–2011 (GRDC, 2016; E-OBS, 2017).

The H-TESSEL land surface model implements four soil layers with an additional snow layer on the top. Interception, infiltration, surface runoff, and evapotranspiration are dealt with by dynamically separating a grid cell into fractions of bare ground, low and high vegetation, intercepted water, and shaded and exposed snow. In contrast, the soil properties of a particular layer are uniformly distributed within one grid cell. Vertical water movement in the soil follows Richards's equation with an additional sink term to allow for water uptake by plants. Runoff per grid cell equals the sum of surface runoff and open drainage at the soil bottom (Balsamo et al., 2009; ECMWF, 2017).

4 Method

The following subsections outline the experiment, which is individually conducted for both the Rhine at Lobith and Basel. Section 4.1 details the predictor combinations and the regression strategy, Sect. 4.2 introduces the variation of the catchment area, and Sect. 4.3 illustrates the validation of the resulting hindcasts.

4.1 Model building

The predictand $y_{i,j}$ denotes observations of mean streamflow at a specific gauging site in $\mathrm{m^3\,s^{-1}}$ for $j = 5, 10, \ldots, 180$ days, starting the first day of each calendar month $i = 1, \ldots, 12$ in the period 1981–2011.

4.1.1 Predictor combinations

The set of predictors consists of variables that either precede or succeed the date of prediction i (Table 2). The first model refRun (reference run) aims to estimate how well the regression works given the best available input data. The combinations named preMet (preceding meteorology) and subMet (subsequent meteorology) are constrained to precipitation and surface air temperature preceding and subsequent to the date of forecast, respectively.

Table 2. Predictor combinations consisting of (with respect to the date of prediction) preceding and subsequent precipitation (p), surface air temperature (t), and runoff (q); the numerical values are either from the E-OBS gridded data set or ECMWF's S4 hindcast archive.

	Preceding		Subsequent		
Model	p^{pre}	t^{pre}	p^{sub}	t^{sub}	q^{sub}
refRun	E-OBS	E-OBS	E-OBS	E-OBS	–
preMet	E-OBS	E-OBS	–	–	–
subMet	–	–	E-OBS	E-OBS	–
S4P	E-OBS	E-OBS	S4	–	–
S4T	E-OBS	E-OBS	–	S4	–
S4PT	E-OBS	E-OBS	S4	S4	–
S4Q	E-OBS	E-OBS	–	–	S4

The S4* combinations constitute the MOS method and consider the seasonal predictions from the S4 hindcast archive, where we use the asterisk as wildcard to refer to any of the S4P, S4T, S4PT, and S4Q models. The S4P and S4T models are used to separate the forecast quality with respect to precipitation and temperature. The S4Q model is tested as H-TESSEL does not implement groundwater dynamics and preceding precipitation and temperature might tap this source of predictability.

4.1.2 Regression

For a particular $y_{i,j}$ we first apply a correlation screening to select the optimal aggregation time $a_{i,j}$ for each predictor according to

$$a_{i,j} = \arg\max_{k} \mid \mathrm{cor}(y_{i,j}, x_{i,k}) \mid, \tag{1}$$

where $x_{i,k}$ is one of the predictors from Table 2 and $k = -10, -20, \ldots, -720$ days in the case of p^{pre} and t^{pre} (backward in time relative to the date of prediction) and $k = 5, 10, \ldots, j$ days in the case of p^{sub}, t^{sub}, and q^{sub} (forward in time relative to the date of prediction). The limit of 720 days is chosen since larger values rarely get selected.

The ordinary least squares hyperplane is then used for prediction without any transformation, basis expansion, or interaction. However, model variance can be an issue: specifically for the preMet model from Table 2 we expect the signal-to-noise ratio to be low for most of the predictands. In combination with the moderate sample size $n = 26$ for model fitting (with respect to the cross-validation; see Sect. 4.1.3), perturbations in the training set can lead to large changes in the predictor's time lengths $a_{i,j}$ and regression coefficients. In order to stabilise model variance, we draw 100 non-parametric bootstrap replicates of the training set, fit the model to these replicates, and combine the predictions by unweighted averaging (Breiman, 1996; Schick et al., 2016).

4.1.3 Cross-validation

Each year with a buffer of 2 years (i.e. the two preceding and subsequent years) is left out and the regression outlined in Sect. 4.1.2 is applied to the remaining years. The fitted models then predict the central years that have been left out. Buffering is used to avoid artificial forecast quality due to hydrometeorological persistence (Michaelsen, 1987).

4.1.4 Lead time

Lead time is introduced by integrating the predicted $\hat{y}_{i,j}$ in time and taking differences with respect to j. For example monthly mean streamflow z_i in July ($i = 7$) is predicted with a lead time of 1 month according to

$$\hat{z}_7 = (\hat{y}_{6,60} \cdot (30+31) \cdot b - \hat{y}_{6,30} \cdot 30 \cdot b)/(31 \cdot b), \quad (2)$$

where $b = 24 \cdot 60 \cdot 60\,\mathrm{s}$ equals the number of seconds of 1 day and both $\hat{y}$ and $\hat{z}$ have unit $\mathrm{m^3\,s^{-1}}$. For 0 lead time, we set $\hat{z}_i = \hat{y}_{i,30}$. Please note that the year 1981 needs to be dropped from the validation (Sect. 4.3) since the length of the streamflow series prevents e.g. January 1981, with a lead time of 1 month, from being forecast.

4.2 Spatial levels

Contrasting the forecast quality of a given model for catchments separated in space inevitably implies a large number of factors, e.g. the geographic location (and thus the grid points of the ESM involved), the orography, or the degree to which streamflow is regulated. In order that these factors are held while screening through a range of catchment areas, we propose to vary the working scale within a particular target catchment.

Following this line of argumentation we apply the model-building procedure from Sect. 4.1 to three distinct sets of subcatchments, which we term "spatial levels" (Table 3). Spatial level 1 simply consists of the target catchment itself, i.e. the Rhine at Lobith and Basel. At spatial levels 2 and 3 we take additional gauging stations from within the Rhine basin, which naturally divide the basin into subcatchments.

Table 3. Subcatchment division of the Rhine at Lobith and Basel. The median area covers 4 orders of magnitude.

	Number of	Area km^2		
	subcatchments	min	median	max
Lobith level 1	1	–	159 700	–
Lobith level 2	5	19 690	33 220	43 550
Lobith level 3	12	8284	13 040	17 610
Basel level 1	1	–	36 000	–
Basel level 2	10	1871	2946	6346
Basel level 3	124	6	187	2654

For these subcatchments we have streamflow observations from the entire upstream area but not the actual subcatchment area itself. To arrive at an estimate of the water volume generated by the subcatchment, we equate the predictand $y_{i,j}$ to the difference of outflow and inflow of that subcatchment. For a particular date of prediction and spatial level, the sum of the resulting subcatchment forecasts $\hat{z}_i$ then constitutes the final forecast for the Rhine at Lobith and Basel.

This procedure implies that we ignore the water travel time: first, when taking the differences of outflows and inflows and second, when summing up the subcatchment forecasts. While the former increases the observational noise, the latter does not affect the regression itself, but it adds a noise term to the final forecast at Lobith and Basel. As the statistical properties of the noise introduced by the water travel time are unknown, we only can argue that the results provide a lower bound of the forecast quality due to this methodological constraint.

4.3 Validation

The forecast quality of the regression models is analysed using the pairs of cross-validated monthly mean streamflow forecasts and observations $(\hat{z}, z)$. These series cover the period 1982–2011 and have a sample size of $n = 360$. In general the validation is based on the mean absolute error (MAE) and Pearson's correlation coefficient (ρ).

The first validation steps focus on the forecasts at Lobith and Basel and thus consider the sum of the subcatchment forecasts $\hat{z}$ per spatial level. The forecasts in the subcatchments itself are addressed in Sect. 4.3.5. Finally, the validation of the 5-day mean streamflow forecasts (Sect. 4.3.6) complements the monthly analysis.

4.3.1 Benchmarks

Climatology and runoff simulated by H-TESSEL serve as benchmarks. The climatology is estimated using the arithmetic mean from the daily streamflow observations. After averaging in time, runoff from H-TESSEL gets post-calibrated via linear regression against the streamflow observations per spatial level. For both benchmarks the cross-validation scheme from Sect. 4.1.3 is applied.

4.3.2 Taylor diagram

Taylor diagrams (Taylor, 2001) provide an instrument to contrast model performances. The plotting position of a particular model has a distance from the origin equal to the standard deviation of its forecasts $\hat{z}$ and is located on the line with an angle of incline $\phi = \arccos(\rho)$. The plotting position of the observations z has a distance from the origin equal to the standard deviation of z and is located on the abscissa. The distance between these two plotting positions equals the root mean squared error with the unconditional bias $E(\hat{Z} - Z)$ removed.

4.3.3 Statistical significance

In the case of the monthly analysis it turns out that the paired differences of absolute errors for a given lead time, spatial level, and reference model r

$$d = \mid \hat{z}^{\mathrm{r}} - z \mid - \mid \hat{z}^{\mathrm{S4*}} - z \mid \tag{3}$$

no longer exhibit serial correlation and approximately follow a Gaussian distribution. Using the mean difference $\bar{d}$, we then report the p values of the two-sided t-test with null hypothesis $\bar{d} = 0$ and alternative hypothesis $\bar{d} \neq 0$. The sample autocorrelation functions and quantile plots against the Gaussian distribution of d for 0 lead time and r being the preMet model are included in the Supplement.

4.3.4 Skill

To evaluate whether a particular model m has skill with respect to a reference model r the MAE ratio

$$s = 1 - \frac{\mathrm{MAE}_m}{\mathrm{MAE}_r} \tag{4}$$

is employed. For example, m could be a S4* model and r the preMet model. $s = 0.1$ means that the model m lowers the MAE of model r by 10 %.

4.3.5 Subcatchments

To help in the interpretation of the forecast quality of the MOS method regarding the spatial levels at Lobith and Basel, we plot, in a qualitative manner, the MAE skill score (Eq. 4) of the S4* and preMet models in space as well as against the subcatchment area, the median of the terrain roughness, the MAE skill score of the subMet with the preMet model as reference, and the MAE skill score of the refRun model with the climatology as reference.

The terrain roughness is included since the atmospheric flow in complex terrain is challenging to simulate and atmospheric general circulation models need to filter the topography according to their spatial resolution (Maraun and Widmann, 2015; Torma et al., 2015). The terrain roughness is defined as the difference of the maximum and minimum elevation value within a 3×3 pixel window (Wilson et al., 2007). It is derived here from the digital elevation model EU-DEM (2016), which has a horizontal resolution of 25 m.

4.3.6 Five-day mean streamflow

In order to predict 5-day mean streamflow, Eq. (2) is used with a step size of 5 days. However, the monthly dates of prediction impose some restrictions to the validation: first, it is not possible to derive regular time series at different lead times as in the monthly analysis. Furthermore, the distributional assumptions required for the statistical test from Sect. 4.3.3 are not valid. The results of the 5-day mean streamflow experiment thus are restricted to a qualitative interpretation.

5 Results

The experiment spans several dimensions (i.e. Lobith versus Basel, dates of prediction, lead times, predictor combinations, spatial levels), so we frequently need to collapse one or several dimensions. The Supplement aims to complete the results as presented below.

5.1 Taylor diagram

Figure 2 shows the Taylor diagrams for Lobith and Basel to get a global overview regarding the lead times, predictor combinations, and spatial levels. Accurate forecasts reproduce the standard deviation of the observations (thus lie on the circle with radius equal to the standard deviation of the observations) and also exhibit high correlation (so travel on this circle towards the observations on the abscissa). At a first glimpse the spatial levels do not introduce clear differences and most of the models mass at the same spots.

The benchmark climatology is outperformed at 0 lead time by all models. At longer lead times the subMet model pops up besides the refRun model and the remaining models approach climatology. For the refRun model we note a correlation of about 0.9 independently of the lead time while the observation's variability generally is underestimated.

For Lobith and 0 lead time we observe an elongated cluster, which comprises all models except the climatology and the refRun model. Some models score a higher correlation – a closer look would reveal that these are the S4P, S4PT, and S4Q models with H-TESSEL standing at the forefront.

5.2 Date of prediction versus lead time

Figure 3 takes a closer look at the clusters in Fig. 2 with the example of the S4PT model and in addition breaks down the prediction skill into the different calendar months. Please note that the ordinate lists the calendar month and not the

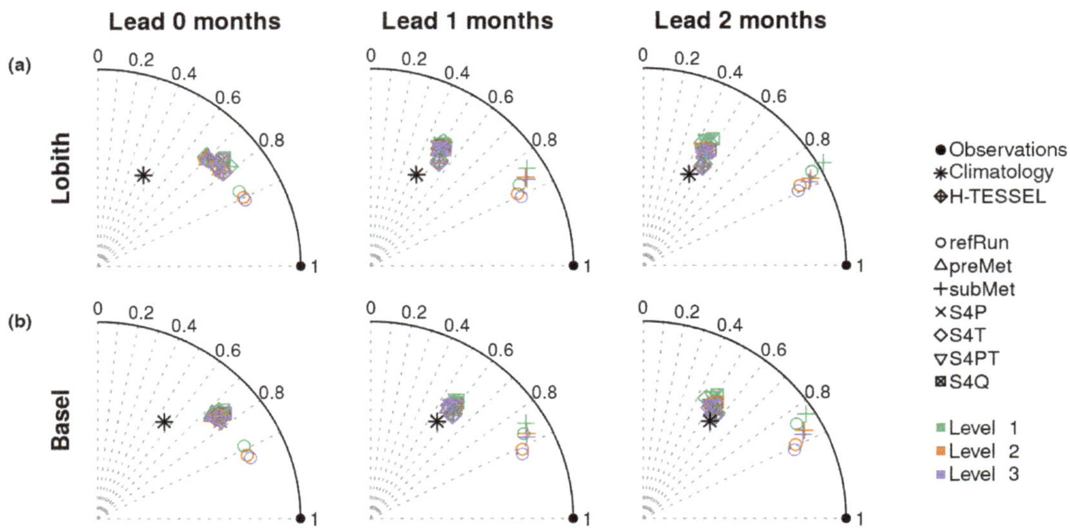

Figure 2. Taylor diagrams for the benchmarks climatology and H-TESSEL and the predictor combinations from Table 2 at Lobith **(a)** and Basel **(b)**; $n = 360$.

date of prediction – e.g. the top rows show the skill in predicting January's mean streamflow for lead times of 0 up to 2 months. Crosses indicate p values smaller than 0.05 when Eq. (3) is applied to the individual calendar months.

In general, the patterns repeat more or less along the spatial levels and the S4PT model only beats the reference models in the denominator of Eq. (4) at 0 lead time. An exception can be observed at Lobith for the month of June, for which the S4PT model most likely outperforms the climatology at 1-month lead time.

While significant differences between the S4PT and the preMet models are rare, the subMet model starts to outperform the S4PT model already at a lead time of 1 month. The comparison against the bias-corrected H-TESSEL runoff shows that the S4PT model might provide more accurate predictions for early summer, but not otherwise.

5.3 Mean absolute error

In order to conclude the analysis of the monthly predictions at Lobith and Basel, Table 4 reports the MAE at 0 lead time. Reading Table 4 along the rows reveals a more or less consistent pattern: the refRun model approximately halves the MAE of the climatology; differences between the preMet, subMet, and S4T models are small; compared to the preMet model, the S4P, S4PT, and S4Q models lower the MAE by about 40 to 50 $m^3 s^{-1}$ for Lobith and by about 10 to 20 $m^3 s^{-1}$ for Basel; and H-TESSEL outperforms the S4* models in the case of Lobith, but not Basel. When reading Table 4 along the columns, we generally note a decreasing MAE when going from spatial level 1 to spatial level 3 at Lobith. In the case of Basel, the MAE remains more or less constant except for the refRun model.

Focusing on the MOS method, Table 5 lists the corresponding MAE skill score (Eq. 4) of the S4* models using the preMet model as the reference. The p values for the null hypothesis "the preMet and S4* models score an equal mean absolute error" are listed in brackets. We see that the S4P, S4PT, and S4Q models score an error reduction ranging from 5 to 12 %. In the case of the S4T model an error reduction is either not existent (Lobith) or small (Basel), which comes along with high p values.

5.4 Subcatchments

Figure 4 depicts the MAE skill score (Eq. 4) for the S4PT model relative to the preMet model for each subcatchment at 0 lead time. If the MAE difference does not exhibit a p value smaller than 0.05 (Eq. 3), the subcatchment is coloured in white. We observe that the MAE skill score takes values in the range of about −0.05 to 0.11 and both the lowest and highest scores occur at Basel and spatial level 3. Negative scores can only be found at Basel and spatial level 3, and positive skill tends to cluster in space.

The same skill scores from Fig. 4 are contrasted in Fig. 5 with the subcatchment area, the median of the terrain roughness, the MAE skill score of the subMet model relative to the preMet model, and the MAE skill score of the refRun model relative to the climatology. If the MAE difference of the S4PT and the preMet models does not exhibit a p value smaller than 0.05, the symbol is drawn with a reduced size. The horizontal lines depict the MAE skill scores from Table 5.

While the first two attributes concern the geography of the subcatchments, the third attribute indicates the relevance of the initial conditions for the subsequent generation of streamflow. The fourth attribute shows how well the S4PT model

Table 4. Mean absolute error at 0 lead time of the benchmarks climatology and H-TESSEL and the predictor combinations from Table 2, rounded to integers. All values have unit $m^3 s^{-1}$; $n = 360$.

	Climatology	H-TESSEL	refRun	preMet	subMet	S4P	S4T	S4PT	S4Q
Lobith level 1	633	421	336	501	502	455	500	453	451
Lobith level 2	633	415	299	484	498	429	482	435	443
Lobith level 3	633	415	285	480	498	424	476	428	444
Basel level 1	239	190	131	197	195	183	195	183	191
Basel level 2	239	185	115	197	193	179	192	178	187
Basel level 3	239	182	110	198	193	179	194	178	188

Table 5. MAE skill score of the S4* models relative to the preMet model (Eq. 4, expressed in %) at 0 month lead time. The p values for the null hypothesis "the preMet and S4* models score an equal mean absolute error" are enclosed in brackets; $n = 360$.

	S4P		S4T		S4PT		S4Q	
Lobith level 1	9	(< 0.01)	0	(0.85)	10	(< 0.01)	10	(< 0.01)
Lobith level 2	11	(< 0.01)	0	(0.78)	10	(< 0.01)	8	(< 0.01)
Lobith level 3	12	(< 0.01)	1	(0.55)	11	(< 0.01)	8	(< 0.01)
Basel level 1	7	(0.01)	1	(0.36)	7	(0.01)	3	(0.07)
Basel level 2	9	(< 0.01)	3	(0.02)	10	(< 0.01)	5	(< 0.01)
Basel level 3	10	(< 0.01)	2	(0.08)	10	(< 0.01)	5	(< 0.01)

performs relative to the climatology as benchmark, when it has access to the best available input data.

The resulting patterns suggest that positive skill does not depend on the subcatchment area. On the other hand, a low terrain roughness and a weak relevance of the initial conditions seem to favour positive skill. The last row finally indicates that positive skill is restricted to subcatchments where the refRun model outperforms climatology. Roughly, a hypothetical relationship appears to strengthen from the top to the bottom plots.

5.5 Five day mean streamflow

Figure 6 shows the correlation coefficient of the 5-day mean streamflow observations and corresponding predictions for all models and benchmarks up to a lead time of 45 days. We observe that the refRun model scores a correlation of about 0.8 with a slowly decreasing tendency towards longer lead times. Furthermore, the subMet model crosses the preMet model approximately in the second week; the preMet model approaches climatology within about 3 weeks; and the subMet model comes close to the refRun model in about 3 weeks.

In addition, we see that the bias-corrected H-TESSEL runoff starts rather cautiously, but it seems to slightly outperform the S4* models at longer lead times. While the S4T model is hardly distinguishable from the preMet model, the S4P, S4PT, and S4Q models appear to outperform the preMet model within the first 20 days (Lobith) and 15 days (Basel).

For the full range of lead times, the spatial levels introduce some clear differences (Fig. 7): the refRun and subMet models are improved at longer lead times along the spatial levels. For lead times longer than about 50 days, the bias-corrected H-TESSEL runoff stays in close harmony with the climatology, while the S4* and preMet models start to score a smaller correlation instead. This effect seems to be mitigated at spatial levels 2 and 3.

6 Discussion

6.1 Model building

In the case of the monthly streamflow, the refRun model ends up with a correlation of about 0.9 for all lead times, spatial levels, and both Lobith and Basel (Fig. 2). Part of this correlation is also the annual cycle (Fig. 1), which already leads to a correlation of about 0.5 when using the climatology as prediction rule. The forecasts from the refRun model do not fully reproduce the observations' variance, which might be improved with a transformation of the predictand (Wang et al., 2012). This option – along with predictors that more explicitly represent the initial conditions, e.g. lake levels, soil moisture content, or snow courses – preferably should be tested in a future study with a small number of catchments and longer time series.

For the 5-day mean streamflow the refRun model gets degraded. At short lead times the correlation amounts to about 0.8, while for longer lead times the correlation exhibits a decreasing trend. Either the present model formulation is less valid (especially for small values of j, say 5 or 10 days, the assumption of linearity might fail) or the scheme to introduce the lead time (Eq. 2) is not appropriate for mean values of

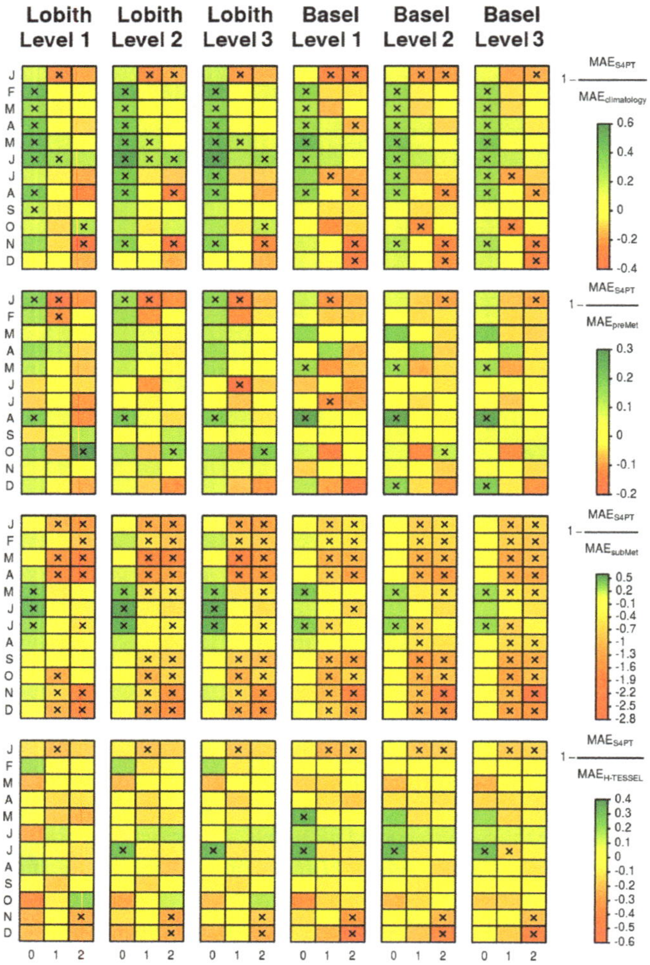

Figure 3. MAE skill score of the S4PT model with respect to the climatology, the preMet and subMet models, and bias corrected H-TESSEL runoff. The ordinate depicts the target calendar month and the abscissa the monthly lead time. Crosses indicate p values smaller than 0.05 for the null hypothesis "the reference model in the denominator and the S4PT model score an equal mean absolute error"; $n = 30$.

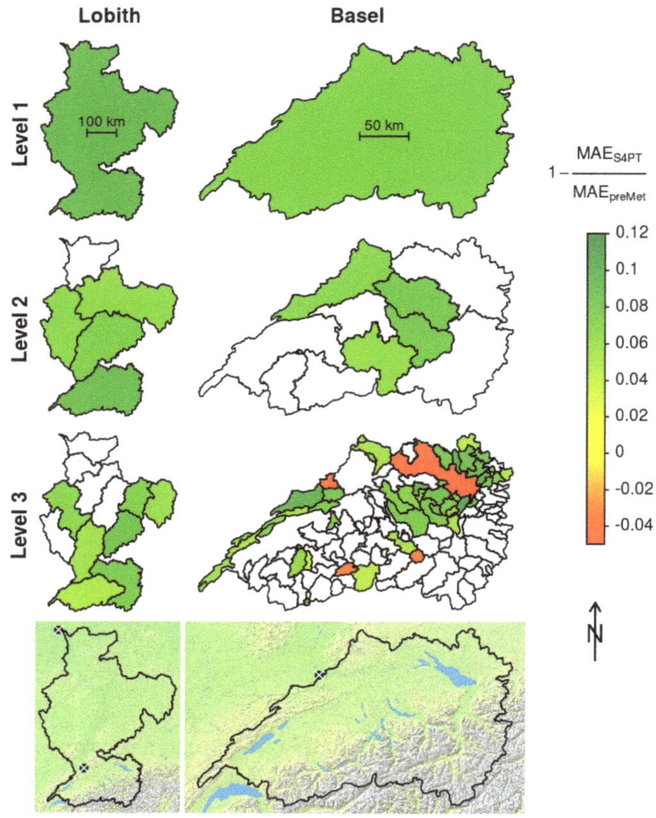

Figure 4. MAE skill score of the S4PT model with respect to the preMet model for each subcatchment and 0 lead time. Subcatchments are only coloured when the p value for the null hypothesis "the preMet and S4PT models score an equal mean absolute error" is smaller than 0.05. In the bottom maps the main outlets at Lobith and Basel are marked with a white cross and open water surfaces are coloured in blue (CORINE, 2016; EU-DEM, 2016); $n = 360$.

small time windows (e.g. the subtraction of streamflow volumes of 155 and 150 days only allows for small prediction errors). Since the final forecast values are not part of the regression equation, it is even possible to perform worse than climatology (Fig. 7).

6.2 Spatial levels

The spatial levels can affect the forecast quality in two ways:

- via the ignorance of the water travel time (Sect. 4.2)
- or by the aggregation of the E-OBS and S4 fields at the catchment scale not being the appropriate spatial resolution (e.g. large-scale grid averages cancel any spatial variability, and for catchment areas below the grid scale a grid point does not necessarily contain information valid at the local scale).

However, clear differences between the spatial levels can only be observed for the 5-day streamflow predictions, where at spatial levels 2 and 3 the forecast quality is improved. Using local information of precipitation, surface air temperature, or runoff appears to compensate for the ignorance of the water travel time.

6.3 preMet-subMet

In Yossef et al. (2013) the ESP-revESP framework is applied to the world's largest river basins using the global hydrological model PCRaster Global Water Balance (PCR-GLOBWB). Considering all calendar months and the Rhine at Lobith, the ESP simulation only outperforms the climatology at 0 lead time; the revESP simulation is outperformed at 0 lead time by both the ESP simulation and climatology; and at longer lead times the revESP simulation clearly outper-

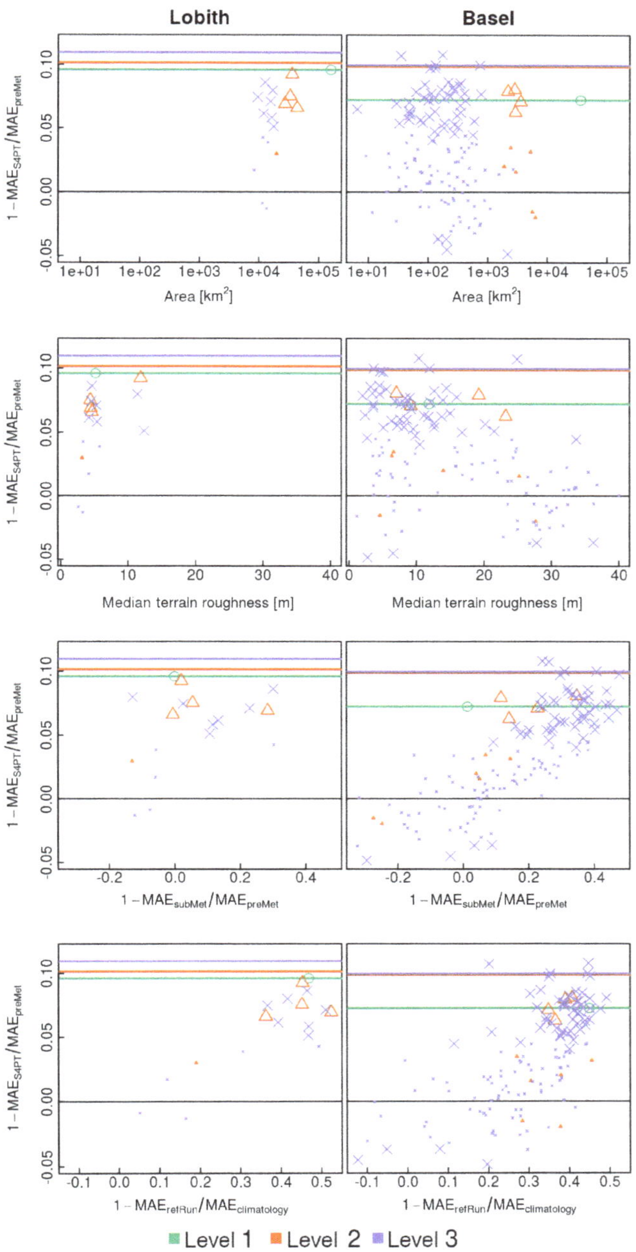

Figure 5. MAE skill score of the S4PT model with respect to the preMet model for each subcatchment and 0 lead time, plotted against subcatchment attributes (see Sect. 4.3.5 for details). Large symbols note a p value smaller than 0.05 for the null hypothesis "the preMet and S4PT models score an equal mean absolute error". The horizontal lines indicate the corresponding skill per spatial level at Lobith and Basel; $n = 360$.

forms both the ESP simulation and climatology. Therefore, the results of Yossef et al. (2013) and those of the present study are mostly in line.

The analysis of the 5-day mean streamflow forecasts (Sect. 5.5) further reveals that the crossover of the preMet and subMet models occurs approximately in the second week. However, this estimate ignores variations within the calendar year and should be considered as a rough guess since the regression method is far from being perfect in the case of the 5-day mean streamflow.

6.4 MOS method

In the case of the monthly mean streamflow forecasts at 0 lead time, the MOS method based on precipitation or runoff provides a smaller mean absolute error than the preMet model (Table 5). Figure 6 suggests that this error reduction at the monthly timescale arises from the predictions of the first 15 to 20 days. Here, it must be stressed that for the present regression strategy temperature subsequent to the date of prediction often is a weak predictor (regression coefficients of the refRun model at spatial level 1 are included in the Supplement). Thus, a rejection of the S4T model does not allow any inference about the forecast quality of surface air temperature itself.

Figure 5 indicates that the subcatchment area is most likely not relevant to score positive skill; rather the S4PT model outperforms the preMet model in subcatchments where the terrain roughness and the relevance of the initial conditions are low. However, the terrain roughness and the relevance of the initial conditions are not independent attributes: Fig. 4 shows that for small subcatchments in the Alpine region positive skill is rarely present (spatial levels 2 and 3 at Basel). These subcatchments generally exhibit a high terrain roughness as well as a high relevance of the initial conditions due to snow accumulation in winter and subsequent melting in spring and summer. A possible explanation could be that errors in the initial condition estimates outweigh the moderate skill contained in the seasonal climate predictions.

6.5 H-TESSEL

Within ECMWF's seasonal forecasting system S4, H-TESSEL aims to provide a lower boundary condition for the simulation of the atmosphere and consequently neither implements streamflow routing nor groundwater storage (Balsamo et al., 2009; ECMWF, 2017). However, H-TESSEL in combination with a linear bias correction often performs best (Table 4).

The S4Q model, which has access to the same input data and in addition conditions on preceding precipitation and temperature, scores a lower forecast accuracy than H-TESSEL in the case of Lobith (Table 4). This is most likely related to overfitting, which is not sufficiently smoothed by the model averaging (Sect. 4.1.2).

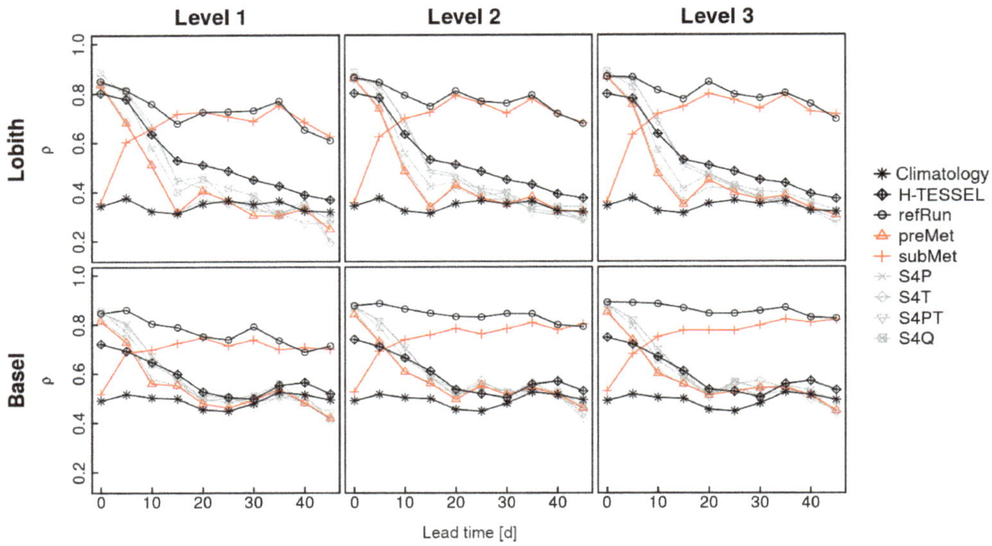

Figure 6. Correlation coefficient of 5-day mean streamflow observations and predictions for lead times up to 45 days; $n = 360$.

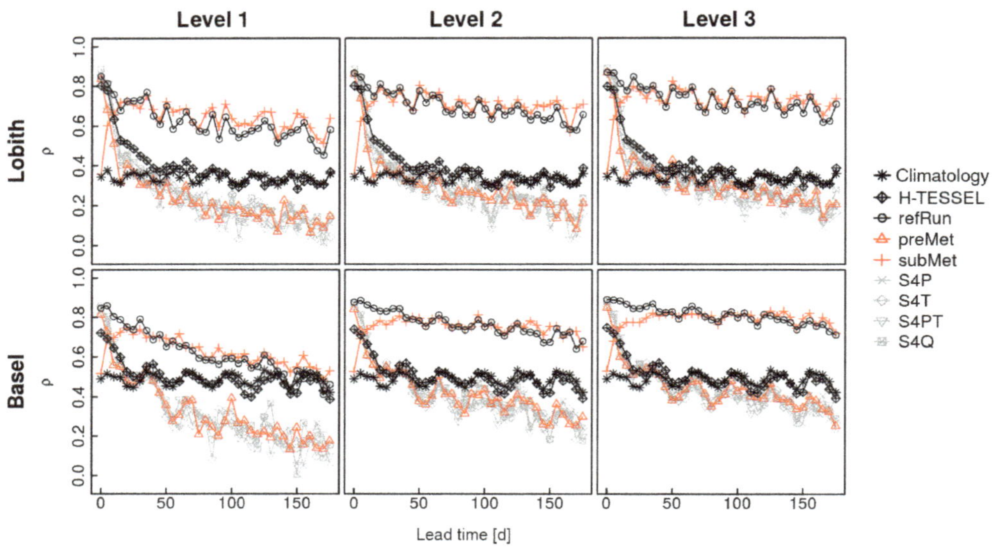

Figure 7. Correlation coefficient of 5-day mean streamflow observations and predictions for lead times up to 175 days; $n = 360$.

7 Conclusions

The present study tests a model output statistics (MOS) method for monthly and 5-day mean streamflow forecasts in the Rhine basin. The method relies on the linear regression model fitted by least squares and uses predictions of precipitation and surface air temperature from the seasonal forecast system S4 of the European Centre for Medium-Range Weather Forecasts. Observations of precipitation and surface air temperature prior to the date of prediction are employed as a surrogate for the initial conditions. In addition, runoff simulated by the S4 land surface component, the H-TESSEL land surface model, is evaluated for its predictive power.

MOS methods often bridge the grid resolution of the dynamical model and the spatial scale of the actual predictand. In order to estimate how the forecast quality depends on the catchment area, a hindcast experiment for the period 1981–2011 is conducted that varies the working scale within the Rhine basin at Lobith and Basel. This variation is implemented by applying the MOS method to subcatchments and combining the resulting forecasts to predict streamflow at the main outlets at Lobith and Basel.

On average, the monthly mean streamflow forecasts based on the initial conditions are skillful with respect to the climatology at 0 lead time for both the Rhine at Lobith and Basel. The MOS method, which in addition has access to the dynamical seasonal predictions, further reduces the mean absolute error by about 5 to 12 % compared to the model that is constrained to the initial conditions. For lead times of 1 and 2 months the forecasts virtually reduce to climatology. These results hold for the entire range of tested subcatchment scales, meaning that effects of a scale mismatch between the horizontal grid resolution and the catchment area do not emerge. Applying the MOS method for 5-day mean streamflow finally results in a rather moderate forecast quality.

We conclude that the present model formulation – in particular the assumption of linearity – is valid for the monthly timescale, catchments with areas up to 160 000 km^2, and water travel times similar to the Rhine river. However, the results also show that a simple linear bias correction of the runoff predicted by the H-TESSEL land surface model is hard to beat. Given the simplicity of a linear bias correction, we think that it could be worth further investigating runoff simulations from land surface components of earth system models for subseasonal to seasonal streamflow forecasting.

Code availability. The regression approach from Sect. 4.1.2 is implemented in an R package maintained on https://github.com/schiggo/SSO.

Competing interests. The authors declare that they have no conflict of interest.

Special issue statement. This article is part of the special issue "Sub-seasonal to seasonal hydrological forecasting". It is not associated with a conference.

Acknowledgements. Streamflow series and catchment boundaries are provided by the following public authorities: the State Institute for the Environment, Measurements and Conservation Baden Wuerttemberg; the Bavarian Environmental Agency; the State of Vorarlberg; the Austrian Federal Ministry of Agriculture, Forestry, Environment and Water; and the Swiss Federal Office for the Environment. Further we acknowledge the E-OBS data set from the EU-FP6 project ENSEMBLES (http://ensembles-eu.metoffice.com/) and the data providers in the ECA&D project (www.ecad.eu) as well as the Copernicus data and information funded by the European Union (EU-DEM and CORINE). We also thank the Global Runoff Data Centre and the European Centre for Medium-Range Weather Forecasts for the access to the data archives. The reviews and comments by Joost Beckers, Kean Foster, and Fredrik Wetterhall substantially improved the manuscript. The study was funded by the Group of Hydrology, which is part of the Institute of Geography at the University of Bern, Bern, Switzerland.

Edited by: Fredrik Wetterhall

References

Balsamo, G., Beljaars, A., Scipal, K., Viterbo, P., van den Hurk, B., Hirschi, M., and Betts, A. K.: A Revised Hydrology for the ECMWF Model: Verification from Field Site to Terrestrial Water Storage and Impact in the Integrated Forecast System, J. Hydrometeorol., 10, 623–643, https://doi.org/10.1175/2008JHM1068.1, 2009.

Beckers, J. V. L., Weerts, A. H., Tijdeman, E., and Welles, E.: ENSO-conditioned weather resampling method for seasonal ensemble streamflow prediction, Hydrol. Earth. Syst. Sc., 20, 3277–3287, https://doi.org/10.5194/hess-20-3277-2016, 2016.

Bennett, J. C., Wang, Q. J., Li, M., Robertson, D. E., and Schepen, A.: Reliable long-range ensemble streamflow forecasts: Combining calibrated climate forecasts with a conceptual runoff model and a staged error model, Water Resour. Res., 52, 8238–8259, https://doi.org/10.1002/2016WR019193, 2016.

Blöschl, G. and Sivapalan, M.: Scale issues in hydrological modelling: A review, Hydrol. Process., 9, 251–290, https://doi.org/10.1002/hyp.3360090305, 1995.

Bosshard, T., Kotlarski, S., Zappa, M., and Schär, C.: Hydrological Climate-Impact Projections for the Rhine River: GCM-RCM Uncertainty and Separate Temperature and Precipitation Effects, J. Hydrometeorol., 15, 697–713, https://doi.org/10.1175/JHM-D-12-098.1, 2014.

Breiman, L.: Bagging Predictors, Mach. Learn., 24, 123–140, https://doi.org/10.1023/A:1018054314350, 1996.

Brunet, N., Verret, R., and Yacowar, N.: An Objective Comparison of Model Output Statistics and "Perfect Prog" Systems in Producing Numerical Weather Element Forecasts, Weather Forecast., 3, 273–283, https://doi.org/10.1175/1520-0434(1988)003<0273:AOCOMO>2.0.CO;2, 1988.

Clark, M. P., Fan, Y., Lawrence, D. M., Adam, J. C., Bolster, D., Gochis, D. J., Hooper, R. P., Kumar, M., Leung, L. R., Mackay, D. S., Maxwell, R. M., Shen, C., Swenson, S. C., and Zeng, X.: Improving the representation of hydrologic processes in Earth System Models, Water Resour. Res., 51, 5929–5956, https://doi.org/10.1002/2015WR017096, 2015.

CORINE: Corine Land Cover 2006 raster data CLC 2006, available at: https://land.copernicus.eu/pan-european/corine-land-cover, (last access: 29 December 2017), 2016.

Crochemore, L., Ramos, M.-H., and Pappenberger, F.: Bias correcting precipitation forecasts to improve the skill of seasonal streamflow forecasts, Hydrol. Earth. Syst. Sc., 20, 3601–3618, https://doi.org/10.5194/hess-20-3601-2016, 2016.

Demirel, M. C., Booij, M. J., and Hoekstra, A. Y.: The skill of seasonal ensemble low-flow forecasts in the Moselle River for three different hydrological models, Hydrol. Earth. Syst. Sc., 19, 275–291, https://doi.org/10.5194/hess-19-275-2015, 2015.

E-OBS: Daily temperature and precipitation fields in Europe V.16, available at: http://www.ecad.eu/download/ensembles/ensembles.php, last access: 29 December 2017.

ECMWF: IFS Documentation, available at: http://www.ecmwf.int/en/forecasts/documentation-and-support/

changes-ecmwf-model/ifs-documentation, last access: 29 December 2017.

EU-DEM: Digital Elevation Model over Europe v1.0, available at: https://land.copernicus.eu/pan-european/satellite-derived-products/eu-dem (last access: 29 December 2017), 2016.

Foster, K. L. and Uvo, C. B.: Seasonal streamflow forecast: a GCM multi-model downscaling approach, Hydrol. Res., 41, 503–507, https://doi.org/10.2166/nh.2010.143, 2010.

Fundel, F., Jörg-Hess, S., and Zappa, M.: Monthly hydrometeorological ensemble prediction of streamflow droughts and corresponding drought indices, Hydrol. Earth. Syst. Sc., 17, 395–407, https://doi.org/10.5194/hess-17-395-2013, 2013.

GRDC: The Global Runoff Data Centre (data base state June 2016), available at: http://www.bafg.de/GRDC/EN/Home/homepage_node.html (last access: 29 December 2017), 2016.

Haylock, M. R., Hofstra, N., Klein Tank, A. M. G., Klok, E. J., Jones, P. D., and New, M.: A European daily high-resolution gridded data set of surface temperature and precipitation for 1950–2006, J. Geophys. Res.-Atmos., 113, D20, https://doi.org/10.1029/2008JD010201, 2008.

Hofstra, N., Haylock, M., New, M., and Jones, P. D.: Testing E-OBS European high-resolution gridded data set of daily precipitation and surface temperature, J. Geophys. Res.-Atmos., 114, D21, https://doi.org/10.1029/2009JD011799, 2009.

Humphrey, G. B., Gibbs, M. S., Dandy, G. C., and Maier, H. R.: A hybrid approach to monthly streamflow forecasting: Integrating hydrological model outputs into a Bayesian artificial neural network, J. Hydrol., 540, 623–640, https://doi.org/10.1016/j.jhydrol.2016.06.026, 2016.

ICPR: Internationally Coordinated Management Plan for the International River Basin District of the Rhine, International Commission for the Protection of the Rhine, 2009.

Jörg-Hess, S., Griessinger, N., and Zappa, M.: Probabilistic Forecasts of Snow Water Equivalent and Runoff in Mountainous Areas, J. Hydrometeorol., 16, 2169–2186, https://doi.org/10.1175/JHM-D-14-0193.1, 2015.

Maraun, D. and Widmann, M.: The representation of location by a regional climate model in complex terrain, Hydrol. Earth. Syst. Sc., 19, 3449–3456, https://doi.org/10.5194/hess-19-3449-2015, 2015.

Marcos, R., Llasat, M. C., Quintana-Segui, P., and Turco, M.: Seasonal predictability of water resources in a Mediterranean freshwater reservoir and assessment of its utility for end-users, Sci. Total Environ., 575, 681–691, https://doi.org/10.1016/j.scitotenv.2016.09.080, 2017.

Michaelsen, J.: Cross-Validation in Statistical Climate Forecast Models, J. Clim. Appl. Meteorol., 26, 1589–1600, https://doi.org/10.1175/1520-0450(1987)026<1589:CVISCF>2.0.CO;2, 1987.

Molteni, F., Stockdale, T., Balmaseda, M., Balsamo, G., Buizza, R., Ferranti, L., Magnusson, L., Mogensen, K., Palmer, T., and Vitart, F.: The new ECMWF seasonal forecast system (System 4), European Centre for Medium Range Weather Forecasts, 2011.

National Academies: Next Generation Earth System Prediction, The National Academies Press, 1st Edn., https://doi.org/10.17226/21873, 2016.

Orth, R. and Seneviratne, S. I.: Predictability of soil moisture and streamflow on subseasonal timescales: A case study, J. Geophys. Res.-Atmos., 118, 10963–10979, https://doi.org/10.1002/jgrd.50846, 2013.

Pappenberger, F., Cloke, H. L., Balsamo, G., Ngo-Duc, T., and Oki, T.: Global runoff routing with the hydrological component of the ECMWF NWP system, Int. J. Climatol., 30, 2155–2174, https://doi.org/10.1002/joc.2028, 2010.

Sahu, N., Robertson, A. W., Boer, R., Behera, S., DeWitt, D. G., Takara, K., Kumar, M., and Singh, R. B.: Probabilistic seasonal streamflow forecasts of the Citarum River, Indonesia, based on general circulation models, Stoch. Env. Res. Risk A., 31, 1–12, https://doi.org/10.1007/s00477-016-1297-4, 2016.

Schick, S., Rössler, O., and Weingartner, R.: Comparison of cross-validation and bootstrap aggregating for building a seasonal streamflow forecast model, Proc. IAHS, 374, 159–163, https://doi.org/10.5194/piahs-374-159-2016, 2016.

Shukla, S., Sheffield, J., Wood, E. F., and Lettenmaier, D. P.: On the sources of global land surface hydrologic predictability, Hydrol. Earth. Syst. Sc., 17, 2781–2796, https://doi.org/10.5194/hess-17-2781-2013, 2013.

Singla, S., Céron, J.-P., Martin, E., Regimbeau, F., Déqué, M., Habets, F., and Vidal, J.-P.: Predictability of soil moisture and river flows over France for the spring season, Hydrol. Earth. Syst. Sc., 16, 201–216, https://doi.org/10.5194/hess-16-201-2012, 2012.

Slater, L. J., Villarini, G., Bradley, A. A., and Vecchi, G. A.: A dynamical statistical framework for seasonal streamflow forecasting in an agricultural watershed, Clim. Dynam., 49, https://doi.org/10.1007/s00382-017-3794-7, 2017.

Taylor, K. E.: Summarizing multiple aspects of model performance in a single diagram, J. Geophys. Res.-Atmos., 106, 7183–7192, https://doi.org/10.1029/2000JD900719, 2001.

Torma, C., Giorgi, F., and Coppola, E.: Added value of regional climate modeling over areas characterized by complex terrain–Precipitation over the Alps, J. Geophys. Res.-Atmos., 120, 3957–3972, https://doi.org/10.1002/2014JD022781, 2015.

van Dijk, A. I. J. M., Peña Arancibia, J. L., Wood, E. F., Sheffield, J., and Beck, H. E.: Global analysis of seasonal streamflow predictability using an ensemble prediction system and observations from 6192 small catchments worldwide, Water Resour. Res., 49, 2729–2746, https://doi.org/10.1002/wrcr.20251, 2013.

Wang, Q. J., Shrestha, D. L., Robertson, D. E., and Pokhrel, P.: A log-sinh transformation for data normalization and variance stabilization, Water Resour. Res., 48, 5, https://doi.org/10.1029/2011WR010973, 2012.

Wilson, M. F. J., O'Connell, B., Brown, C., Guinan, J. C., and Grehan, A. J.: Multiscale Terrain Analysis of Multibeam Bathymetry Data for Habitat Mapping on the Continental Slope, Mar. Geod., 30, 3–35, https://doi.org/10.1080/01490410701295962, 2007.

Wood, A. W. and Lettenmaier, D. P.: An ensemble approach for attribution of hydrologic prediction uncertainty, Geophys. Res. Lett., 35, 14, https://doi.org/10.1029/2008GL034648, 2008.

Yossef, N. C., Winsemius, H., Weerts, A., van Beek, R., and Bierkens, M. F. P.: Skill of a global seasonal streamflow forecasting system, relative roles of initial conditions and meteorological forcing, Water Resour. Res., 49, 4687–4699, https://doi.org/10.1002/wrcr.20350, 2013.

Yuan, X., Roundy, J. K., Wood, E. F., and Sheffield, J.: Seasonal Forecasting of Global Hydrologic Extremes: System Development and Evaluation over GEWEX Basins, B. Am. Meteorol. Soc., 96, 1895–1912, https://doi.org/10.1175/BAMS-D-14-00003.1, 2015a.

Yuan, X., Wood, E. F., and Ma, Z.: A review on climate-model-based seasonal hydrologic forecasting: physical understanding and system development, Wiley Interdisciplinary Reviews: Water, 2, 523–536, https://doi.org/10.1002/wat2.1088, 2015b.

14

Streamflow characteristics from modeled runoff time series–importance of calibration criteria selection

Sandra Pool[1], **Marc J. P. Vis**[1], **Rodney R. Knight**[2], **and Jan Seibert**[1,3,4]

[1]Department of Geography, University of Zurich, Zurich, Switzerland
[2]U.S. Geological Survey Lower Mississippi-Gulf Water Science Center, 640 Grassmere Park, Suite 100, Nashville, TN 37211, USA
[3]Department of Earth Sciences, Uppsala University, Uppsala, Sweden
[4]Department of Physical Geography, Stockholm University, Stockholm, Sweden

Correspondence to: Sandra Pool (sandra.pool@geo.uzh.ch)

Abstract. Ecologically relevant streamflow characteristics (SFCs) of ungauged catchments are often estimated from simulated runoff of hydrologic models that were originally calibrated on gauged catchments. However, SFC estimates of the gauged donor catchments and subsequently the ungauged catchments can be substantially uncertain when models are calibrated using traditional approaches based on optimization of statistical performance metrics (e.g., Nash–Sutcliffe model efficiency). An improved calibration strategy for gauged catchments is therefore crucial to help reduce the uncertainties of estimated SFCs for ungauged catchments. The aim of this study was to improve SFC estimates from modeled runoff time series in gauged catchments by explicitly including one or several SFCs in the calibration process. Different types of objective functions were defined consisting of the Nash–Sutcliffe model efficiency, single SFCs, or combinations thereof. We calibrated a bucket-type runoff model (HBV – Hydrologiska Byråns Vattenavdelning – model) for 25 catchments in the Tennessee River basin and evaluated the proposed calibration approach on 13 ecologically relevant SFCs representing major flow regime components and different flow conditions. While the model generally tended to underestimate the tested SFCs related to mean and high-flow conditions, SFCs related to low flow were generally overestimated. The highest estimation accuracies were achieved by a SFC-specific model calibration. Estimates of SFCs not included in the calibration process were of similar quality when comparing a multi-SFC calibration approach to a traditional model efficiency calibration. For practical applications, this implies that SFCs should preferably be estimated from targeted runoff model calibration, and modeled estimates need to be carefully interpreted.

1 Introduction

Reliable runoff information is fundamental for many water resources-related tasks such as flood prevention, drought mitigation, management of drinking water supply and hydropower, or river restoration. Runoff modeling is a tool that can be used to create runoff time series when observed time series are not available. Runoff simulations usually focus on either representing the general shape of the hydrograph or on accurately simulating specific streamflow characteristics relevant to a respective application. However, the extraction of streamflow characteristics (SFCs) from a simulated time series may produce poor estimates when these characteristics were not included in model calibration. Ecologically relevant SFCs are properties of the annual streamflow hydrograph defining the structure and functioning of aquatic and riparian biodiversity (Richter et al., 1996; Poff et al., 1997). The accurate prediction of streamflow characteristics is a core determinate to defining how streamflow and aquatic communities relate. A large number of SFCs have been suggested to characterize ecologically relevant aspects of the flow regime (Tharme, 2003) and have become the basis for decision-support systems integrating resource management with ecological response (Cartwright et al., 2017).

Multivariate regression or runoff models are used to estimate SFCs when observed streamflow time series data are not available (Hailegeorgis and Alfredsen, 2016). The estimation of SFCs with linear regression usually relates a single SFC to catchment characteristics such as climate, land cover, and geographic and geologic variables (e.g., Sanborn and Bledsoe, 2006; Carlisle et al., 2010; Knight et al., 2012). This approach is inflexible in a sense that the regression is SFC-specific and does not allow for analysis of potential water-use and land management (Murphy et al., 2013). These disadvantages can be partially overcome by applying runoff models. Simulated streamflow time series from runoff models can be used to calculate any SFC and, by changing model input and parameters, different scenarios such as climate change, groundwater withdrawals, land use, and riverine change can be simulated (Poff et al., 2010; Murphy et al., 2013; Olsen et al., 2013; Shrestha et al., 2014). While statistical models such as multiple linear regressions often provide greater accuracy (Murphy et al., 2013), runoff models provide opportunities for also evaluating climate or land-use change scenarios.

Runoff models are used in both ecohydrology and hydrological modeling as tools to simulate specific aspects of the runoff regime. The terms, SFCs or ecological flow indices, are often used to refer to such specific aspects of the flow regime in ecohydrology studies, whereas the more recently introduced term, hydrological signatures, has been used in hydrological modeling (Jothityangkoon et al., 2001; Wagener et al, 2007). Hydrological signatures can often support a physical interpretation of the way a catchment functions and are seen as valuable metrics especially for modeling ungauged catchments (Jothityangkoon et al., 2001), for selecting appropriate model structures (Euser et al., 2013) or guiding model parameter selection in a meaningful way (Yilmaz et al., 2008), and for classifying catchments (Wagener et al., 2007; Sawicz et al., 2011). Regardless of the terminology and the ultimate goal, the basic goal is the quantification of certain aspects of a streamflow time series. In this paper, we use the term SFC as equivalent to hydrological signature, but generally prefer the term SFC to emphasize their ecological relevance.

Estimated streamflow characteristics are prone to significant errors when calculated from simulated time series (Murphy et al., 2013; Shrestha et al., 2014; Vis et al., 2015). This is due in part to the objective functions used for evaluating the model error such as the commonly used model efficiency (Nash and Sutcliffe, 1970) or volume error, which do not ensure that a model reproduces particular streamflow characteristics. These objective functions subsequently guide model parameter calibration, which strongly influences the simulated hydrograph (for an overview, see Pfannerstill et al., 2014) in terms of annual, seasonal, and monthly volumes and magnitudes. For example, Vis et al. (2015) compared model simulation from calibrations based on only the model efficiency with calibrations based on the combination of multiple objectives such as model efficiency, model efficiency of log-transformed flow, volume error, and Spearman rank correlation. All these calibration approaches tended to overestimate low flows and underestimate medium and high-flow-related SFCs. Estimation accuracy varied greatly between SFCs, with absolute biases between 3 and 33 %. Large differences in estimation accuracy are also reported by Shrestha et al. (2014) and Ryo et al. (2015). Their multi-objective calibration approach resulted in runoff simulations favoring high flows at the expense of the estimation accuracy of low flows. The large variability in estimated SFC accuracy as well as the bias in the estimates can generally be observed independently of the model used to simulate the runoff time series (Caldwell et al., 2015). A remedy to this large variability and bias is to incorporate SFCs into model calibration schemes. For example, Westerberg et al. (2011) and Pfannerstill et al. (2014) focused on specific evaluation points or segments of the flow-duration curve (FDC) during model calibration. Both studies report better overall performance for the simulated hydrograph with a FDC-based calibration compared to a more traditional calibration approach using, for example, the model efficiency (Nash and Sutcliffe, 1970). However, runoff models calibrated using FDC have to be constrained by additional SFCs if one is interested in the exact timing of events or when snow-related runoff processes are of importance (Westerberg et al., 2011). Yilmaz et al. (2008) combined information on different segments of the FDC with the runoff ratio and the rainfall–runoff lag time to guide model parameter selection in terms of primary catchment functions. These hydrologically meaningful signatures generally improved hydrograph simulation, but their value was limited for the process of vertical redistribution of excess rainfall in the catchment. In a recent study, Kiesel et al. (2017) compared estimates of ecologically relevant SFCs simulated from model calibrations using different objective functions including SFCs and the Kling–Gupta efficiency (Gupta et al., 2009). They found that including all SFCs of interest in the model calibration resulted in better SFC estimates than a calibration using the Kling–Gupta efficiency. Instead of aiming at a well-simulated, general hydrograph, Hingray et al. (2010) and Olsen et al. (2013) focused on certain aspects of the streamflow regime that were considered most important. Their results, which are echoed by Murphy et al. (2013), suggest that the runoff model performs reasonably well for the aspects on which it is calibrated, whereas it only modestly represents other runoff characteristics. Hence, developing an approach to increase the accuracy of estimated SFCs from runoff model time series continues to be an open challenge in hydrological modeling.

This study expands on the study of Vis et al. (2015) where various combinations of traditionally used objective functions were evaluated with respect to a suite of ecologically relevant SFCs. Their model calibrations with the model efficiency (R_{eff}) outperformed multi-objective model calibrations (different combinations of R_{eff}, log-transformed flow, volume error, and Spearman rank correlation) for the investi-

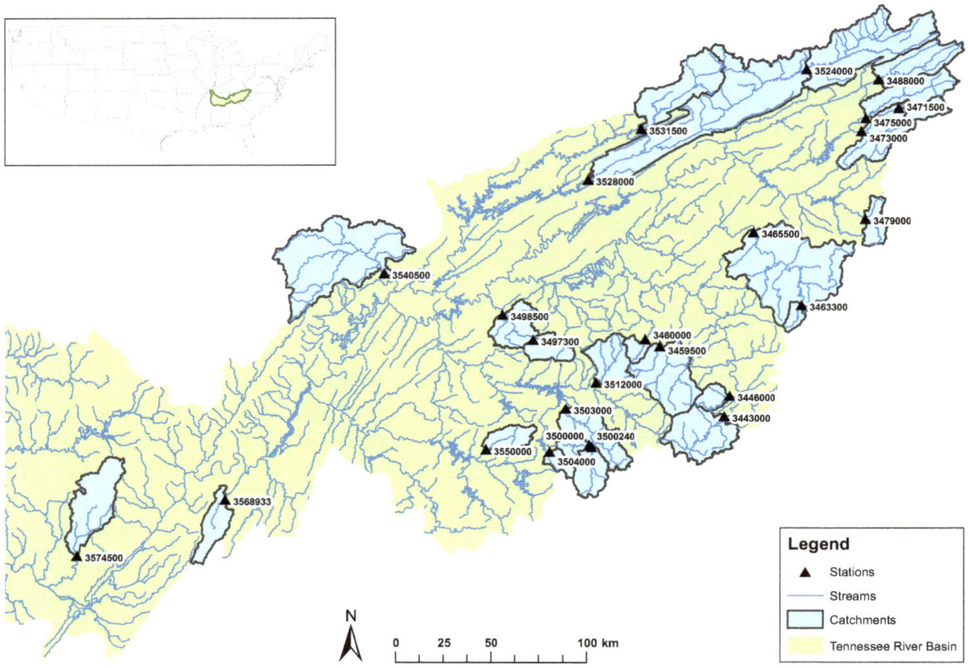

Figure 1. Location of the 25 study catchments in the Tennessee River basin (Table 1 in Vis et al., 2015, for more information).

gated SFCs. It was furthermore hypothesized that the explicit consideration of SFCs in runoff model calibration could reduce bias in estimated SFCs. The main objective of this study was therefore to assess the potential for a runoff model calibrated using specific aspects of the flow regime to more accurately estimate a suite of SFCs as compared to using a model efficiency-based calibration approach. The general approach was based on the idea that most information essential for estimating SFCs is preserved in the simulated hydrograph by including selected SFCs in model calibration. Our modeling approach relies on catchments with observed runoff time series and therefore does not answer the question of how to simulate SFCs in ungauged or altered catchments. However, the prediction of runoff for ungauged catchments benefits from an improved and informed calibration strategy for gauged catchments, which is used in the subsequent regionalization. For regionalization approaches we refer to studies such as Yadav et al. (2007), Viglione et al. (2013), or Westerberg et al. (2016).

The following questions are addressed in this paper:

1. How well is a single SFC simulated when that SFC is used as the model objective function? (Objective function is the SFC of interest.)

2. How well is a single SFC simulated when the model objective function contains one or multiple other SFCs? (Objective function can include the SFC of interest, but generally contains one or multiple other SFCs.)

3. How does the accuracy of estimated SFCs vary between traditional calibration approaches and those where the SFCs of interest are included? (Objective functions are different combinations of SFC(s) and the model efficiency.)

Throughout this study, we refer to traditional and "SFC-based" objective functions. Traditional objective functions were defined as efficiency criteria based on statistical performance metrics computed from (transformed) model residuals (e.g., R_{eff} or volume error). In contrast, "SFC-based" objective functions evaluate specific hydrograph aspects, such as event frequencies, timing, or variability of runoff, that are of ecological relevance in our study region.

2 Materials and methods

2.1 Catchment locations and characteristics

The study catchments are all located in the 106 000 km^2 Tennessee River basin in the southeastern United States (Fig. 1), which is one of the most diverse temperate freshwater ecosystems in the world (Abell et al., 2000). A large number of endemic fish species and a unique assemblage of mussels, crayfish, and salamanders make the Tennessee River basin an excellent area for ecohydrological studies (Abell et al., 2000). From a study published by Knight et al. (2008), 25 catchments in the Tennessee River basin with observed streamflow time series (U.S. Geological Survey, 2016b), precipitation (U.S. Department of Commerce, 2007a), temperature (U.S. Department of Commerce, 2007b), and potential evaporation data (Rotstayn et al., 2006) were selected. The catchment areas range between 100 and 4800 km^2 with elevations ranging from 174 to 937 m (U.S. Geological Survey, 2016a). Land cover for the study catchments is predominantly hardwood forest and pasture. Air temperature and precipitation vary between catchments according to both catchment elevation and longitude. Mean annual air temperature in the 25 catchments varies between 9.3 and 14.7 °C, and annual precipitation varies from 1500 to 2020 mm, with fall being slightly drier and less than 8 % of annual precipitation falling as snow. Runoff is highest in winter and lowest in summer, ranging from 400 to 1300 mm a^{-1} (millimeters per year). Variability in soil thickness (Omernik, 1987), regolith thickness, karst development, and topographic slope (Hoos, 1990; Wolfe et al., 1997; Law et al., 2009) are documented as asserting the most influence on runoff.

2.2 Selection of SFCs

Thirteen SFCs assessed in this study were chosen for use in model scenarios based on discernible functional connections with fish community diversity (Knight et al., 2008, 2014). This set of 13 SFCs represents each of the major flow regime components commonly used in ecological studies (e.g., Olden and Poff, 2003; Arthington et al., 2006; Caldwell et al., 2015): magnitude, ratio, frequency, variability, and date (Table 1). For this study the SFCs were additionally grouped according to flow conditions (mean, low, and high flow), because different aspects of the hydrograph have been shown to be sensitive to the objective function used for model calibration (for an overview, see Pfannerstill et al., 2014). The SFCs were calculated using the U.S. Geological Survey (2014) EflowStats R package. Please note that some of the tested SFCs (DH13, ML20, MA26, DH16, and FL2) are defined as scaled with the median, mean, or total runoff. The scaling leads to SFC values that are dependent on flow magnitudes. The magnitude of the simulation error for DH13, ML29, MA26, DH16, and FL2 is therefore dependent on runoff magnitudes, whereas the sign of the simulation error is not affected by the normalization.

2.3 The runoff model

The HBV (Hydrologiska Byråns Vattenavdelning) model (Bergström, 1976; Lindström et al., 1997) is a bucket-type hydrologic model for simulating continuous runoff series. Model inputs are daily rainfall and air temperature, as well as daily potential evaporation values. Hydrologic processes are represented by four different routines corresponding to snow, soil water, groundwater, and runoff routing, with a combined total of 16 parameters. In the snow routine, snow accumulation and snowmelt are calculated by a degree-day method. Snowmelt together with rainfall and potential evaporation are input to the soil-water routine, where the actual evaporation and the groundwater recharge are computed based on the soil-moisture storage. The groundwater (or response) routine consists of a connected shallow and deep groundwater reservoir and simulates peak flow, intermediate runoff, and baseflow. These three runoff components are taken together and transformed by a triangular weighting function during the routing process to calculate the runoff at the catchment outlet. Runoff can be modeled in a semi-distributed way by separating a catchment into elevation bands. Thereby, the snow and soil-water routines are calculated for each elevation band, whereas the groundwater storage and the runoff routing routines are treated as a lumped representation of the entire catchment. HBV exists in different versions, whereby the general structure of the model remains the same. The version applied in this study is HBV-light (Seibert and Vis, 2012). Like for all bucket-type models, parameters in the HBV model cannot be determined a priori: they are identified by model calibration instead. More detailed information on the HBV model can be found in Bergström (1976), Lindström et al. (1997), and Seibert and Vis (2012).

2.4 Modeling approach

2.4.1 Model setup

For each of the 25 catchments the number of elevation bands was defined by splitting the catchment into elevation zones of 200 m. Elevation zones covering less than 5 % of the catchment area were merged with the adjacent elevation zone. For the resulting elevation bands, air temperature and rainfall were computed with a lapse rate of 0.6 °C per 100 m and 10 % per 100 m, respectively. Potential evaporation was assumed to be uniform over the whole catchment.

Model simulations were run for two time periods, one lasting from the hydrological years (1 October until 30 September) 1984 to 1996 and the other lasting from 1997 to 2009. The approximately 3 years preceding each simulation period (January 1982 to September 1984 and January 1995 to September 1997, respectively) served to establish state variables of the model. A warm-up period was needed to ensure that the different state variables at the beginning of the simulation period were consistent with the preceding meteorological conditions and parameter values. The two simulation periods were used for model calibration and validation. For calibration, a genetic algorithm (Seibert, 2000) was used and the range of possible parameter values was specified based on previous studies (Lindström et al., 1997; Seibert, 1999; Table 2 in Vis et al., 2015). The 100 independent calibration trials allowed us to account for parameter uncertainty or equifinality (Beven and Freer, 2001) and resulted in a set of 100 calibrated parameter sets for each objective function (Fig. 2).

Table 1. Description of streamflow characteristics used to calibrate the runoff model (adapted from Knight et al., 2014; U.S. Geological Survey, 2014) ($mm\,d^{-1}$: millimeters per day; –: no units; a^{-1}: per annum; %: percent).

Streamflow characteristic	Abbreviation	Further explanation	Flow condition	Unit
		Magnitude		
Mean annual runoff	MA41	Mean annual daily runoff	Mean flow	($mm\,d^{-1}$)
Maximum October runoff	MH10	Mean of October runoff maxima for each year	High flow	($mm\,d^{-1}$)
Lowest 15 % of daily runoff	E85	Daily mean runoff that is exceeded 85 % of the time for the period of record	Low flow	($mm\,d^{-1}$)
Rate of runoff recession	RA7	Median change in log of runoff for days in which the change is negative across the period of record	Mean flow	($mm\,d^{-1}$)
		Ratio		
Average 30-day maximum runoff	DH13	Mean annual maximum of a 30-day moving average runoff divided by the median for the entire record	High flow	(–)
Baseflow	ML20	Ratio of total baseflow to total flow. Baseflow is the minimum flow magnitude in a 5-day window if 90 % of that minimum flow magnitude is less than the minimum flow magnitude of the 5 day window before and after the considered window	Low flow	(–)
Stability of runoff	TA1	Measure of the constancy of a flow regime by dividing daily flows into predetermined flow classes. The 11 flow classes capture flow ranging from flow less than 0.1 times the logarithmic mean flow to flow more than 2.25 times the logarithmic mean flow.	Mean flow	(–)
		Frequency		
Frequency of moderate floods	FH6	Average number of high-flow events per year that are equal to or greater than 3 times the median annual flow for the period of record	High flow	(a^{-1})
Frequency of larger floods	FH7	Average number of high-flow events per year that are equal to or greater than 7 times the median annual flow for the period of record	High flow	(a^{-1})
		Variability		
Variability of March runoff	MA26	Standard deviation for March runoff over the period of record divided by the mean runoff for March over the period of record	Mean flow	(%)
Variability in high-flow pulse duration	DH16	Standard deviation for the yearly average high-flow pulse duration (daily flow greater than the 75th percentile) divided by the mean of the yearly average high-flow pulse duration multiplied by 100	High flow	(%)
Variability of low-flow pulse count	FL2	Standard deviation for the average number of yearly low-flow pulses (daily flow less than the 25th percentile) divided by the mean low-flow pulse counts multiplied by 100	Low flow	(%)
		Date		
Timing of annual minimum runoff	TL1	Julian date of annual minimum flow occurrence	Low flow	(Julian day)

2.4.2 Choice of objective functions for model calibration

The complete model calibration process was conducted for 25 catchments and using data from all five different types of objective functions (see Table 2 for the exact equations) that focused on different aspects of the hydrograph. In the first step, model parameters were constrained by maximizing the model efficiency (R_{eff}, Nash and Sutcliffe, 1970). The model efficiency is the most widely used objective function

Table 2. Objective functions used in model calibration. Objective functions were calculated with observed (obs) and simulated (sim) runoff (Q) or SFCs (I).

Objective function	Abbreviation	Definition	Optimal value
Model efficiency	R_{eff}	$1-\frac{\sum(Q_{\mathrm{obs}}-Q_{\mathrm{sim}})^2}{\sum(Q_{\mathrm{obs}}-\overline{Q_{\mathrm{obs}}})^2}$	1
Efficiency for each individual SFC[1]	I_{Single}	$1-\frac{\lvert I_{\mathrm{obs}}-I_{\mathrm{sim}}\rvert}{I_{\mathrm{obs}}}$	1
SFC and model efficiency	$I_{\mathrm{Single\_Reff}}$	$0.5\left(I_{\mathrm{Single}}+R_{\mathrm{eff}}\right)$	1
Efficiency for the selected SFCs[2]	I_{Multi}	$\frac{1}{n}\left(I_{\mathrm{Single}_1}+\ldots+I_{\mathrm{Single}\_n}\right)$	1
SFCs and model efficiency	$I_{\mathrm{Multi\_Reff}}$	$\frac{n-1}{n}I_{\mathrm{Multi}}+\frac{1}{n}R_{\mathrm{eff}}$	1

[1] For each of the 13 SFCs a specific I_{Single} exists.

[2] I_{Multi} consists of the n most robust and informative SFCs.

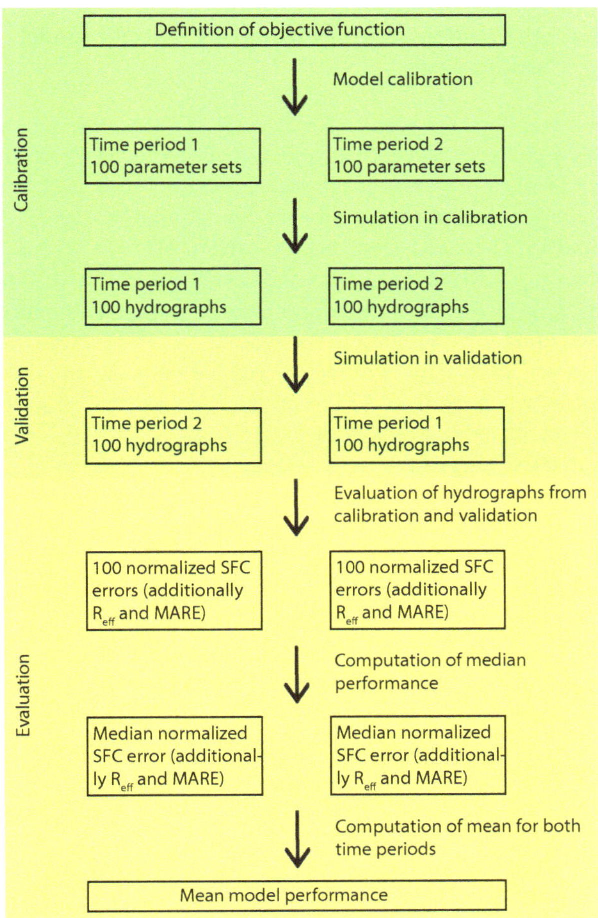

Figure 2. Flow chart of the modeling approach consisting of calibration, validation, and evaluation in time period 1 (1984–1996) and time period 2 (1997–2009) and completed for each of the five objective function types R_{eff}, I_{Sinlge}, $I_{\mathrm{Single\_Reff}}$, I_{Multi}, and $I_{\mathrm{Multi\_Reff}}$.

in hydrological modeling, and it served as a benchmark for the objective functions that included SFCs. Model calibration with R_{eff} tends to reduce simulation errors in magnitude and timing of high-flow conditions at the expense of errors in low-flow conditions (Legates and McCabe, 1999; Krause et al., 2005).

Next, a new efficiency measure that consisted of one single SFC (I_{Single}) was defined to explicitly incorporate individual SFCs into model calibration (Table 2). Each of the 13 selected SFCs was used separately for model calibration, resulting in 13 versions of I_{Single}. Additionally, each SFC efficiency measure was combined with R_{eff}, whereby both metrics were equally weighted ($I_{\mathrm{Single\_Reff}}$). The use of a single SFC as the objective function allowed calibration to focus on a specific aspect of the hydrograph, while adding R_{eff} helped to improve the overall shape of the hydrograph, including the magnitude and timing of events.

Based on the results from the individual SFCs, an objective function consisting of equally weighted normalized SFCs was defined (I_{Multi}, Table 2). This "SFC-based" efficiency measure was again combined with R_{eff} ($I_{\mathrm{Multi\_Reff}}$). For the resulting combined objective function, the same weights were assigned to each metric to make sure the individual SFCs had sufficient influence on the model calibration and were not dominated by R_{eff}. The number of SFCs constituting I_{Multi} was not previously fixed. Instead, a minimum number of SFCs was selected so that the resulting objective function was both robust and informative. These two requirements for the objective function could be achieved by only including SFCs that are robust and informative. A SFC was considered robust when the SFC calculated from a model simulation with I_{Single} had relatively small errors over the full range of catchments in both validation time periods compared to other SFCs. A SFC was regarded as being informative when it also yielded relatively good simulations for other SFCs. The robustness and information value of a SFC were therefore assessed relative to other SFCs, enabling acceptable trade-off solutions for all SFCs, with a minimum

Table 3. Performance measures used in model evaluation. Performance measures were calculated with observed (obs) and simulated (sim) runoff (Q) or SFCs (I).

Performance measure	Abbreviation	Definition	Optimal value
Model efficiency	R_{eff}	$1-\frac{\sum(Q_{obs}-Q_{sim})^2}{\sum(Q_{obs}-\overline{Q_{obs}})^2}$	1
Mean absolute relative error[1]	MARE	$1-\frac{1}{n}\sum\frac{\lvert Q_{obs}-Q_{sim}\rvert}{Q_{obs}}$	1
Normalized SFC error[2]	nSFC	$\frac{I_{obs}-I_{sim}}{R_{obs}}$	0

[1] n is the number of days.
[2] R is the range of possible values of a SFC for the respective catchment.

number of SFCs being potentially representative for (most of) the 13 SFCs.

2.4.3 Evaluation of model performance

Model performance in calibration and validation was evaluated by means of normalized SFC error, R_{eff}, and mean absolute relative error (MARE) (see Table 3 for the exact equations). These evaluation criteria were calculated for all 100 runoff simulations based on the five different types of objective functions in both validation time periods and for all 25 catchments. For the interpretation of the results, the median model efficiency of each objective function, validation period, and catchment was selected as the representative value for the model efficiency distribution. Simulation uncertainty stemming from the 100 parameter sets was assessed by a two-sided binomial test with the null hypothesis that the probability for overestimation and underestimation of a SFC is equal to 50 %.

As there are significant differences in the SFC ranges, a normalization was needed that allowed comparison of the different SFCs. Instead of normalizing in terms of relative error, an approach was applied that normalizes the SFC estimation error. The normalization of a SFC was computed as the absolute simulation error divided by the range of possible values for that SFC in the respective catchment (Table 3). To calculate these SFC ranges, 10 000 Monte Carlo simulations were run for each respective catchment using randomly chosen parameter values from the previously identified parameter space (Lindström et al., 1997; Seibert, 1999; Table 2 in Vis et al., 2015). The Monte Carlo simulations represented the potential variation in a certain SFC if no information was available to constrain the runoff model. The range was then calculated as the difference between the 10th and 90th percentiles of the simulated SFC values.

3 Results

The HBV model was capable of reproducing the observed runoff for the study catchments reasonably well. Model calibration on R_{eff} resulted in R_{eff} values between 0.68 and 0.89 with a median of 0.79. The corresponding R_{eff} values in validation ranged from 0.62 to 0.86 with a median of 0.77.

3.1 The use of single SFCs as objective functions in model calibration

3.1.1 Estimation accuracy using SFC-specific model calibrations

Model calibration results for the 13 SFCs confirmed that HBV-light is capable of estimating different SFCs with a high level of precision if the respective SFC was used as an objective function (I_{Single}) for model calibration (the 13 absolute nSFCs varied between 0.000 and 0.005 for calibrations with I_{Single}). Both I_{Single} and the combined objective function I_{Single_Reff} clearly outperformed model calibrations based on R_{eff} with regard to the estimation of SFCs (Fig. 3a). However, calibration with I_{Single} yielded poor model performances when evaluated in terms of R_{eff}, whereas R_{eff} efficiencies of calibrations with either I_{Single_Reff} or R_{eff} were comparable (Fig. 3a).

Validation results (Fig. 3b) exhibited a similar pattern in model performance to the calibration results. The median absolute normalized error of the 13 SFCs was relatively low for model runs based on the objective functions I_{Single} and I_{Single_Reff} compared to model calibration with R_{eff}. The comparable SFC estimation accuracy of I_{Single} and I_{Single_Reff} that often outperformed model simulations with R_{eff} confirms the value of SFCs for model calibration aiming at a respective SFC. An exceptional behavior can be observed for MH10, where the estimation accuracy was negatively affected by a calibration based on the SFC itself (Fig. 5a–c).

3.1.2 How informative is a SFC for estimating any SFC?

The calibrations for all 13 versions of I_{Single} and I_{Single_Reff} resulted in a total in 26 different runoff simulations that were evaluated by calculating the normalized SFC error for the calibration and validation periods. The SFC TA1 (stability of runoff; Fig. 4a and b) was selected as a representative example to illustrate that the use of SFCs as a single objective

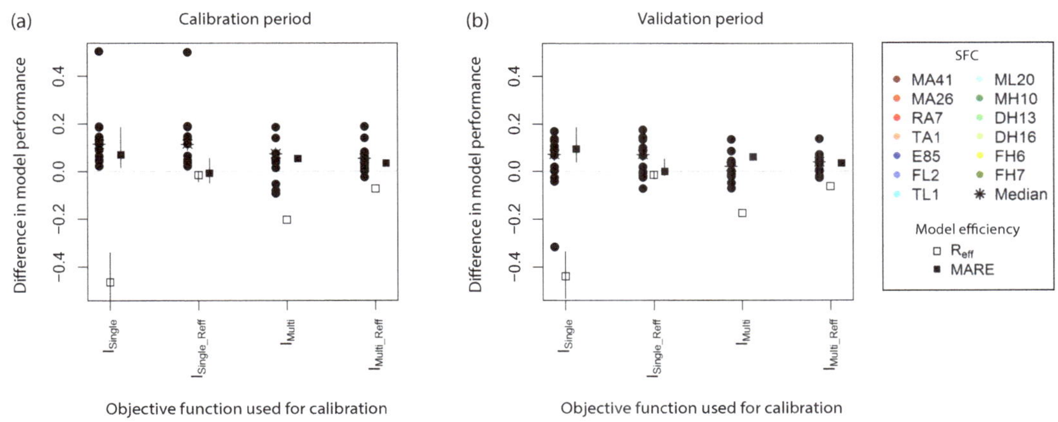

Figure 3. Model performance in **(a)** calibration and **(b)** validation in terms of absolute normalized SFC errors (nSFC), R_{eff}, and MARE depending on the objective function used in calibration. Model performance is shown as the difference between a model calibration with R_{eff} and model calibrations with I_{Single}, $I_{\text{Single\_Reff}}$, I_{Multi}, or $I_{\text{Multi\_Reff}}$ (positive values indicate that model calibration with I_{Single}, $I_{\text{Single\_Reff}}$, I_{Multi}, or $I_{\text{Multi\_Reff}}$ resulted in better model performance than model calibration with R_{eff}; negative values indicate that model calibration with I_{Single}, $I_{\text{Single\_Reff}}$, I_{Multi}, or $I_{\text{Multi\_Reff}}$ resulted in poorer model performance than model calibration with R_{eff}). Model performance values correspond to the median of the 25 catchments and the mean of both modeling time periods.

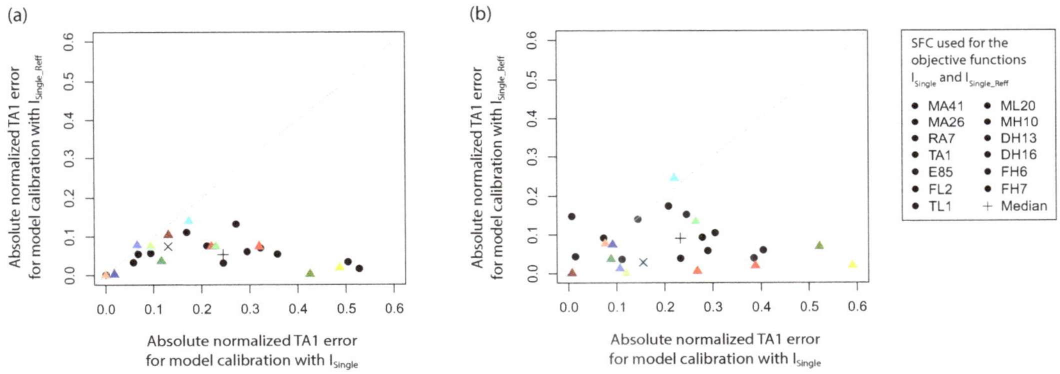

Figure 4. Absolute normalized TA1 error (nSFC) in **(a)** calibration and **(b)** validation calculated from model calibrations with the objective functions I_{Single} and $I_{\text{Single\_Reff}}$. Absolute normalized SFC errors correspond to the median of the 25 catchments and are shown separately for both modeling time periods (triangles for period 1, 1984–1996, and circles for period 2, 1997–2009). The x and plus symbols represent the median of periods 1 and 2, respectively. (Absolute normalized TA1 error for model calibrations with the objective function R_{eff} was 0.08, period 1, and 0.05, period 2, in calibration and 0.002, period 1, and 0.15, period 2, in validation.)

function (I_{Single}) generally resulted in poor SFC estimates for those SFCs not included in I_{Single} in both model calibration and validation when compared to model calibrations with $I_{\text{Single\_Reff}}$ or R_{eff}. Estimation accuracies from calibrations with $I_{\text{Single\_Reff}}$ and R_{eff} were often of comparable magnitude. Error magnitudes from the three described objective function types (I_{Single}, $I_{\text{Single\_Reff}}$, and R_{eff}) could vary considerably between time periods (illustrated by triangles and circles, respectively, in Fig. 4a and b).

3.2 The use of multiple SFCs for model calibration

Figure 6a shows simulation results for the objective function I_{Single} for all 25 catchments and both modeling time periods. The five SFCs with the highest robustness (less variability in error; Fig. 6a) were RA7, ML20, FH6, E85, and MA41. All five of these SFCs could be used for the objective function I_{Multi}; however, E85 (lowest 15 % of daily runoff) was discarded as potential SFC for I_{Multi} because of its redundant information with ML20 (baseflow). The information value of the remaining 4 SFCs for each of the 13 SFCs is presented in Fig. 6b. All 13 SFCs were relatively well simulated by model calibrations with I_{Single} of either RA7, ML20, FH6, or MA41 (colored circles in Fig. 6b) compared to calibrations with other SFCs.

Median estimates of the 13 SFCs in the calibration period were slightly lower when the model was calibrated with

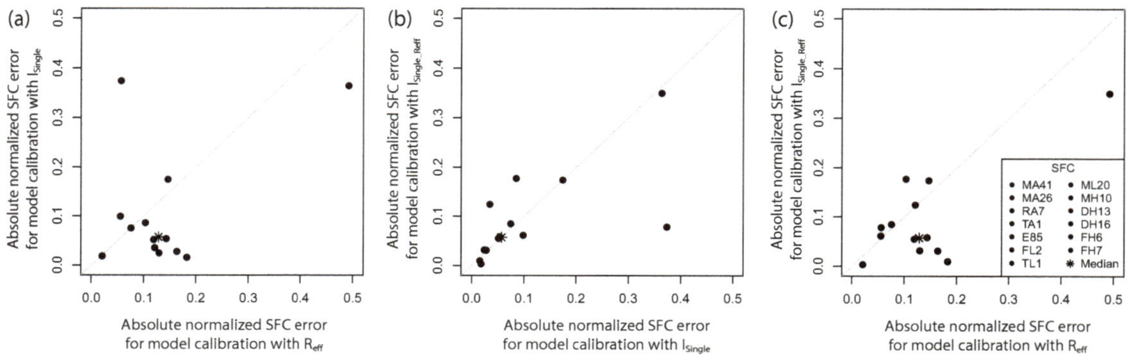

Figure 5. Comparison of absolute normalized SFC errors (nSFC) in validation calculated from model calibrations with the objective functions R_{eff}, I_{Single}, and $I_{\mathrm{Single\_Reff}}$. Absolute normalized SFC errors correspond to the median of the 25 catchments and the mean of both modeling time periods.

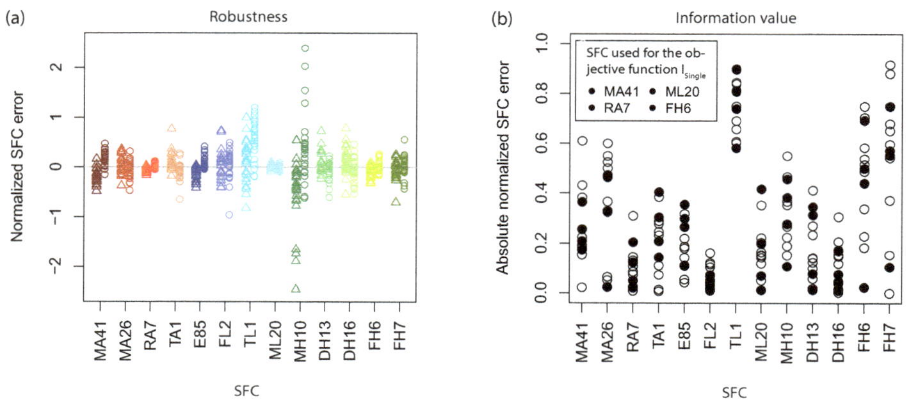

Figure 6. **(a)** Robustness: normalized SFC errors (nSFC) in validation calculated from model calibrations with the objective function I_{Single} for the respective SFC. Values are shown for all 25 catchments and both modeling time periods (triangles for period 1, 1984–1996, and circles for period 2, 1997–2009). **(b)** Information value: absolute normalized SFC errors (nSFC) in validation calculated from model calibrations with all 13 objective functions I_{Single}. Model performance values correspond to the median of the 25 catchments and the mean of both modeling time periods. Each open circle represents 1 of the 13 SFCs used for I_{Single}. The colored circles refer to the final selection of SFCs for the objective function I_{Multi}.

I_{Multi} rather than $I_{\mathrm{Multi\_Reff}}$. Both of these objective functions led to better model performance for SFCs than calibrating with R_{eff} alone (Fig. 3a). Model performance for the validation period with $I_{\mathrm{Multi\_Reff}}$ had a lower median error for SFCs than the error associated with using I_{Multi} as an objective function (Fig. 3b). The comparison of I_{Multi} and $I_{\mathrm{Multi\_Reff}}$ for all SFCs separately (Fig. 7a) revealed that for most SFCs both objective functions resulted in similar estimates. $I_{\mathrm{Single\_Reff}}$ was better for estimating SFCs than $I_{\mathrm{Multi\_Reff}}$, especially for SFCs not included in the $I_{\mathrm{Multi\_Reff}}$ objective function (Fig. 7b). Comparing simulations from $I_{\mathrm{Multi\_Reff}}$ and R_{eff} revealed a smaller median error of the SFCs when calibrating with $I_{\mathrm{Multi\_Reff}}$ (Figs. 3b and 7c). Yet, for most SFCs not explicitly incorporated into the objective function $I_{\mathrm{Multi\_Reff}}$, the objective function R_{eff} performed equally well or slightly better than $I_{\mathrm{Multi\_Reff}}$ (Fig. 7c).

3.3 Estimation accuracy for SFCs

Figure 8 provides an overview (median of all 25 catchments) of how well SFCs were simulated by presenting the results for both modeling time periods and all five objective function types. Error magnitudes ranged between −25 and 25 % for the majority of SFCs. Considerably higher estimation accuracy was achieved for ML20 (−5 to 2 %), whereas estimation accuracies were lowest for MH10 and TL1, with error magnitudes up to 40 and 77 %, respectively. For some SFCs (e.g., MA26 and TL1) the error tended to be higher in one of the two modeling time periods, whereas for other SFCs (e.g., RA7 and MH10) the objective function had a distinct influence on the error magnitude. There was no evidence that the estimation accuracy depends on flow components (magnitude, ratio, frequency, variability, and date) or flow conditions (low, medium, and high flow).

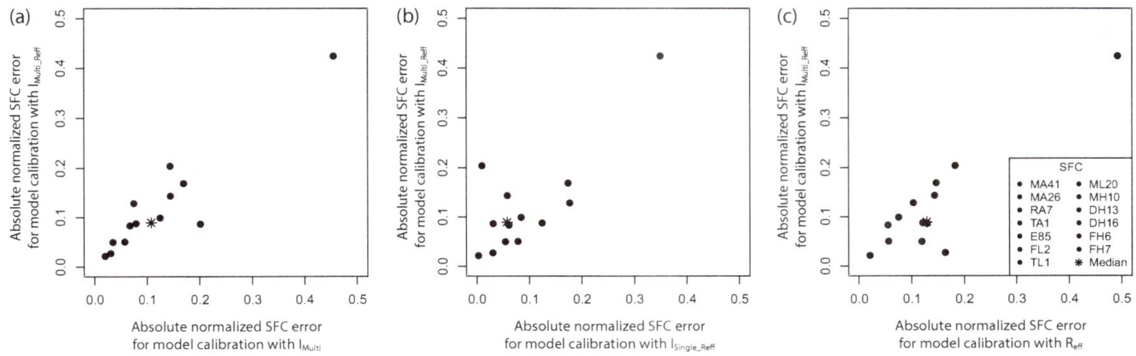

Figure 7. Comparison of absolute normalized SFC errors (nSFC) in validation calculated from model calibrations with the objective functions R_{eff}, $I_{\mathrm{Single\_Reff}}$, I_{Multi}, and $I_{\mathrm{Multi\_Reff}}$. Absolute normalized SFC errors correspond to the median of the 25 catchments and the mean of both modeling time periods.

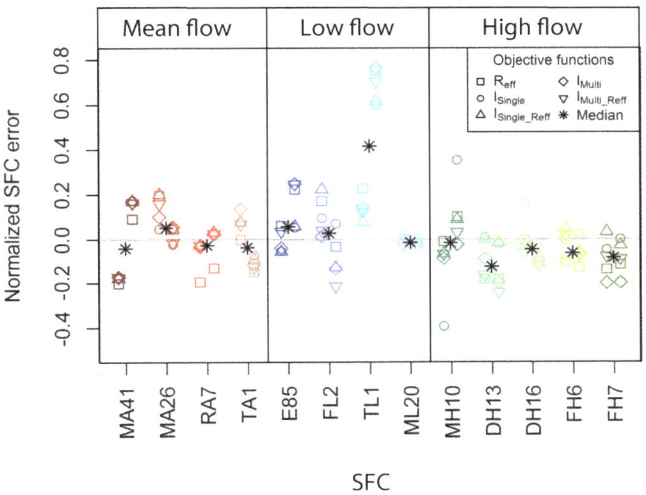

Figure 8. Normalized SFC errors (nSFC) in validation depending on the objective function used in calibration. Model performance values correspond to the median of the 25 catchments and are shown for both modeling time periods (period 1, 1984–1996, on the left side and period 2, 1997–2009, on the right side).

The median error (illustrated by stars in Fig. 8) was used for the evaluation of the underestimation or overestimation of SFCs. Among the tested SFCs, an underestimation was observed for all five SFCs representing high-flow conditions as well as for three of four mean-flow-related SFCs. With one exception, low-flow SFCs were overestimated. This overall pattern was less evident when evaluating each objective function and time period separately (Figs. 8 and 9). The SFCs DH16 and MH10 indicate two typically observed deviations in the overall pattern. DH16 is an example of a SFC that could be regarded as being clearly underestimated by the model, because of its negative bias in 9 out of 10 cases (median values in Fig. 9a). However, for objective functions or modeling time periods with a low magnitude in the median bias, the underestimation of the SFC was not statistically significant. Even in the case of a median pointing to statistically significant underestimation, there might be a substantial number of catchments for which DH16 was overestimated. A second commonly observed phenomenon is shown by the SFC MH10 (Fig. 9b). While MH10 had mostly small but statistically significant median errors, there were many catchments with considerably higher errors. Although MH10 was the most extreme example, it illustrates that small median errors do not guarantee good results for all catchments.

4 Discussion

4.1 On the importance of the choice of the objective function

The results demonstrated that the objective function used for model calibration strongly influences the estimation accuracy of SFCs. This finding confirms the findings of previous studies (e.g., Hingray et al., 2010; Westerberg et al., 2011; Murphy et al., 2013; Olsen et al., 2013; Pfannerstill et al., 2014; Shrestha et al., 2014; Caldwell et al., 2015; Vis et al., 2015) and points out the importance of making a careful choice of the objective function for model calibration. The benefit of optimizing one specific SFC lies in the relatively accurate estimation of the respective SFC compared to a calibration with R_{eff} or a multi-SFC objective function. Model calibration on one single SFC clearly emphasizes the hydrograph aspects of the selected SFC possibly neglecting an adequate representation of other hydrograph characteristics. This implies that calibrations with I_{Single} can lead to poor model performance for SFCs not included in the objective function. The fact that a calibration with R_{eff} and a calibration with multiple SFCs lead to comparable estimates for most SFCs indicates that the main hydrological processes of the catchments are similarly well represented with the two approaches. Considering that SFCs not incorporated into the objective function

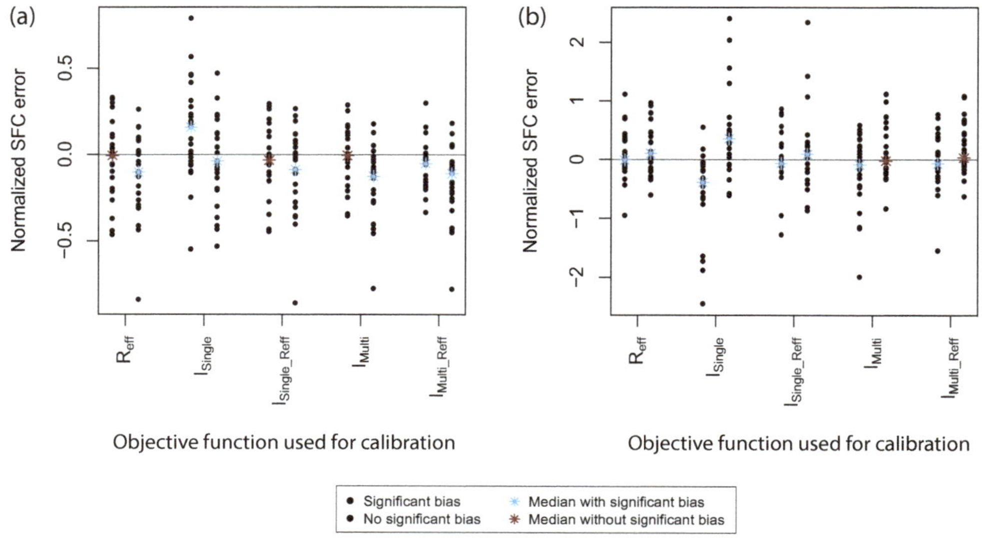

Figure 9. **(a)** Normalized DH16 errors (nSFC) and **(b)** normalized MH10 errors (nSFC) in validation depending on the objective function used in calibration. Normalized SFC errors are shown for all 25 catchments and for both modeling time periods (period 1, 1984–1996, on the left side and period 2, 1997–2009, on the right side). Colors indicate the significance of the results assessed by a two-sided binomial test at a confidence level of 0.95. Note the difference in the *y* axis.

I_{Multi} showed little change compared to calibrations with R_{eff} brings into question the benefit of including SFCs in model calibration instead of applying a traditional calibration approach when aiming at estimating a suite of SFCs. This is surprising because the SFCs selected for I_{Multi} or $I_{\text{Multi\_Reff}}$ provide information on high flows, recession rate, percentage of baseflow, and annual runoff volume, and therefore should help in constraining the model with respect to different important runoff processes. These results are different from those of Yilmaz et al. (2008) and Pfannerstill et al. (2014), whose multi-metric runoff model calibration resulted in an improved general shape of the hydrograph. Although their calibration approach was mainly based on various segments of the flow duration curve, it is unclear why the conclusions differ that much. From the above discussion it becomes evident that calibrating a runoff model for estimating many different SFCs from one single hydrograph is a trade-off between finding a parameterization that is general enough to represent different aspects of the hydrograph and that simultaneously emphasizes specific SFCs. These trade-off situations are common as perfect model parameterizations are usually not possible due to a variety of uncertainty sources, such as model structural uncertainty and input and runoff data uncertainty (Beven, 2016).

A noticeable result from the current study is the distinct difference in model performance in calibration and validation when using the objective function I_{Single}. While almost perfect fits are achieved in calibration for all catchments and SFCs, model errors tend to be much higher in validation, with a considerable spread between catchments as well as a clear difference depending on the SFC. This observation confirms that the model is able to simulate the SFCs well, but also outlines that a good model calibration does not imply robust simulations in validation. In general, it seems that SFCs that are strongly related to physical catchment properties (e.g., rate of streamflow recession) are the most robust, followed by SFCs representing an average flow condition with a moderate robustness. SFCs that are a measure of more extreme high-flow conditions are the least robust, possibly because these conditions are subject to inter-annual weather changes and are more difficult to model due to their dynamic behavior. A low robustness could also indicate that the model structure might be suboptimal for some catchments.

The two least robust SFCs are MH10 and TL1. MH10 simulations with I_{Single} yield by far the poorest results of all objective function types, with very large normalized error in both positive and negative directions. In comparison, the high estimation errors for TL1 depend on the modeling time period. The high estimation errors for TL1 in period 2 stem from years where the minimum runoff was simulated in late winter while the observed minimum was in late fall. By visually analyzing the temperature and runoff time series, it can be hypothesized that such model simulations mainly happened in years with successive weeks of continuously little precipitation during late winter. Such prolonged drier periods occurred more often in one of the two modeling time periods and thus evoked the distinct bias in model accuracy depending on the simulation period. Both TL1 and MH10 are calculated from a single value per year, as opposed to, e.g., RA7, which is based on all recessions. In model calibration, many parameter sets are derived that perfectly simulate this single value. However, a good simulation of either TL1 or

MH10 is not so much dependent on an accurate representation of dominant runoff processes. Thus, model results for the validation period using input data of identical quality can fail to accurately simulate either SFC because of parameter sets "tuned" to the data as opposed to being based on modeling the process.

4.2 Model performance regarding SFCs

The runoff model tends to underestimate SFCs related to mean and high-flow conditions, while SFCs representing low-flow conditions are generally overestimated. These results are consistent with those of Olsen et al. (2013), Caldwell et al. (2015), Vis et al. (2015), and Kiesel et al. (2017) and can partly be explained by the model behavior characterized by a less pronounced runoff response to precipitation events but increased groundwater discharge to the stream during drier periods compared to the observed data (Vis et al., 2015). The observations that average flow conditions are better simulated than extremes (Caldwell et al., 2015; Vis et al., 2015) or that high-flow-related SFCs are more accurately estimated than those related to low flow (Shrestha et al., 2014; Ryo et al., 2015) cannot be confirmed with our results. None of these earlier studies explicitly included SFCs in model calibration and the deviating results could be attributed to the differing approaches to defining the objective function(s). This presumption is supported by the previously described differences in results of Vis et al. (2015), although they applied the same runoff model, catchments, and SFCs.

4.3 How to select SFCs for a multi-index calibration approach

The current study supports the assumption that including SFCs in model calibration helps to preserve most hydrograph aspects relevant to those SFCs. Thus, an objective function based on several SFCs is expected to result in a hydrograph from which a suite of SFCs can be calculated. Not knowing which SFCs will be relevant for a given study, a guideline as to which SFCs the model calibration could be based on would be helpful. The first step towards a guideline consists of selecting SFCs that are potentially valuable for model calibration. This selection was based on the concept of robustness and information value of SFCs, which is comparable to the approach used by Euser et al. (2013), who assessed the realism of model structures. Like Euser et al. (2013), results from the current study indicated that high robustness was not necessarily related to high information value, emphasizing the importance of selecting SFCs by jointly evaluating robustness and information value. The concept of information value and robustness favors simulations that preserve important hydrograph characteristics, as can be seen from the slightly improved median estimation accuracy of SFCs with the objective functions I_{Multi} or $I_{\mathrm{Multi\_Reff}}$ compared to estimations with R_{eff} only.

A model calibrated on certain flow conditions (low, medium, and high flow) is beneficial for SFCs representing these flow conditions (see, e.g., Murphy et al., 2013), so it was hypothesized that the information value of the selected SFCs is highest for SFCs belonging to the same group of flow conditions. The confirmation of this hypothesis would allow us to draw general conclusions about a minimum number of SFCs required for model calibration. Surprisingly the results did not reveal any pattern related to flow conditions and thus no recommendation for the final selection of SFCs can be made. It seems that the selection of SFCs for an informative and robust objective function depends on the type and the combination of SFCs one is interested in. Since this study was based on a limited number of SFCs it could be interesting to test the hypothesis by analyzing a greater number of SFCs. Testing a larger number of SFCs might reveal relations that are difficult to see with a small sample. Furthermore, more knowledge about the effect of single SFCs or the combination of SFCs used as objective functions on runoff simulations could be gained by using synthetic data and a modeling approach where an excellent hydrograph fit is possible (e.g., "HBV-land" in Seibert and Vis, 2012).

4.4 Objective functions, their estimation accuracy, and consequences for practical applications

The emphasis of SFC-related modeling studies changed from estimating single SFCs to simulating a suite of SFCs (Olden and Poff, 2003). The modeling design of this study combined both approaches for the same SFCs and catchments and thus enabled a direct comparison of the results. Ideally, the runoff model could be calibrated to simulate a hydrograph for each catchment from which any SFC can be calculated. Such an approach ensures a relatively small calibration effort, which is especially valuable if one is interested in modeling many catchments and/or various scenarios. However, results indicate that SFCs related to a more generally calibrated model (e.g., R_{eff}, I_{Multi}, or $I_{\mathrm{Multi\_Reff}}$) are less accurate than when they are estimated from hydrographs based on targeted model calibrations (e.g., I_{Single} or $I_{\mathrm{Single\_Reff}}$). This fact has substantial implications for the later application of simulated SFCs in decision-support systems for integrated resource management. As stated by Carlisle et al. (2010), with high errors in SFC estimates, only considerable flow departures from natural conditions can be detected. Also, inaccurate SFC values can impede the generation of more robust flow alteration–ecosystem change relationships that are ultimately needed for sustainable flow management guidelines (Arthington et al., 2006; Poff and Zimmermann, 2010; Gillespie et al., 2015; Cartwright et al., 2017).

As with regional statistical approaches, incorporating SFCs into model objective functions implies that a modeler knows which SFCs are relevant and that the model must be recalibrated if one is interested in additional SFCs. The advantage of runoff models over multivariate regressions and observed streamflow series includes their use for climate scenario analysis or for simulating runoff in ungauged catchments, with the latter being one of the ultimate aims in the ELOHA framework (Poff et al., 2010). Modeling SFCs gets even more challenging when moving from a gauged to an ungauged catchment. An appropriate calibration strategy targeted to the main simulation goal is crucial for any subsequent regionalization.

4.5 Choice of the runoff model for estimating SFCs

When comparing SFCs estimated from simulations of different runoff models, the question can be raised whether the results depend on the selected model. This question is especially important for resource managers who need to make decisions based on model results from different studies (Caldwell et al., 2015). A comparison of runoff models with different spatial scales that rely on different data inputs was conducted by Caldwell et al. (2015). Their results do not indicate that a certain runoff model is more suited for predicting SFCs than others, but rather that the calibration process probably has as much influence as the model structure. Thus, it can be assumed that the conclusions of this study would be similar if a different calibrated runoff model was applied.

5 Conclusions

In this study, we evaluated the value of using SFCs for the calibration of a runoff model used to estimate SFCs. The results suggest that the choice of the objective function used for model calibration strongly influences the estimation accuracy of SFCs. While the model was capable of correctly simulating any of the tested SFCs, a good reproduction of a particular SFC was generally achieved when this SFC was included in the objective function. SFC estimates from model simulations with an objective function consisting of a representative selection of SFCs resulted in comparable accuracies to the estimates from model runs based on the commonly used model efficiency when evaluated against SFCs not included in the objective function. Estimates of SFCs that are less dependent on the short-term weather input or SFCs representing average flow conditions were more robust than other SFCs. Since the results imply that one has to consider significant uncertainties when simulated time series are used to derive SFCs that were not included in the calibration, we strongly recommend calibrating the runoff model explicitly for the SFCs of interest.

Author contributions. SP, MV, RK, and JS designed this study based on a previous collaboration; MV performed the runoff simulations; SP analyzed the results that were discussed with all co-authors. Writing of the paper was led by SP with contribution of all co-authors.

Competing interests. The authors declare that they have no conflict of interest.

Acknowledgements. This paper is a product of discussions and activities that took place at the U.S. Geological Survey John Wesley Powell Center for Analysis and Synthesis as part of the workgroup focusing on Water Availability for Ungauged Rivers (https://powellcenter.usgs.gov/). Funding for this research was provided by the U.S. Geological Survey Cooperative Water Program and the University of Zurich. Any use of trade, firm, or product names is for descriptive purposes only and does not imply endorsement by the U.S. Government. We thank the reviewers Björn Guse, Oddbjørn Bruland, and Sjur Kolberg for their constructive and detailed comments that helped to improve the quality of our manuscript.

Edited by: Dimitri Solomatine

References

Abell, R. A., Olson, D. M., Dinerstein, E., Hurley, P. T., Diggs, J. T., Eichbaum, W., Walters, S., Wettengel, W., Allnutt, T., Loucks, C. J., and Hedao, P. (Eds.): Freshwater ecoregions of North America: A conservation assessment, Island Press, Washington, DC, USA, 2000.

Arthington, A. H., Bunn, S. E., Poff, N. L., and Naiman, R. J.: The challenge of providing environmental flow rules to sustain river ecosystems, Ecol. Appl., 16, 1311–1318, https://doi.org/10.1890/1051-0761(2006)016[1311:TCOPEF]2.0.CO;2, 2006.

Bergström, S.: Development and application of a conceptual runoff model for Scandinavian catchments, SMHI, Norrköping, Sweden, No. RHO 7, 134 pp., 1976.

Beven, K.: Facets of uncertainty: Epistemic uncertainty, non-stationarity, likelihood, hypothesis testing, and communication, Hydrolog. Sci. J., 61, 1652–1665, https://doi.org/10.1080/02626667.2015.1031761, 2016.

Beven, K. and Freer, J.: Equifinality, data assimilation, and uncertainty estimation in mechanistic modelling of complex environmental systems using the GLUE methodology, J. Hydrol. 249, 11–29, https://doi.org/10.1016/S0022-1694(01)00421-8, 2001.

Caldwell, P. V., Kennen, J. G., Sun, G., Kiang, J. E., Butcher, J. B., Eddy, M. C., Hay, L. E., LaFontaine, J. H., Hain, E. F., Nelson, S. A. C., and McNulty, S. G.: A comparison of hydrologic models for ecological flows and water availability, Ecohydrology, 8, 1525–1546, https://doi.org/10.1002/eco.1602, 2015.

Carlisle, D. M., Falcone, J., Wolock, D. M., Meador, M. R., and Norris, R. H.: Predicting the natural flow regime: models for assessing hydrological alteration in streams, River Res. Appl., 26, 118–136, https://doi.org/10.1002/rra.1247, 2010.

Cartwright, J., Caldwell, C., Nebiker, S., and Knight, R.: Putting flow–ecology relationships into practice: A decision-support system to assess fish community response to water-management scenarios, Water, 9, 196, https://doi.org/10.3390/w9030196, 2017.

Euser, T., Winsemius, H. C., Hrachowitz, M., Fenicia, F., Uhlenbrook, S., and Savenije, H. H. G.: A framework to assess the realism of model structures using hydrological signatures, Hydrol. Earth Syst. Sci., 17, 1893–1912, https://doi.org/10.5194/hess-17-1893-2013, 2013.

Gillespie, B. R., Desmet, S., Kay, P., Tillotson, M. R., and Brown, L. E.: A critical analysis of regulated river ecosystem responses to managed environmental flows from reservoirs, Freshwater Biol., 60, 410–425, https://doi.org/10.1111/fwb.12506, 2015.

Gupta, H. V., Kling, H., Yilmaz, K. K., and Martinez, G. F.: Decomposition of the mean squared error and NSE performance criteria: Implications for improving hydrological modelling, J. Hydrol., 377, 80–91, https://doi.org/10.1016/j.jhydrol.2009.08.003, 2009.

Hailegeorgis, T. T. and Alfredsen, K.: Regional statistical and precipitation-runoff modelling for ecological applications: Prediction of hourly streamflow in regulated rivers and ungauged basins, River Res. Appl., 33, 233–248, https://doi.org/10.1002/rra.3006, 2016.

Hingray, B., Schaefli, B., Mezghani, A., and Hamdi, Y.: Signature-based model calibration for hydrological prediction in mesoscale Alpine catchments, Hydrolog. Sci. J., 55, 1002–1016, https://doi.org/10.1080/02626667.2010.505572, 2010.

Hoos, A. B.: Recharge rates and aquifer hydraulic characteristics for selected drainage basins in middle and east Tennessee, U.S. Geological Survey, Nashville, Tennessee, USA, Water Resources Investigations Report 90–4015, 39 pp., 1990.

Jothityangkoon, C., Sivapalan, M., and Farmer, D. L.: Process controls of water balance variability in a large semi-arid catchment: Downward approach to hydrological model development, J. Hydrol., 254, 174–198, https://doi.org/10.1016/S0022-1694(01)00496-6, 2001.

Kiesel, J., Guse, B., Pfannerstill, M., Kakouei, K., Jähnig, S. C., and Fohrer, N.: Improving hydrological model optimization for riverine species, Ecol. Indic., 80, 376–385, https://doi.org/10.1016/j.ecolind.2017.04.032, 2017.

Knight, R. R., Gregory, M. B., and Wales, A. K.: Relating streamflow characteristics to specialized insectivores in the Tennessee River Valley: A regional approach, Ecohydrology, 1, 394–407, https://doi.org/10.1002/eco.32, 2008.

Knight, R. R., Gain, W. S., and Wolfe, W. J.: Modelling ecological flow regime: an example from the Tennessee and Cumberland River basins, Ecohydrology, 5, 613–627, https://doi.org/10.1002/eco.246, 2012.

Knight, R. R., Murphy, J. C., Wolfe, W. J., Saylor, C. F., and Wales, A. K.: Ecological limit functions relating fish community response to hydrologic departures of the ecological flow regime in the Tennessee River basin, United States, Ecohydrology, 7, 1262–1280, https://doi.org/10.1002/eco.1460, 2014.

Krause, P., Boyle, D. P., and Bäse, F.: Comparison of different efficiency criteria for hydrological model assessment, Adv. Geosci., 5, 89–97, https://doi.org/10.5194/adgeo-5-89-2005, 2005.

Law, G. S., Tasker, G. D., and Ladd, D. E.: Streamflow-characteristic estimation methods for unregulated streams of Tennessee, U.S. Geological Survey, Reston, Virginia, USA, Scientific Investigations Report 2009–5159, 212 pp., 2009.

Legates, D. R. and McCabe, G. J.: Evaluating the use of "goodness-of-fit" measures in hydrologic and hydroclimatic model validation, Water Resour. Res., 35, 233–241, https://doi.org/10.1029/1998WR900018, 1999.

Lindström, G., Johansson, B., Persson, M., Gardelin, M., and Bergström, S.: Development and test of the distributed HBV-96 hydrological model, J. Hydrol., 201, 272—288, https://doi.org/10.1016/S0022-1694(97)00041-3, 1997.

Murphy, J. C., Knight, R. R., Wolfe, W. J., and Gain, W. S.: Predicting ecological flow regime at ungauged sites: A comparison of methods, River Res. Appl., 29, 660–669, https://doi.org/10.1002/rra.2570, 2013.

Nash, J. E. and Sutcliffe, J. V.: River flow forecasting through conceptual models part I – A discussion of principles, J. Hydrol., 10, 282–290, https://doi.org/10.1016/0022-1694(70)90255-6, 1970.

Olden, J. D. and Poff, N. L.: Redundancy and the choice of hydrologic indices for characterizing streamflow regimes, River Res. Appl., 19, 101–121, https://doi.org/10.1002/rra.700, 2003.

Olsen, M., Troldborg, L., Henriksen, H. J., Conallin, J., Refsgaard, J. C., and Boegh, E.: Evaluation of a typical hydrological model in relation to environmental flows, J. Hydrol., 507, 52–62, https://doi.org/10.1016/j.jhydrol.2013.10.022, 2013.

Omernik, J. M.: Ecoregions of the Conterminous United States, Ann. Assoc. Am. Geogr., 77, 118–125, https://doi.org/10.1111/j.1467-8306.1987.tb00149.x, 1987.

Pfannerstill, M., Guse, B., and Fohrer, N.: Smart low flow signature metrics for an improved overall performance evaluation of hydrological models, J. Hydrol., 510, 447–458, https://doi.org/10.1016/j.jhydrol.2013.12.044, 2014.

Poff, N. L. and Zimmerman, J. K.: Ecological responses to altered flow regimes: A literature review to inform the science and management of environmental flows, Freshwater Biol., 55, 194–205, https://doi.org/10.1111/j.1365-2427.2009.02272.x, 2010.

Poff, N. L., Allan, J. D., Bain, M. B., Karr, J. R., Prestegaard, K. L., Richter, B. D., Sparks, R. E., and Stromberg, J. C.: The natural flow regime, BioScience, 47, 769–784, https://doi.org/10.2307/1313099, 1997.

Poff, N. L., Richter, B. D., Arthington, A. H., Bunn, S. E., Naiman, R. J., Kendy, E., Acreman, M., Apse, C., Bledsoe, B. P., Freeman, M. C., Henriksen, J., Jacobson, R. B., Kennen, J. G., Merritt, D. M., O'Keeffe, Y. H., Olden, J. D., Rogers, K., Tharme, R. E., and Warner, A.: The ecological limits of hydrologic alteration (ELOHA): A new framework for developing regional environmental flow standards, Freshwater Biol., 55, 147–170, https://doi.org/10.1111/j.1365-2427.2009.02204.x, 2010.

Richter, B. D., Baumgartner, J. V., Powell, J., and Braun, D. P.: A method for assessing hydrologic alteration within ecosystems, Conserv. Biol., 10, 1163–1174, https://doi.org/10.1046/j.1523-1739.1996.10041163.x, 1996.

Rotstayn, L. D., Roderick, M. L., and Farquhar, G. D.: A simple pan-evaporation model for analysis of climate simulations: Evaluation over Australia, Geophys. Res. Lett., 33, L7715, https://doi.org/10.1029/2006GL027114, 2006.

Ryo, M., Iwasaki, Y., and Yoshimura, C.: Evaluation of spatial pattern of altered flow regimes on a river network using a distributed hydrological model, PloS ONE, 10, e0133833, https://doi.org/10.1371/journal.pone.0133833, 2015.

Sanborn, S. C. and Bledsoe, B. P.: Predicting streamflow regime metrics for ungauged streams in Colorado, Washington, and Oregon, J. Hydrol., 325, 241–261, https://doi.org/10.1016/j.jhydrol.2005.10.018, 2006.

Sawicz, K., Wagener, T., Sivapalan, M., Troch, P. A., and Carrillo, G.: Catchment classification: empirical analysis of hydrologic similarity based on catchment function in the eastern USA, Hydrol. Earth Syst. Sci., 15, 2895–2911, https://doi.org/10.5194/hess-15-2895-2011, 2011.

Seibert, J.: Regionalization of parameters for a conceptual rainfall-runoff model, Agr. Forest Meteorol., 98–99, 279–293, https://doi.org/10.1016/S0168-1923(99)00105-7, 1999.

Seibert, J.: Multi-criteria calibration of a conceptual runoff model using a genetic algorithm, Hydrol. Earth Syst. Sci., 4, 215–224, https://doi.org/10.5194/hess-4-215-2000, 2000.

Seibert, J. and Vis, M. J. P.: Teaching hydrological modeling with a user-friendly catchment-runoff-model software package, Hydrol. Earth Syst. Sci., 16, 3315–3325, https://doi.org/10.5194/hess-16-3315-2012, 2012.

Shrestha, R. R., Peters, D. L., and Schnorbus, M. A.: Evaluating the ability of a hydrologic model to replicate hydroecologically relevant indicators, Hydrol. Process., 28, 4294–4310, https://doi.org/10.1002/hyp.9997, 2014.

Tharme, R. E.: A global perspective on environmental flow assessment: Emerging trends in the development and application of environmental flow methodologies for rivers, River Res. Appl., 19, 397–441, https://doi.org/10.1002/rra.736, 2003.

U.S. Department of Commerce: Divisional normals and standard deviations of temperature, precipitation, and heating and cooling degree days 1971–2000 (and previous normals periods), Section 2 precipitation, United States Department of Commerce, Washington, DC, USA, Climatography of the United States No. 85, 2007a.

U.S. Department of Commerce: Divisional normals and standard deviations of temperature, precipitation, and heating and cooling degree days 1971–2000 (and previous normals periods), Section 1 temperature, United States Department of Commerce: Washington, DC, USA, Climatography of the United States No. 85, 2007b.

U.S. Geological Survey: EflowStats R-package, available at: https://github.com/USGS-R/EflowStats (last access: July 2016), 2014.

U.S. Geological Survey: The National Map, 3D Elevation Program Products and Services Web page, available at: http://nationalmap.gov/3DEP/3dep_prodserv.html (last access: November 2015), 2016a.

U.S. Geological Survey: National Water Information System – Web interface, https://doi.org/10.5066/F7P55KJN, 2016b.

Viglione, A., Parajka, J., Rogger, M., Salinas, J. L., Laaha, G., Sivapalan, M., and Blöschl, G.: Comparative assessment of predictions in ungauged basins – Part 3: Runoff signatures in Austria, Hydrol. Earth Syst. Sci., 17, 2263–2279, https://doi.org/10.5194/hess-17-2263-2013, 2013.

Vis, M., Knight, R., Pool, S., Wolfe, W., and Seibert, J.: Model calibration criteria for estimating ecological flow characteristics, Water, 7, 2358–2381, https://doi.org/10.3390/w7052358, 2015.

Wagener, T., Sivapalan, M., Troch, P., and Woods, R.: Catchment classification and hydrologic similarity, Geography Compass, 1, 901–931, https://doi.org/10.1111/j.1749-8198.2007.00039.x, 2007.

Westerberg, I. K., Guerrero, J.-L., Younger, P. M., Beven, K. J., Seibert, J., Halldin, S., Freer, J. E., and Xu, C.-Y.: Calibration of hydrological models using flow-duration curves, Hydrol. Earth Syst. Sci., 15, 2205–2227, https://doi.org/10.5194/hess-15-2205-2011, 2011.

Westerberg, I. K., Wagener, T., Coxon, G., McMillan, H. K., Castellarin, A., Montanari, A., and Freer, J.: Uncertainty in hydrological signatures for gauged and ungauged catchments, Water Resour. Res., 52, 1847–1865, https://doi.org/10.1002/2015WR017635, 2016.

Wolfe, W., Haugh, C., Webbers, A., and Diehl, T.: Preliminary conceptual models of the occurrence, fate, and transport of chlorinated solvents in karst regions of Tennessee, U.S. Geological Survey, Nashville, Tennessee, USA, Water Resources Investigations Report 97–4097, 88 pp., 1997.

Yadav, M., Wagener, T., and Gupta, H.: Regionalization of constraints on expected watershed response behavior for improved predictions in ungauged basins, Adv. Water Resour., 30, 1756–1774, https://doi.org/10.1016/j.advwatres.2007.01.005, 2007.

Yilmaz, K. K., Gupta, H. V., and Wagener, T.: A process-based diagnostic approach to model evaluation: Application to the NWS distributed hydrologic model, Water Resour. Res., 44, W09417, https://doi.org/10.1029/2007WR006716, 2008.

15

Benchmarking ensemble streamflow prediction skill in the UK

Shaun Harrigan[1,2], **Christel Prudhomme**[1,2,3], **Simon Parry**[2], **Katie Smith**[2], **and Maliko Tanguy**[2]

[1]European Centre for Medium-Range Weather Forecasts (ECMWF), Shinfield Park, Reading, RG2 9AX, UK
[2]Centre for Ecology & Hydrology, Wallingford, Oxfordshire, OX10 8BB, UK
[3]Department of Geography, Loughborough University, Loughborough, Leicestershire, LE11 3TU, UK

Correspondence: Shaun Harrigan (shaun.harrigan@ecmwf.int)

Abstract. Skilful hydrological forecasts at sub-seasonal to seasonal lead times would be extremely beneficial for decision-making in water resources management, hydropower operations, and agriculture, especially during drought conditions. Ensemble streamflow prediction (ESP) is a well-established method for generating an ensemble of streamflow forecasts in the absence of skilful future meteorological predictions, instead using initial hydrologic conditions (IHCs), such as soil moisture, groundwater, and snow, as the source of skill. We benchmark when and where the ESP method is skilful across a diverse sample of 314 catchments in the UK and explore the relationship between catchment storage and ESP skill. The GR4J hydrological model was forced with historic climate sequences to produce a 51-member ensemble of streamflow hindcasts. We evaluated forecast skill seamlessly from lead times of 1 day to 12 months initialized at the first of each month over a 50-year hindcast period from 1965 to 2015. Results showed ESP was skilful against a climatology benchmark forecast in the majority of catchments across all lead times up to a year ahead, but the degree of skill was strongly conditional on lead time, forecast initialization month, and individual catchment location and storage properties. UK-wide mean ESP skill decayed exponentially as a function of lead time with continuous ranked probability skill scores across the year of 0.75, 0.20, and 0.11 for 1-day, 1-month, and 3-month lead times, respectively. However, skill was not uniform across all initialization months. For lead times up to 1 month, ESP skill was higher than average when initialized in summer and lower in winter months, whereas for longer seasonal and annual lead times skill was higher when initialized in autumn and winter months and lowest in spring. ESP was most skilful in the south and east of the UK, where slower responding catchments with higher soil moisture and groundwater storage are mainly located; correlation between catchment base flow index (BFI) and ESP skill was very strong (Spearman's rank correlation coefficient $= 0.90$ at 1-month lead time). This was in contrast to the more highly responsive catchments in the north and west which were generally not skilful at seasonal lead times. Overall, this work provides scientific justification for when and where use of such a relatively simple forecasting approach is appropriate in the UK. This study, furthermore, creates a low cost benchmark against which potential skill improvements from more sophisticated hydro-meteorological ensemble prediction systems can be judged.

1 Introduction

Skilful hydrological forecasts at sub-seasonal to seasonal lead times would provide a valuable tool for improved decision making for wide range of sectors such as water resources management (Anghileri et al., 2016), hydropower operations (Hamlet et al., 2002), and agriculture (Letcher et al., 2004), particularly in times of slow onset events such as drought (Simpson et al., 2016). One of the earliest operational hydrological forecasting methods is ensemble streamflow prediction (ESP). ESP was pioneered in the US at the National Weather Service (NWS) during the 1970s and 1980s as a means of providing ensemble forecasts of streamflow for a variety of lead times from 1 day to seasonal and beyond (Day, 1985; Twedt et al., 1977; originally stood for Extended Streamflow Prediction). Two years of severe drought in California in 1976 and 1977 provided the motivation for such hydrological forecasting developments at the time (Wood et al., 2016b). In the UK, the 2010–2012 drought in England and Wales provided the impetus for the establishment of the first operational seasonal hydrological forecasting service, the Hydrological Outlook UK (HOUK), which went live in June 2013 (Prudhomme et al., 2017; forecasts available at: http://www.hydoutuk.net/). ESP is used as one of three hydrological forecasting methods in HOUK and also feeds into the Environment Agency's monthly "Water Situation Report for England" (operational for groundwater levels in March 2012), providing forward look ESP forecasts of streamflow for 29 catchments out to a 12-month lead time (https://www.gov.uk/government/collections/water-situation-reports-for-england).

In the traditional formulation of ESP, as used in this paper, historical sequences of climate data (precipitation, potential evapotranspiration, and/or temperature) at the time of forecast are used to force hydrological models, providing a plausible range of representations of the future streamflow states. The source of ESP skill is therefore due to initial hydrologic conditions (IHCs) from antecedent stores of soil moisture, groundwater, snowpack, and channel streamflow itself (Wood et al., 2016a; Wood and Lettenmaier, 2008) which can be detectable up to a year ahead (Staudinger and Seibert, 2014), rather than from skilful atmospheric forecasts. The original operational concept of the NWS ESP forecasting system was that it was flexible, easy to use, and could be run efficiently using simple conceptual hydrological models (Day, 1985). Traditional ESP, while simple, is still widely used today in operational seasonal hydrological forecasting (e.g. US NWS and HOUK) and as a low cost forecast against which to benchmark potential skill improvements from more sophisticated hydro-meteorological ensemble prediction systems (e.g. Arnal et al., 2017; Crochemore et al., 2017; Pappenberger et al., 2015; Thober et al., 2015; Wood et al., 2005).

Several studies have established the skill of the ESP method for catchments in particular regions based on carefully constructed hindcast experiments. For example, in the western US, Franz et al. (2003) found ESP forecasts in 14 snow dominated catchments were, on average, skilful (compared to benchmark climatology forecasts) with a lead time up to 7 months, particularly when initialized early in the spring snowmelt season. Wood and Lettenmaier (2008) found that information about IHCs was more important than climate information during the transition between wet and dry seasons in two western US catchments up to a 5-month lead time. For non-snow dominated catchments in the southeast of the US, Li et al. (2009) showed that harnessing the long memory of soil moisture and groundwater stores can provide skilful ESP forecasts, as the impact of anomalous dry or wet conditions can take weeks or months to dissipate. Wang et al. (2011) found simple conceptual rainfall-runoff models were able to reliably estimate conditional catchment IHCs in two east Australian catchments, subsequently producing ESP forecasts of comparable skill to the current operational Bayesian Joint Probability statistical forecast system (BJP, Wang et al., 2009) at 1- and 3-month lead times. More recently, Singh (2016) assessed the potential for long-range ESP forecasting for integrated water management in four catchments (two rainfall dominated and two snowfall dominated) on the South Island of New Zealand and found ESP to be skilful out to a 3-month lead time, with greatest improvements over climatology forecasts in summer. The previous studies demonstrate that the traditional ESP method is skilful at both short and long lead times in many regions around the world and, given its relative ease of application and low computational cost, remains a valuable ensemble hydrological forecasting approach. Although ESP is being used operationally within the UK, its skill has not yet been investigated at the catchment scale within a rigorous hindcast experiment and is therefore the focus of this paper.

By definition, a forecast can only be considered *skilful* if it is more accurate against observations than some simpler and/or cheaper reference or *benchmark* forecast (Jolliffe and Stephenson, 2003; Wilks, 2011). Pappenberger et al. (2015) identified three classes of benchmark forecasts commonly used in hydrological forecasting: (i) climatology, used for seasonal forecasting, (ii) persistence, used for short range forecasting, and (iii) simplified hydrology models, for testing whether more complex models provide useful skill gains. We define the process of *benchmarking* as establishing the skill of a forecasting system (here ESP) against a simpler benchmark forecast across various lead times, forecast initialization months, and for a large sample of diverse catchments within the study domain. Consequently, the aim of this paper is to establish the skill of the traditional ESP method for forecasting streamflow in the UK at the catchment scale using (streamflow) climatology as the benchmark forecast within a rigorous 50-year hindcast study design. Three key research questions emerge:

1. When is ESP skilful, in terms of a wide range of lead times and forecast initialization months?
2. Where is ESP skilful, in terms of spatial distribution of skilful forecasts both regionally and at the individual catchment scale across the UK?
3. Why is ESP skilful, in terms of individual catchment storage capacity?

Section 2 describes the hydroclimatic data used and the selection of catchments, Sect. 3 outlines the methods leading to the generation of ESP hindcasts. Results are presented in Sect. 4 and discussed in Sect. 5, before key conclusions and avenues for further work are offered in Sect. 6. Details about how to access the ESP hindcast archive used in this study as well as supplementary data and figures are given in the "Data availability" section at the end of the article.

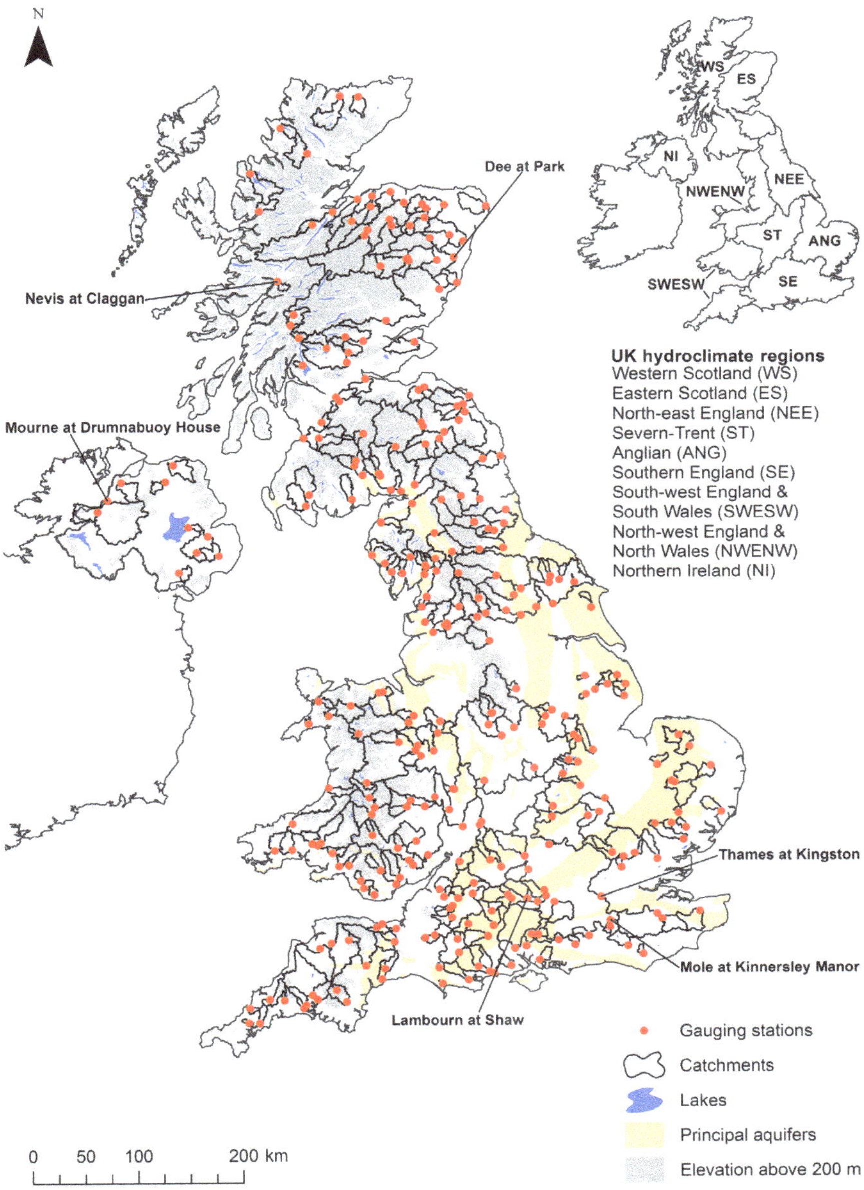

Figure 1. Location of 314 gauging stations (red dots) and catchment boundaries (black lines) with upland areas (shaded in grey) and principal aquifers (shaded in pale yellow). UK Hydroclimate Regions, derived from grouping smaller UK hydrometric areas, are shown inset.

2 Data

We selected a set of 314 catchments for our ESP evaluation from the UK National River Flow Archive (NRFA; http://nrfa.ceh.ac.uk/), chosen to be representative of the range of UK hydroclimatic conditions and ensuring good spatial coverage (Fig. 1). These catchments include those used for routinely assessing the current and future UK hydrological status (e.g. National Hydrological Monitoring Programme, 2017) as well as 128 catchments that are part of the new version of the UK Benchmark Network (UKBN2; Harrigan et al., 2017) that can be considered relatively free from human disturbances such as water abstractions, urbanization, and reservoir impacts. Individual details of all 314 catchments are given in the Supplement Table S1.

Observed catchment average daily mean streamflow Q ($m^3 s^{-1}$), daily precipitation P ($mm\,d^{-1}$), and daily potential evapotranspiration ET_p ($mm\,d^{-1}$) were extracted for each catchment and are needed for three tasks: (i) as input to the hydrological model calibration (Q, P, and ET_p; Sect. 3.1); (ii) to generate historic climate sequences (P and ET_p, Sect. 3.2) used as forcing to the ESP method; and (iii) as forcing to the reference simulation (P and ET_p; i.e. proxy observations in Sect. 3.3).

Q was retrieved from the NRFA over the longest possible period of observed Q across the 314 stations, 32 water years from 1983 to 2014 (water year from 1 October to 30 September referred to by the calendar year in which it ends). P was retrieved from the 1 km gridded CEH-GEAR dataset (Keller et al., 2015; Tanguy et al., 2016) between 1961 and 2015 for the UK. ET_p according to Penman–Monteith for FAO-defined well-watered grass was retrieved from the 1 km gridded CHESS-PE dataset (Robinson et al., 2016, 2017) between 1961 and 2015 for catchments in Great Britain. CHESS-PE does not cover Northern Ireland, so an alternative 5 km ET_p dataset for the UK based on the temperature-based McGuinness–Bordne equation was used for these 10 catchments instead (Tanguy et al., 2017, 2018).

Catchment characteristics are summarized in Table 1 for the UK and nine hydroclimate regions as shown in Fig. 1 inset. The nine UK Hydroclimate Regions were derived by merging contiguous UK hydrometric areas (National River Flow Archive, 2014) that reflect broad hydrological and climatological similarity across the UK and are used for aiding interpretation of results. The distribution of the 314 catchments within the nine regions varies between 10 in Northern Ireland (NI) and 59 in Southern England (SE). Catchment areas range from 4.4 to 9948 km^2 with a median area of 181 km^2. There is a distinctive hydroclimatic gradient in the UK with wetter more responsive upland catchments in the north and west, and drier lowland catchments in the south and east, many of which drain the principal Chalk and Limestone aquifers. The slow flow contribution from groundwater and other delayed sources, such as lakes, snow, and soil water storage, was characterized using the base flow index (BFI; Gustard et al., 1992) obtained from UK NRFA metadata. BFI ranges between 0 and 1 with values ~ 0.15–0.35 representative of more responsive rainfall-runoff regimes in the north and west whereas many Chalk rivers in the south east have a BFI ≥ 0.9. Three regions (Severn-Trent (ST), Anglian (ANG), and SE) have median runoff ratios (RR) < 0.5 meaning more precipitation is lost to evaporation than runoff in the majority of these catchments. Less than 5 % of catchments have a significant amount of snowfall, defined here following Berghuijs et al. (2014) as catchments with a long-term mean fraction of precipitation falling as snow $\overline{F}_s > 0.15$, and are mainly situated in Eastern Scotland (ES). The range of these hydroclimatic characteristics provide a large and diverse set of catchments to benchmark ESP skill.

3 Methods

3.1 Hydrological modelling

The first of four key methodological steps was to calibrate and evaluate the GR4J (Génie Rural à 4 paramètres Journalier) model (Perrin et al., 2003) used for the generation of streamflow series. It is a daily lumped catchment rainfall-runoff model with a parsimonious structure consisting of four free parameters that require calibration against streamflow observations using daily P and ET_p as input. GR4J has been shown to reliably simulate the hydrology of a diverse set of catchments (Perrin et al., 2003) including temporal transition between wet and dry periods (Broderick et al., 2016), and for the generation of ESP forecasts (e.g. Pagano et al., 2010). The GR4J structure includes a soil moisture accounting reservoir (capacity controlled with parameter X1 [mm]), a water exchange function (rate controlled by parameter X2 [mm d^{-1}]), and a non-linear routing store to represent baseflow (capacity determined by parameter X3 [mm]), with rainfall-runoff time lags (set in days by parameter X4 [d]) controlled by two unit hydrographs.

GR4J was calibrated using the open source "airGR" package v1.0.2 in R (Coron et al., 2016, 2017) with the inbuilt calibration optimization algorithm based on a steepest descent local search procedure and default parameter ranges. The modified Kling–Gupta efficiency (KGE_{mod}; Gupta et al., 2009; Kling et al., 2012) applied to root squared transformed flows KGE_{mod}[sqrt] was used as the objective function for automatic fitting, thus placing weight on mid-range flows, rather than high or low flows. This was decided given ESP forecasts are made across the year during both dry and wet conditions. A split sample test (Klemeš, 1986) was used by dividing the 32-year complete period (CP; water years 1983–2014) of available streamflow observations into two equal 16-year segments for calibration and evaluation: period 1 (P1; water years 1983–1998) and period 2 (P2; water years 1999–2014). Three calibrated GR4J parameter sets were created for each catchment using data from P1, P2, and CP, thus allowing testing of parameter stability between P1 and P2. Model performance against streamflow observations was evaluated using KGE_{mod}[sqrt], the Nash–Sutcliffe efficiency (NSE; Nash and Sutcliffe, 1970), and percent bias (PBIAS; Gupta et al., 1999) to assess water balance errors.

The UK-wide median (5th and 95th percentile) KGE_{mod}[sqrt] is 0.94 (0.83, 0.97) for calibration (CP) and for evaluation 0.92 (0.80, 0.96) and 0.92 (0.78, 0.96) for P1 and P2, respectively (Table 2). Median PBIAS across all catchments over CP is low, −0.1 % (−3.7, 0.7 %). Overall, GR4J performs well against streamflow observations and parameter sets remain stable across P1 and P2 with comparable performance to Crochemore et al. (2017) and Poncelet et al. (2017) using GR6J for catchments across France, Germany, and Austria. For completeness and comparison with other works, the NSE was calculated as it is the most universally used metric. Spatial maps and summary statistics for KGE_{mod}[sqrt] and NSE are provided in Fig. S1 in the Supplement and, notwithstanding differences in study design, results for GR4J are on par with other large sample catchment modelling studies in the UK (e.g. Crooks et al., 2009, using the probability distributed

Table 1. Summary statistics of eight catchment characteristics for the UK and nine hydroclimate regions shown in Fig. 1. The median across n catchments within each region is given with the 5th and 95th percentile ranges in parentheses. Area, median elevation, and base flow index (BFI) were retrieved from the UK NRFA. Mean annual Q, P, and ET_P were calculated over water years 1983–2014 using data in Sect. 2. RR is the runoff ratio and $\overline{F}_s$ * is the long term (water years 1983–2014) mean fraction of precipitation that has fallen as snow.

Region	n	Area (km^2)	Median elevation (m a.s.l.)	BFI (–)	Mean annual Q ($mm\,yr^{-1}$)	Mean annual P ($mm\,yr^{-1}$)	Mean annual ET_p ($mm\,yr^{-1}$)	RR $\overline{Q}/\overline{P}$ (–)	$\overline{F}_s$ (–)
UK	314	181 (27, 1844)	179 (60, 437)	0.5 (0.27, 0.89)	595 (162, 1839)	1031 (648, 2202)	504 (400, 542)	0.59 (0.24, 0.87)	0.03 (0.01, 0.14)
WS	35	229 (64, 1745)	268 (146, 468)	0.33 (0.20, 0.61)	1115 (554, 2847)	1460 (998, 3145)	428 (391, 476)	0.74 (0.58, 0.90)	0.06 (0.03, 0.12)
ES	43	289 (70, 2759)	303 (100, 596)	0.51 (0.34, 0.67)	693 (338, 1498)	1040 (783, 1970)	432 (387, 481)	0.63 (0.44, 0.84)	0.09 (0.06, 0.21)
NEE	30	344 (11, 1910)	264 (88, 449)	0.43 (0.26, 0.82)	559 (344, 1054)	1037 (757, 1462)	486 (455, 516)	0.57 (0.44, 0.83)	0.07 (0.04, 0.09)
ST	25	198 (48, 6345)	145 (87, 312)	0.56 (0.41, 0.79)	392 (209, 844)	858 (670, 1311)	511 (493, 528)	0.46 (0.31, 0.68)	0.03 (0.02, 0.05)
ANG	33	99 (23, 1540)	80 (33, 132)	0.56 (0.25, 0.88)	183 (128, 254)	655 (600, 716)	535 (528, 551)	0.27 (0.21, 0.36)	0.03 (0.03, 0.04)
SE	59	134 (18, 1091)	105 (43, 178)	0.64 (0.23, 0.96)	356 (146, 568)	856 (654, 1033)	529 (520, 541)	0.42 (0.20, 0.64)	0.02 (0.01, 0.03)
SWESW	47	174 (29, 915)	207 (77, 377)	0.51 (0.37, 0.67)	979 (507, 1549)	1372 (1002, 1971)	519 (495, 537)	0.69 (0.51, 0.83)	0.01 (0.00, 0.03)
NWENW	32	112 (30, 1094)	210 (108, 360)	0.35 (0.27, 0.58)	1154 (390, 2102)	1529 (884, 2429)	478 (457, 514)	0.75 (0.44, 0.91)	0.04 (0.02, 0.05)
NI	10	230 (68, 1235)	140 (90, 172)	0.38 (0.33, 0.50)	688 (533, 1206)	1111 (917, 1565)	475 (466, 488)	0.63 (0.57, 0.77)	0.01 (0.00, 0.02)

* $\overline{F}_s$ calculated using the CemaNeige snow-accounting module (Valéry et al., 2014) within the airGR package (Coron et al., 2016, 2017) applied to the GR4J model (Perrin et al., 2003).

model (PDM; Moore, 2007) for 120 catchments). All streamflow simulations (proxy observations, and benchmark and ESP forecasts) were generated using model parameter sets calibrated over CP and with KGE_{mod}[sqrt] as objective function; median and ranges of calibrated parameter values for GR4J X1, ..., X4 across the UK and nine hydroclimate regions are given in Table 2 and for individual catchments in Table S1 along with respective performance metrics.

3.2 Generation of ESP hindcasts from historic climate data

In step 2, initial hydrologic conditions (IHCs) were estimated for each catchment and forecast initialization date by forcing the calibrated GR4J model with 4 years of observed P and ET_p previous to the forecast initialization date, over the 1961–2015 period, thus the first usable forecast date after model spin up is 1 January 1965. Secondly, a 51-member ensemble m of streamflow hindcasts was generated for each forecast initialization date (first of each month) by forcing GR4J with 51 historic climate sequences (P and ET_p pairs) extracted from 1961 to 2015 out to a 12-month lead time at a daily time step. Each of the 51 generated hindcast time series were then temporally aggregated to provide a forecast of mean streamflow over seamless lead times of 1 day to 12 months, resulting in 365 lead times per forecast (leap days were removed). Following convention in the HOUK, lead time (LT) in this paper refers to the streamflow (expressed as mean daily streamflow) over the period from the forecast initialization date to n days (or months) ahead in time. So a January ESP forecast with 1-month lead time is the mean daily streamflow from 1 January to the end of January and a January forecast with 2-month lead time is the mean daily streamflow from 1 January to the end of February.

Although it is not possible to create a hindcast experiment under exactly the same conditions experienced in operational mode, effort was made to ensure historic climate sequences did not artificially inflate skill (see Robertson et al., 2016) by using leave-three-years-out cross-validation (L3OCV) whereby the 12-month forecast window and the two succeeding years were not used as climate forcings. This was done to account for persistence from known large-scale climate–streamflow teleconnections such as the North Atlantic Oscillation with influences lasting from several seasons to years (Dunstone et al., 2016). Because this climate information could be an advantage, but is not available in

Table 2. Summary statistics of GR4J calibrated parameters and performance metrics for the UK and nine hydroclimate regions shown in Fig. 1. The median across n catchments within each region is given with the 5th and 95th percentile ranges in parentheses. Calibration (Cal) was over the complete period (CP; water years 1983–2014) and evaluation (Eval) for both period 1 (P1; water years 1983–1998) and period 2 (P2; 1999–2014).

Region	n	GR4J X1 (mm)	GR4J X2 (mm d^{-1})	GR4J X3 (mm)	GR4J X4 (d)	Cal (CP) KGE_{mod}[sqrt] (–)	Eval (P1) KGE_{mod}[sqrt] (–)	Eval (P2) KGE_{mod}[sqrt] (–)	Cal (CP) PBIAS (%)
UK	314	250 (78, 955)	−0.1 (−4.2, 0.8)	40 (12, 380)	1.3 (1.0, 2.6)	0.94 (0.83, 0.97)	0.92 (0.80, 0.96)	0.92 (0.78, 0.96)	−0.1 (−3.7, 0.7)
WS	35	130 (46, 438)	0.0 (−0.6, 0.6)	27 (14, 130)	1.2 (1.1, 2.1)	0.93 (0.83, 0.96)	0.92 (0.82, 0.95)	0.91 (0.81, 0.95)	0.1 (−2.2, 1.2)
ES	43	296 (112, 523)	0.0 (−0.7, 0.8)	43 (18, 104)	1.2 (1.1, 1.8)	0.90 (0.74, 0.94)	0.88 (0.74, 0.94)	0.88 (0.71, 0.94)	−0.5 (−2.2, 0.4)
NEE	30	277 (79, 499)	0.0 (−1.1, 0.7)	24 (12, 109)	1.3 (1.1, 2.3)	0.92 (0.87, 0.95)	0.91 (0.83, 0.94)	0.90 (0.78, 0.93)	−0.2 (−7.1, 0.4)
ST	25	345 (142, 1169)	−0.5 (−1.0, 0.5)	44 (18, 153)	1.4 (1.1, 2.7)	0.96 (0.88, 0.97)	0.93 (0.83, 0.96)	0.92 (0.80, 0.96)	0.2 (−1.6, 0.7)
ANG	33	286 (128, 773)	−0.8 (−4.5, −0.1)	28 (5, 371)	1.5 (1.2, 2.7)	0.92 (0.86, 0.95)	0.88 (0.82, 0.94)	0.88 (0.81, 0.94)	−0.2 (−8.7, 1.4)
SE	59	411 (160, 1877)	−0.7 (−17.2, 1.0)	77 (6, 703)	1.4 (1.0, 9.5)	0.95 (0.88, 0.97)	0.92 (0.82, 0.96)	0.92 (0.8, 0.96)	−0.1 (−5.0, 0.4)
SWESW	47	205 (83, 459)	0.1 (−1.0, 0.9)	81 (29, 182)	1.2 (0.9, 2.0)	0.97 (0.94, 0.97)	0.94 (0.86, 0.97)	0.94 (0.85, 0.96)	−0.3 (−1.2, 0.3)
NWENW	32	141 (60, 480)	0.2 (−0.6, 0.8)	36 (19, 134)	1.2 (1.1, 1.8)	0.95 (0.93, 0.97)	0.95 (0.88, 0.96)	0.94 (0.87, 0.96)	0.0 (−0.5, 0.4)
NI	10	146 (70, 244)	0.2 (−0.1, 0.3)	23 (16, 37)	1.4 (1.1, 1.9)	0.93 (0.91, 0.96)	0.93 (0.86, 0.95)	0.93 (0.86, 0.95)	−0.1 (−1.0, 0.9)

operational forecasting, it was not used in the hindcast experiment. Using the first forecast on 1 January 1965 as an example, 51 sequences of P and ET_p pairs of length 365 days (from 1 January to 31 December) were extracted from observed P and ET_p records between 1961 and 2015, but not for 1965, 1966, or 1967. To keep a 51-member ensemble across all hindcast years, forecasts made in 2013 and 2014 did not have enough data for L3OCV so in these cases climate sequences from 1961, and 1961 and 1962, respectively, were instead removed. The skill of ESP was evaluated over a 50-year hindcast period N between 1965 and 2015 for each of 12 initialization months i (January to December) and all 365 LTs. In total, 600 hindcasts were generated ($N \times i$) with 51 ensemble members each at 365 LTs across 314 catchments resulting in over 3.5×10^9 forecast values of streamflow in the ESP hindcast archive.

3.3 Creation of proxy streamflow observation series

In step 3, a proxy streamflow observation series was produced by forcing the calibrated GR4J model with observed P and ET_p over 1961–2015 with a 4-year model spin-up. A 4-year model spin up ensures model states are appropriately stabilized, especially important for slower responding catchments (e.g. in Southern England and Anglian regions). The proxy observation series, the best estimate of streamflow observations given current model and observed meteorological data, is used to evaluate ESP forecasts against. It is common to use this approach instead of using direct streamflow observations as it has the advantage of isolating loss of skill to IHCs rather than from model errors and biases (e.g. Alfieri et al., 2014; Pappenberger et al., 2015; Wood et al., 2016a; Yossef et al., 2013).

3.4 Evaluation of ESP skill

In step 4, forecast skill is presented as a skill score, which is the improvement over the benchmark forecast using some measure of accuracy A, given generically by Wilks (2011) in Eq. (1):

$$\text{skill score} = \frac{A_{\text{fc}} - A_{\text{bench}}}{A_{\text{perf}} - A_{\text{bench}}}, \tag{1}$$

where A_{fc} is the accuracy measure of the hydrological forecasting system Q_{fc} (here ESP) against observations $Q_{\text{obs*}}$ (here *proxy observations); A_{bench} is the accuracy measure of the benchmark forecast Q_{bench} against $Q_{\text{obs*}}$, and A_{perf} is the value of A in the case of a perfect forecast (typically

1 or 0 depending on metric). For each forecast made over the hindcast period the probabilistic skill of the full ESP 51-member ensemble forecast Q_{fc} was evaluated against a probabilistic climatology benchmark forecast Q_{bench}. Q_{bench} was calculated as the full sample climatological distribution of proxy streamflow observations over 1965–2015 for the forecast period. Similar to the historic climate forcing sequences in Sect. 3.2, the probabilistic climatology benchmark forecast was also created using L3OCV to account for persistence known to occur for several years in streamflow, particularly during drought (Wilby et al., 2015). In testing, we performed the skill evaluation with and without cross-validation of ESP forecasts and streamflow climatology benchmark forecasts. It was found that cross-validation was important, as in some cases failing to cross-validate ESP forecasts inflated skill scores whereas failing to cross-validate climatological benchmark forecasts deflated skill scores (i.e. the benchmark forecast was advantaged thereby disadvantaging ESP forecasts); in some cases skill scores were advantaged/disadvantaged by $\pm 15\,\%$.

The continuous ranked probability score (CRPS) (Hersbach, 2000) accuracy measure A, and corresponding skill score (CRPSS), was used for evaluating the probabilistic skill of ESP. The CRPS penalizes biased forecasts and those with low sharpness (Wilks, 2011). The Ferro et al. (2008) ensemble size correction for CRPS was applied to account for differences between the number of members in Q_{fc} (period 1961–2015 → L3OCV → $n = 51$) and Q_{bench} (period 1965–2015 → L3OCV → $n = 47$), as done in evaluation of hydrological ensemble forecasting elsewhere (e.g. Crochemore et al., 2017). Calculation of skill scores was undertaken using the open source “easyVerification” package v0.4.2 in R (MeteoSwiss, 2017).

The CRPS is one of the most recommended scores for evaluation of overall hydrological ensemble forecast performance (Pappenberger et al., 2015). However, several commonly used metrics were also calculated for evaluation of deterministic ESP performance (using the ESP ensemble mean): Pearson correlation coefficient (Cor.), the mean squared error skill score (MSESS), and the deterministic equivalent to CRPSS, the mean absolute error skill score (MAESS). The pattern of results in terms of where and when ESP is most/least skilful was found to be independent of chosen metric, with virtually identical results between probabilistic (using CRPSS) and deterministic (using MAESS) results (see Fig. S2), and so for brevity the remainder of paper is based on CRPSS only. A skill score of 1 indicates a perfect forecast, a skill score > 0 shows the ESP forecast is more skilful than the benchmark, a skill score $= 0$ shows ESP is only as accurate as the benchmark, and a skill score < 0 warns that ESP is inferior to the benchmark forecast. The CRPSS was applied to the 314 catchments for the 12 initialization months and 365 lead times for each year over the 50-year hindcast period.

4 Results

Results are presented in the following order: First, ESP skill is shown for all 365 lead times (LT), then by forecast initialization month for a sample of eight representative LTs commonly used in operational hydrological forecasting (i.e. short (1 and 3 days), extended (1 and 2 weeks), monthly (1 month), seasonal (3 and 6 months), and annual (12 months)). Second, the spatial distribution of ESP skill is shown, both averaged across the UK and each of the nine hydroclimate regions, then for individual catchments to explore sub-region heterogeneity. Third, the relationship between catchment storage and ESP skill is assessed.

Reducing accuracy of a forecast to a numeric skill metric value is abstract and difficult to interpret. Throughout the results and discussion sections skill score values are assigned qualitative descriptions according to degree of skill based on the CRPSS: very high [0.75, 1]; high [0.5, 0.75); moderate [0.25, 0.5); low (0, 0.25); no skill $= 0$, and negative skill < 0; CRPSS values which are near zero, defined between ± 0.05, are regarded as “neutrally skilful” (after Bennett et al., 2017). Five example 1965–2015 hindcast time series with skills ranging from very high to negative skill are visualized in Fig. 2 and act as a graphical reference in the remainder of the paper to aid interpretation of skill.

4.1 Timing of ESP skill

4.1.1 Lead time

UK-wide mean ESP skill across all catchments and initialization months decays exponentially as a function of lead time (Fig. 3). Mean CRPSS values from short (1-day) to extended (2-week) lead times range from 0.75 to 0.30, and across monthly, seasonal (3-month), and annual lead times from 0.20, 0.11, to 0.04, respectively. There is large spread around mean skill scores for any lead time, depicted by the semi-transparent 5th and 95th percentile bands across the 314 catchments in Fig. 3. For example, at a 2-week lead time CRPSS values are bound between 0.11 and 0.71, and for monthly lead times between 0.06 and 0.59. Skill scores for the deterministic ESP ensemble mean (measured by MAESS) are virtually the same as those for probabilistic forecasts (measured by CRPSS) for all lead times and regions (see Fig. S2c and d).

4.1.2 Initialization month

ESP skill varies depending on forecast initialization month (IM) and the time of year, with highest and lowest skill conditional on lead time. Figure 4 shows skill scores for initialization months January to December for short and extended lead times (LTs) as summarized by boxplots across all catchments. Skill scores for these four sample LTs (1-day, 3-day, 1-week, and 2-week) are highest in summer months (June, July, and August) with August the most skilful forecast IM

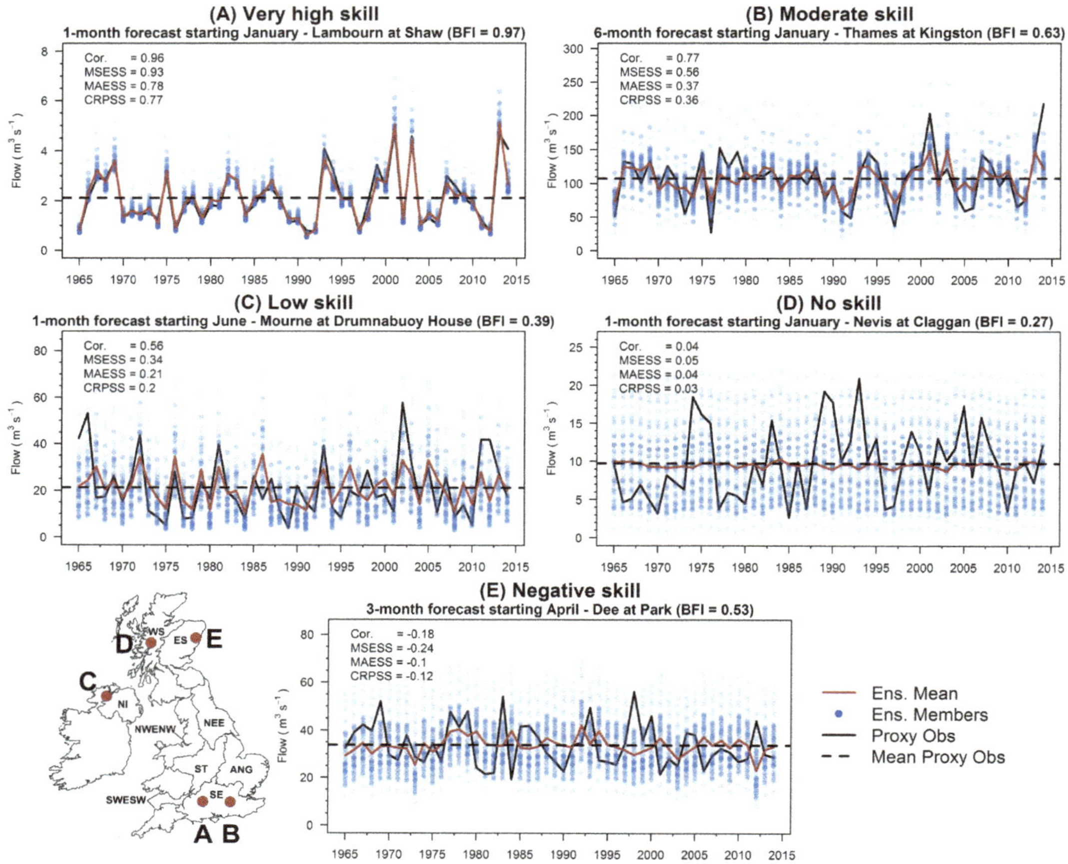

Figure 2. Five example 1965–2015 hindcast time series in which skill metrics range from very high **(a)** to negative skill **(e)**. The red line is the 51-member ESP ensemble mean, black line the proxy observed streamflow (also known as a perfect forecast), semi-transparent blue dots show the ensemble spread for each hindcast year, and the dashed horizontal black line shows mean proxy observed streamflow (analogous to a deterministic climatology benchmark forecast, although not cross-validated here as was done in calculation of skill scores (i.e. simply the same value repeated each year)).

on average, whereas skill is lower for winter months (December, January, and February) with January the least skilful forecast IM. Skill scores across IMs for the four sample monthly to annual LTs are shown in Fig. 5. Skill is also highest for the 1-month forecasts when initialized in August, however for 3-month, 6-month, and 12-month LTs, forecast skill is generally higher for autumn (September, October, and November) and winter IMs, with October the most skilful on average. All four monthly, seasonal, and annual LTs have lowest skill scores when initialized in spring months, particularly April, which in the UK is a transition month between winter months with lowest soil moisture deficits (SMDs) and warmer summer months with highest SMDs.

The decay in skill with LT as shown in Fig. 3 also occurs across all initialization months (Figs. 4 and 5). Whilst mean ESP skill tends towards zero for longer LTs, there are many catchments with much higher than average skill scores. For example, for 1-month LT ESP forecasts initialized in August the average UK-wide ESP skill is moderate (CRPSS = 0.30), but 36 catchments have high skill (CRPSS ≥ 0.5), and a CRPSS as high as 0.91 is achieved for the Lambourn at Shaw in Southern England.

4.2 Spatial distribution of ESP skill

4.2.1 UK hydroclimate regions

Figure 6 shows a heatmap of mean ESP skill across initialization months for the UK and for nine hydroclimate regions using the CRPSS metric. The same patterns are found for Cor., MSESS, and MAESS (Fig. S2). ESP skill has a prominent spatial pattern across the UK consistent over shorter and longer LTs. Least skilful UK regions are Western Scotland (WS), North-west England & North Wales (NWENW), and Northern Ireland (NI), whereas Severn-Trent (ST), Anglian (ANG), and Southern England (SE) are most skilful. Using a 1-week LT as an example, ESP is over twice as skilful in SE (CRPSS = 0.57) than in WS (CRPSS = 0.25). All regions are, on average, skilful out to 1-month LT, but by 3-month LT WS, NWENW, and NI are only neutrally skilful; at LTs up

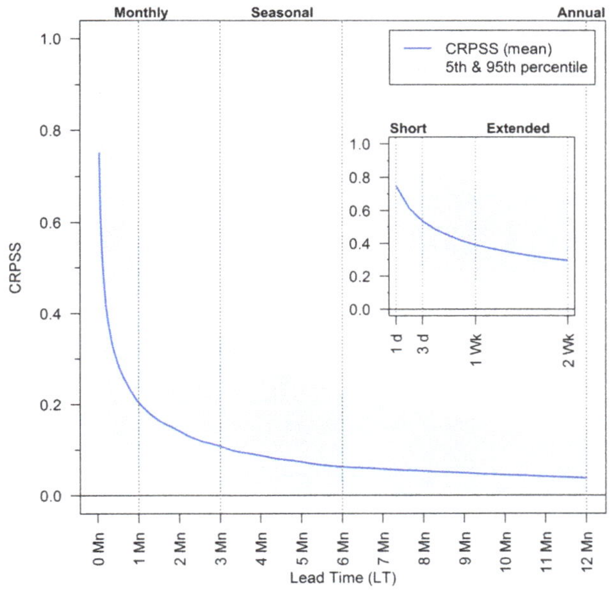

Figure 3. UK-wide mean ESP CRPSS values across all 314 catchments and 12 forecast initialization months for all 365 lead times (LTs) with short and extended lead times also shown inset for readability. The range of skill scores across catchments at each LT is shown by the semi-transparent 5th and 95th percentile band. Vertical lines represent eight commonly used operational forecasting LTs from short (1 and 3 days), extended (1 and 2 weeks), monthly (1 month), seasonal (3 and 6 months), to annual (12 months).

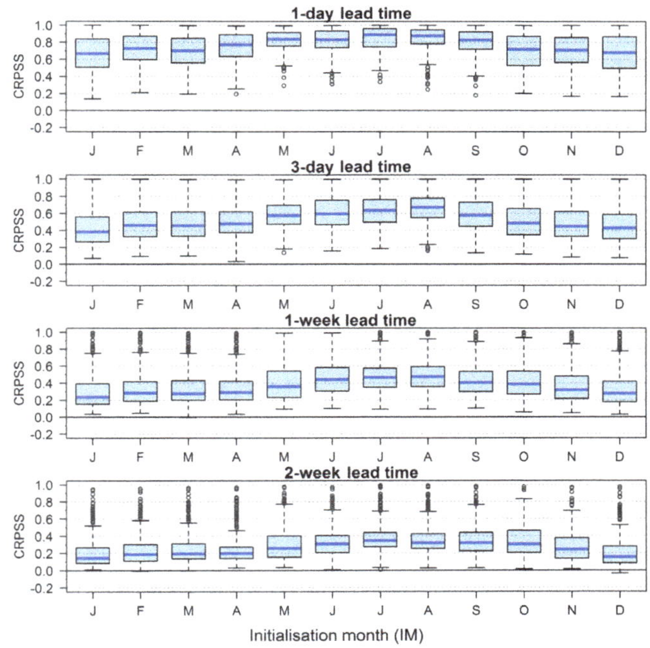

Figure 4. UK-wide ESP skill scores across 314 catchments for each of the 12 forecast initialization months for four short and extended lead times. Boxplots summarize CRPSS values with the blue line representing the median, and boxes the interquartile range (IQR); whiskers extend to the most extreme data point, which is no more than 1.5 times the IQR from the box, and grey circles are outliers beyond this range.

to 6 and 12 months ST, ANG, and SE are the only regions to remain skilful, as a whole.

4.2.2 Catchment scale

There is considerable sub-region heterogeneity when skill scores for individual forecasts at the catchment scale are examined. CRPSS values are mapped in Fig. 7 for all 314 catchment locations for a sample of four LTs (ranging from extended to annual) and three initialization months (January, April, and August). Although WS is considered a low skill region overall at a 1-week LT in Fig. 6 (i.e. CRPSS = 0.248), moderate to high skill ESP forecasts can be made for some catchments at different times of the year. For example, August 1-week LT forecasts (Fig. 7c) in WS are moderately skilful (CRPSS $\geq$ 0.25) for over 80 % of the 35 catchments or even highly skilful (CRPSS $\geq$ 0.5) for 20 % of catchments. In all regions, almost all individual catchments are more skilful than the benchmark climatological forecast for up to extended LTs (i.e. Fig. 7a–c).

Sub-region heterogeneity is much more apparent for monthly, seasonal, and annual LTs (Fig. 7d–l). As in Fig. 6, skill decays at different rates depending on region and lead time, but also initialization month. However, the finer spatial information in Fig. 7 shows that skill decays towards zero at vastly different rates for individual catchments even within

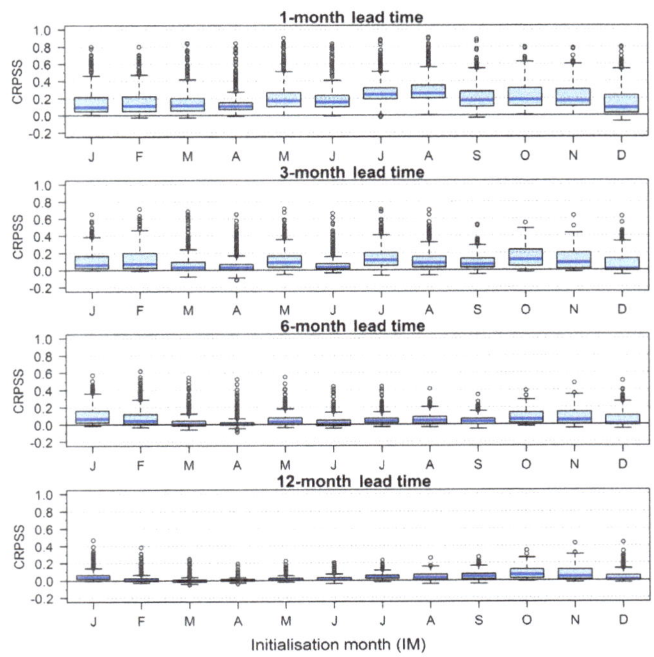

Figure 5. As in Fig. 4 but for the four monthly, seasonal, and annual lead times.

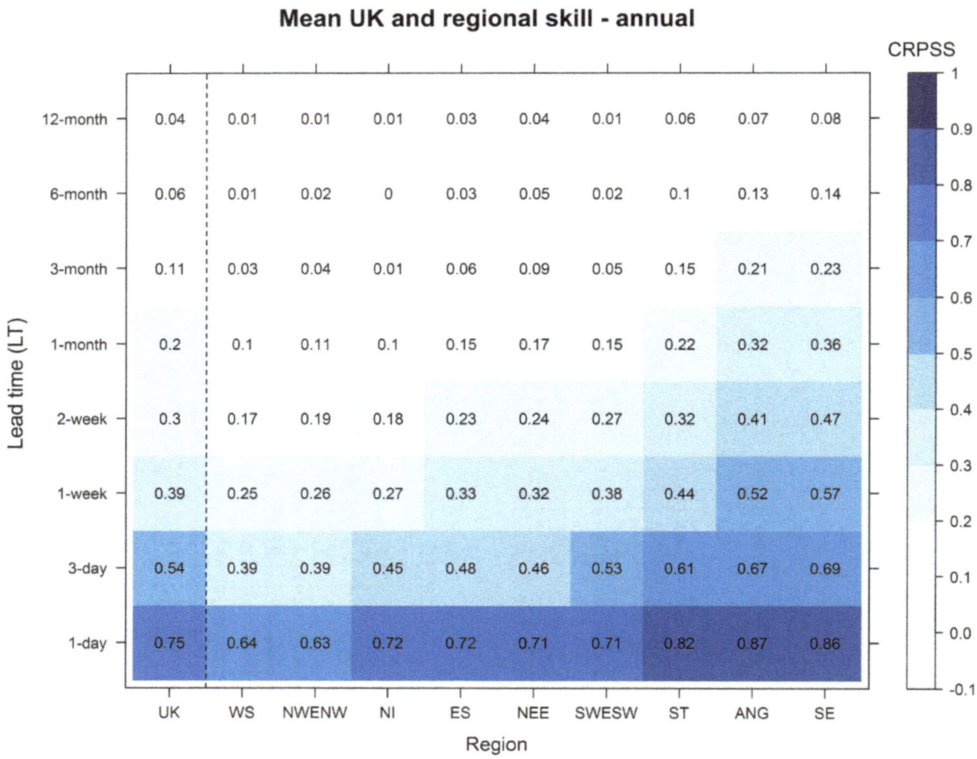

Figure 6. Mean ESP skill across all 12 forecast initialization months for the UK and for each of the nine hydroclimate regions ordered from least to most skilful (horizontal axis) at eight sample lead times (vertical axis). Skill is given by the CRPSS with darker and lighter shades showing higher and lower skill, respectively; mean skill score values are shown within each cell.

the same region. For example, despite low average skill of January 12-month LT forecasts in SE (CRPSS = 0.14), nearly 20 % of catchments have moderate skill. In April, when UK-wide forecasts at longer LTs are least skilful (i.e. Fig. 5), skilful forecasts can still be made at monthly and seasonal LTs for the majority of catchments in ST, ANG, and SE (Fig. 7e and h). Sub-region heterogeneity is perhaps most prominent for the Thames basin in SE. The April 3-month LT forecast for the Thames at Kingston has low skill (CRPSS = 0.22, size = 9948 km^2), but two of its sub-catchments have contrasting skills; the Lambourn at Shaw is highly skilful (CRPSS = 0.65, size = 234 km^2) whereas the forecast made for the Mole at Kinnersley Manor has effectively no skill (CRPSS = 0.02, size = 142 km^2).

4.3 Relationship between catchment storage and ESP skill

The relationship between the two calibrated GR4J catchment storage parameters, X1 (soil moisture store capacity [mm]) and X3 (groundwater store capacity [mm]), BFI, and ESP skill (CRPSS) for $n = 314$ individual catchments is shown in the scatterplot matrix in Fig. 8 using the non-parametric Spearman's rank correlation coefficient ρ. It is difficult to link X1 and X3 specifically to soil moisture and groundwater storage capacity, respectively, as GR4J is not a physically based hydrological model. However, their sum (X1 + X3) can be considered an estimate of total catchment storage (excluding water in the river channel and snowpack). Total catchment storage (X1 + X3) is strongly positively (non-linearly) correlated with BFI ($\rho = 0.87$); catchments with high BFIs tend to have much higher than average catchment storage capacity. The BFI is also very strongly positively correlated with ESP skill ($\rho = 0.90$). The 1-month LT forecast skill (based on CRPSS) averaged across all 12 initialization months is used to demonstrate this, but similar results are found over the range of lead times, individual initialization months, and skill metrics (not shown). Forecasts in the most responsive catchments (BFI ≤ 0.35, 20 % of catchments) have on average low skill (CRPSS = 0.08) whereas the slowest responding catchments (BFI ≥ 0.9, 5 % of catchments) have high skill (CRPSS = 0.66).

5 Discussion

Overall, the ESP method is found to be skilful when benchmarked against climatology in the UK, but the degree of skill is dependent on lead time, initialization month, and individual catchment location and storage properties.

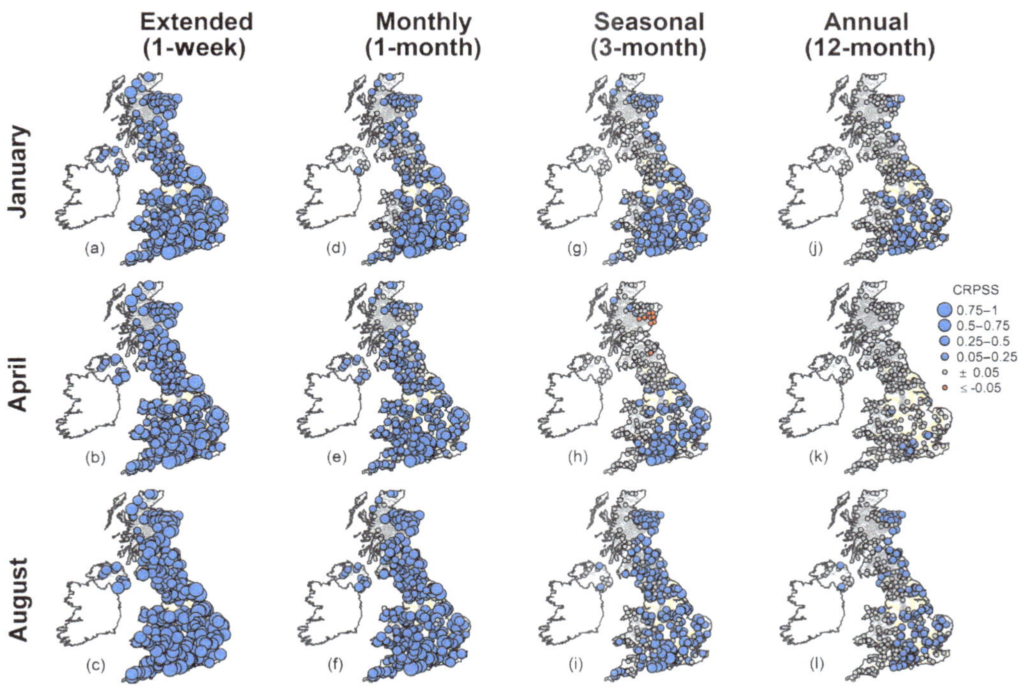

Figure 7. ESP skill for individual forecasts made at each of the 314 catchment locations for four sample lead times (columns) and three initialization months (rows). Larger and smaller circles represent higher and lower skill from CRPSS, respectively, with blue circles when ESP is more skilful than benchmark climatology and red when ESP is less skilful. Grey circles represent neutrally skilful forecasts (i.e. CRPSS values between ± 0.05).

5.1 When is ESP skilful?

UK-wide ESP forecasts for short lead times (out to 3 days) are on average highly skilful (CRPSS ≥ 0.5) and for extended lead times (out to 2 weeks) moderately skilful (CRPSS ≥ 0.25). Mean ESP skill decays exponentially with increasing lead time so skill is on average much lower for monthly, seasonal, and annual lead times, as expected. However, the magnitude of skill is not uniform across the 12 forecast initialization months. ESP skill for short, extended, and monthly lead times is higher than average when initialized in summer months and lower than average for winter months. Svensson (2016) also found higher skill across the UK when initialized in summer (highest also for August forecasts at a 1-month lead time) using the statistical persistence forecasting method. This is consistent with Li et al. (2009) and Shukla and Lettenmaier (2011), who found soil moisture initial hydrologic conditions (IHCs) contributed to greater skill for forecasts initialized in the warmer summer season than the cold winter season in the south-east of the US, up to a 1-month lead time; this was said to be due to drier initial soil moisture states in summertime. Similarly, Staudinger and Seibert (2014) found drier initial soil moisture was connected to longer persistence in all seasons except winter in Switzerland. Soil moisture deficits (SMDs) are also highest in summer in the UK, peaking in July, and lowest in winter (based on UK Met Office MORECS dataset (Hough and Jones, 1997) over 1961–2015). This could help explain why up to 1-month LT hydrological forecasts initialized in summer months using IHCs alone (e.g. ESP) are more skilful than if initialized in winter in the UK. Higher summer ESP forecast skill could be capitalized upon operationally given seasonal climate predictability over northern Europe is notoriously challenging for summer rainfall (e.g. Weisheimer and Palmer, 2014).

In contrast, ESP skill at seasonal to annual lead times is generally higher than average when initialized in winter and autumn months, and lowest in April. However, these higher skills occur in catchments with higher BFIs, suggesting that perhaps groundwater from large slowly responding aquifers is the source of ESP skill at these longer lead times. This is supported by Wood and Lettenmaier (2008), who found that baseflow dominates hydrological persistence in winter in the Rio Grande River in the US. Staudinger and Seibert (2014) also found for simulations initialized in winter, wetter initial conditions lead to longer persistence, although they note it was difficult to separate the relative influences from snow and aquifer memory. Lower longer-range skill for forecasts initialized in spring months was also found by Svensson (2016) for a 3-month LT based on statistical streamflow persistence forecasts. However, there are limited seasonal hydrological hindcast studies for the UK that have also assessed skill at

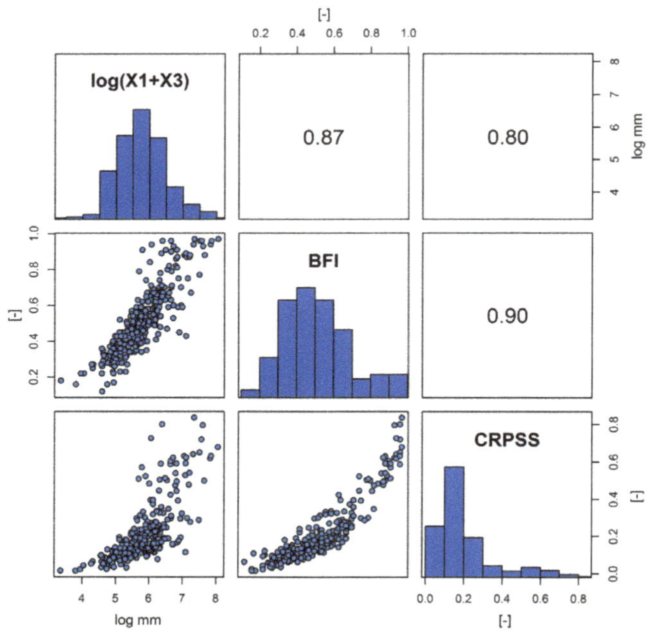

Figure 8. Scatterplot matrix between catchment storage capacity (X1 soil moisture store capacity [mm] + X3 groundwater store capacity [mm]), BFI, and ESP skill (CRPSS) with $n = 314$ using the non-parametric Spearman's rank correlation coefficient ρ. Skill is the 1-month CRPSS skill score magnitude averaged across all 12 initialization months. Catchment storage capacity (X1 + X3) was re-expressed by taking the natural log, as raw values are heavily positively skewed.

longer than 3-month lead times to compare results. Spring in the UK is characterized as a transition season between lowest (winter) and highest (summer) SMDs, in which groundwater recharge no longer occurs and baseflow begins its recession. Factors that might contribute to lower skilled forecasts initialized in spring, and indeed to differences in skill across all initialization months, include: potentially higher variability in IHC storage states, changing variability in rainfall across the forecast window (e.g. from late spring to early autumn), and differences in model performance for different months over the year due to the global calibration of GR4J. Given the answer is likely a combination of many of these factors, among others, further work should endeavour to attribute differences in skill during different times of the year, but this is outside the scope of this paper.

5.2 Where is ESP skilful?

The skill of ESP is also not uniformly distributed in space. Least skilful hydroclimate regions within the UK are situated in the north and west (WS, NWENW, and NI) whereas the most skilful are situated in the south and east (ST, ANG, and SE) across all lead times studied. This prominent spatial pattern was also noted, among others, by Svensson et al. (2015) and Svensson (2016) using statistical persistence forecasting and Bell et al. (2017) using a gridded national-scale hydrological model. These space–time patterns are also apparent in skill maps of individual catchments (i.e. Fig. 7), although there is marked sub-region heterogeneity, as demonstrated using the Thames basin: the slow responding Lambourn at Shaw sub-basin (BFI = 0.97) was highly skilful whereas the fast responding Mole at Kinnersley Manor catchment (BFI = 0.39) had virtually no ESP skill.

5.3 Why is ESP skilful?

The most skilful ESP regions of the UK are also those that are underlain by the UK's principal aquifers (Fig. 1). Catchments with larger calibrated soil moisture and groundwater storage capacity parameters in GR4J (i.e. X1 and X3) are also situated in ST, ANG, and SE, and tend to have a higher base flow index (BFI) (Table 2). The BFI is therefore broadly interpreted here as an integrated index of catchment storage capacity and is inferred to be responsible for modulating ESP skill – catchments with higher storage are more skilful with skill decaying at a much slower rate as lead time increases, compared to catchments with low storage capacity. For example, forecasts for the Lambourn remains on average moderately skilful (i.e. CRPSS ≥ 0.25) until a lead time of 306 days, but the Mole drops below the moderately skilful threshold at a lead time of just 10 days.

These findings are consistent with the current physical understanding of sources of ESP skill in non-snow dominated catchments in the literature. Water storage within the soil introduces a memory effect whereby anomalously dry or wet conditions can take weeks or months to be "forgotten" (Ghannam et al., 2016; Li et al., 2009), and the slow transformation of precipitation to streamflow in catchments with highly permeable aquifers in the south east of the UK leads to temporal streamflow dependence for up to a season ahead, and longer (Chiverton et al., 2015). Although it is encouraging that GR4J storage parameter values (X1 and X3) appear to show some physical realism, a note of caution is needed as GR4J is not a physically based hydrological model, nor is it guaranteed that these results are directly transferable to any lumped catchment hydrological model. It has also been noted that the BFI in the UK is influenced by many other factors such as lake and snow storage (Parry et al., 2016), therefore a more detailed examination of the physical hydrogeological controls on catchment BFI, such as in Bloomfield et al. (2009) for the Thames, is needed at a national scale.

The ESP method was originally developed and tested in the snow dominated catchments of the western US with particular strength in forecasting spring snow melt driven streamflow (e.g. Franz et al., 2003; Wood and Lettenmaier, 2008). Because the source of ESP is from IHCs, and because individual catchments will have different relative contributions of IHC sources (e.g. snow, soil moisture, and groundwater), ESP skill must be assessed using a large sample of diverse catchment types and sizes for each region it is being applied in (e.g. Yossef et al., 2013). The present study adds to the broader international literature on benchmarking ESP skill in non-snow dominated catchments. In particular, results show that IHCs in catchments with large soil moisture and groundwater storage provide skill up to a year ahead. It must however be acknowledged that the UK is not completely snow-free. Just under 5 % of catchments studied have a significant snow contribution (i.e. $\overline{F}_s > 0.15$) located mainly in upland areas of Eastern Scotland (ES) (see Fig. 1). In the present experimental set-up, snow accumulation and melt processes were not represented within the GR4J model. This would explain why ES has the lowest GR4J model performance for the reference simulation of all regions (Table 2). In addition, the worst performing forecast in the entire ESP hindcast archive is the 3-month LT April forecast for the Dee at Park with a negative CRPSS $= -0.12$ (see Fig. 2e). In this instance both the ESP forecast and the proxy streamflow observations (or perfect model) which the forecast was evaluated against was not a good enough representation of reality.

ESP in its traditional form as used here provides the *lower limit* of streamflow forecasting skill in the absence of skilful atmospheric forecasts (Pagano et al., 2010) or improved hydrological process representation (e.g. snow). As such, ESP assumes near total uncertainty about future rainfall; when there is limited to no influence of IHCs on streamflow prediction (e.g. highly responsive catchments), the ESP ensemble mean and spread defaults to climatology (see Fig. 2d). Given the known influence of the NAO on rainfall and therefore streamflow in the UK, particularly in the north and west for winter (e.g. Svensson et al., 2015), there is potential for an NAO-conditioned ESP method to be developed. This would involve sub-sampling historic climate sequences used to force ESP based on years most similar to NAO conditions at the time of forecast. Beckers et al. (2016) developed an ENSO-conditioned ESP method for three test sites in the US Pacific Northwest and found skill improvements in the order of 5–10 %; the study also presented the added value of including a weather resampling technique to account for the unavoidable reduction in ensemble size. Overall, low ESP forecast performance and sharpness in highly responsive catchments in the north and west would be expected to improve with the incorporation of information that reduces rainfall forcing uncertainty at all lead times but particularly seasonal, whether from ensemble sub-sampling or inclusion of skilful atmospheric forecasts.

6 Conclusions

Ensemble streamflow prediction (ESP) has a rich history internationally as a low cost and efficient ensemble hydrological forecasting system used operationally across a range of lead times. The ESP method using simple lumped conceptual hydrological models is currently one of three methods used within the operational Hydrological Outlook UK (HOUK) seasonal hydrological forecasting service and also feeds into the Environment Agency's monthly "Water Situation Report for England". However, the skill of ESP at the catchment scale under a rigorous hindcast experiment for a large sample of diverse catchments across the UK had not previously been investigated.

We conclude that ESP is skilful against a climatology benchmark forecast in the majority of catchments across all lead times up to a year ahead, but the degree of skill is strongly conditional on lead time, forecast initialization month, and individual catchment location and storage properties. In summary:

- ESP skill decayed exponentially with increasing lead time but catchments with larger storage capacity decayed at a much slower rate, resulting in the possibility of low to moderate skill forecasts based on initial hydrologic conditions (IHCs) alone even at a 12-month lead time for some catchments.

- For short (1–3 days), extended (1–2 weeks), and monthly forecasts, skill was highest when initialized in summer months and lowest in winter months.

- For seasonal (3–6 months) to annual forecasts, skill was highest when initialized in winter and autumn months, but only for catchments with high storage capacity (i.e. high base flow index). Longer range forecast skill was lowest when initialized in spring, particularly April, which is likely due to the complex interplay of hydrological and climatological processes involved during the transition from lower winter to higher summer soil moisture deficit conditions and needs to be explored further.

- ESP is most skilful in the south and east of the UK, where slower responding catchments with higher storage are mainly located. This is in contrast to the more highly responsive catchments in the north and west which are generally not skilful at seasonal lead times. However, substantial sub-region heterogeneity was observed and skilful ESP forecasts are still possible at the individual catchment scale despite when the region as a whole has low skill.

We show that simple lumped conceptual rainfall-runoff models (here using GR4J) are able to be used to produce skilful ESP forecasts at short to annual lead times in the UK. This hindcast experiment provides scientific justification for when (lead time and initialization month) and where (region and catchment types) use of such a relatively simple forecasting approach is appropriate. Currently, ESP is only used operationally in the UK at seasonal and annual lead times in England and Wales. This skill evaluation has shown that much higher skills are possible for short (1–3 days) and extended (1–2 weeks) lead times in all regions across the UK and opens the potential for applying ESP as a low cost and efficient catchment-scale ensemble hydrological forecasting system in a wider context.

Finally, most ensemble hydrological forecasting systems are benchmarked against an arguably too simplistic climatology benchmark forecast which is not particularly challenging to beat. Pappenberger et al. (2015) calls this "naïve skill" and argues that a forecasting system can only be classified as having "real skill" when it performs better than a "tough to beat" lower cost benchmark forecast system. The ESP hindcast archive derived and presented in this study provides such a "tough to beat" simplified hydrology model benchmark in which the potential value of improvements from more sophisticated forms of ESP (e.g. incorporation of snow processes, sub-sampling historic climate) or more complex and expensive hydro-meteorological ensemble forecasting systems can be judged. When and where ESP cannot provide skilful streamflow forecasts provides an opportunity to benchmark the degree to which recent improvements in seasonal prediction of UK regional rainfall (e.g. Baker et al., 2017) leads to improvements over using IHCs alone (i.e. our ESP method), and is the focus of future work.

Competing interests. The authors declare that they have no conflict of interest.

Acknowledgements. This work was funded by NERC National Capability funding to CEH and the NERC funded Improving Predictions of Drought for User Decision-Making (IMPETUS) project (NE/L010267/1). Statistical analyses and graphics were implemented in the open-source R programming language. Streamflow data and metadata are from the NRFA and MORECS dataset from the UK Met Office. We thank Cecilia Svensson for fruitful discussions about this work and Nuria Bachiller-Jareno for help in designation of the UK Hydroclimate Regions. Finally, we thank Guillaume Thirel and two anonymous referees for their constructive feedback that has greatly improved this paper.

Edited by: Quan J. Wang

References

Alfieri, L., Pappenberger, F., Wetterhall, F., Haiden, T., Richardson, D., and Salamon, P.: Evaluation of ensemble streamflow predictions in Europe, J. Hydrol., 517, 913–922, https://doi.org/10.1016/j.jhydrol.2014.06.035, 2014.

Anghileri, D., Voisin, N., Castelletti, A., Pianosi, F., Nijssen, B., and Lettenmaier, D. P.: Value of long-term streamflow forecasts to reservoir operations for water supply in snow-dominated river catchments, Water Resour. Res., 52, 4209–4225, https://doi.org/10.1002/2015WR017864, 2016.

Arnal, L., Cloke, H. L., Stephens, E., Wetterhall, F., Prudhomme, C., Neumann, J., Krzeminski, B., and Pappenberger, F.: Skilful seasonal forecasts of streamflow over Europe?, Hydrol. Earth Syst. Sci. Discuss., https://doi.org/10.5194/hess-2017-610, in review, 2017.

Baker, L. H., Shaffrey, L. C., and Scaife, A. A.: Improved seasonal prediction of UK regional precipitation using atmospheric circulation, Int. J. Climatol., https://doi.org/10.1002/joc.5382, 2017.

Beckers, J. V. L., Weerts, A. H., Tijdeman, E., and Welles, E.: ENSO-conditioned weather resampling method for seasonal ensemble streamflow prediction, Hydrol. Earth Syst. Sci., 20, 3277–3287, https://doi.org/10.5194/hess-20-3277-2016, 2016.

Bell, V. A., Davies, H. N., Kay, A. L., Brookshaw, A., and Scaife, A. A.: A national-scale seasonal hydrological forecast system: development and evaluation over Britain, Hydrol. Earth Syst. Sci., 21, 4681–4691, https://doi.org/10.5194/hess-21-4681-2017, 2017.

Bennett, J. C., Wang, Q. J., Robertson, D. E., Schepen, A., Li, M., and Michael, K.: Assessment of an ensemble seasonal streamflow forecasting system for Australia, Hydrol. Earth Syst. Sci., 21, 6007–6030, https://doi.org/10.5194/hess-21-6007-2017, 2017.

Berghuijs, W. R., Woods, R. A., and Hrachowitz, M.: A precipitation shift from snow towards rain leads to a decrease in streamflow, Nat. Clim. Change, 4, 583–586, https://doi.org/10.1038/nclimate2246, 2014.

Bloomfield, J. P., Allen, D. J., and Griffiths, K. J.: Examining geological controls on baseflow index (BFI) using regression analysis: An illustration from the Thames Basin, UK, J. Hydrol., 373, 164–176, https://doi.org/10.1016/j.jhydrol.2009.04.025, 2009.

Broderick, C., Matthews, T., Wilby, R. L., Bastola, S., and Murphy, C.: Transferability of hydrological models and ensemble averaging methods between contrasting climatic periods, Water Resour. Res., 52, 8343–8373, https://doi.org/10.1002/2016WR018850, 2016.

Chiverton, A., Hannaford, J., Holman, I., Corstanje, R., Prudhomme, C., Bloomfield, J., and Hess, T. M.: Which catchment characteristics control the temporal dependence structure of daily river flows?, Hydrol. Process., 29, 1353–1369, https://doi.org/10.1002/hyp.10252, 2015.

Coron, L., Perrin, C., and Michel, C.: airGR: Suite of GR hydrological models for precipitation-runoff modelling, R package version 1.0.2, available at: https://webgr.irstea.fr/airGR/?lang=en (last access: 14 July 2017), 2016.

Coron, L., Thirel, G., Delaigue, O., Perrin, C., and Andréassian, V.: The suite of lumped GR hydrological models in an R package, Environ. Model. Softw., 94, 166–171, https://doi.org/10.1016/j.envsoft.2017.05.002, 2017.

Crochemore, L., Ramos, M.-H., Pappenberger, F., and Perrin, C.: Seasonal streamflow forecasting by conditioning climatology with precipitation indices, Hydrol. Earth Syst. Sci., 21, 1573–1591, https://doi.org/10.5194/hess-21-1573-2017, 2017.

Crooks, S. M., Kay, A. L., and Reynard, N. S.: Regionalised Impacts of Climate Change on Flood Flows: Hydrological Models, Catchments and Calibration, Centre for Ecology & Hydrology, Environment Agency, Defra, London, 2009.

Day, G. N.: Extended Streamflow Forecasting Using NWSRFS, J. Water Resour. Plan. Manag., 111, 642–654, 1985.

Dunstone, N., Smith, D., Scaife, A., Hermanson, L., Eade, R., Robinson, N., Andrews, M., and Knight, J.: Skilful predictions of the winter North Atlantic Oscillation one year ahead, Nat. Geosci., 9, 809, https://doi.org/10.1038/ngeo2824, 2016.

Ferro, C. A. T., Richardson, D. S., and Weigel, A. P.: On the effect of ensemble size on the discrete and continuous ranked probability scores, Meteorol. Appl., 15, 19–24, https://doi.org/10.1002/met.45, 2008.

Franz, K. J., Hartmann, H. C., Sorooshian, S., and Bales, R.: Verification of National Weather Service Ensemble Streamflow Predictions for Water Supply Forecasting in the Colorado River Basin, J. Hydrometeorol., 4, 1105–1118, https://doi.org/10.1175/1525-7541(2003)004<1105:VONWSE>2.0.CO;2, 2003.

Ghannam, K., Nakai, T., Paschalis, A., Oishi, C. A., Kotani, A., Igarashi, Y., Kumagai, T., and Katul, G. G.: Persistence and memory timescales in root-zone soil moisture dynamics, Water Resour. Res., 52, 1427–1445, https://doi.org/10.1002/2015WR017983, 2016.

Gupta, H. V., Sorooshian, S., and Yapo, P. O.: Status of Automatic Calibration for Hydrologic Models: Comparison with Multilevel Expert Calibration, J. Hydrol. Eng., 4, 135–143, https://doi.org/10.1061/(ASCE)1084-0699(1999)4:2(135), 1999.

Gupta, H. V., Kling, H., Yilmaz, K. K., and Martinez, G. F.: Decomposition of the mean squared error and NSE performance criteria: Implications for improving hydrological modelling, J. Hydrol., 377, 80–91, https://doi.org/10.1016/j.jhydrol.2009.08.003, 2009.

Gustard, A., Bullock, A., and Dixon, J. M.: Low flow estimation in the United Kingdom, Institute of Hydrology, Wallingford, available at: http://nora.nerc.ac.uk/6050/ (last access: 3 November 2016), 1992.

Hamlet, A. F., Huppert, D., and Lettenmaier, D. P.: Economic Value of Long-Lead Streamflow Forecasts for Columbia River Hydropower, J. Water Resour. Plan. Manag., 128, 91–101, https://doi.org/10.1061/(ASCE)0733-9496(2002)128:2(91), 2002.

Harrigan, S., Hannaford, J., Muchan, K., and Marsh, T. J.: Designation and trend analysis of the updated UK Benchmark Network of river flow stations: The UKBN2 dataset, Hydrol. Res., https://doi.org/10.2166/nh.2017.058, 2017.

Hersbach, H.: Decomposition of the Continuous Ranked Probability Score for Ensemble Prediction Systems, Weather Forecast., 15, 559–570, https://doi.org/10.1175/1520-0434(2000)015<0559:DOTCRP>2.0.CO;2, 2000.

Hough, M. N. and Jones, R. J. A.: The United Kingdom Meteorological Office rainfall and evaporation calculation system: MORECS version 2.0 – an overview, Hydrol. Earth Syst. Sci., 1, 227–239, https://doi.org/10.5194/hess-1-227-1997, 1997.

Jolliffe, I. T. and Stephenson, D. B.: Forecast verification: a practitioner's guide in atmospheric science, John Wiley & Sons, England, 2003.

Keller, V. D. J., Tanguy, M., Prosdocimi, I., Terry, J. A., Hitt, O., Cole, S. J., Fry, M., Morris, D. G., and Dixon, H.: CEH-GEAR: 1 km resolution daily and monthly areal rainfall estimates for the UK for hydrological and other applications, Earth Syst. Sci. Data, 7, 143–155, https://doi.org/10.5194/essd-7-143-2015, 2015.

Klemeš, V.: Operational testing of hydrological simulation models, Hydrol. Sci. J., 31, 13–24, https://doi.org/10.1080/02626668609491024, 1986.

Kling, H., Fuchs, M., and Paulin, M.: Runoff conditions in the upper Danube basin under an ensemble of climate change scenarios, J. Hydrol., 424–425, 264–277, https://doi.org/10.1016/j.jhydrol.2012.01.011, 2012.

Letcher, R. A., Chiew, F. H. S., and Jakeman, A. J.: An Assessment of the Value of Seasonal Forecasts in Australian Farming Systems, Int. Congr. Environ. Model. Softw., available at: http://scholarsarchive.byu.edu/iemssconference/2004/all/79 (last access: 2 February 2017), 2004.

Li, H., Luo, L., Wood, E. F., and Schaake, J.: The role of initial conditions and forcing uncertainties in seasonal hydrologic forecasting, J. Geophys. Res.-Atmos., 114, D04114, https://doi.org/10.1029/2008JD010969, 2009.

MeteoSwiss: easyVerification: Ensemble Forecast Verification for Large Data Sets, R package version 0.4.2, available at: http://CRAN.R-project.org/package=easyVerification, last access: 14 July 2017.

Moore, R. J.: The PDM rainfall-runoff model, Hydrol. Earth Syst. Sci., 11, 483–499, https://doi.org/10.5194/hess-11-483-2007, 2007.

Nash, J. E. and Sutcliffe, J. V.: River flow forecasting through conceptual models part I – A discussion of principles, J. Hydrol., 10, 282–290, https://doi.org/10.1016/0022-1694(70)90255-6, 1970.

National Hydrological Monitoring Programme: Hydrological summary for the United Kingdom: April 2017, NERC/Centre for Ecology & Hydrology, Wallingford, UK, available at: http://nrfa.ceh.ac.uk/monthly-hydrological-summary-uk, last access: 20 May 2017.

National River Flow Archive: Integrated Hydrological Units of the United Kingdom: Hydrometric Areas with Coastline, available at: https://doi.org/10.5285/1957166d-7523-44f4-b279-aa5314163237 (last access: 10 June 2017), 2014.

Pagano, T., Hapuarachchi, P., and Wang, Q. J.: Continuous rainfall-runoff model comparison and short-term daily streamflow forecast skill evaluation, CSIRO, available at: https://publications.csiro.au/rpr/pub?pid=csiro:EP103545 (last access: 1 July 2017), 2010.

Pappenberger, F., Ramos, M. H., Cloke, H. L., Wetterhall, F., Alfieri, L., Bogner, K., Mueller, A., and Salamon, P.: How do I know if my forecasts are better? Using benchmarks in hydrological ensemble prediction, J. Hydrol., 522, 697–713, https://doi.org/10.1016/j.jhydrol.2015.01.024, 2015.

Parry, S., Wilby, R. L., Prudhomme, C., and Wood, P. J.: A systematic assessment of drought termination in the United Kingdom, Hydrol. Earth Syst. Sci., 20, 4265–4281, https://doi.org/10.5194/hess-20-4265-2016, 2016.

Perrin, C., Michel, C., and Andréassian, V.: Improvement of a parsimonious model for streamflow simulation, J. Hydrol., 279, 275–289, https://doi.org/10.1016/S0022-1694(03)00225-7, 2003.

Poncelet, C., Merz, R., Merz, B., Parajka, J., Oudin, L., Andréassian, V., and Perrin, C.: Process-based interpretation of conceptual hydrological model performance using a multinational catchment set, Water Resour. Res., 53, 7247–7268, https://doi.org/10.1002/2016WR019991, 2017.

Prudhomme, C., Hannaford, J., Harrigan, S., Boorman, D., Knight, J., Bell, V., Jackson, C., Svensson, C., Parry, S., Bachiller-Jareno, N., Davies, H., Davis, R., Mackay, J., McKenzie, A., Rudd, A., Smith, K., Bloomfield, J., Ward, R., and Jenkins, A.: Hydrological Outlook UK: an operational streamflow and groundwater level forecasting system at monthly to seasonal time scales, Hydrol. Sci. J., 62, 2753–2768, https://doi.org/10.1080/02626667.2017.1395032, 2017.

Robertson, D., Bennett, J., and Schepen, A.: How good is my forecasting method? Some thoughts on forecast evaluation using cross-validation based on Australian experiences, HEPEX Blog, available at: https://hepex.irstea.fr/how-good-is-my-forecasting (last access: 13 March 2017), 2016.

Robinson, E. L., Blyth, E., Clark, D. B., Comyn-Platt, E., Finch, J., and Rudd, A. C.: Climate hydrology and ecology research support system potential evapotranspiration dataset for Great Britain (1961–2015) [CHESS-PE], https://doi.org/10.5285/8baf805d-39ce-4dac-b224-c926ada353b7, 2016.

Robinson, E. L., Blyth, E. M., Clark, D. B., Finch, J., and Rudd, A. C.: Trends in atmospheric evaporative demand in Great Britain using high-resolution meteorological data, Hydrol. Earth Syst. Sci., 21, 1189–1224, https://doi.org/10.5194/hess-21-1189-2017, 2017.

Shukla, S. and Lettenmaier, D. P.: Seasonal hydrologic prediction in the United States: understanding the role of initial hydrologic conditions and seasonal climate forecast skill, Hydrol. Earth Syst. Sci., 15, 3529–3538, https://doi.org/10.5194/hess-15-3529-2011, 2011.

Simpson, M., James, R., Hall, J. W., Borgomeo, E., Ives, M. C., Almeida, S., Kingsborough, A., Economou, T., Stephenson, D., and Wagener, T.: Decision Analysis for Management of Natural Hazards, Annu. Rev. Environ. Resour., 41, 489–516, https://doi.org/10.1146/annurev-environ-110615-090011, 2016.

Singh, S. K.: Long-term Streamflow Forecasting Based on Ensemble Streamflow Prediction Technique: A Case Study in New Zealand, Water Resour. Manag., 30, 2295–2309, https://doi.org/10.1007/s11269-016-1289-7, 2016.

Staudinger, M. and Seibert, J.: Predictability of low flow – An assessment with simulation experiments, J. Hydrol., 519, 1383–1393, https://doi.org/10.1016/j.jhydrol.2014.08.061, 2014.

Svensson, C.: Seasonal river flow forecasts for the United Kingdom using persistence and historical analogues, Hydrol. Sci. J., 61, 19–35, https://doi.org/10.1080/02626667.2014.992788, 2016.

Svensson, C., Brookshaw, A., Scaife, A. A., Bell, V. A., Mackay, J. D., Jackson, C. R., Hannaford, J., Davies, H. N., Arribas, A., and Stanley, S.: Long-range forecasts of UK winter hydrology, Environ. Res. Lett., 10, 064006, https://doi.org/10.1088/1748-9326/10/6/064006, 2015.

Tanguy, M., Dixon, H., Prosdocimi, I., Morris, D. G., and Keller, V. D. J.: Gridded estimates of daily and monthly areal rainfall for the United Kingdom (1890–2015) [CEH-GEAR], https://doi.org/10.5285/33604ea0-c238-4488-813d-0ad9ab7c51ca, 2016.

Tanguy, M., Prudhomme, C., Smith, K., and Hannaford, J.: Historic Gridded Potential Evapotranspiration (PET) based on temperature-based equation McGuinness-Bordne calibrated for the UK (1891–2015), https://doi.org/10.5285/17b9c4f7-1c30-4b6f-b2fe-f7780159939c, 2017.

Tanguy, M., Prudhomme, C., Smith, K., and Hannaford, J.: Historical gridded reconstruction of potential evapotranspiration for the UK, Earth Syst. Sci. Data Discuss., https://doi.org/10.5194/essd-2017-137, in review, 2018.

Thober, S., Kumar, R., Sheffield, J., Mai, J., Schäfer, D., and Samaniego, L.: Seasonal Soil Moisture Drought Prediction over Europe Using the North American Multi-Model Ensemble (NMME), J. Hydrometeorol., 16, 2329–2344, https://doi.org/10.1175/JHM-D-15-0053.1, 2015.

Twedt, T. M., Schaake Jr., J. C., and Peck, E. L.: National Weather Service extended streamflow prediction, in Proceedings of the 45th Annual Western Snow Conference, 52–57, Albuquerque, New Mexico, available at: https://westernsnowconference.org/node/1106 (last access: 30 June 2017), 1977.

Valéry, A., Andréassian, V., and Perrin, C.: "As simple as possible but not simpler": What is useful in a temperature-based snow-accounting routine? Part 2 – Sensitivity analysis of the Cemaneige snow accounting routine on 380 catchments, J. Hydrol., 517, 1176–1187, https://doi.org/10.1016/j.jhydrol.2014.04.058, 2014.

Wang, E., Zhang, Y., Luo, J., Chiew, F. H. S., and Wang, Q. J.: Monthly and seasonal streamflow forecasts using rainfall-runoff modeling and historical weather data, Water Resour. Res., 47, W05516, https://doi.org/10.1029/2010WR009922, 2011.

Wang, Q. J., Robertson, D. E., and Chiew, F. H. S.: A Bayesian joint probability modeling approach for seasonal forecasting of streamflows at multiple sites, Water Resour. Res., 45, W05407, https://doi.org/10.1029/2008WR007355, 2009.

Weisheimer, A. and Palmer, T. N.: On the reliability of seasonal climate forecasts, J. R. Soc. Interface, 11, 20131162, https://doi.org/10.1098/rsif.2013.1162, 2014.

Wilby, R. L., Prudhomme, C., Parry, S., and Muchan, K. G. L.: Persistence of Hydrometeorological Droughts in the United Kingdom: A Regional Analysis of Multi-Season Rainfall and River Flow Anomalies, J. Extreme Events, 2, 1550006, https://doi.org/10.1142/S2345737615500062, 2015.

Wilks, D. S.: Statistical methods in the atmospheric sciences, Academic press, Oxford, 2011.

Wood, A. W. and Lettenmaier, D. P.: An ensemble approach for attribution of hydrologic prediction uncertainty, Geophys. Res. Lett., 35, L14401, https://doi.org/10.1029/2008GL034648, 2008.

Wood, A. W., Kumar, A., and Lettenmaier, D. P.: A retrospective assessment of National Centers for Environmental Prediction climate model–based ensemble hydrologic forecasting in the western United States, J. Geophys. Res.-Atmos., 110, D04105, https://doi.org/10.1029/2004JD004508, 2005.

Wood, A. W., Hopson, T., Newman, A., Brekke, L., Arnold, J., and Clark, M.: Quantifying Streamflow Forecast Skill Elasticity to Initial Condition and Climate Prediction Skill, J. Hydrometeorol., 17, 651–668, https://doi.org/10.1175/JHM-D-14-0213.1, 2016a.

PERMISSIONS

All chapters in this book were first published in HESS, by Copernicus Publications; hereby published with permission under the Creative Commons Attribution License or equivalent. Every chapter published in this book has been scrutinized by our experts. Their significance has been extensively debated. The topics covered herein carry significant findings which will fuel the growth of the discipline. They may even be implemented as practical applications or may be referred to as a beginning point for another development.

The contributors of this book come from diverse backgrounds, making this book a truly international effort. This book will bring forth new frontiers with its revolutionizing research information and detailed analysis of the nascent developments around the world.

We would like to thank all the contributing authors for lending their expertise to make the book truly unique. They have played a crucial role in the development of this book. Without their invaluable contributions this book wouldn't have been possible. They have made vital efforts to compile up to date information on the varied aspects of this subject to make this book a valuable addition to the collection of many professionals and students.

This book was conceptualized with the vision of imparting up-to-date information and advanced data in this field. To ensure the same, a matchless editorial board was set up. Every individual on the board went through rigorous rounds of assessment to prove their worth. After which they invested a large part of their time researching and compiling the most relevant data for our readers.

The editorial board has been involved in producing this book since its inception. They have spent rigorous hours researching and exploring the diverse topics which have resulted in the successful publishing of this book. They have passed on their knowledge of decades through this book. To expedite this challenging task, the publisher supported the team at every step. A small team of assistant editors was also appointed to further simplify the editing procedure and attain best results for the readers.

Apart from the editorial board, the designing team has also invested a significant amount of their time in understanding the subject and creating the most relevant covers. They scrutinized every image to scout for the most suitable representation of the subject and create an appropriate cover for the book.

The publishing team has been an ardent support to the editorial, designing and production team. Their endless efforts to recruit the best for this project, has resulted in the accomplishment of this book. They are a veteran in the field of academics and their pool of knowledge is as vast as their experience in printing. Their expertise and guidance has proved useful at every step. Their uncompromising quality standards have made this book an exceptional effort. Their encouragement from time to time has been an inspiration for everyone.

The publisher and the editorial board hope that this book will prove to be a valuable piece of knowledge for researchers, students, practitioners and scholars across the globe.

LIST OF CONTRIBUTORS

Colby M. Steelman, Celia S. Kennedy, Donovan C. Capes and Beth L. Parker
G360 Institute for Groundwater Research, College of Engineering and Physical Sciences, University of Guelph, Guelph, Ontario, Canada

Ana Clara Santos
School of Architecture, Civil and Environmental Engineering (ENAC), École Polytechnique Fédérale de Lausanne (EPFL), Lausanne, Switzerland
Instituto Superior Técnico, Technical University of Lisbon, Lisbon, Portugal

Andrea Rinaldo
School of Architecture, Civil and Environmental Engineering (ENAC), École Polytechnique Fédérale de Lausanne (EPFL), Lausanne, Switzerland
Dipartimento di Ingegneria Civile Edile e Ambientale, Universitá degli studi di Padova, Padua, Italy

Bettina Schaefli
School of Architecture, Civil and Environmental Engineering (ENAC), École Polytechnique Fédérale de Lausanne (EPFL), Lausanne, Switzerland
Faculty of Geosciences and Environment, University of Lausanne, Lausanne, Switzerland

Maria Manuela Portela
Instituto Superior Técnico, Technical University of Lisbon, Lisbon, Portugal

Simon Höllering, Jan Wienhöfer, Jürgen Ihringer and Erwin Zehe
Karlsruhe Institute of Technology (KIT), Karlsruhe, Germany

Luis Samaniego
Helmholtz Centre for Environmental Research, Leipzig, Germany

Paweł Wilk, Paulina Orlińska-Woźniak and Joanna Gębala
Section of Modelling Water Quality Surface - Institute of Meteorology and Water Management, National Research Institute, 01-673 Warsaw, 61 Podleśna St., Poland

Qiang Li, Xiaohua Wei and Krysta Giles-Hansen
Department of Earth, Environmental and Geographic Sciences, University of British Columbia (Okanagan campus), 1177 Research Road, Kelowna, British Columbia, V1V 1V7, Canada

Xin Yang
Key Laboratory of Virtual Geographic Environment, Nanjing Normal University, Ministry of Education, Nanjing, 210023, China
State Key Laboratory Cultivation Base of Geographical Environment Evolution, Nanjing, 210023, China
Jiangsu Center for Collaborative Innovation in Geographical Information Resources Development and Application, Nanjing, 210023, China

Mingfang Zhang
School of Resources and Environment, University of Electronic Science and Technology of China, Chengdu, 611731, China

Wenfei Liu
Institute of Ecology and Environmental Science, Nanchang Institute of Technology, Nanchang, 330099, China

Daniel J. Karran and Cherie J. Westbrook
Department of Geography and Planning, Centre for Hydrology, University of Saskatchewan, Saskatoon, SK S7N 5C8, Canada

Joseph M. Wheaton
Department of Watershed Sciences, Utah State University, Logan, UT 84322-5210, USA

Carol A. Johnston
Department of Natural Resource Management, South Dakota State University, Brookings, SD 57007, USA

Angela Bedard-Haughn
Department of Soil Science, University of Saskatchewan, Saskatoon, SK S7N 5A8, Canada

Aurélien Beaufort, Nicolas Lamouroux, Hervé Pella, Thibault Datry and Eric Sauquet
Irstea, UR RiverLy, centre de Lyon-Villeurbanne, 5 rue de la Doua CS 20244, 69625 Villeurbanne, France

Terri S. Hogue
Civil and Environmental Engineering and Hydrologic Science and Engineering, Colorado School of Mines, Golden, Colorado, USA

Samuel Saxe
Civil and Environmental Engineering and Hydrologic Science and Engineering, Colorado School of Mines, Golden, Colorado, USA
National Research Program, United States Geological Survey, Lakewood, Colorado, USA

Lauren Hay
National Research Program, United States Geological Survey, Lakewood, Colorado, USA

Yongqiang Zhang and David Post
CSIRO Land and Water, Acton 2601, Canberra, Australia

David McInerney, Greer Humphrey, Mark A. Thyer, Holger R. Maier, Graeme C. Dandy and Dmitri Kavetski
School of Civil, Environmental and Mining Engineering, The University of Adelaide, North Terrace, Adelaide, South Australia, 5005, Australia

Matthew S. Gibbs
School of Civil, Environmental and Mining Engineering, The University of Adelaide, North Terrace, Adelaide, South Australia, 5005, Australia
Department of Environment, Water and Natural Resources, Government of South Australia, Adelaide, 5000, Australia

Firas Saleh, Venkatsundar Ramaswamy, Nickitas Georgas, Alan F. Blumberg and Julie Pullen
Stevens Institute of Technology, Davidson Laboratory, Department of Civil, Environmental and Ocean Engineering, Hoboken NJ, 07030, USA

Simon Etter, Barbara Strobl and H. J. Ilja van Meerveld
Department of Geography, University of Zurich, Winterthurerstrasse 190, 8057 Zurich, Switzerland

Jan Seibert
Department of Geography, University of Zurich, Winterthurerstrasse 190, 8057 Zurich, Switzerland
Department of Aquatic Sciences and Assessment, Swedish University of Agricultural Sciences, 75007 Uppsala, Sweden

Simon Schick, Ole Rössler and Rolf Weingartner
Institute of Geography, University of Bern, Bern, Switzerland
Oeschger Centre for Climate Change Research, University of Bern, Bern, Switzerland

Sandra Pool and Marc J. P. Vis
Department of Geography, University of Zurich, Zurich, Switzerland

Rodney R. Knight
U.S. Geological Survey Lower Mississippi-Gulf Water Science Center, 640 Grassmere Park, Suite 100, Nashville, TN 37211, USA

Jan Seibert
Department of Geography, University of Zurich, Zurich, Switzerland
Department of Earth Sciences, Uppsala University, Uppsala, Sweden
Department of Physical Geography, Stockholm University, Stockholm, Sweden

Shaun Harrigan
European Centre for Medium-Range Weather Forecasts (ECMWF), Shinfield Park, Reading, RG2 9AX, UK
Centre for Ecology & Hydrology, Wallingford, Oxfordshire, OX10 8BB, UK

Christel Prudhomme
European Centre for Medium-Range Weather Forecasts (ECMWF), Shinfield Park, Reading, RG2 9AX, UK
Centre for Ecology & Hydrology, Wallingford, Oxfordshire, OX10 8BB, UK
Department of Geography, Loughborough University, Loughborough, Leicestershire, LE11 3TU, UK

Simon Parry, Katie Smith and Maliko Tanguy
Centre for Ecology & Hydrology, Wallingford, Oxfordshire, OX10 8BB, UK

Index

S

T

V

W

CPSIA information can be obtained
at www.ICGtesting.com
Printed in the USA
LVHW060852030122
707669LV00002B/104

9 781639 872848